The Herpesviruses

Volume 1

THE VIRUSES

Series Editors
HEINZ FRAENKEL-CONRAT, *University of California*
Berkeley, California
ROBERT R. WAGNER, *University of Virginia School of Medicine*
Charlottesville, Virginia

THE HERPESVIRUSES, Volumes 1, 2, and 3
Edited by Bernard Roizman

The Herpesviruses

Volume 1

Edited by
BERNARD ROIZMAN
University of Chicago
Chicago, Illinois

SPRINGER SCIENCE+BUSINESS MEDIA, LLC

Library of Congress Cataloging in Publication Data

Main entry under title:

The Herpesviruses.

(The Viruses)
Includes bibliographical references and index.
1. Herpesvirus diseases. 2. Herpesviruses. I. Roizman, Bernard, 1929- . II. Series. [DNLM: 1. Herpesviridae. QW 165.5.H3 H5637]
RC147.H6H57 1982 616.9′25 82-15034

ISBN 978-1-4684-4165-9 ISBN 978-1-4684-4163-5 (eBook)

DOI 10.1007/978-1-4684-4163-5

Originally published by Plenum Press, New York in 1982
Softcover reprint of the hardcover 1st edition 1982

Contributors

Christopher Beisel, Division of Biological Sciences, The University of Chicago, Chicago, Illinois 60637
Andrew Cheung, Ph.D., Division of Biological Sciences, The University of Chicago, Chicago, Illinois 60637
Timothy Dambaugh, Ph.D., Division of Biological Sciences, The University of Chicago, Chicago, Illinois 60637
Ronald C. Desrosiers, Ph.D., New England Regional Primate Research Center, Harvard Medical School, Southborough, Massachusetts 01772
Guy de-Thé, M.D., Ph.D., Faculty of Medicine Alexis Carrel, Lyon; IRSC—CNRS, Villejuif, France
Susan Fennewald, Division of Biological Sciences, The University of Chicago, Chicago, Illinois 60637
Bernard Fleckenstein, M.D., Institut für Klinische Virologie, University of Erlangen-Nürnberg, D-8520 Erlangen, West Germany
Mark Heller, Ph.D., Division of Biological Sciences, The University of Chicago, Chicago, Illinois 60637
Gertrude Henle, M.D., The Joseph Stokes, Jr., Research Institute of The Children's Hospital of Philadelphia, University of Pennsylvania School of Medicine, Philadelphia, Pennsylvania 19104
Werner Henle, M.D., The Joseph Stokes, Jr., Research Institute of The Children's Hospital of Philadelphia, University of Pennsylvania School of Medicine, Philadelphia, Pennsylvania 19104
Mary Hummel, Ph.D., Division of Biological Sciences, The University of Chicago, Chicago, Illinois 60637
Elliott Kieff, M.D., Ph.D., Division of Biological Sciences, The University of Chicago, Chicago, Illinois 60637
Walter King, Ph.D., Division of Biological Sciences, The University of Chicago, Chicago, Illinois 60637
George Miller, M.D., Department of Epidemiology and Public Health, Yale University School of Medicine, New Haven, Connecticut 06510

Meihan Nonoyama, Ph.D., Department of Virology, Showa University Research Institute for Biomedicine in Florida, Clearwater, Florida 33520

Laurence Noel Payne, Ph.D., Houghton Poultry Research Station, Houghton, Huntingdon, Cambridgeshire PE17 2DA, England

James E. Robinson, M.D., Department of Pediatrics, Yale University School of Medicine, New Haven, Connecticut 06510

Bernard Roizman, Ph.D., Marjorie B. Kovler Viral Oncology Laboratories, The University of Chicago, Chicago, Illinois 60637

Vicky van Santen, Ph.D., Division of Biological Sciences, The University of Chicago, Chicago, Illinois 60637

Foreword

The first volume of the nineteen-volume series entitled *Comprehensive Virology* was published in 1974 and the last is yet to appear. We noted in 1974 that virology as a discipline had passed through its descriptive and phenomenological phases and was joining the molecular biology revolution. The volumes published to date were meant to serve as an in-depth analysis and standard reference of the evolving field of virology. We felt that viruses as biological entities had to be considered in the context of the broader fields of molecular and cellular biology. In fact, we felt then, and feel even more strongly now, that viruses, being simpler biological models, could serve as valuable probes for investigating the biology of the far more complex host cell.

During the decade-long compilation of a series of books like *Comprehensive Virology*, some of the coverage will obviously not remain up-to-date. The usual remedy to this aspect of science publishing is to produce a second edition. However, in view of the enormous increase in knowledge about viruses, we felt that a new approach was needed in covering virology in the 1980s and 1990s. Thus we decided to abandon the somewhat arbitrary subgrouping of the subject matter under the titles Reproduction, Structure and Assembly, Regulation and Genetics, Additional Topics, and Virus–Host Interactions. Instead we have organized a new series entitled *The Viruses*. This series will consist of individual volumes or groups of volumes, each to deal with a single virus family or group, each to be edited with full responsibility by an acknowledged authority on that topic, and each to cover all aspects of these viruses, ranging from physicochemistry to pathogenicity and ecology. Thus, over the next several years we plan to publish single volumes or multiple-volume sets devoted to each of the following virus families: Herpesviridae, Adenoviridae, Papovaviridae, Parvoviridae, Poxviridae, Reoviridae, Retroviridae, Picornaviridae, Togaviridae, Rhabdoviridae, Myxoviridae, and Paramyxoviridae, as well as hepatitis viruses, plant viruses, bacterial viruses, insect viruses, and perhaps other groups of viruses if and when they are deemed appropriate for comprehensive coverage and analysis.

This first volume of *The Viruses* begins a three- or four-volume set that will provide comprehensive coverage of herpesviruses. The editor is Bernard Roizman, who also deserves much of the credit for conceiving and promoting the idea of a second but independent series to succeed *Comprehensive Virology.* The Herpesviridae comprise a family of viruses widespread throughout the animal kingdom, many of which are extremely important pathogens. Diseases caused by herpesviruses are of ever-increasing significance as serious medical problems. In addition, research on the molecular biology, genetics, pathogenicity, and immunology of these complex viruses has in recent years undergone a veritable metamorphosis, which promises to continue for some years to come.

Heinz Fraenkel-Conrat
Robert R. Wagner

Contents

Chapter 7

The Molecular Biology of Marek's Disease Herpesvirus

Meihan Nonoyama

Chapter 8

Biology of Marek's Disease Virus and the Herpesvirus of Turkeys

Laurence Noel Payne

CHAPTER 1

The Family Herpesviridae: General Description, Taxonomy, and Classification

BERNARD ROIZMAN, Ph.D.

I. DESCRIPTION OF THE FAMILY

In the last 50 years, more than 80 distinct herpesviruses have been isolated from a wide variety of animal species. These viruses, comprising the family Herpesviridae, share structural features of their virions. As listed in Table I, the herpesvirus virion is made of four architectural elements: (1) a core consisting of a fibrillar spool on which the DNA is wrapped; (2) an icosadeltahedral capsid containing 12 pentameric and 150 hexameric capsomeres; (3) variable amounts of globular material asymmetrically arranged around the capsid and designated as the tegument; and (4) a membrane or envelope surrounding the entire structure (Wildy *et al.*, 1960; Roizman and Furlong, 1974). The virions of the various herpesviruses cannot be differentiated by electron-microscopic examination. The viruses are, however, readily differentiated on the basis of biological properties, the immunological specificity of their virions, and the size, base composition, and arrangement of their genomes. In fact, few virus families exhibit as much variation as the members of the family Herpesviridae.

A brief description listing the main characteristics of the family Herpesviridae is presented in Table I. These characteristics are common to all herpesviruses examined to date. Table II identifies the viruses that belong to the family Herpesviridae.

BERNARD ROIZMAN, Ph.D. • Marjorie B. Kovler Viral Oncology Laboratories, The University of Chicago, 910 East 58th Street, Chicago, Illinois 60637.

TABLE I. Characteristics of the Family Herpesviridae

Taxon	Vernacular name	Approved name
Family	Herpesvirus group	Herpesviridae

Main characteristics

A. Properties of the virus particles
   1. Nucleic acid: DNA, double-stranded, linear 32–75 G+C mole %, molecular weight approximately 80–150 × 10^6.
   2. Proteins: More than 20 structural polypeptides with molecular weights from 12,000 to 220,000.
   3. Lipid: Exact percentage of total weight unknown, probably variable; located in virion envelope.
   4. Carbohydrate: Exact percentage of total weight unknown, identified largely as covalently linked to envelope proteins.
   5. Physicochemical characteristics: Buoyant density (CsCl) of virion 1.20–1.29 g/cm^3. Molecular weight 10^9 or higher.
   6. Morphology: The virion (120–200 nm in diameter) consists of four structural components. The core consists of a fibrillar spool on which the DNA is wrapped. The ends of the fibers are anchored to the underside of the capsid shell. The capsid (100–110 nm) is an icosahedron with 5 capsomeres on each edge; it contains 150 hexameric and 12 pentameric capsomeres. The hexameric capsomeres contain a hole running halfway through the long axis. The tegument surrounding the capsid consists of globular material that is often distributed asymmetrically and may be variable in amount. The envelope, a bilayered membrane surrounding the tegument, has surface projections. The intact envelope is impermeable to negative stain.
   7. Antigenic properties: Neutralizing antibody reacts with major viral glycoproteins located in the viral envelope. An Fc receptor for immunoglobulin G may be present in the virion envelope.
   8. Effects of virus suspensions on cells: Fusion and agglutination occur rarely or only under very special conditions in the absence of replication.

B. Virus replication
   1. Entry: The viral envelope adsorbs to receptors on the plasma membrane of the host cell, ultimately fuses with the membrane, and releases the capsid into the cytoplasm. A DNA–protein complex is then translocated into the nucleus.
   2. Replication: Viral DNA is transcribed in the nucleus. Messenger RNAs generated from the transcripts are translated in the cytoplasm. Viral DNA is replicated in the nucleus and is spooled into preformed, immature nucleocapsids.
   3. Maturation and egress: The ability to infect cells is acquired as capsids become enveloped by budding through the inner lamella of the nuclear membrane and, in some instances, through other membranes of the cell. Virus particles accumulate in the space between the inner and outer lamellae of the nuclear membrane and in cysternae of the endoplasmic reticulum. Virus particles are released by transport to the cell surface through the modified endoplasmic reticulum.

C. Biological aspects
   1. Host range: Each virus has its own host range; this range may vary considerably both in nature and in the laboratory. Herpesviruses occur in both warm- and cold-blooded vertebrates and invertebrates. Some herpesviruses have been reported to induce neoplasia in their natural hosts and/or in experimental animals or in both. In cell culture, herpesviruses have been reported to convert cell strains into continuous cell lines that may cause invasive tumors in appropriate experimental hosts.

(Continued)

TABLE I. (*Continued*)

2. Transmission: For many herpesviruses, transmission is by contact between moist mucosal surfaces. Some herpesviruses can be transmitted transplacentally, intrapartum, via breast milk, or by blood transfusions, and some are probably also transmitted by airborne and waterborne routes.
3. Association of virus with host: Herpesviruses may remain latent in their primary host for the lifetime of those hosts; cells harboring latent virus may vary depending on the virus.

II. NAMING OF HERPESVIRUSES

Herpesviruses derive their name from the Greek herpes, ερπειν, meaning "creep." The term herpes was used in medicine to describe a variety of skin conditions and diseases for at least 25 centuries, but its meaning changed considerably during that time. Beswick (1962) and Wildy (1973) provided a fascinating history of the usage of the terms, and selected usage was illustrated by Onions (1971), but of the many clinical conditions, only a few, i.e., herpes simplex, herpes labialis, herpes genitalis, and herpes zoster are still designated by this term.

The earliest herpesviruses were named after the clinical conditions or the diseases they cause. Herpes simplex virus, herpes zoster virus, Aujeszky's disease virus (pseudorabies), and Marek's disease virus are examples of such designations. Subsequently, herpes viruses have also been designated after their hosts (e.g., herpesvirus hominis), the cell pathology they cause (e.g., cytomegaloviruses), or their discoverers (e.g., Epstein–Barr virus). Although many of these designations are firmly entrenched in the scientific literature, they either contravene the rules established by the International Committee for the Taxonomy of Viruses (ICTV) (e.g., the naming of viruses after a disease or their discoverers) or are misleading because the animal species from which the virus was initially isolated is not in fact the sole host. To preclude haphazard and frequently contradictory designations, the Herpesvirus Study Group appointed by the ICTV recommended a provisional system for the labeling of herpesviruses. The recommendations are as follows (Roizman *et al.*, 1981):

1. Each herpesvirus should be named after the taxonomic unit—the family or subfamily—to which its primary natural host belongs. The subfamily name should be applied to viruses isolated from the family Bovidae and from primates; in all other instances, the family name should be used. When family names are used, the name should end in *id* (e.g., Equid, Caviid). Names based on the

TABLE II. Provisional Designation and Classification of Viruses in the Family Herpesviridae

Provisional designation[a]	Common name (synonyms)	Subfamily[b]	G + C (mole %)	Genome properties: group[c] and (mol wt. × 10^6)	References
Viruses of humans					
Human herpesvirus 1	Herpes simplex virus type 1	α	67	E (96)	Gruter (1924), Schneweis (1962), Roizman (1979)
Human herpesvirus 2	Herpes simplex virus type 2	α	69	E (96)	Schneweis (1962), Roizman (1979)
Human herpesvirus 3	Varicella–zoster virus	α	46	D? (79–100)	Weller (1953), Iltis *et al.* (1977), Dumas *et al.* (1980), Ludwig *et al.* (1972)
Human herpesvirus 4	Epstein–Barr virus	γ	59	C (114)	Epstein *et al.* (1965), Schulte-Holthausen and Zur Hausen (1970)
Human herpesvirus 5	Cytomegalovirus	β	57	E (145)	Smith (1956), Stinski (1982), Plummer *et al.* (1969), Westrate *et al.* (1980), Sheldrick, Berthelot, Laithier, and Fleckenstein (personal communication)
Viruses of nonhuman primates					
Aotine herpesvirus 1	*Herpesvirus aotus* type 1	β	56	E (145)	Daniel *et al.* (1971), Sheldrick, Berthelot, Laithier, and Fleckenstein (personal communication)
Aotine herpesvirus 2	*Herpesvirus aotus* type 2	γ	—	B (100)	Barahona *et al.* (1973a), Fleckenstein, Daniel, and Mulder (personal communication)

Aotine herpesvirus 3	*Herpesvirus aotus* type 3	β	56	E (145)	Daniel *et al.* (1973), Sheldrick, Berthelot, Laithier, and Fleckenstein (personal communication)
Aotine herpesvirus 4	Owl monkey cytomegalovirus	β	—	—	Ablashi *et al.* (1972)
Ateline herpesvirus 1	Spider monkey herpesvirus	α	72	—	Hull *et al.* (1972), Goodheart and Plummer (1974)
Ateline herpesvirus 2	*Herpesvirus ateles* strain 810	γ	48	B (90)	Melendez *et al.* (1972), Fleckenstein *et al.* (1978)
Ateline herpesvirus 3	*Herpesvirus ateles* strain 73	γ	—	—	Deinhardt *et al.* (1973)
Callitrichine herpesvirus 1[d]	*Herpesvirus sanguinus*	—	—	—	Melendez *et al.* (1970)
Callitrichine herpesvirus 2	SSG, marmoset cytomegalovirus	β	—	—	Nigida *et al.* (1979)
Cebine herpesvirus 1	Capuchin herpesvirus (AL-5)	β	—	—	Lewis *et al.* (1974)
Cebine herpesvirus 2	Capuchin herpesvirus (AP-18)	β	—	—	Sabin and Wright (1934)
Cercopithecine herpesvirus 1	Simian herpesvirus, B virus, herpesvirus B	α	75	—	Malherbe and Harwin (1958)
Cercopithecine herpesvirus 2	SA8	α	67	E (100)	Goodheart and Plummer (1974), Mouchel, Berthelot, Desrosiers, and Sheldrick (personal communication)
Cercopithecine herpesvirus 3	SA6	β	51	—	Goodheart and Plummer (1974), Malherbe *et al.* (1963)
Cercopithecine herpesvirus 4	SA-15	β	—	—	Malherbe *et al.* (1963)
Cercopithecine herpesvirus 5	African green monkey cytomegalovirus (AGM-CMV)	β	—	—	Black *et al.* (1963)
Cercopithecine herpesvirus 6	Liverpool vervet monkey virus (LVMV or LVV)	α	52	—	Clarkson *et al.* (1967)
Cercopithecine herpesvirus 7	Patas monkey herpesvirus (MMV or PHV), delta herpesvirus	α	—	—	McCarthy *et al.* (1968)
Cercopithecine herpesvirus 8	Rhesus monkey cytomegalovirus	β	52	—	Asher *et al.* (1969)

(Continued)

TABLE II. (*Continued*)

Provisional designation[a]	Common name (synonyms)	Subfamily[b]	G+C (mole %)	Genome properties: group[c] and (mol wt. × 10^6)	References
Cercopithecine herpesvirus 9	Medical Lake macaque herpesvirus, simian varicella virus	α	—	—	Blakely *et al.* (1973)
Cercopithecine herpesvirus 10	Rhesus leukocyte-associated herpesvirus strain I (551 I)	γ	—	—	Frank *et al.* (1973)
Cercopithecine herpesvirus 11	Rhesus leukocyte-associated herpesvirus strain II (553 I)	γ	—	—	Frank *et al.* (1973)
Cercopithecine herpesvirus 12	Herpesvirus papio, baboon herpesvirus	γ	—	C (114)	Heller and Kieff (1981), Falk *et al.* (1976)
Cercopithecine herpesvirus 13	Herpesvirus cyclopsis	—	—	—	Jackman *et al.* (1977)
Cercopithecine herpesvirus 14	African green monkey EBV-like virus (AGM-EBV)	γ	—	—	Bocker *et al.* (1980)
Pongine herpesvirus 1[d]	Chimpanzee herpesvirus, herpesvirus pan	γ	—	C (114)	Landon *et al.* (1968), Heller and Kieff (personal communication)
Pongine herpesvirus 2[d]	Orangutan herpesvirus	γ	—	—	Rasheed *et al.* (1977)
Pongine herpesvirus 3	Herpesvirus gorilla	γ	—	—	Neubaner *et al.* (1979)
Saimiriine herpesvirus 1	Marmoset herpesvirus, herpesvirus M, herpes T, *Herpesvirus tamarinus*, *Herpesvirus platyrrhinae* type 1	α	67	D (100)	Goodheart and Plummer (1974), Holmes *et al.* (1964), Mouchel, Berthelot, Desrosiers, and Sheldrick (personal communication)
Saimiriine herpesvirus 2	Squirrel monkey herpesvirus, *Herpesvirus saimiri*	γ	46	B (103)	Melendez *et al.* (1968)
Tupaiine herpesvirus 1[d]	Tree shrew herpesvirus	—	65–66	—	Mirkovic *et al.* (1970), Darai *et al.* (1979, 1981), Ludwig (1972)

Viruses of other mammals					
Acelaphine herpesvirus 1	Malignant catarrhal fever virus of wildebeest	γ (?)	—	—	Plowright *et al.* (1960)
Acelaphine herpesvirus 2	Hartebeest herpesvirus	γ (?)	—	—	Reid and Rowe (1973)
Bovine herpesvirus 1	Infectious bovine rhinotracheitis virus, infectious pustular vulvovaginitis virus	α	72	D :	Plummer *et al.* (1969), Gibbs and Rweyemamu (1977), Madin, *et al.* (1956)
Bovine herpesvirus 2	Bovine mammillitis virus, Allerton virus, pseudo-lumpy skin disease virus	α	64	E (88)	Buchman and Roizman (1978a,b), Gibbs and Rweyemamu (1977), Martin *et al.* (1966)
Bovine herpesvirus 3	Bovine "orphan" herpesvirus	α	—	—	Gibbs and Rweyemamu (1977), Bartha *et al.* (1967)
Canid herpesvirus 1	Dog herpesvirus	α	32	—	Plummer *et al.* (1969), Strandberg and Carmichael (1965)
Caprine herpesvirus 1	Sheep herpesvirus	?	—	—	Mackay (1969)
Caprine herpesvirus 2	Domestic goat herpesvirus	α	—	—	Saito *et al.* (1974)
Caviid herpesvirus 1	Guinea pig herpesvirus (Hsuing–Kaplow virus)	—	—	—	Hsuing and Kaplow (1969), Bhatt *et al.* (1971)
Caviid herpesvirus 2	Guinea pig cytomegalovirus	β	57	—	Hartley *et al.* (1971), Middelkamp *et al.* (1967), Nayak (1971)
Cricetid herpesvirus (?)	Hamster herpesvirus	β	—	—	Smith (1959)
Elephantid herpesvirus	Elephant (loxodontal) herpesvirus	—	—	—	Basson *et al.* (1971)
Equid herpesvirus 1	Equine rhinopneumonitis virus, equine abortion virus	α	57	D (94)	Randall *et al.* (1953), Plummer *et al.* (1973)
Equid herpesvirus 2	Slow growing "cytomegalovirus"-like virus	β (?)	—	—	Plummer *et al.* (1973), Plummer and Waterson (1963)
Equid herpesvirus 3	Coital-exanthema virus	α	66	—	Bagust (1971), Ludwig *et al.* (1971)

(*Continued*)

TABLE II. (*Continued*)

Provisional designation[a]	Common name (synonyms)	Subfamily[b]	G+C (mole %)	Genome properties: group[c] and (mol wt. × 10^6)	References
Felid herpesvirus 1	Cat herpesvirus, infectious rhinotracheitis virus	α	46	—	Plummer *et al.* (1969), Ditchfield and Grinyer (1965)
Felid herpesvirus 2	Cat cytomegalovirus	β (?)	—	—	Fabricant and Gillespie (1974)
Leporid herpesvirus 1	Cottontail herpesvirus, *Herpesvirus sylvilagus*	γ	33	—	Goodheart and Plummer (1974), Hinze (1971)
Leporid herpesvirus 2	*Herpesvirus cuniculi*	—	—	—	Nesburn (1969)
Lorisine herpesvirus 1[d]	Kinkajou herpesvirus, *Herpesvirus pottos*	—	—	—	Barahona *et al.* (1973b)
Macropodid herpesvirus 1	Parma wallaby herpesvirus 1	—	71	—	Whalley and Webber (1979)
Murid herpesvirus 1	Mouse cytomegalovirus	β	59	(132)	Smith (1954), Mosmann and Hudson (1973)
Murid herpesvirus 2	Rat cytomegalovirus	β	47	—	Berezesky *et al.* (1971)
Murid herpesvirus 3	Mouse thymic virus	—	—	—	Parker *et al.* (1973), Rowe and Capps (1961)
Sciurid herpesvirus 1	European ground squirrel cytomegalovirus	β	—	—	Diosi *et al.* (1967)
Suid herpesvirus 1	Pseudorabies virus, Aujeszky's disease virus	α	74	D (91)	Gustafsohn (1970)
Suid herpesvirus 2	Inclusion-body rhinitis virus, pig cytomegalovirus	β	—	—	L'Ecuyer and Corner (1966), Valicek *et al.* (1970)
Viruses of birds					
Anatid herpesvirus 1	Duck plaque herpesvirus	—	—	—	Baudet (1928), Breese and Dardiri (1968)
Ciconiid herpesvirus 1	Black stork herpesvirus	—	—	—	Kaleta *et al.* (1980b)

Columbid herpesvirus 1	Pigeon herpesvirus 1	—	59	—	Cornwell *et al.* (1970), Lee *et al.* (1972)
Falconid herpesvirus 1	Falcon inclusion-body disease virus	—	—	—	Mare and Graham (1973)
Gallid herpesvirus 1	Infectious laryngotracheitis virus	α	45	—	Plummer *et al.* (1969), Lee *et al.* (1972), May and Tittsler (1925), Cruickshank *et al.* (1963)
Gallid herpesvirus 2	Marek's disease herpesvirus	γ	46	E (110)	Churchill and Biggs (1967), Lee *et al.* (1971) (see also Chapter 7), Sheldrick *et al.* (1982)
Gruid herpesvirus 1	Crane herpesvirus	—	—	—	Burtscher and Grunberg (1979)
Meleagrid herpesvirus 1	Turkey herpesvirus	γ	48	E (104)	Lee *et al.* (1972), Kawamura *et al.* (1969), Sheldrick *et al.* (1982)
Perdicid herpesvirus 1	Bobwhite quail herpesvirus	—	—	—	Kaleta *et al.* (1980a)
Phalacrocoracid herpesvirus 1	Cormorant herpesvirus	—	58	—	Lee *et al.* (1972), French *et al.* (1973)
Psittacid herpesvirus 1	Parrot herpesvirus; recently rediscovered Pacheco's disease virus	—	—	—	Simpson *et al.* (1975)
Strigid herpesvirus 1	Virus of hepatosplenitis of owls	—	61	—	Lee *et al.* (1972), Schetter (1970), Burki *et al.* (1973)
Viruses of reptiles					
Elapid herpesvirus (?)	Indian cobra, banded krait, Siamese cobra virus	—	—	—	Monroe *et al.* (1968), Lunger and Clark (1978)
Chelonid herpesvirus 1	Gray patch disease agent of green sea turtle	—	—	—	Rebell *et al.* (1975), Haines and Kleese (1977)
Chelonid herpesvirus (2?)	Pacific pond turtle virus	—	—	—	Frye *et al.* (1977)
Chelonid herpesvirus (3?)	Painted turtle virus	—	—	—	Cox *et al.* (1980)
Iguanid herpesvirus 1	Green iguana virus	—	—	—	Zeigel and Clark (1972), Clark and Karzon (1972)
Lacertid herpesvirus (?)	Green lizard virus	—	—	—	Raynaud and Adrian (1970)

(*Continued*)

TABLE II. (*Continued*)

Provisional designation[a]	Common name (synonyms)	Subfamily[b]	G+C (mole %)	Genome properties: group[c] and (mol wt. × 10^6)	References
Viruses of amphibians					
Ranid herpesvirus 1	Lucke virus	—	46	—	Lucke (1938), Wagner *et al.* (1970), Gravell (1971)
Ranid herpesvirus 2	Frog virus 4	—	56	—	Gravell (1971), Rafferty (1965)
Viruses of bony fishes					
Cyprinid herpesvirus (?)	Carp pox virus	—	—	—	Schubert (1964)
Ictalurid herpesvirus 1	Channel catfish virus (CCV)	α	56	A (86)	Chousterman *et al.* (1979), Wolf and Darlington (1971)
Percid herpesvirus 1	Wally epidermal hyperplasia virus	—	—	—	Kelly *et al.* (1980)
Pleuronectid herpesvirus	*Herpesvirus scophthalmus*, turbot herpesvirus	—	—	—	Buchanan *et al.* (1978)
Salmonid herpesvirus 1	*Herpesvirus salmonis*	—	—	—	Wolf *et al.* (1978)
Salmonid herpesvirus 2	*Oncorhynchus masou* virus	—	—	—	Kimura *et al.* (1980)

[a] A question mark denotes a provisional listing pending isolation and characterization of the virus.
[b] Subfamily classification is based on criteria listed in Table III.
[c] Genome classification is based on criteria listed in the text and Table IV.
[d] The formal classification does not include subfamilies. In these instances, it was assumed that the subfamily name would be the same as the family name.

subfamily to which the host belongs should end in *ine* (e.g., Bovine, Cebine). The only exceptions are viruses isolated from man; because humans are the sole living members of the family Hominidae, the term human rather than hominid shall be used. Because virus names include terms that are at the hierarchial level of "species" and the ICTV has not yet considered official species designations, the label for each of the herpesviruses may be considered vernacular and therefore may be transliterated into the language in which the publication is printed. As species nomenclature is made official by the ICTV, the question of transliteration may be reconsidered.

2. The herpesviruses within each host family or subfamily will be given arabic numbers. The numbers should not be preceded by the word "type." Each new herpesvirus will receive the next available number.

The rationale for some of the proposed rules was published in the first report of the Herpesvirus Study Group (Roizman *et al.*, 1973). Some of the objectives of those proposals are as follows:

1. Inasmuch as herpesviruses have been isolated from nearly every species investigated to date, the naming of viruses after the species or genus of the host is impractical; in addition to the large number of "groupings" such labeling would create, it is often difficult to determine which species is the main natural host. The naming of herpesviruses after the host's family creates a more manageable number of groups, but presents particular problems for viruses isolated from the family Bovidae and from nonhuman primates. Specifically, the selection of host family name as the basis for the provisional nomenclature was based on the notion that the number of equivalent hierarchical groupings should be in some balance with the number of viruses placed in these groupings. At this time, only the number of viruses isolated from nonhuman primates is out of balance with the number of groups in which they are placed. In addition, in the case of the herpesviruses isolated from the family Bovidae, it is convenient to differentiate among the viruses of sheep, domestic cattle, and wildebeest, and this can be readily accomplished by grouping viruses according to the subfamily to which the host belongs. Conceivably, it may be necessary to apply subfamily rather than family names to herpesviruses isolated from other families when the number of herpesviruses escalates.

2. The proposal to use transliterated names based on the family or subfamily to which the primary host belongs is designed to avoid the use of latinized names, which would be confused with an official binomial nomenclature. The latter will depend on ICTV proceedings in the future.

3. The arabic number with which a particular herpesvirus is labeled should carry no implied meaning about the properties of the virus.

The herpesviruses the properties of which are sufficiently well defined to merit designation are listed in Table II. The current names of some of these viruses, particularly those from humans, are so well established that they are unlikely to be totally abandoned in favor of the new labels; however, the naming of new herpesviruses should follow the recommendations listed herein.

III. CLASSIFICATION OF HERPESVIRUSES

Extensive studies carried out in the past decade have made it apparent that in the course of their evolution, herpesviruses have undergone considerable diversification with respect to (1) virion antigenic properties, (2) biological properties, and (3) composition, size, and arrangement of their deoxynucleotide sequences, even though the morphology of the virion has been preserved. Of these three major sets of properties that could be used for classification of the family Herpesviridae into major subdivisions, the virion antigenic properties appear to be least useful. Specifically, serological tests have identified clusters of closely related herpesviruses that also by other criteria can be placed in the same or closely related genera; however, serology has not provided a firm foundation for subdivision of the viruses into a small number of subfamily groups. Before I consider a classification based on the remaining two major sets of properties, it seems worthwhile to list the minimum number of subdivisions permitted by each set.

A. Grouping of Herpesviruses on the Basis of Biological Properties

With respect to biological properties, the herpesviruses that have been studied in some detail can be clustered into three subfamilies on the basis of host range, duration of reproductive cycle, cytopathology, and characteristics of latent infection (Table III). These characteristics do not have equal weight in each subfamily. For example, the differentiation between subfamilies Alphaherpesvirinae and Gammaherpesvirinae is based on host range *in vitro* and characteristics of latent infection. The differentiation between subfamily Betaherpesvirinae and the others is based primarily on the length of the reproductive cycle and slow development of cytopathology in cell cultures. Cell association was specifically excluded because this characteristic often conflicts with surmised and, in some instances, known properties of viruses *in vivo*. Thus, although Gallid herpesvirus 2 and human herpesvirus 3 are strongly cell-associated in cell culture, cell-free virus can be found in high titer at the sites of multiplication, and their spread in nature indicates dissemination by cell-free virus rather than by infected cells. Furthermore, in the case of Gallid herpesvirus 2, dissemination by cell-free virus has been proven.

TABLE III. Subdivision of Herpesviruses According to Their Biological Properties

Subfamily	Examples
Alphaherpesvirinae	
Host range: *In vivo* variable from very wide to very narrow; *in vitro* also variable.	Human herpesvirus 1 Human herpesvirus 2 Equid herpesvirus 1 Bovine herpesvirus 2 Suid herpesvirus1
Duration of reproductive cycle: Short.	
Cytopathology: Rapid spread of infection in cell culture resulting mass destruction of susceptible cells. Establishment of carrier cultures of susceptible cells harboring nondefective genomes difficult to accomplish.	
Latent infections: Frequently, but not exclusively, in ganglia.	
Betaherpesvirinae	
Host range: *In vivo* narrow, frequently restricted to the species or genus to which the host belongs; *in vitro* replicates best in fibroblasts, although exceptions exist.	Human herpesvirus 5 Murid herpesvirus 1 Suid herpesvirus 2 Caviid herpesvirus 2
Duration of reproductive cycle: Relatively long.	
Cytopathology: Slowly progressing lytic foci in cell culture. The infected cells frequently become enlarged (cytomegalia) both *in vitro* and *in vivo*. Inclusions containing DNA frequently present in both nuclei and cytoplasm. Carrier cultures easily established.	
Latent infections: Possibly in secretory glands, lymphoreticular cells, and kidneys and other tissues.	
Gammaherpesvirinae	
Host range: *In vivo* usually limited to the same family or order as the host it naturally infects. *In vitro* all members of this subfamily replicate in lymphoblastoid cells, and some also cause lytic infections in some types of epithelioid and fibroblastoid cells. Viruses in this group are specific for either B or T lymphocytes. In the lymphocyte, infection is frequently arrested either at a prelytic stage with persistence and minimum expression of the viral genome or at a lytic stage, causing cell death without production of complete virions.	Human herpesvirus 4 Gallid herpesvirus 2
Duration of reproductive cycle: Variable.	
Cytopathology: Variable.	
Latent infections: Latent virus is frequently demonstrated in lymphoid tissue.	
Leporid herpesvirus 1	
Meleagrid herpesvirus 1	

Because a particular herpesvirus may clearly belong to a particular subfamily by virtue of its overall characteristics even though it may not meet all criteria, it seemed appropriate to designate the subfamilies by neutral names (Alpha, Beta, Gamma), rather than according to a prominent member or characteristic of the group.

B. Grouping of Herpesviruses on the Basis of the Structure of Their Genomes

Base composition, size of the genome, and arrangement of the reiterated sequences within the genome are three characteristics of the structure of herpesvirus genomes that can be considered.

The average base composition is not helpful for a hierarchical classification inasmuch as it appears to vary continuously from 32 to 75 G+C moles% (see Table II); subdivision according to base composition would therefore be arbitrary and, as in the case of Gallid herpesvirus 2 and human herpesvirus 1, would tend to separate viruses with similar genome sizes and sequence arrangements into different subfamilies.

Although herpesviruses differ significantly in the sizes of their genomes, this characteristic is more suitable for identification than for classification inasmuch as the molecular weight of most herpesvirus genomes is not known. There is in addition the possibility that the presently discernible groupings of small, medium, and large herpesvirus genomes may become less discrete as additional information on other herpesviruses becomes available.

The most useful characteristic for the classification of herpesviruses is the arrangement of reiterated sequences. For purposes of classification, it is convenient to focus on reiteration of sequences containing at least 100 nucleotides. On this basis, the viral genomes examined to date fall into five groups (Table IV) as follows:

Group A: Characterized by single reiteration of one set of sequences in the same orientation at the termini (Chousterman *et al.*, 1979).

Group B: Characterized by multiple reiterations of one set of sequences in the same orientation at both termini (Bornkamm *et al.*, 1976).

Group C: Characterized by multiple reiterations of one set of sequences in the same orientation at both termini as well as by internal tandem reiterations of other sets of sequences (Raab-Traub *et al.*, 1980).

Group D: Characterized by inverted reiteration of a set of sequences from one terminus internally, as well as by reiteration of a subset of these sequences in the same orientation at the other terminus (Ben-Porat *et al.*, 1979).

Group E: Characterized by inverted reiteration of sets of sequences for both termini internally, as well as by reiteration of a subset of these sequences at both termini in the same orientation (Sheldrick and Berthelot, 1975; Wadsworth *et al.*, 1975).

C. Selection of Viral Properties for a Hierarchical Classification

One basic problem confronting the Study Group has been that no two of the three major sets of properties described above can be used for

TABLE IV. Grouping of Herpesviruses on the Basis of the Properties of Their Genomes

Group	Arrangement of reiterated sequences	Number of isomeric arrangements[a]	Examples
A	Single set reiterated at termini in the same orientation	1	Ictalurid herpesvirus 1
B	Numerous reiterations of the same set of sequences at both termini in the same orientation	1	Saimiriine herpesvirus 2 Ateline herpesvirus 2
C	(1) Numerous reiterations of the same set of sequences at both termini in the same orientation; (2) variable number of tandem reiterations of different sequences internally	1	Human herpesvirus 4
D	(1) Single set of sequences from terminus reiterated internally; (2) subset of terminal sequences reiterated at both termini in the same orientation	2	Suid herpesvirus 1 Equid herpesvirus 1
E	(1) Single set of sequences from both termini reiterated in inverted form internally; (2) subset of terminal sequences reiterated at both termini in the same orientation	4	Human herpesvirus 1 Human herpesvirus 2 Bovine herpesvirus 2 Gallid herpesvirus 2[b] Human herpesvirus 5[b]

[a] Defined by the number of genome populations differing in the location of sequences in the unique regions relative to the termini.
[b] The presence of a terminal sequence reiterated at both termini is surmised, but not yet proved.

clustering all herpesviruses into hierarchically equivalent subdivisions. For example, although Saimiriine herpesvirus 2, human herpesvirus 4, and Gallid herpesvirus 2 can readily be placed into the subfamily Gammaherpesvirinae, their genomes fall into groups B, C, and E, respectively (Bornkamm *et al.*, 1976; Raab-Traub *et al.*, 1980) (see also Chapter 7). Even if the differences in the sizes of the genomes and the presence of internal reiteration of human herpesvirus 4 (4) were ignored, these viruses would still constitute two distinct groups on the basis of their genome structure. Because it is desirable to reduce the number of subfamilies to a minimum, we propose a hierarchical classification based on the dominance of one set of properties over all others. Inasmuch as biological properties are more readily established following isolation of a new herpesvirus, these properties have been selected as the dominant criterion for classification of herpesvirus into subfamilies—the major, hierarchically equivalent, subdivisions of the family. Accordingly, herpesviruses have been subdivided into three subfamilies designated Alphaherpesvirinae, Betaherpesvirinae, and Gammaherpesvirinae, by criteria listed in Table III. The provisional classification of herpesviruses into subfamilies is presented in Table II.

The classification of herpesviruses into genera should reflect phy-

TABLE V. Classification of Selected Herpesviruses into Subfamilies and Genera

Taxon	Approved name[a]	Common name
Family	Herpesviridae	Herpesvirus group
Subfamily	Alphaherpesvirinae	Rapidly growing highly cytolytic herpesviruses
Genus	*Simplexvirus*	Herpes simplex-like viruses
Species	Human herpesvirus 1	Herpes simplex virus 1
Species	Human herpesvirus 2	Herpes simplex virus 2
Species	Bovine herpesvirus 2	Bovine mammillitis virus
(Species[b])	Cercopithecine herpesvirus 2	SA8
(Species[b])	Cercopithecine herpesvirus 1	B virus
Genus	*Poikilovirus*	Pseudorabieslike viruses
Species	Suid herpesvirus 1	Pseudorabies viruses
(Species[c])	Equid herpesvirus 1	Equine rhinopneumonitis virus
Genus	*Varicellavirus*	Varicella--zoster-like viruses
Species	Human herpesvirus 3	Varicella–zoster virus
Subfamily	Betaherpesvirinae	Slow growing, cytomegalic viruses
Genus	*Cytomegalovirus*	Human cytomegaloviruses
Species	Human herpesvirus 5	Human cytomegalovirus
Genus	*Muromegalovirus*	Murine cytomegaloviruses
Species	Murid herpesvirus 1	Murine cytomegalovirus
Subfamily	Gammaherpesvirinae	Lymphocyte-associated viruses
Genus	*Lymphocrytovirus*	Epstein–Barr-like viruses
Species	Human herpesvirus 4	Epstein–Barr virus
Species	Cercopithecine herpesvirus 12	Baboon herpesvirus
Species	Pongine herpesvirus 1	Chimpanzee herpesvirus
Genus	*Thetalymphocryptovirus*	Marek's disease-like herpesviruses
Species	Gallid herpesvirus 2	Marek's disease herpesvirus
Sepcies	Meleagrid herpesvirus 1	Turkey herpesvirus
Genus	*Rhadinovirus*	Saimiri ateles-like herpesviruses
Species	Ateline herpesvirus 2	*Herpesvirus ateles* (strain 810)
Species	Ateline herpesvirus 3	*Herpesvirus ateles* (strain 73)
Species	Saiminiine herpesvirus 2	*Herpesvirus saimiri*

[a] Only the family and subfamily names have been approved by the ICTV. The remainder are proposed names.

[b] Parentheses indicate that the species is a candidate for that genus on the basis of serological data. Final placement will be determined on the basis of the genome structure.

[c] Parentheses indicate that the species is a candidate for that genus based on genome structure and weak serological relationship. Final placement deferred until nucleotide sequence homology is determined.

logenetic relationship and shall be on the basis of genome structure, serological relationship, and, when appropriate, sequence homology. A proposed initial classification of herpesviruses into genera is shown in Table V. Definitive assignments will be consistent with the criteria listed above. Thus, the genus *Simplexvirus* in subfamily Alphaherpesvirinae includes human herpesvirus 1, human herpesvirus 2, and Bovine herpesvirus 2, which have similar genome structures (Buchman and Roizman, 1978a,b) and are closely related by serology and nucleotide sequence homology (Sterz *et al.*, 1974). The two candidate members are related serologically (Honess and Watson, 1977), but their sequence homology with the other members of the genus is not yet known. Similar criteria were employed for the erection of several genera in the Gammaherpesvirinae. In several instances involving the Alphaherpesvirinae and Betaherpesvirinae subfamilies, it was clear that some individual herpesviruses will constitute sole or initial members of new genera, and these were designated accordingly.

REFERENCES

Ablashi, D.V., Chopra, H.C., and Armstrong, G.R., 1972, A cytomegalovirus isolated from an owl monkey, *Lab. Anim. Sci.* **22**:190–195.

Asher, D.M., Gibbs, C.J., and Long, D.J., 1969, Rhesus monkey cytomegalovirus: Persistent asymptomatic viruria Abstract, *Bacteriol. Proc.* 191.

Bagust, T.J., 1971, The equine herpesviruses (a review), 1971, *Vet. Bull.* **41**:79–92.

Barahona, H.H., Melendez, L.V., King, N.W., Daniel, M.D., Fraser, C.E.O., and Preville, A.E., 1973a, Herpesvirus aotus type 2: A new viral agent from owl monkeys (*Aotus trivirgatus*), *J. Infect. Dis.* **127**:171–178.

Barahona, H.H., Trum, B.F., Melendez, L.V., Garcia, F.G., King, N.W., Daniel, M.D., and Jackman, D.A., 1973b, A new herpes virus isolated from kinkajou (*Botos flavus*), *Lab. Anim. Sci.* **23**:830–836.

Bartha, A., Juhasz, M., Liebermann, H., Hantschel, H., and Schulze, P., 1967, Isolierung und Eigenschaften eines bovinen Herpesvirus von einem Kalb mit respiratorischer Krankheit und Keratokonjunktivitis, *Arch. Exp. Vet. Med.* **21**:616–623.

Basson, P.A., McCully, R.M., De Voss, B., Young, E., and Kruger, S.P., 1971, Some parasitic and other natural diseases of the African Elephant in the Kruger National Park, *Onderstepoort. J. Vet. Res.* **38**:239–254.

Baudet, A.E.R.F., 1928, Mortality in ducks in the Nederlands caused by a filterable virus: Fowl plague, *Tijdschr. Diergeneeskd.* **50**:455–459.

Ben-Porat, T., Rixon, F. J., and Blankenship, M.L., 1979, Analysis of the structure of the genome of pseudorabies virus, *Virology* **95**:285–294.

Berezesky, I.K., Grimley, P.M., Tyrell, S.A., and Rabson, A.S., 1971, Ultrastructure of a rat cytomegalovirus, *Exp. Mol. Pathol.* **14**:337–349.

Beswick, T.S.L., 1962, The origin and the use of the word herpes, *Med. Hist.* **6**:214–232.

Bhatt, P.N., Percy, D.H., Craft, J.L., and Jonas, A.M., 1971, Isolation and characterization of a herpes-like (Hsiung–Kaplow) virus from guinea pigs, *J. Infect. Dis.* **123**:178–179.

Black, P.H., Hartley, J., and Rowe, W.P., 1963, Isolation of cytomegalovirus from African green monkey, *Proc. Soc. Exp. Biol. Med.* **112**:601–605.

Blakely, G.A., Lourie, B., Morton, W.G., Evans, H.H., and Kaufman, A.F., 1973, A varicella-like disease in macaque monkeys, *J. Infect. Dis.* **127**:617–625.

Bocker, J.F., Tiedemann, K.-H., Bornkamm, G.W., and Zur Hausen, M., 1980, Characterization of an EBV-like virus from African green monkey lymphoblasts, *Virology* **101**:291–295.

Bornkamm, G.W., Delius, H., Fleckenstein, B., Werner, F.J., and Mulder, C., 1976, The structure of herpesvirus saimiri genomes: Arrangement of heavy and light sequences within the M genome, *J. Virol.* **19**:154–161.

Breese, S.S., Jr., and Dardiri, A.H., 1968, Electron microscopic characterization of duck plague virus, *Virology* **34**:160–169.

Buchanan, J.S., Richards, R.H., Sommerville, C., and Madeley, C.R., 1978, A herpestype virus from turbot (*Scophthalmus maximus* L), *Vet Rec.* **102**:527–528.

Buchman, T.G., and Roizman, B., 1978a, Anatomy of bovine mammillitis virus DNA. I. Restriction endonuclease maps of four populations of molecules that differ in the relative orientation of their long and short components, *J. Virol.* **25**:395–407.

Buchman, T.G., and Roizman, B., 1978b, Anatomy of bovine mammillitis DNA. II. Size and arrangements of the deoxynucleotide sequences, *J. Virol.* **27**:239–254

Burki, F., Burtscher, H., and Sibalin, M., 1973, Herpesvirus strigis: A new avian herpesvirus. I. Biological properties, *Arch. Gesamte Virusforsch.* **43**:14–24.

Burtscher, H., and Grunberg, W., 1979, Herpesvirus-hepatitis bei Kränchen (Aves—Gruidae). I. Pathomorphologische Befunde, *Zentralbl. Veterinaermed. Reihe B* **26**:561–569.

Cebrian, J., Kaschka-Dierich, C., Berthelot, N., and Sheldrick, P. 1982, Inverted repeat nucleotide sequences in the genomes of Marek's disease virus and the herpesvirus of the turkey *Proc. Natl. Acad. Sci. USA* **79**:555–558.

Chousterman, S., Lacasa, M., and Sheldrick, P., 1979, Physical map of the channel catfish virus genome: Location of sites for restriction endonucleases *Eco*RI, *Hind*III, *Hpa*I, and *Xba*I, *J. Virol.* **31**:73–85.

Churchill, A.E., and Biggs, P.M., 1967, Agent of Marek's disease in tissue culture, *Nature (London)* **215**:528–530.

Clark, H.F., and Karzon, D.T., 1972, Iguana virus, a herpes-like virus isolated from cultured cells of a lizard, *Iguana iguana*, *Infect. Immun.* **5**:559–569.

Clarkson, M.J., Thorpe, E., and McCarthy, K., 1967, A virus disease of captive vervet monkeys (*Cercopithecus aethiops*) caused by a new herpesvirus, *Arch. Gesamte Virusforsch.* **22**:219–234.

Cornwell, H.J.C., Wright, N.G., and McCusker, H.B., 1970, Herpesvirus infection of pigeons. II. Experimental infection of pigeons and chicks, *J. Comp. Pathol.* **80**:229–232.

Cox, W.R., Rapley, W.A., and Barker, I.K., 1980, Herpesvirus-like infection in the painted turtle (*Chrysemys picta*), *J. Wildl. Dis.* **16**:445–449.

Cruickshank, J.O., Berry, D.M., and Hay, B., 1963, The fine structure of infectious laryngotracheitis virus, *J. Virol.* **20**:376–378.

Daniel, M.D., Melendez, L.V., King, N.W., Fraser, C.E.O., Barahona, H.H., Hung, R.D., and Garcia, F.G., 1971, Herpesvirus aotus: A latent herpesvirus from owl monkeys (*Aotus trivirgatus*)—Isolation and characterization, *Proc. Soc. Exp. Biol. Med.* **138**:835–845.

Daniel, M.D., Melendez, L.V., King, N.W., Barahona, H.H., Fraser, C.E.O., Garcia, F.G., and Silva, D., 1973, Isolation and characterization of a new virus from owl monkeys: Herpesvirus aotus type 3, *Am. J. Phys. Anthropol.* **38**:497–500.

Darai, G., Matz, B., Schroder, C.H., Flugel, R.H., Berger, U., Munk, K., and Gelderblom, H., 1979, Characterization of a tree shrew herpesvirus isolated from a lymphosarcoma, *J. Gen. Virol.* **43**:541–551.

Darai, G., Flugel, R.M., Matz, B., and Delius, H., 1981, DNA of Tupaia herpesviruses, in: *Herpesvirus DNA* (Y. Becker, ed.), p. 345–361, Martinus Nijhoff Publishers, The Hague 1981.

Deinhardt, F., Falk, L.A., and Wolfe, L.G., 1973, Simian herpesviruses, *Cancer Res.* **33**:1424–1426.

Diosi, P., Babusceac, L., and David, C., 1967, Recovery of cytomegalovirus from the submaxillary glands of ground squirrels, *Arch. Gesamte Virusforsch.* **20**:383–386.

Ditchfield, J., and Grinyer, I., 1965, Feline rhinotracheitis virus: A feline herpesvirus, *Virology* **26**:504–506.

Dumas, A.M., Geelen, J.L.M.C., Maris, W., and Van der Noordaa, J., 1980, Infectivity and molecular weight of varicella-zoster virus DNA, *J. Gen. Virol.* **47**:233–235.

Epstein, M.A., Henle, W., Achong, B.G., and Barr, Y.M., 1965, Morphological and biological studies on a virus in cultured lymphoblasts from Burkitt's lymphoma, *J. Exp. Med.* **121**:761–770.
Fabricant, C.G., and Gillespie, J.H., 1974, Identification and characterization of a second feline herpesvirus, *Infect. Immun.* **9**:460–466.
Falk, L., Deinhardt, F., Nonoyama, M., Wolfe, L.G., Bergholz, C., Lapin, B., Yakovleva, L., Agrba, V., Henle, G., and Henle, W., 1976, Properties of a baboon lymphotropic herpesvirus related to Epstein–Barr virus, *Int. J. Cancer* **18**:798–807.
Fleckenstein, B., Bornkamm, G.W., Mulder, C., Werner, F.-J., Daniel, M.D., Falk, L.A., and Delius, H., 1978, Herpesvirus ateles DNA and its homology with herpesvirus saimiri nucleic acid, *J. Virol.* **25**:361–373.
Frank, A.L., Bissell, J.A., Rowe, D.S., Dunnick, N.R., Mayner, R.E., Hopps, H.E., Parkman, P.D., and Meyer, H.M., Jr., 1973, Rhesus leucocyte-associated herpesvirus. I. Isolation and characterization of a new herpesvirus recovered from rhesus monkey leukocytes, *J. Infect. Dis.* **128**:618–629.
French, E.L., Purchase, H.G., and Nazerian, K., 1973, A new herpevirus isolated from a nestling cormorant (*Phalacrocorax melanoleucos*), *Avian Pathol.* **2**:3–15.
Frye, F.L., Oshiro, L.O., Dutra, F.R., and Carney, J.D., 1977, Herpesvirus-like infection in two Pacific pond turtles, *J. Am. Vet. Med. Assoc.* **171**:882–884.
Gibbs, E.P.J., and Rweyemamu, M.M., 1977, Bovine herpesviruses, *Vet. Bull.* **47**:317–425.
Goodheart, C., and Plummer, G., 1974, The densities of herpes viral DNAs, in: *Progress in Medical Virology*, Vol. 19 (J.L. Melnick, ed.), p. 324, S. Karger, Basel.
Gravell, M., 1971, Viruses and renal carcinoma of *Rana pipiens*. X. Comparison of herpes-type viruses associated with Lucke tumor-bearing frogs, *Virology* **43**:730–733.
Gruter, W., 1924, Das Herpesvirus, seine aetiologische und klinische Bedeutung, *Muench. Med. Wochenschr.* **71**:1058–1060.
Gustafsohn, D.P., 1970, Pseudorabies in: *Diseases of Swine*, 3rd ed. (H.W. Dunne, ed.), pp. 337–355, Iowa State Univ. Press.
Haines, H., and Kleese, W.C., 1977, Effect of water temperature on a herpesvirus infection of sea turtles, *Infect. Immun.* **15**:756–759.
Hartley, H.W., Rowe, W.P., and Huebner, R.J., 1971, Serial propagation of the guinea pig salivary gland virus in tissue culture, *Proc. Soc. Exp. Biol. Med.* **96**:281–285.
Heller, M., and Kieff, E.D., 1981, Colinearity between the DNAs of Epstein–Barr virus and herpes virus papio, *J. Virol.* **37**:698–709.
Hinze, H.C., 1971, New member of the herpesvirus group isolated from wild cottontail rabbits, *Infect. Immun.* **3**:350–354.
Holmes, A.W., Caldwell, R.G., Dedmon, R.E., and Deinhardt, F., 1964, Isolation and characterization of a new herpes virus, *J. Immunol.* **92**:602–610.
Honess, R.W., and Watson, D.H., 1977, Unity and diversity in the herpesviruses, *J. Gen. Virol.* **37**:15–38.
Hsiung, G.D., and Kaplow, L.S., 1969, Herpes like virus isolated from spontaneously degenerated tissue culture derived from leukemia-susceptible guinea pigs, *J. Virol.* **3**:355–357.
Hull, R.N., Dwyer, A.C., Holmes, A.W., Nowakowski, E., Deinhardt, F., Lennette, E.H., and Emmons, R.W., 1972, Recovery and characterization of a new simian herpesvirus from a fatally infected spider monkey, *J. Natl. Cancer Inst.* **49**:225–230.
Iltis, J.P., Oakes, J.E., Hyman, R.W., and Rapp, F., 1977, Comparison of the DNAs of varicella-zoster viruses isolated from clinical cases of varicella and herpes zoster, *Virology* **82**:345–352.
Jackman, D.A., King, N.W., Daniel, M.D., Sehgal, D.K., and Fraser, C.E.O., 1977, *M. cyclopsis*: A new herpesvirus isolated from *Macaca cyclopsis*, Abstracts of the Annual Meeting of the American Society for Microbiology, Abstract V348.
Kaleta, E.F., Marschall, H.J., Glunder, G., and Stiburek, B., 1980a, Isolation and serological differentiation of a herpesvirus from bobwhite quail (*Colinus virginianus*, L. 1758), *Arch. Virol.* **66**:359–364.

Kaleta, E.F., Mikami, T., Marschall, H.J., Heffels, U., Heidenreich, M., and Stiburek, B., 1980b, A new herpesvirus isolated from black storks (*Ciconia nigra*), *Avian Pathol.* 9:301–310.

Kawamura, H., King, D.J., and Anderson, D.P., 1969, A herpesvirus isolated from kidney cell culture of normal turkeys, *Avian Dis.* **13**:853–863.

Kelly, R.K., Nielsen, O., and Yamamoto, T., 1980, A new herpes-like virus (HLV) of fish (*Stizostedion vitreum-vitreum*): Abstract, Proceedings of the Meeting of the Tissue Culture Association, *In Vitro* **16**:225.

Kimura, T., Ioshimizu, M., and Tanaka, M., 1980, Salmonid viruses: Effect of *Oncorhynchus masou* virus (OMV) in fry of chum salmon (*Oncorhynchus keta*), *Fish Health News* 9:2–3.

Landon, J.E., Ellis, L.B., Zeve, V.H., and Fabrizio, D.P., 1968, Herpes-type virus in cultured leukocytes from chimpanzees, *J. Natl. Cancer Inst.* **40**:181–192.

L'Ecuyer, C., and Corner, A.H., 1966, Propagation of porcine cytomegalic inclusion disease virus in cell cultures: Preliminary report, *Can. J. Comp. Med. Vet. Sci.* **30**:321–326.

Lee, L.F., Kieff, E.D., Bachenheimer, S.L., Roizman, B., Spear, P.G., Burmester, B.R., and Nazerian, K., 1971, Size and composition of Marek's disease virus deoxyribonucleic acid, *J. Virol.* 7:289–294.

Lee, L.F., Armstrong, R.L., and Nazerian, K., 1972, Comparative studies of six avian herpesviruses, *Avian Dis.* **16**:799–805.

Lewis, M.A., Frye, L.D., Gibbs, C.J., Jr., Chou, S.M., Cutchins, E.C., and Gajdusek, D.C., 1974, Isolation and characterization of two new unrelated herpesviruses from capuchin monkeys, Abstracts of the Annual Meeting of the American Society for Microbiology, Abstract 336.

Lucke, B., 1938, Carcinoma of the leopard frog: Its probable causation by a virus, *J. Exp. Med.* **68**:457–466

Ludwig, H.O., 1972, Untersuchungen am genetischen Material von Herpesviren. I. Biophysikalisch–chemische Charakterisierung von Herpesvirus-Desoxyribonucleinsauren, *Med. Microbiol. Immunol* **157**:186–211.

Ludwig, H., Biswal, N., Bryans, J.T., and McCombs, R.M., 1971, Some properties of the DNA from a new equine herpesvirus, *Virology* **45**:534–537.

Ludwig, H., Haines, H.G., Biswal, N., and Benyesh-Melnick, M., 1972, The characterization of varicella-zoster virus DNA, *J. Gen. Virol.* **14**:111–114.

Lunger, P.D., and Clark, H.F., 1978, Reptilia-related viruses, *Adv. Virus Res.* **23**:159–204.

Mackay, J.M.K., 1969, Tissue culture studies of sheep pulmonary adenomatosis (Faastekle). I. Direct culture of affected lungs, *J. Comp. Pathol.* **79**:141–146.

Madin, S.H., York, C.J., and McKercher, D.G., 1956, Isolation of the infectious bovine rhinotracheitis virus, *Science* **124**:721–722.

Malherbe, H., and Harwin, R., 1958, Neurotropic virus in African monkeys, *Lancet* **2**:530.

Malherbe, M., Harwin, R., and Ulrich, M., 1963, The cytopathic effects of vervet monkey viruses, *S. Afr. Med. J.* **37**:407–411.

Mare, C.J., and Graham, D.L., 1973, Falcon herpesvirus, the etiologic agent of inclusion body disease of falcons, *Infect. Immun.* **8**:118–126.

Martin, W.B., Hay, D., Crawford, L.V., Lebouvier, G.L., and Crawford, E.M., 1966, Characteristics of bovine mammillitis virus, *J. Gen. Microbiol.* **45**:325–332.

May, H.G., and Tittsler, R.P., 1925, Tracheo-laryngitis in poultry, *J. Am. Vet. Med. Assoc.* **67**:229–231.

McCarthy, K., Thorpe, E., Laursen, A.C., Heymann, C.S., and Beale, J.A., 1968, Exanthematous disease in patas monkeys caused by a herpes virus, Lancet **2**:856–857.

Melendez, L.V., Daniel, M.D., Hunt, R.D., and Garcia, F.G., 1968, An apparently new herpesvirus from primary kidney cultures of the squirrel monkey (*Saimiri sciureus*), *Lab. Anim. Care* **18**:374–381.

Melendez, L., Hunt, R.D., Daniel, M.D., and Trum, B.F., 1970, New World monkeys, herpesviruses and cancer, in: *Infections and Immunosuppression in Subhuman Primates*, pp. 111–117, Munksgaard, Copenhagen.

Melendez, L.V., Hunt, R.D., King, N.W., Barahona, H.H., Daniel, M.D., Fraser, C.E.O., and Garcia, F.G., 1972, Herpesvirus ateles, a new lymphoma virus of monkeys, *Nature (London) New Biol.* **235:**182–184.

Middelkamp, J.N., Patrizi, G., and Reed, C.A., 1967, Light and electron microscopic studies of the guinea pig cytomegalovirus, *J. Ultrastruct. Res.* **18:**85–101.

Mirkovic, R., Voss, W.R., and Benyesh-Melnick, M., 1970, Characterization of a new herpes-type virus indigenous for tree shrews, Proceedings of the 10th International Congress of Microbiology, Mexico City, pp. 181–189.

Monroe, J.H., Shibley, G.P., Schidlovsky, G., Nakai, T., Howalson, A.F., Wivel, N.W., and O'Connor, T.E., 1968, Action of snake venom on Rauscher virus, *J. Natl. Cancer Inst.* **40:**135–145.

Mosmann, T.R., and Hudson, J.B., 1973, Some properties of the genome of murine cytomegalovirus (MCV), *Virology* **54:**135–149.

Nayak, D.P., 1971, Isolation and characterization of a herpesvirus from leukemic guinea pigs, *J. Virol.* **8:**579–588.

Nesburn, A.B., 1969, Isolation and characterization of a herpes-like virus from New Zealand albino rabbit kidney cultures: A probable reisolation of virus III of Rivers, *J. Virol.* **3:**59–69.

Neubaner, R.H., Rabin, H., Strnad, B.C., Nonoyama, M., and Nelson-Rees, W.A., 1979, Establishment of a lymphoblastoid cell line and isolation of an Epstein–Barr-related virus of gorilla origin, *J. Virol.* **31:**845–848.

Nigida, S.M., Falk, L.A., Wolfe, G., and Deinhardt, F., 1979, Isolation of a cytomegalovirus from salivary glands of white-lipped marmosets (*Saguinus fuscicollis*), *Lab. Anim. Sci.* **29:**53–60.

Onions, C.T., 1971, *The Compact Edition of the Oxford English Dictionary*, Oxford University Press, Oxford.

Parker, J.C., Vernon, M.L., and Cross, S.S., 1973, Classification of mouse thymic virus as a herpesvirus, *Infect. Immun.* **7:**305–308.

Plowright, W., Ferris, R.D., and Scott, G.R., 1960, Blue wildebeest and the aetiological agent of bovine malignant catarrhal fever, *Nature (London)* **188:**1167–1169.

Plummer, G., and Waterson, A.P., 1963, Equine herpesviruses, *Virology* **19:**412–416.

Plummer, G., Goodheart, C.R., Henson, D., and Bowling, C.P., 1969, A comparative study of the DNA density and behavior in tissue cultures of fourteen different herpesviruses, *Virology* **39:**134–137.

Plummer, G., Goodheart, C.R., and Studdert, M.J., 1973, Equine herpesviruses: Antigenic relationships and DNA densities, *Infect. Immun.* **8:**621–627.

Raab-Traub, N., Dambaugh, T., and Kieff, E., 1980, DNA of Epstein–Barr virus. VIII. B95-8, the previous prototype, is an unusual deletion of derivative, Cell **22:**257–267.

Rafferty, K.A., Jr., 1965, The cultivation of inclusion-associated virus from Lucke tumor frogs, *Ann. N. Y. Acad. Sci.* **126:**3–21.

Randall, C.C., Ryden, F.W., Doll, E.R., and Shell, F.S., 1953, Cultivation of equine abortion virus in fetal horse tissue *in vitro*, *Am. J. Pathol.* **29:**139–153.

Rasheed, S., Rongey, R.W., Bruszweski, J., Nelson-Rees, W.A., Rabin, H., Neubaner, R.H., Esra, G., and Gardner, M.B., 1977, Establishment of a cell line with associated Epstein–Barr-like virus from a leukemic orangutan, *Science* **198:**407–409.

Raynaud, A., and Adrian, M., 1970, Lesions cutañees a structure papillomatouse associées a des virus chez le lezard vert (*Lacerta virelis laur.*), *C. R. Acad. Sci. Ser. D* **283:**845–847.

Rebell, G., Rywlin, A., and Haines, H., 1975, A herpesvirus-type agent associated with skin lesions of green sea turtle in aquaculture, *Am. J. Vet. Res.* **36:**1221–1224.

Reid, H.W., and Rowe, L.W., 1973, The attenuation of a herpesvirus (malignant catarrhal fever virus) isolated from hartebeest (*Alcelaphus buselaphus cokei* Gunther), *Res. Vet. Sci.* **15:**144–146.

Roizman, B., 1979, The structure and isomerization of herpes simplex virus genomes, *Cell* **16:**481–494.

Roizman, B., and Furlong, F., 1974, The replication of Herpesviruses, in: *Comprehensive Virology*, Vol. 3 (H. Fraenkel-Conrad and R.R. Wagner, eds.), pp. 229–403, Plenum Press, New York.

Roizman, B., Bartha, A., Biggs, P.M., Carmichael, L.E., Granoff, A., Hampar, B., Kaplan, A.S., Melendez, L.V., Munk, K., Nahmias, A., Plummer, G., Rajcani, J., Terni, M., de Thé, G., Watson, D.H., and Wildy, P., 1973, Provisional labels for herpesviruses, *J. Gen. Virol.* **20**:417–419.

Roizman, B., Carmichael, L.E., Deinhardt, F., de Thé, G.B., Nahmias, A.J., Plowright, W., Rapp, F. Sheldrick, P., Takahashi, M., and Wolf, K., 1981, Herpes-viridae: Definition, provisional nomenclature and taxonomy *Intervirology* **16**:201–217.

Rowe, W.P., and Capps, W.I., 1961, A new mouse virus causing necrosis of the thymus in newborn mice, *J. Exp. Med.* **113**:831–844.

Sabin, A.B., and Wright, A.M., 1934, Acute ascending myelitis following a monkey bite with the isolation of a virus capable of producing the disease, *J. Exp. Med.* **59**:115–136.

Saito, J.K., Gribble, D.H., Berrios, P.E., and Knight, H.D., 1974, A new herpesvirus isolate from goats: Preliminary report, *Am. J. Vet. Res.* **35**:847–848.

Schetter, C.H., 1970, *In vitro* Untersuchungen über die Eigenschaften des Virus der Hepatitis et Splenitis Infectiosa Strigum, Verhandlungsbericht XII Int. Symp. Erkr. Zootiere, Budapest, pp. 205–209.

Schneweis, K.E., 1962, Serologische Untersuchungen zur Typendifferenzierung des herpesvirus hominis, *Z. Immunitaetsforsch. Exp. Ther.* **124**:24–48.

Schubert, von G., 1964, Electronenmikroskopische Untersuchungen zur Pockenkrankheit des Karpfens, *Z. Naturforsch.* **19**:675–682.

Sheldrick, P., and Berthelot, N., 1975, Inverted repetitions in the chromosome of herpes simplex virus, *Cold Spring Harbor Symp. Quant. Biol.* **39**:667–678.

Shulte-Holthausen, H., and Zur Hausen, H., 1970, Purification of the Epstein–Barr virus and some properties of its DNA, *Virology* **40**:776–779.

Simpson, C.F., Hanley, J.E., and Gaskin, J.M., 1975, Psittacine herpesvirus infection resembling Pacheco's parrot disease, *J. Infect. Dis.* **13**:390–396.

Smith, M.G., 1954, Propagation of salivary gland virus of the mouse in tissue cultures, *Proc. Soc. Exp. Biol. Med.* **86**:435–440.

Smith, M.G., 1956, Propagation in tissue cultures of a cytopathogenic virus from human salivary gland virus (SGV) disease (22498), *Proc. Soc. Exp. Biol. Med.* **92**:424–430.

Smith, M.G., 1959, The salivary gland viruses of man and animals (cytomegalic inclusion disease), *Prog. Med. Virol.* **2**:171–202.

Sterz, H., Ludwig, H., and Rott, R., 1974, Immunologic and genetic relationship between herpes simplex virus and bovine herpes mammillitis virus, *Intervirology* **2**:1–13.

Stinski, M., 1982, Molecular biology of cytomegaloviruses, in: *The Herpesviruses*, Vol. 2 (B. Roizman, ed.), pp. Plenum Press, New York (in press).

Strandberg, J.D., and Carmichael, L.E., 1965, Electron microscopy of a canine herpesvirus, *J. Bacteriol.* **90**:1790–1791.

Valicek, L., Smid, B., Pleva, V., and Mensik, J., 1970, Porcine cytomegalic inclusion disease virus, *Arch. Gesamte Virusforsch.* **32**:19–30.

Wadsworth, S., Jacob, R.J., and Roizman, B., 1975, II. Size, composition and arrangement of inverted terminal repetitions, *J. Virol* **15**:1487–1497.

Wagner, E.K., Roizman, B., Savage, T., Spear, P.G., Mizell, M., Durr, F.E., and Sypowicz, D., 1970, Characterization of the DNA of herpesviruses associated with Lucke adenocarcinoma of the frog and Burkitt lymphoma of man, *Virology* **42**:257–261.

Weller, T.H., 1953, Serial propagation *in vitro* of agents producing inclusion bodies derived from varicella and herpes zoster, *Proc. Soc. Exp. Biol. Med.* **83**:3440–346.

Westrate, M.W., Geelen, J.L.M.C., and van der Noordaa, J., 1980, Human cytomegalovirus DNA: Physical maps for the restriction endonucleases Bgl II, Hind III, and Xba I, *J. Gen. Virol.* **49**:1–21.

Whalley, J.M., and Webber, C.E., 1979, Characteristics of Parma wallaby herpesvirus grown in marsupial cells, *J. Gen. Virol.* **45**:423–430.

Wildy, P., 1973, Herpes: History and classification, in: *The Herpesviruses* (A.S. Kaplan, ed.), pp. 1–25, Academic Press, New York.
Wildy, P., Russell, W.C., and Horne, R.W., 1960, The morphology of herpes viruses, *Virology* **12**:204–224.
Wolf, K., and Darlington, R.W., 1971, Channel catfish virus: A new herpesvirus of ictalurid fish, *J. Virol.* **8**:525–533.
Wolf, K., Darlington, R.W., Taylor, W.G., Quimby, M.C., and Nagabayashi, T., 1978, Herpesvirus salmonis: Characterization of a new pathogen of rainbow trout, *J. Virol.* **27**:659–666.
Zeigel, R.F., and Clark, H.F., 1972, Electron microscopic observation on a new herpes-type virus isolated from *Iguana iguana* and propogated in reptilian cells *in vitro, Infec. Immun.* **5**:570–582.

CHAPTER 2

Epidemiology of Epstein–Barr Virus and Associated Diseases in Man

GUY DE-THÉ, M.D., Ph.D.

To investigate the origins of diseases, one should first study the effects of the seasons, of the hot or cold winds, of the soil, naked or wooded, of the rains. Secondly, one will consider most attentively the way the inhabitants live, drink, eat, are indolent or active in exercise and labor.

—Hippocrates (460–377 B.C.)

I. PROVINCE OF EPIDEMIOLOGY

A. Epidemiology: The Study of the Distribution and of the Determinants of Diseases

This definition includes two main areas: (1) descriptive epidemiology, covering the distribution of diseases in place and time; and (2) analytical epidemiology, aimed at characterizing possible causal factors.

The term epidemic was until recently restricted to acute outbreaks of infectious diseases. Now it covers any occurrence of any illness in a community that is clearly in excess of normal expectation. Thus, non-infectious diseases, such as coronary heart diseases or certain cancers (e.g., lung carcinomas), do exhibit epidemic patterns in certain societies

---

GUY DE-THÉ, M.D., Ph.D. • Faculty of Medicine Alexis Carrel, Rue G. Paradin, 69372, Lyon; IRSC—CNRS, Villejuif, France.

(Western countries). For many public-health workers however, epidemiology should not only be concerned with diseases but also embrace the study of markers and origin of health in populations, health being, according to the World Health Organization, a state of physical, mental, and social well-being. The relationship between epidemiology and other disciplines has evolved in recent years. In classic epidemiology, the contributions of clinical medicine, of pathology, and of biostatistics were dominant. Today, microbiology, biochemistry, immunology, experimental pathology, and molecular biology have become complementary disciplines to epidemiology, which is now embracing a multidisciplinary approach to the understanding of normal or pathological processes in populations. In return, epidemiological data can now identify subject areas within which experimental investigations have the greatest chances of providing critical new information on fundamental aspects of biology.

B. Epidemiological Concept of Cause

Two events following each other can only be suspected of being related as cause and effect. The term "causal relationship," being abstract, cannot be universal, and epidemiology having a practical purpose in searching risk factors, which offers the possibility of disease prevention, a causal association is epidemiologically accepted only when changes in the first event lead to modifications in the second. Thus, the ideal proof of causality for an epidemiologist is the prevention of a disease through intervention against a putative causal factor. In most cases, the causal nature of an association is not easy to demonstrate, and in the absence of experimental proof, the association must be strong, consistent with existing knowledge, specific, and temporal (USPHS, 1964). The notion of "causal association" is further complicated by the fact that degenerative diseases (cardiovascular, neurological, or malignant) are multifactorial in origin and result from a multistep pathogenic process.

Ideally, experimentalists search for the "necessary cause" without which the disease does not occur, even if this factor is not sufficient. But often it is not easy to distinguish between "necessary" and "contributory" factors. As an example, *Mycobacterium tuberculosis* is considered as the "necessary," if not sufficient, cause of tuberculosis, but an illness clinically indistinguishable from and histopathologically similar to tuberculosis can be caused by different agents. The notion of disease "entity" must precede the search for causes. Another example is Burkitt lymphoma (BL). It represents an epidemiological, histopathological, and clinical entity in the equatorial belt, but one can observe lymphomas that are histopathologically similar, but clinically different, in temperate climates. Causal factors associated with endemic BL in Africa are absent in nonendemic BL-type lymphomas observed in temperate climates. Epidemiologically, clinical entities represent the main target for prevention, whereas in fundamental research, pathogenesis at the cellular level represents the main goal.

Since Pasteur, many diseases have been found to be caused by microbiological agents, with one disease corresponding to one agent. The postulates elaborated by Jacob Henle and formulated by Robert Koch (1890) refer to parasitic and acute infectious diseases. They imply that parasites should occur in every case of the disease and in no other disease. The isolation of the parasite, to be cultured in the laboratory, should reinduce the disease in the susceptible host.

C. Viral Diseases and Criteria for Causation

With regard to viral diseases, Rivers (1937), as early as 1937, objected that the presence of the virus cannot be demonstrated in every case of the disease produced by it and that nondiseased individuals may be viral carriers. He stressed the importance of the absence of antibody to the putative causative virus in the patients' sera at the onset of illness and its appearance during recovery. He noted, however, that the disease could result from a reinfection or reactivation of a viral infection and therefore that the absence of antibody prior to the illness might not be a proper criterion. The virologist's "dilemma" was further discussed by Huebner (1957), who, in reviewing the different postulates, proposed that the clinical entity suspected to be virally induced should be regularly associated with the virus, experimentally reproduced in animals and in human volunteers inoculated with the agent in a controlled study, and preventable by a specific vaccine. Epidemiological cross-sectional and longitudinal studies may be necessary when highly prevalent human viruses are involved. The final proof that the Epstein–Barr virus (EBV) was causally related to infectious mononucleosis rested on prospective serological investigations, establishing the absence of antibodies to the virus prior to the disease and their regular appearance during illness (Evans and Niederman, 1976).

Reactivation of latent or persistent viral infections may cause specific diseases such as zoster in immunodepressed or cancerous patients. After a primary infection, the varicella virus will hide latent in spinal neural ganglia. The immunodepression that accompanies certain degenerative diseases will reactivate the latent varicella virus and induce clinical zoster in the territory of the infected spinal ganglion. Similarly, a reactivation of a latent infection by the EBV may be causally related to the clinical onset of nasopharyngeal carcinoma (NPC). Herpesviruses, papovaviruses, measles virus, rubella virus, and adenoviruses appear to be good candidates for causing clinical conditions associated with reactivation of latent viral infections. The best demonstration of a causal relationship between degenerative diseases, including cancers, and a reactivation of a latent viral infection will come from a successful antiviral intervention in this reactivation, which could prevent the occurrence of disease.

D. Carcinogenesis: A Multistep Process

The occurrence of a cancerous cell appears to imply a succession of heritable changes at the cellular level. Beremblum (1940) proposed the concept of a two stage initiation–promotion carcinogenesis, each step being caused by different classes of chemical agents. Recently, Peto (1977) proposed that carcinomas related to environmental factors develop as a result of a multistage process involving possibly four to six independent steps, each with specific causes. A tumor would arise from the proliferation of a clone derived from a cell that had undergone all the necessary changes. Reviewing this field, Beremblum (1978) stressed the differences between oncogenic agents, either initiators or promoters, and the cocarcinogenic agents, which have a permissive influence on carcinogenesis by modifying the oncogenic agent or the host, but are not directly involved in the cellular carcinogenic process.

What is the point of action of certain viruses in the chain of events leading to malignancies? As discussed by de-Thé (1980a), viruses could act as initiators, promoters, or possibly cocarcinogens for certain tumors such as endemic primary liver carcinoma and hepatitis B virus, endemic BL, NPC, and EBV, epidemic bovine esophageal cancers, and papilloma viruses (Jarrett *et al.*, 1980). All these viruses being ubiquitous, their involvement in causing geographically restricted malignancies must result from extreme conditions of infection by the virus such as very early age of the host or simultaneously with malaria. That a virus may represent a risk factor without representing the necessary agent for the tumor is a difficult notion to accept in our Pasteurian era. Nevertheless, EBV is causally related to endemic BL in Africa (de-Thé *et al.*, 1978a), but does not appear to be involved in histopathologically similar tumors of temperate climates (see Sections IV.D and IV.F). EBV would act as a cellular initiator in the process, leading to BL in endemic areas (de-Thé, 1977, 1978b, 1980b), but in nonendemic areas, the process would be initiated differently, with a frequency 20–40 times lower.

E. Epidemiological Strategy

Three epidemiological methods are used in medicine:

1. Descriptive epidemiology characterizes the diseases in terms of place and time, as well as the agent, the host, and the environment. Following are definitions of some terms of descriptive epidemiology that will be used in this chapter:

Incidence: Number of new cases of a disease occurring in a population in a unit of time (usually a year).

Incidence rate: Number of new cases over the total population at risk within a period of time (year). The denominator in such a case represents the population at risk or under surveillance.

Prevalence: Number of cases of diseases existing at one time; the *prevalence rate* gives the number of such cases divided by the population at risk. The time period is usually a year or a given instant of time (point prevalence). In serological surveys, the prevalence is represented by the presence of antibody to a specific antigen, of the antigen itself, or of any other blood marker in a given population at the time of collection. The prevalence rate is the number of sera having marker divided by the number of individuals tested.

The end product of descriptive epidemiology should be the formulation of etiological hypotheses and the characterization of risk factors.

2. Analytical epidemiology is aimed at weighing risk factors or at testing hypotheses of causation formulated above. Two types of analytical studies are commonly employed, the retrospective and the prospective.

Case–control studies compare the presence or absence of the suspected causal factors in patients with the disease under study to their presence or absence in individuals without the disease but similar to the patients in such characteristics as age, sex, and environment. Retrospective studies usually lead to further characterization of risk factors, but rarely do they bring the final proof of causality. This is particularly true when the putative causal factor and the disease are separated by a long period of time. In such a case, the observed situation at the time of the disease and control is very difficult to assess, since it results from a long and complicated series of events occurring between the exposure to the suspected agent and the disease.

The prospective approach is a more dynamic way of looking at risk factors, but it must be supported by well-defined and testable hypotheses. The prospective method usually consists in following up a characterized cohort of individuals at risk for a given disease. The longitudinal study of such a cohort will indicate the individuals who are at high risk, and by studying the information obtained at different times before the disease process, one could prove or disprove hypotheses regarding the role of a virus, for example, in the pathogenesis of a disease. Prospective studies are generally more expensive than retrospective studies, but bring more valuable answers.

3. Experimental epidemiology utilizes the epidemiological models using either animals or human volunteers. In the past, human volunteers participated in studies not only of infectious mononucleosis, but also of yellow fever, malaria, hepatitis, and even syphilis, raising important moral as well as technical and medical issues. From the technical viewpoint, susceptible volunteers were sometimes difficult to characterize. Medically, the concern is that the seriousness of the induced disease

could lead to permanent disability. Finally, the moral and ethical right to use human subjects is now a subject of great debate in Western democracies, and it is now a prerequisite that volunteers sign an informed consent.

II. NATURAL HISTORY OF EPSTEIN–BARR VIRUS

When Epstein described in 1964 the presence of herpes virions in lymphoblasts cultured from Burkitt lymphoma (BL) biopsies, it was difficult to visualize how herpesviruses could be implicated in oncogenesis. This was the time when RNA tumor viruses (oncornaviruses) were believed to represent the "final common way" of oncogenesis in animal and man (Huebner and Todaro, 1969).

The belief that the discovery was trivial stemmed from the finding that this virus, named after its discoverers, Epstein and Barr infected practically everybody in East Africa and even in the United States, where BL was unknown. The great challenge was to determine how an extremely common environmental agent (infecting 80–100% of individuals around the world) could become an oncogenic factor in certain circumstances. This having been partially resolved, the EBV is now emerging as a unique model for viral and chemical carcinogenesis, since this virus appears to act in a way similar to that proposed for chemical carcinogens.

A. Natural History of Epstein–Barr Virus in the Organism

The major route of transmission of EBV appears to be the saliva, breast feeding being a potential but not proven route and blood transfusion being an accidental but proven one (W. Henle *et al.*, 1968; Gerber *et al.*, 1969). Within 2 weeks after severe primary infection, such as infectious mononucleosis (IM), transforming EBV, as tested on cord-blood lymphocytes, can be found in saliva and throat washings of IM patients (Bender, 1962; Gerber *et al.*, 1972), sometimes for years (Miller *et al.*, 1973). In fact, 15–20% of young adults in the United States are virus shedders (Gerber *et al.*, 1972), some remaining so for years. The exact site of viral replication in the oropharyngeal area is not entirely known, although EBV has been repeatedly found in saliva from the parotid gland (Miller *et al.*, 1973; Morgan *et al.*, 1979). The questions as to which cells replicate EBV in this salivary gland and whether tonsils or nasopharyngeal mucosa or both also participate in the process of viral replication remain open.

The serological and cell-mediated immune responses that follow the primary infection by EBV will be described in Section III. It is important to note here that antibodies do not influence the shedding of EBV in saliva. Hence, a chronic source of infectious virus is available for nonimmune individuals, particularly children.

Table I lists the different EBV antigens and the serological techniques used in seroepidemiology. The serological profiles and the sequences of events from primary infection to tumors are given in Table II. It is not understood why antibodies to viral capsid antigen (VCA) always precede antibodies to the early antigen (EA) complex, which at the cellular level is synthesized before the VCA, and diffuse outside the cell. The great variability observed in the delay for developing antibodies to EBV-induced nuclear antigen (EBNA) after primary infection may reflect a genetically controlled cell-mediated immune response. In certain hereditary immune deficiencies, antibodies to EBNA never develop (see below). It would be informative to investigate the HLA profile, especially D and D-R, of the various types of poor EBNA responders. At an individual level, the VCA titer reached at primary infection is stable throughout life, except in case of reactivation. In contrast, antibodies to EA are specifically transient. Antibodies to EBNA tend to be stable, but not to the degree observed for VCA.

Then follows a long period during which the EBV infection is latent in B lymphocytes present in the circulating blood, lymph nodes, and spleen. Latency is the process by which a virus is present intracellularly in an unexpressed or partially expressed state. This latency can be reactivated *in vitro* when such lymphocytes are cultivated, leading to the establishment of long-term lymphoblastoid-cell lines (LCLs). *In vivo*, reactivation can occur, for example, in immunosuppressed, kidney-transplanted patients (Chang *et al.*, 1978); EBV is then excreted in their saliva, and their serological profiles exhibit significant rises in antibody titers against EA and VCA (see Table II). As will be discussed in Section V.D, such a reactivation is observed in pre-nasopharyngeal carcinoma (NPC) conditions and may represent a critical oncogenic event.

When B lymphocytes from IM patients or even from normal individuals are cultured *in vitro*, these lymphocytes are induced to produce EB virions, which in turn infect new lymphocytes, resulting in their immortalization and the establishment of LCLs (Rickinson *et al.*, 1974). Many compounds have been shown to induce EB viral expression in latently infected cells: halogenated pyrimidines [5-iodo-2-deoxyuridine (IUdR) or 5-bromodeoxyuridine (BUDR)] (Gerber, 1972; Hampar *et al.*, 1972; Glaser *et al.*, 1976); tumor promoters, such as the phorbol ester 12-*O*-tetradecanoyl phorbol-13 acetate (TPA) (zur Hausen *et al.*, 1978), antibodies to surface immunoglobulin M (IgM) (Tovey *et al.*, 1978); heat (Vonka and Kutinova, 1973); and even platinum compounds (Vonka *et al.*, 1972). TPA represents the most active compound of croton oil, extracted from seeds of *Euphorbia*. It is of interest to note here that extracts from many other Euphorbiaceae exhibit *in vitro* inducing activity of EA in Raji cells (Ito *et al.*, 1981).

Host factors that control latency *in vivo* are poorly known. Age and

TABLE I. Epstein–Barr Virus Antigens and Epstein–Barr Virus Serological Tests

Viral antigens	Source of antigens	Localization of antigens in test cells	Techniques and references[a]
Viral capsid antigen (VCA)	Acetone-fixed EBV producer cell lines (P_3HR-1, EB-3)	Total	Immunofluorescence[1,2] Immunoperoxidase[3–5]
Early antigen (EA)			
Diffuse (D)	Acetone or methanol-fixed (D component) activated Raji cells (see below)	Cytoplasmic	Immunofluorescence[6] Immunoperoxidase[4] Radioimmunoassay[7]
Restricted (R)	Methanol-fixed (R component) abortively superinfected or BUdR- or IUdR- or PHA- or TPA-activated nonproducer cell lines (Raji)	Golgi area	
Membrane antigens (MA)	Unfixed viable EBV producer cell lines		Immunofluorescence[2,8,9] Radioimmunoassay[10] Immunoperoxidase[4]

Antibody-dependent cellular cytotoxicity (ADCC)	—		ADCC[11–14]
Lymphocyte-defined membrane antigens (LYDMAS)	Viable EBV producer and nonproducer cell lines	Cell surface	Lymphocytotoxicity[15]
EBV-induced nuclear antigen (EBNA)	Acetone methanol-fixed nonproducer cell lines		Anticomplementary immunofluorescent (ACIF) test[16] Anticomplementary immunoenzymatic (ACIE) method[17] Radioimmunoassay[18]
Complement-fixation soluble (CF/S)	Cell-free extracts of nonproducer cell lines	Nuclear	Complement fixation[19–23] Radial immunodiffusion[24]
Neutralizing (Neut)	Infectious EBV		Inhibition of EA induction[25] Inhibition of colony formation[26,27] Inhibition of transformation[28]

[a] (1) G. Henle and W. Henle (1966); (2) Klein *et al.* (1968); (3) Granlund and Andrese (1977); (4) Stephens *et al.* (1977); (5) Liu and Zeng (1979); (6) Henle *et al.* (1970b); (7) Moar *et al.* (1977); (8) Goldstein *et al.* (1969); (9) Klein *et al.* (1967); (10) Hewetson *et al.* (1972); (11) Pearson and Qualtière (1978); (12) Jondal (1976); (13) Pearson *et al.* (1975); (14) Pearson and Orr (1976); (15) Svedmyr and Jondal (1975); (16) Reedman and Klein (1973); (17) Zeng *et al.* (1980b); (18) Brown *et al.* (1974); (19) Armstrong *et al.* (1966); (20) Baron *et al.* (1975); (21) Gerber and Deal (1970); (22) Pope *et al.* (1969b); (23) Vonka *et al.* (1970); (24) Matsuo *et al.* (1976); (25) Pearson *et al.* (1970); (26) Rocchi *et al.* (1973a); (27) Rocchi *et al.* (1973b); (28) Miller *et al.* (1972a).

TABLE II. Serological Profiles Observed at Different Periods of Life and in Different Conditions

Condition	VCA IgM (IgA)	VCA IgG	EA	EBNA	Transforming EBV in saliva (IgA)
Newborn (maternal antibodies)	−	+	−	+	−
6 months until primary infection	−	−	−	−	−
Primary infection (preclinical)	+	−	−	−	−
Recent primary infection (clinical IM)	+	+/++	+/−	−	+/−
Immune status (latent infection)	−	+	−	+	+/−
Reactivation (immunodepression)	−	++	+/++	+	+
Reactivation (pre-NPC conditions)	(IgA +)	++	+	+	(IgA +)
NPC clinical onset	(IgA ++)	+++	++	+/++	(IgA ++)
Burkitt lymphoma	−	++	−/++	+	+

species of individuals from whom lymphocytes are immortalized by EBV influence the state of latency of the derived cell lines: neonatal (cord-blood) lymphocytes usually give nonproducer LCLs, whereas adult lymphocytes give more producer LCLs; human immortalized lymphocytes are as a rule less productive than New World monkey (squirrel, cottontop, or marmoset) lymphocytes (Miller *et al.*, 1972b, 1974). Cell type and level of differentiation are other important parameters in latency and induction. Somatic-cell hybrids between lymphoid immortalized cells (Raji) and Hela epithelial-cell lines have shown that the producer status is dominant over the nonproducer one and that inducibility is under the control of multiple host factos (Glaser and Rapp, 1972; Glaser and Nonoyama, 1974).

Many questions remain unanswered regarding the natural history of EBV at the organism level: What are the sites and cell type(s) involved in viral replication at time of primary infection (parotid gland, nasopharynx, nervous tissues) and in viral reactivation in immunodepressed patients or in pre-NPC conditions? What are the cell types infected during latency? What are the host factors that govern immunovirological control of EBV primary infection, latency, and reactivation?

B. Natural History of Epstein–Barr Virus in Populations

The knowledge of the comparative epidemiological behavior of EBV is a prerequisite to hypothesize how this ubiquitous agent can be etiologically associated with rare or geographically restricted diseases, such as IM (Western countries), BL (equatorial belt), or NPC (South East Asia, North and East Africa).

Soon after Epstein *et al.* (1964) discovered the virus that now bears their names, deception came when Old *et al.* (1966) observed that precipitating antibodies to EBV were not restricted to BL patients, but were found in all age/sex-matched African control patients and normal children. The differences between both tumorous and normal groups were the geometric mean titers (GMTs) of various antibodies to EBV (G. Henle *et al.*, 1969, 1971). Numerous epidemiological surveys showed that EBV infection prevailed the world over, even in isolated groups, such as Aleutian or Melanesian Islanders (Lang *et al.*, 1977; Tischendorf *et al.*, 1970), or in the Indians of the Amazonian plateau, where measles and certain respiratory viral infections are absent (Black *et al.*, 1970).

1. Variations in Age-Specific Prevalence of EBV Infection

Significant differences exist, however, in the prevalence and in the age-specific incidence of primary infection among geographic areas and socioeconomic groups in any given country. Figure 1 gives the prevalence of EBV infection in the 4- to 6-year age group and in areas as different as the island of Barbados and the state of Connecticut. The observed differences are enormous: nearly 100% of children are infected by this age in Barbados (where IM does not exist), whereas fewer than 30% of children of the same age have been infected in Connecticut (where IM is prevalent). Figure 2 gives an idea of the age-specific prevalence of EBV infection in these two areas (Barbados and Connecticut), with Hawaii being intermediate. These data suggest that the age of primary infection depends on the socioeconomic development. The age-specific prevalence of EBV infection is socioeconomically related in the United States, as can be seen in Fig. 3. At 2–4 years of age, 50–60% of black Americans are already infected by EBV, vs. around 10% of white Americans. By the age of 16–18, the difference disappears (maximum incidence of IM among whites). Evans (1974) has compared young adult males in various settings in the United States, and as can be seen in Fig. 4, variations are important: 93%

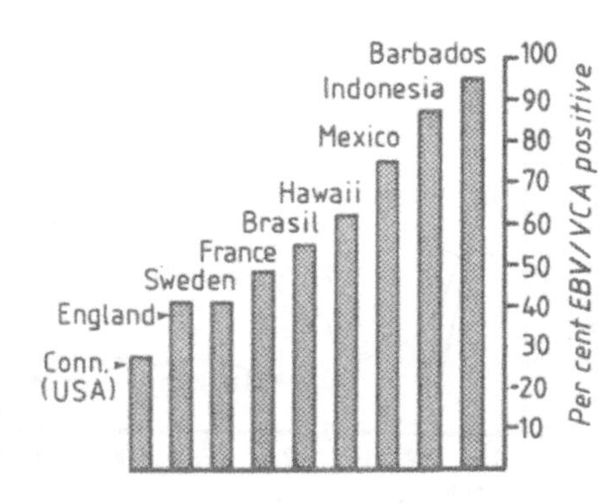

FIGURE 1. EBV prevalence within the 4- to 6-year age group in different geographic areas. Adapted from Evans (1974).

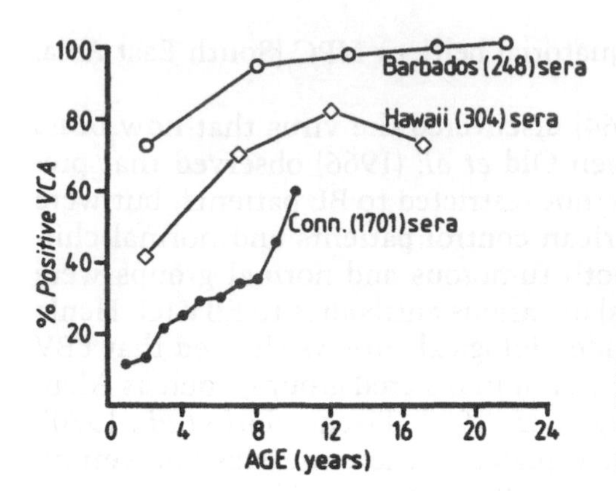

FIGURE 2. Prevalence of VCA antibodies, according to age, in Barbados, Hawaii, and Connecticut. Adapted from Evans (1974) and Jennings (1973).

of military recruits are EBV-positive, but only 26% of Yale freshmen in the 1958–1963 classes.

Changes in time of the age-specific prevalence of infection by EBV in a given population can be observed. As can be seen in Fig. 4, in 11 years' time (1958–1968), the prevalence of EBV infection in Yale freshmen increased from 26 to 51%, suggesting sociocultural changes (such as sexual permissivity). In contrast, Japanese medical-school freshmen in Kamamoto showed a regular increase in the proportion of EBV-negative students from 3% in 1973 to 6.5% in 1979 (Hinuma, personal communication).

2. Variations in the Immune Response to Primary Infection

The results cited above indicate that the age of the primary EBV infection varies with geographic areas and sociocultural groups. Figures 5 and 6 give the age-specific prevalence of VCA-positive individuals in the West Nile district of Uganda (high-BL area) and among Chinese Singaporians (at high risk for NPC) together with age-specific prevalence of EA and GMTs of VCA- and EA-positive individuals (de-Thé *et al.*, 1975b). The VCA prevalence curves show that before 3 years of age, 95% of Ugandan children are VCA-positive, while only 23% of Chinese children

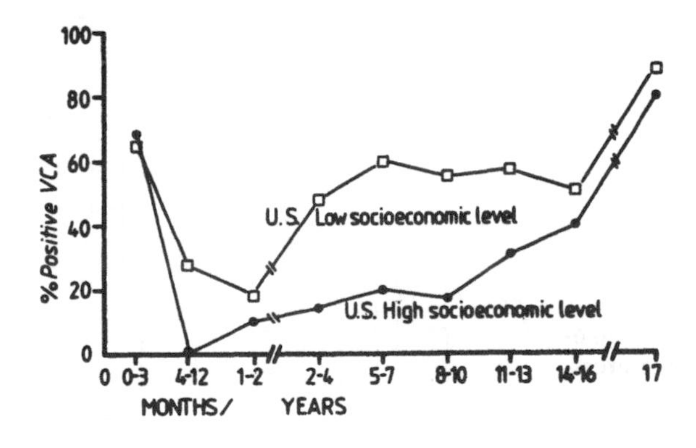

FIGURE 3. Age distribution in high and low socioeconomic groups in the United States. Adapted from G. Henle *et al.* (1969).

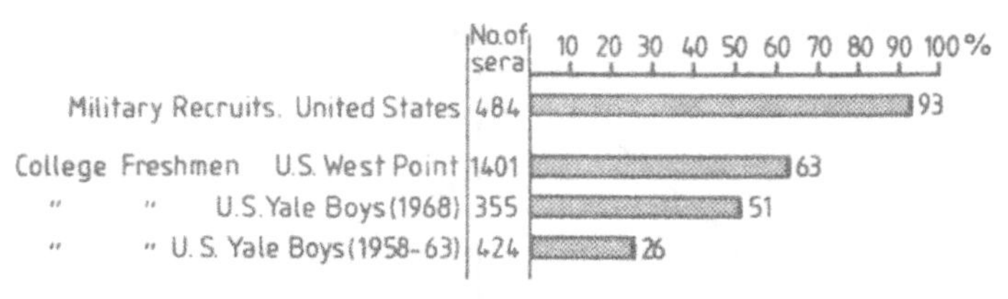

FIGURE 4. Percentage of EBV-VCA-positive sera in young adults. Adapted from Evans (1974).

of the same age in Singapore are positive. Thus, during the first 2 years of life, a massive infection takes place in Ugandan children, but not in Singapore, or for that matter in any socioeconomically developed country. To estimate the level of immune response of these two populations to primary EBV infection, one can consider the age-specific EA prevalence and GMTs of VCA and EA. It can be seen in Figs. 5 and 6 that in Uganda there is a high peak of VCA GMT between 1 and 2 years of age at a level (GMT = 421) similar to that of BL patients, a peak that is not observed in Singapore, or anywhere in economically developed countries.

The curve of GMT VCA in Uganda (Fig. 5) shows a sharp decline from the ages of 1 to 5 years, while on an individual basis, VCA titers are stable. Since the GMTs were calculated using the positive sera, this curve might suggest that at 1 year of age, most of the positive sera have a very high VCA titer, whereas at 5 years of age, for example, the GMT of positive VCA sera reflects a mixture of children with high titers (those infected early in age) and of children infected later who responded to primary infection with a lower VCA titer. In fact, in the West Nile district of Uganda, at 1 year of age, 83% of positive infants had VCA titers of 1280 or higher, whereas at the age of 3 years, only 25% had VCA titers of 1280 or higher (de-Thé, unpublished). This point is relevant to interpret the results of the Ugandan BL prospective study (see Section IV.F).

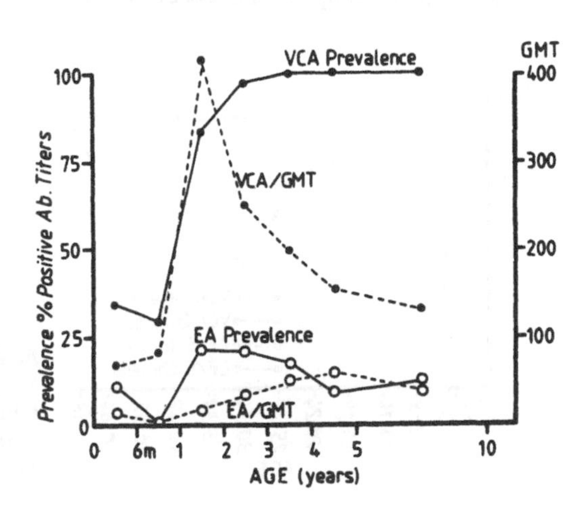

FIGURE 5. Prevalence and GMTs of positive sera for antibodies to VCA and EA according to age in the West Nile district of Uganda. Adapted from de-Thé *et al.* (1975b) and de-Thé (1977).

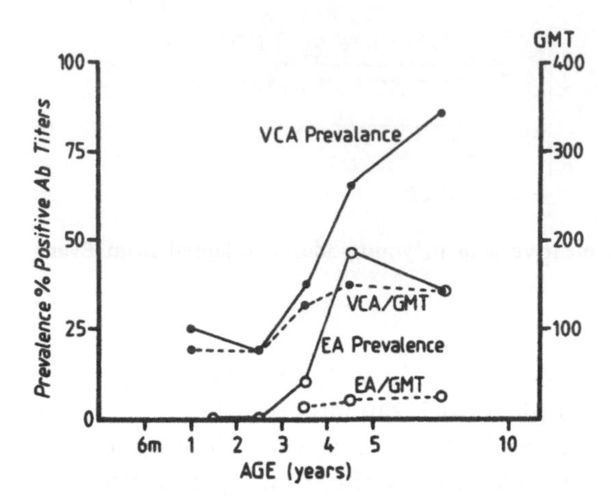

FIGURE 6. Prevalence and GMTs of positive sera for antibodies to VCA and EA, according to age, among Chinese Singaporians. Adapted from de-Thé *et al.* (1975b) and de-Thé (1977).

Other markers of the immune response to primary infection are antibodies to EA: the age-specific prevalence rate is significantly lower and earlier in Ugandans than in Singaporian Chinese (Fig. 5 and 6), as though EBV infection too early in life would result in a poor EA response.

NPC is prevalent among Chinese Singaporians, but not among Indo-Pakistanese Singaporians. It was therefore interesting to compare the age-specific prevalence of infection of these two populations as well as their age-specific response to EA and EBNA. Figures 7 and 8 show that there is no difference between the two groups in VCA age-specific prevalence or in prevalence and GMTs of antibodies to EA. In contrast, the prevalence of complement-fixation soluble (CF/S) antibodies according to age was significantly higher in Chinese than in Indian communities of Singapore.

For comparison, Figs. 9 and 10, from the People's Republic of China, give the prevalence rates and GMTs of CF/S antibodies according to age of Chinese in Kwangchow, Guangxi Autonomous Region (at high risk for NPC), and in Peking (at low risk for NPC) (Gu and Zeng, 1978). As observed in Singapore between Chinese and Indo-Pakistanese, the Can-

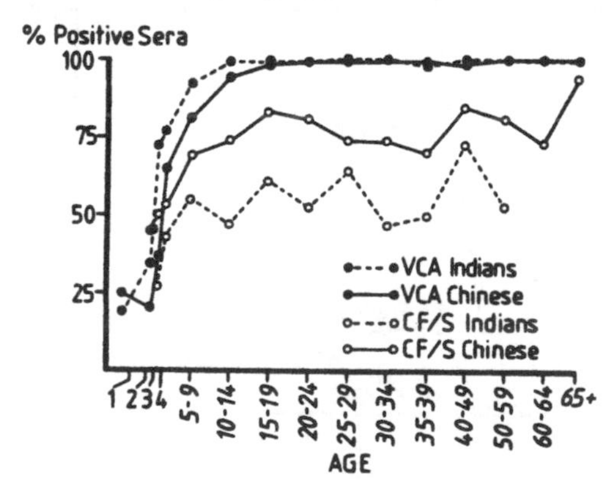

FIGURE 7. Prevalence of VCA and CF/S antibodies according to age among Chinese (at high risk for NPC) and Indians (at low risk for NPC) in Singapore. From de-Thé *et al.* (1975b).

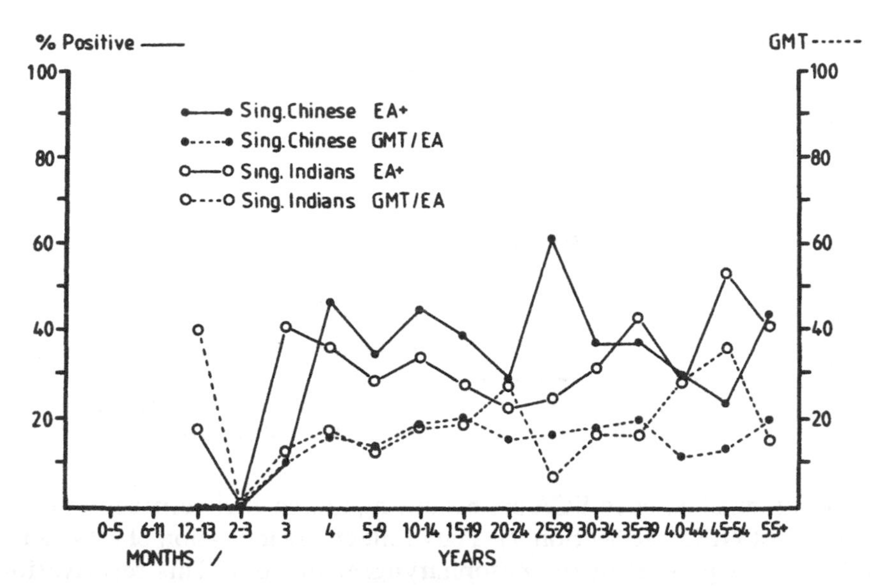

FIGURE 8. Prevalence of EA (D + R) antibodies and GMTs of positive sera among Chinese and Indian Singaporians (de-Thé and colleagues, unpublished data).

tonese at high risk for NPC maintain a higher CF/S (or EBNA) antibody level as compared to northern Chinese at low risk for this EBV-associated tumor.

3. Reactivation of Latent Epstein–Barr Virus Infection in Certain Diseases

An interesting EBV serological marker, associated with NPC, was observed by Zeng *et al.* (1979) in Cantonese Chinese population groups. As can be seen in Fig. 11, the prevalence of IgA VCA in the normal population around Kwangchow increases with age and seems to precede by a few years the evolution of the NPC prevalence in the same popu-

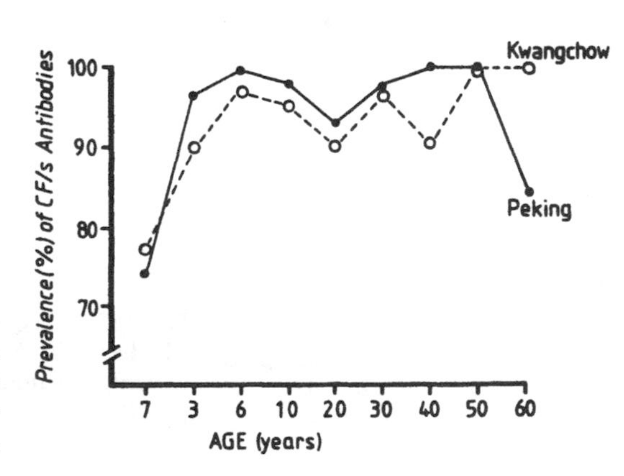

FIGURE 9. Prevalence of CF/S antibodies in Chinese from Kwangchow (high NPC risk) and Peking (low NPC risk). From Gu and Zeng (1978).

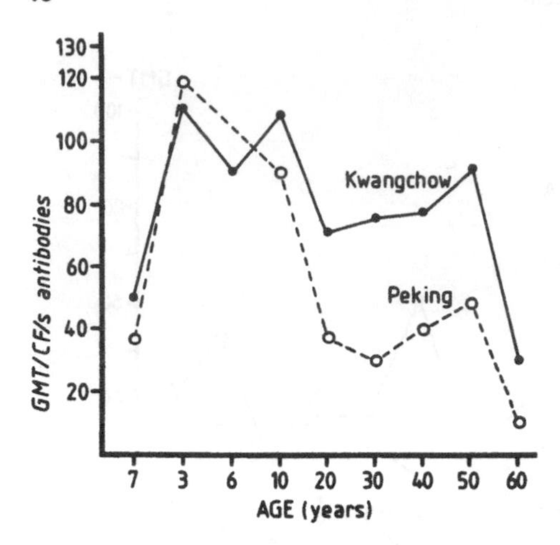

FIGURE 10. GMTs of CF/S antibodies in Chinese populations at high risk (Kwangchow) or low risk (Peking) for NPC. From Gu and Zeng (1978).

lation. IgA antibodies to VCA are becoming increasingly critical for early diagnosis of NPC (see Section V). They reflect a reactivation of EBV, with viral markers present in the nasopharyngeal mucosa. This reactivation appears to be due to sociocultural factors and to exposure to certain chemical carcinogens that have EBV-inducing potential (see Section V).

This reactivation, which reflects EBV activity in the nasopharyngeal mucosa, might be causally associated with NPC development (de-Thé, 1980b; de-Thé and Zeng, 1981), in contrast to the "passive" reactivation observed in a number of diseases associated with natural or induced depressed cell-mediated immunity, such as systemic lupus erythematosus (Stancek and Robensky, 1979), renal transplantation (Marker *et al.*, 1979; Cheeseman *et al.*, 1980), and Hodgkin's disease (Lange *et al.*, 1978). In the study of Cheeseman and colleagues, leukocyte interferon treatment of renal-transplant recipient aimed at controlling reactivation of latent infections of the herpes group (mostly cytomegalovirus and EBV) resulted in a reduced EBV excretion rate in the saliva, whereas antithy-

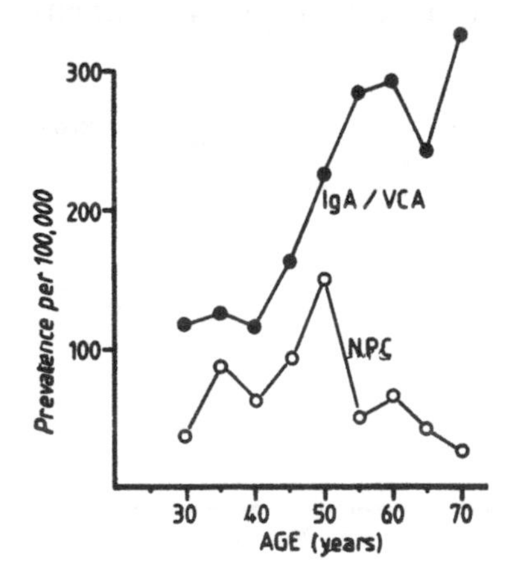

FIGURE 11. Age-specific prevalence rates of IgA antibodies to VCA in six communes of Zangwu County (Guangxi Autonomous Region). From Zeng *et al.* (1979).

mocyte immunoglobulin increased this excretion and increased VCA titers by 4-fold (Cheeseman *et al.*, 1980).

III. DISEASE ASSOCIATED WITH PRIMARY INFECTION BY EPSTEIN–BARR VIRUS: INFECTIOUS MONONUCLEOSIS

Common viral diseases are usually the result of primary infection by the causative agent. Epstein–Barr virus (EBV) primary infections are most of the time silent or unrecognized clinically; a few days with fever and transient enlarged cervical lymph nodes in children will not cause a doctor's consultation or a visit to a specialized hospital, and even if this were the case, an EBV serology would not be done. In contrast, infectious mononucleosis (IM), being often a dramatic event, is now well characterized clinically and epidemiologically. Historically, it is Emil Pfeiffer (1889) in Wiesbaden who, in 1889, first described a condition that he named "glandular fever." The classic description in 1920 and the name of infectious mononucleosis are due to Sprunt and Evans (1920) from Johns Hopkins University.

A. Clinical, Histopathological, and Epidemiological Characterization of Infectious Mononucleosis

With an insidious onset, the typical form of IM involves a sore throat, fever, severe fatigue, nausea, and sweating. At physical examination, enlarged cervical lymph nodes and splenomegaly are the main findings. If the acute syndrome is relatively short (1–2 weeks), convalescence is usually long with lasting fatigue. Severe headache is common, but the Guillain–Barré syndrome, suggesting a direct involvement of EBV infection in the CNS, is rare (Grose and Feorino, 1972). Minor jaundice is frequent and liver functions altered as a rule.

The blood count shows first a mild leukopenia due to a decrease in the number of polymorphs followed by a sharp increase in total leukocyte count (10,000–15,000 WBC/mm^3), due to the presence of large and atypical mononuclear leukocytes. These atypical cells consist of hyperbasophilic, vacuolated blastic cells, often with kidney-shaped nuclei, sometimes so pleomorphic that the diagnosis of leukemia is evoked.

The diagnosis of IM rests on the Paul–Bunnell–Davidsohn test detecting IgM heterophil antibodies agglutinating sheep erythrocytes (Paul and Bunnell, 1932; Davidsohn and Nelson, 1969). IM is accompanied by the transient development of a large number of heterophil antibodies, mostly of the IgM class, directed against sheep, horse, and bovine RBC, but also to the Ii blood group, nuclear factors, proteins OX19, and other factors, this being probably due to the EBV mitogenic activity on committed B lymphocytes. In asymptomatic primary EBV infections, the Paul–Bunnell test is usually negative (Sumaya, 1977).

Epidemiologically, IM is restricted to temperate climates and to high socioeconomic groups. It occurs sporadically with minor outbreaks in Western military camps and educational institutions in late winter or summer, but never in characterized epidemics. The age distribution in the United States, as shown in Fig. 12, has a peak at 15 years for girls and 17–18 years for boys, reflecting difference in sexual maturation (Heath *et al.*, 1972). The highest incidence rate in Atlanta was found to be 345 for 100,000 teenagers (15–19 years old). In contrast to the general belief, IM [with IgM EBV viral capsid antigen (VCA) antibodies] can be observed in the elderly (Pickens and Murdoch, 1979). IM can follow blood transfusions or bone-marrow transplantation in EBV-negative recipients (Sullivan *et al.*, 1978). The frequency of such events is not known, but blood transfusion centers should avoid using blood samples that have early antigen (EA) antibodies reflecting an active EBV infection of the donor.

In contrast to temperate countries with high socioeconomic and hygienic development, IM is exceptional in tropical and developing countries. As an example, not a single case of IM was recorded in 5000 admissions in the infirmary of a college in the Philippines (Evans and Campos, 1971). Repeated requests for information on the existence of typical IM syndromes with the Paul–Bunnell–Davidsohn test in Africa, China, Hong Kong, and Singapore have regularly indicated the rarity of IM among natives (de Thé, unpublished data). This absence is believed to be related to the early age of EBV infection in these areas. A very low incidence of IM was observed at the University of Hawaii, as compared to the University of California at Davis, the Asian student population being less susceptible than whites to IM (Chang *et al.*, 1979). It should be noted that Addy *et al.* (1978) found 5% of Paul–Bunnell–Davidsohn-positive sera among 715 jaundiced patients in Ghana, but no EBV serology was performed.

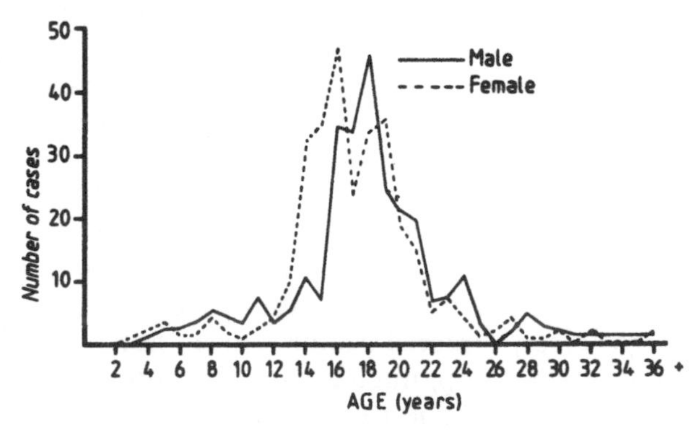

FIGURE 12. Age distribution of IM in Georgia, 1978. From Evans and Niederman (1976).

TABLE III. Summary of 11 Prospective Studies on Epstein–Barr Infection and Infectious Mononucleosis[a]

EBV antibody status at time of survey	Number of sera collected	EBV infection after 1 year (%)	IM after 1 year (%)
With VCA antibodies	3733	0	0
Lacking VCA antibodies	1547	16	7

[a] From Evans and Niederman (1976).

B. Evidence of Causality between Epstein–Barr Virus and Infectious Mononucleosis

The proof of causality between EBV and IM is based on seroepidemiological evidence and on partial success in the experimental transmission of the disease to monkeys and men.

The first suggestion that EBV might be etiologically related to IM came when a laboratory technician in the laboratory of Drs. W. and G. Henle who previously lacked EBV VCA antibodies seroconverted in the course of IM, while her circulating lymphocytes, which failed to grow *in vitro* during the acute phase of illness, gave rise to permanent cultures when collected during her convalescence (G. Henle *et al.*, 1968) [see Chapter 5 (Section I)]. The next step was the finding that all pre-IM sera collected at Yale University lacked antibodies to EBV, while the corresponding acute-phase and convalescent sera exhibited EBV antibodies (Niederman *et al.*, 1968). In addition, lymphoblastoid-cell lines could be obtained from nearly all IM blood leukocytes, but not from leukocytes of individuals who lacked antibodies to EBV (Diehl *et al.*, 1968). Finally, a number of prospective studies confirmed that the absence of VCA antibodies indicated susceptibility to IM, whereas the presence of these antibodies indicated immunity to the disease (see Table III) (Evans *et al.*, 1968; Niederman *et al.*, 1968 University Health Physicians and PHS Laboratories, 1971; Sawyer *et al.*, 1971; Hallee *et al.*, 1974). It is estimated that in highly developed countries, up to 25% of EBV-negative adolescents eventually develop IM syndromes. It is further estimated that if 90% of IM syndromes are caused by EBV, 5–7% are due to cytomegalovirus 1% to *Toxoplasma gondii*, and exceptional cases to viral hepatitis, rubella, herpes simplex, or adenovirus (Evans, 1978).

C. Epstein–Barr Virus Serological Profile during Infectious Mononucleosis

Figure 13 shows the evolution of the IgM and IgG antibodies to VCA, EA, and EBV-induced nuclear antigen (EBNA) during clinical IM. The essential characteristics are the lack of antibodies during incubation and

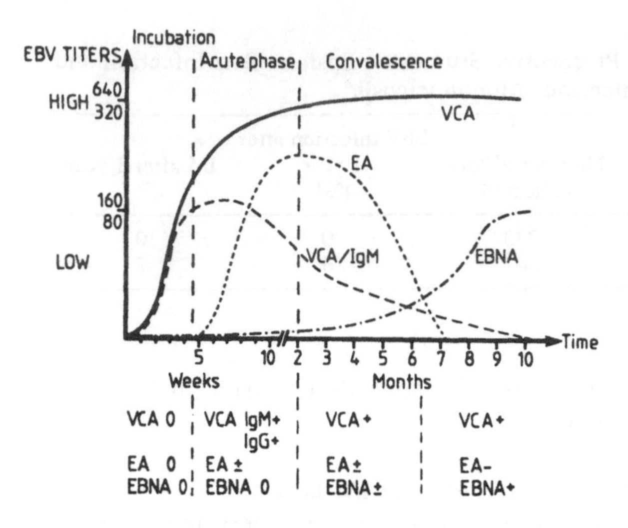

FIGURE 13. Evolution of EBV reactivities during IM syndromes. Adapted from de-Thé and Lenoir (1977).

the progressive development of VCA antibodies observed during the acute phase, in both the IgM and IgG classes. Antibodies to EA are directed against the diffuse (D) components (as in the case of nasopharyngeal carcinoma and in contrast to Burkitt lymphoma). The lack of antibodies to EBNA is the rule during the clinical illness and early convalescence. The "VCA+ IgM+ EA-D+ EBNA−" profile is pathognomonic for IM and can be used in the diagnosis of IM. In the first few weeks of convalescence, IgM antibodies to VCA disappear and IgG to EA/D tend to decrease. EA antibodies can be used as a marker of the evolution of the disease: if they increase in titer, a close clinical surveillance is necessary, since some fatal cases of IM are associated with increased EA titers (Sohier *et al.*, 1981). Another useful clinical marker is the amount of time necessary to mount an immune response to EBNA or to complement-fixation EBV soluble (CF/S) antigen after IM (Sohier, 1971; Sohier and de-Thé, 1978). When the recovery is rapid, EBNA antibodies appear within 6–10 weeks after the acute IM phase. But sometimes, after 6–10 months, or even years after primary infection, no EBNA or CF/S antibodies are detectable, suggesting that there is a poor cell-mediated immune response to the EBV infection (G. Henle *et al.*, 1974).

D. Deadly Lymphoid Proliferations following Epstein–Barr Virus Primary Infection

The clinical syndrome observed during acute IM is generally considered as the result of an immunopathological process in which induced killer T cells are specifically targeted to hit B lymphocytes infected with EBV (Svedmyr and Jondal, 1975). As a matter of fact, the majority of atypical lymphocytes during the late phase of IM are sheep-erythrocyte-rosette-forming cells that are cytotoxic for autologous EBV-carrying cells *in vitro*. The clinical recovery of IM depends on proper T-cell-mediated immune response. It is not surprising, therefore, that certain immune deficiencies are associated with severe EBV infection, leading to deadly lymphoid proliferations. The polyclonal B-cell proliferation, or "lymphoma," that developed after primary EBV infection as reported by Robinson *et al.* (1980) was deadly within a few days, while blood and every organ were heavily infiltrated with EBNA-positive B lymphocytes. The question whether such a proliferation should be considered as a lymphoma or not is a matter for debate (Wright *et al.*, 1980). Another case of fatal diffuse malignant lymphoma following primary EBV infection was recently reported (Thestrup-Pedersen *et al.*, 1980), wherein T-cell deficiencies concerning subpopulations of natural killer (NK) cells and T cells with receptors for IgG (Tγ cells) were substantiated with evidence of EBV-positive circulating B lymphocytes. A possible connection between an interferon-induced chromosomal defect (16q22) and the reduced level of NK- and Tγ-cell activities, leading to EBV-positive B-lymphocyte profileration, was envisaged by the authors.

That EBV can cause fatal IM in certain genetically susceptible individuals is now well documented (Bar *et al.*, 1974; Britton *et al.*, 1978; Virelizier *et al.*, 1978; Crawford *et al.*, 1979), but the immunopathological process involved does not appear to be the same in all cases (complex humoral and cell-mediated immunodeficiency in the case of Britton and colleagues, failure to generate immune interferon in the case of Virelizier and colleagues).

One report that appears to be of particular interest is the X-linked recessive lymphoproliferative syndrome of Purtilo *et al.* (1975, 1979), consisting of an ineffective response to EBV primary infection followed by a great frequency (up to 40%) of different types of malignant lymphoproliferations, including Burkitt-type lymphomas, B immunoblastic sarcomas, plasmocytomas, and others, appearing at variable times after EBV primary infection. The "Purtilo syndrome" may also involve cardiac and neurological congenital anomalies in blood relatives. The extent to which atypical IM leads to polyclonal, fatal lymphoproliferations has not yet been determined, and the registry set up by Purtilo (1980) should be helpful for both fundamental and applied research on EBV-associated diseases.

E. Infectious Mononucleosis as a Risk Factor for Hodgkin's Disease

The possible relationship between IM and the subsequent development of lymphoproliferative malignancies, especially Hodgkin's disease (HD), was first suggested by Dameshek (1969). A number of epidemiological studies supported this view (Carter and Penman, 1969; Miller and

TABLE IV. Frequency of Hodgkin's Disease in Individuals Who Have Experienced Infectious Mononucleosis

Authors	Cohort size	Obs.	Exp.	RR	Remarks
Carter and Penman (1969)	2,779	3	1.3	2.3	—
Miller and Beebe (1973)	2,437	2	1	2	—
Connelly and Christine (1974)	4,429	5	1	5	—
Rosdahl *et al.* (1974)	17,073	17	6	2.8	HD mostly in males
Muñoz *et al.* (1978)	9,454	7	1.8	4	HD mostly in males
Kvale *et al.* (1979)	5,840	6	2	3	—
TOTALS	42,012	40	13.1	19.1	

Beebe, 1973; Connelly and Christine, 1974; Rosdahl *et al.*, 1974; Munoz *et al.*, 1978; Kvale *et al.*, 1979), and as can be seen in Table IV, the increased relative risk (RR) of developing Hodgkin-type lymphoma varies from 2 to 5 for people who have had IM as compared to the general population.

Worth noting is the relatively short time interval elapsing between IM and HD: 2–11 years, with a mean of approximately 3 years, 95% of the cases occurring within 7 years. Munoz *et al.* (1978) raised the possibility of misdiagnosis between IM and HD, since some HD patients exhibit heterophil antibodies with a positive Paul–Bunnell–Davidsohn test, without having clinical IM (Wolf *et al.*, 1969). Progression from IM to monocytic leukemias was recently described by Hehlmann *et al.* (1980). A prospective follow-up of severe IM cases with collection of sequential serum specimens would be helpful to assess the sequence of events and the role of EBV in malignant lymphoproliferations.

IV. ROLE OF EPSTEIN–BARR VIRUS IN THE PATHOGENESIS OF BURKITT LYMPHOMA

In this section, we shall see how Burkitt lymphoma (BL), as described in equatorial Africa in the late 1950s, represents a clinical, pathological, and unique epidemiological entity.

The causal role of Epstein–Barr virus (EBV) in the development of such a tumor is now established and its mode of action partly clarified (de-Thé, 1979). Outside the endemic areas of the equatorial belt, lymphomas with a pathology very similar or identical to that observed in endemic BL do exist, although at a much lower frequency. It appears as though the EBV would have little to do with the development of these latter lymphomas. The comparison between these two groups of lymphomas should elucidate the unique role of EBV in the endemic entity.

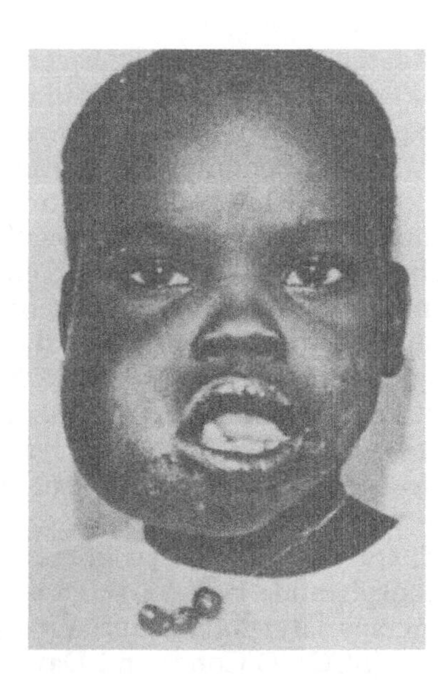

FIGURE 14. Jaw tumor in a Ugandan BL.

A. Clinical and Pathological Entity of Burkitt Lymphoma

The disease described by Dennis Burkitt (1958) had unique clinical features that stimulated surgeons at Mulago Hospital of Makarere University in Kampala, Uganda, to uncover it as a specific entity. Figure 14 shows a characteristic jaw tumor, now recognized the world over, as the typical aspect of BL. Jaw tumors can occur in any of the four quadrants of the maxillae, superior and inferior. As can be seen in Fig. 15, this localization is present in 100% of cases around 3 years of age and in decreasing proportion in older age groups or in areas where tumor incidence is lower (Burkitt, 1970). The radiological aspect of early lesions in the upper jaw or mandible involves the papillae of developing teeth

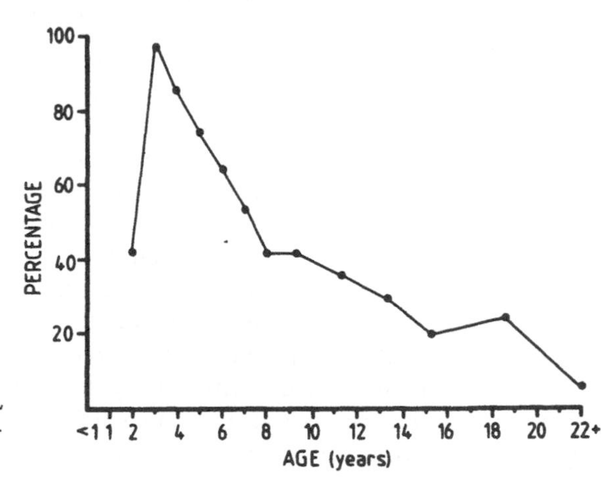

FIGURE 15. Percentage of BL patients with jaw tumors according to age. From Burkitt (1970).

(Adatia, 1970). Tooth buds corresponding to feather follicles in chickens, a known site of replication of the Marek disease herpesvirus, merit futher experimental and clinical investigation in BL genesis.

Other common localizations of BL involve the testicles, ovaries (in older girls), breast, liver, retroperitoneum, and eventually stomach and intestine, but very exceptionally lymph nodes and spleen, this being in sharp contrast to the lymphomas of similar histopathology as observed in temperate climates. Invasion of bone marrow is observed in terminal phases of the disease. Thus, the absence of lesion of the lymph nodes and spleen in this entity is a remarkable and unique feature for BL. Peripheral nerves are not affected in BL, in contrast to the frequency with which they are affected in other lymphomas involving the jaws. Intracranial invasion and peridural deposit of lymphomatous tumor cells are common at a late stage of the disease, resulting in hemi-, para-, or quadraplegia. There was exceptional immediate response of this lymphoma to cyclophosphamide that attracted worldwide attention in the 1960s. Chemotherapy later included vincristine, methotrexate, cytosine Arabinoside, and (BCG) (O'Conor and Davies, 1960; O'Conor, 1961; Clifford, 1970b; Ziegler, 1972, 1977). The 5-year disease-free survival is at present around 25% (Olweny *et al.*, 1980).

The original histopathological description of this tumor was made by O'Conor and Davies (O'Conor and Davies, 1960; O'Conor, 1961), who referred to it as a poorly differentiated lymphocytic lymphoma with the presence of large histiocytes giving the classic "starry-sky" picture. Fialkow *et al.* (1970) later provided evidence for the monoclonal origin of this B-cell lymphoma.

B. Geographic Distribution

Even now, the best epidemiological data are those that were collected by Dennis Burkitt covering thousands of miles in East, Central, and South Africa in the late 1950s. In a remarkably short period of time, he was able to characterize the climatic conditions that determine the geographic distribution of the tumor (Burkitt, 1962a,b).

Twenty years later, his maps are still fully valid. The lymphoma belt, as can be seen in Fig. 16, is limited in West Africa by the 10–15° north latitude and by the Ethiopian highlands and northern Kenya in East Africa. The southern limits in Central Africa are the French-speaking countries and the subtropical coastal plain of Mozambique. The low incidence of BL in the islands of Zanzibar and Pemba, 20 miles off the shores of the endemic coast of Tanzania, is still unexplained (Burkitt, 1969). But there are blanks within the African BL belt, and the distribution of the tumor is strictly determined by altitude, decreasing from 1500 m above sea level at the equator level to 300 m at the coast of South Africa. Figure 17 is a map of the mountainous region of Southeast Africa: the

FIGURE 16. World distribution of endemic BL lymphomas. Black areas indicate endemic areas. Crosses indicate places where nonendemic lymphomas with BL histopathology were described. From Burkitt (1970).

BL-free regions correspond precisely to the areas that are higher than 1200 m in altitude (Burkitt, 1970). It is pertinent to recall here that the high plateau in East Africa is more densely populated than the lower valleys, where the temperature, humidity, and mosquitoes make life much more difficult. Denis Burkitt showed that the tumors developed where rainfall was more than 60 cm a year and the temperature always above 16°C. Outside Africa, endemic BL occurs in holoendemic malaria areas of New Guinea, Malaysia, Colombia, and Brazil (Burkitt, 1967; Collected reports, 1967; Carvalho *et al.*, 1973).

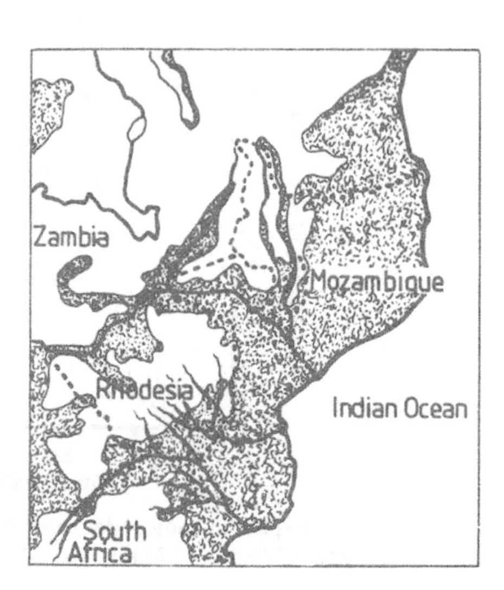

FIGURE 17. Distribution of BL endemic zones in Southeast Africa. BL is endemic in the shadowed areas (below 1200 m altitude) and absent in the highly populated plateau (white areas). Adapted from Burkitt (1970).

C. Age and Sex Distribution

The Burkitt tumor has never been observed under 2 years of age in equatorial Africa, suggesting that the oncogenic event must take place after birth. As can be seen in Fig. 18, its incidence rises very steeply from age 3, and peaks are observed around 5–7 years of age. In Central and West French-speaking Africa, the age distribution is wider than in East Africa, but the peaks remain at the same age.

Adult cases of BL have been observed in East Africa in immigrants from a low- to a high-incidence area (Burkitt and Wright, 1966). In reviewing this topic, Morrow *et al.* (1976) noted that about 50% of BL immigrant patients were over 15 years of age, suggesting that adults were susceptible to a factor present in the high-incidence area but absent in the tumor-free area. In this context, it is worth recalling that the mean age of occurrence is around 8 years in the high-BL areas, but about 16 years in the low-incidence areas. Compared to other childhood tumors, BL occurs at a slightly older age than tumors such as neuroblastomas, nephroblastomas, or retinoblastomas, but earlier than malignant lymphomas and acute leukemias as seen in temperate climates.

The influence of sex has been noted in all areas, with a male/female ratio with jaw tumor of about 3.2 : 1 and for patients without jaw lesions of 2.1 : 1. Thus, the highest incidence for BL is seen in boys aged 5–9, in whom this incidence reaches 15–20/100,000 in high-risk areas (Pike *et al.*, 1967). Genetic, racial, or tribal factors, which would include soci-

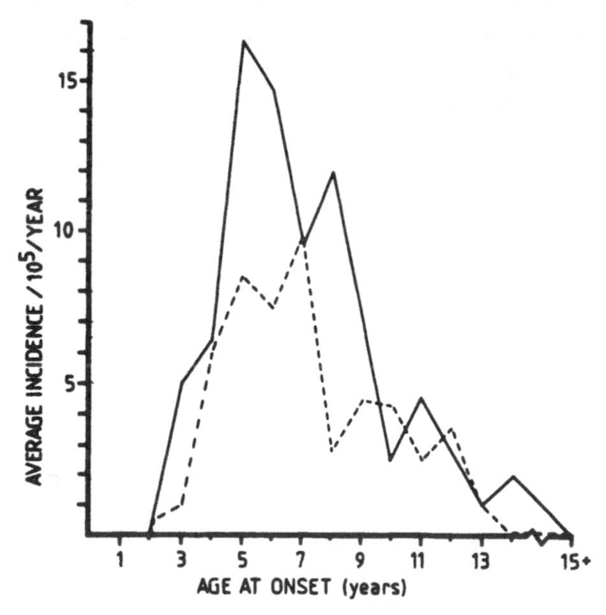

FIGURE 18. Age-specific indicendence of BL in the West Nile district of Uganda during the period 1961–1975. (——) Males; (- - -) females. From Williams *et al.* (1978).

cultural factors, do not appear to be critical for the development of BL. In Uganda, for example, a few tumors have been observed in Indian communities that accounted for less than 1% of the Ugandan population in the 1960s when this evaluation was made.

D. Space and Time Clustering

Space and time clustering of BL represents a most uncommon and surprising epidemiological characteristic for a cancer and at the same time supports strongly the intervention of an environmental factor. The phenomenon was first described by Pike *et al.* (1967) and analyzed by Williams *et al.* (1969) for the BL cases as observed in the West Nile district of Uganda during the period 1961–1965. Thereafter, clustering was not seen with cases observed for the next 5 years (1966–1971) in the same area, but later on strong evidence of clustering was again reported for the two years 1972–1973 (Williams *et al.*, 1978). When all cases observed during the period 1961–1975 were analyzed, the existence of space–time clustering was confirmed and further indicated that patients involved in such clusters were significantly older than patients not involved in clusters. No evidence of space–time clustering for BL could be observed, however, in other areas of Uganda, Tanzania, or Ghana (Brubaker *et al.*, 1973; Morrow *et al.*, 1976, 1977; Biggar and Nkrumah, 1979), thus making the West Nile district a very unique epidemiological niche.

Seasonal variations in the incidence of BL were also observed in the West Nile district of Uganda, where BL were diagnosed twice as often in the second half of the year as in the first (Williams *et al.*, 1974). Finally, it should be noted here that a decline in the overall incidence of BL is being observed in Uganda and Tanzania (Morrow *et al.*, 1976; International Agency for Research on Cancer, Annual Report, 1978).

E. Association between Epstein–Barr Virus and Burkitt Lymphoma

1. Viral Markers in Tumor Cells

Zur Hausen *et al.* (1970) were the first to show that BL biopsies in endemic areas contained EBV genome detectable by nucleic acid hybridization. Later on, a number of studies confirmed this original description, indicating that up to 96% of BL biopsies originating from endemic areas contained an average of 30 EBV genome equivalents per tumor cell (Nonoyama and Pagano, 1973; Nonoyama *et al.*, 1973; Pagano *et al.*, 1973; Lindahl *et al.*, 1974). Viral RNA transcripts corresponding to 3–6% of the EBV DNA were found in large quantity in BL biopsies (Dambaugh *et al.*, 1979).

The detection of the EBV-induced nuclear antigen (EBNA) in tumor

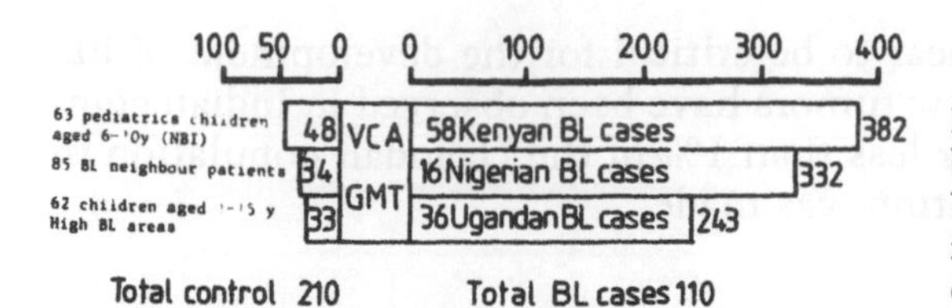

FIGURE 19. GMTs of VCA antibodies of BL cases from East and West Africa as compared to various control groups. Prepared from data of G. Henle *et al.* (1969).

cells was successful in a lower proportion of cases than the detection of EBV genomes (Pagano *et al.*, 1973; Reedman *et al.*, 1974; Ziegler, 1977; de-Thé *et al.*, 1978a). As an example, in the study of Olweny *et al.* (1977), 27 of the 34 DNA-positive BL biopsies were found to be EBNA-positive. In the same study, none of the 25 non-BL African lymphomas exhibited viral markers in the tumor cells.

2. EBV-Specific Immune Response

The serological profile of patients with BL was first studied by G. Henle *et al.* (1969), who, studying viral capsid antigen (VCA) antibodies, found that there was no difference between BL and various controls in the frequency of individuals who had experienced EBV infection, but observed significant differences in the geometric mean titers (GMTs) of VCA antibodies in BL patients as compared with various controls.

Figure 19, which was prepared from the data of G. Henle *et al.* (1969), shows a dramatic difference in VCA GMTs between BL patients from Kenya, Nigeria, and Uganda and various control groups, such as normal children, neighbors of BL patients, and sick patients from pediatric wards of the Nairobi hospital.

Regarding antibodies to early antigen (EA), G. Henle *et al.* (1971), testing 156 BL cases from Kenya and Uganda, found that around 75% of BL patients exhibited antibodies to the restricted component of EA, but at a level usually much lower than that obtained with VCA antibodies. In fact, as can be seen in Fig. 20, a direct relationship exists between the level of the VCA titers and that of the EA response.

200 Control		VCA titers	156 BL patients	
EA+/VCA+	GMT/EA		GMT/EA	EA+/VCA+
		1280	350	14/14
2/4	53	640	230	35/37
8/33	26	320	84	37/40
9/47	24	160	44	18/25
3/63	10	80	32	12/27
0/47		40	24	3/12
0/35		20	0	Total 119/156
0/70		10	0	
Total 22/200				

100 40 | 100 200 300 340

FIGURE 20. Prevalence and GMTs of EA antibodies in BL patients. Prepared with data from G. Henle *et al.* (1971).

The recently developed EBV-specific antibody-dependent cellular cytoxicity (ADCC) was applied to BL sera (Pearson *et al.*, 1979). These antibodies, directed against membrane antigen (MA) components, appear to have a prognostic value for BL patients, as can be seen in Table V. The nature of the ADCC-directed MA is not yet fully defined, but is related to quite an extent to the MAs as previously defined by Klein *et al.* (1969) and Svedmyr *et al.* (1970).

That BL patients regularly exhibited a specific immune response to the EBV and that their tumor cells regularly contained EBV viral markers was suggestive, but did not prove, that EBV is causally related to BL. The fact that this virus exhibited *in vitro* transforming activity on human and simian B lymphocytes (immortalization) and was able to induce *in vivo* lymphomas in some New World monkey species (Epstein *et al.*, 1973; Shope *et al.*, 1973) further favors the hypothesis that this virus represents the oncogenic agent of BL. But direct demonstration could be obtained only by epidemiological means, either by studying the sequence of events that occurs between primary infection and tumor development or by successful antiviral intervention resulting in the prevention of the tumor.

F. Ugandan Prospective Study

After the discovery of the causal role of EBV in infectious mononucleosis (IM) (see Section III.B), the main question for BL etiology was whether BL represents a malignant IM in African children immunodepressed by heavy parasitic infections, such as malaria. The question then arose whether a prospective study aimed at identifying BL candidates was feasible and could help establish the role of EBV in BL (International Agency for Research on Cancer, Annual Report, 1968). If BL were to develop as a consequence of a primary infection taking place shortly before tumor development in very few children escaping from the early infection taking place in Uganda (see Section II.B.2), then the sera col-

TABLE V. Response to Therapy of Burkitt Lymphoma Patients Grouped According to Antibody-Dependent Cellular Cytotoxicity Titers[a]

ADCC group[b]	Number tested	GMT			Clinical regression (%)[c]
		ADCC	VCA	EA	
High	19	10,255	484	70	74
Medium	28	1,046	334	12	36
Low	7	<240	324	25	29

[a] From Pearson *et al.* (1979).
[b] High-titered group, higher than 3840; medium-titered group, 240–3840; low-titered group, less than 240.
[c] Based on complete or partial regression following treatment.

lected prior to the incubating period should lack EBV antibodies. An alternate hypothesis was considered in which a long and heavy exposure to EBV was a necessary condition for BL development; in such a case, sera of future BL cases would exhibit high antibodies to EBV. Finally, if no causal relationship existed between EBV and BL development, then the EBV profile of children who later on will develop BL should not differ from that observed in age/sex- and locality-matched controls (Geser and de-Thé, 1972).

Before a long-term prospective study could be undertaken, it was necessary to evaluate the stability of EBV titer in the African population. A feasibility study in 1968 and 1969 showed a remarkable stability of individual antibody titers to VCA over an 18-month period of time (Kafuko *et al.*, 1972). It was estimated that a 5-year follow-up of approximately 37,000 children aged 4–8 would then be necessary to test the proposed hypotheses (Geser and de-Thé, 1972).

From February 1972 to the end of 1974, serum samples were obtained from 42,000 children, aged 0–8 years, in four counties of the West Nile district of Uganda (Fig. 21), with excellent cooperation from the population (85% of the eligible children were in fact bled) (de-Thé *et al.*, 1978a). Then an extensive follow-up of these children was conducted to detect every BL case, up to late 1979, when the survey of the population had to be stopped because of the political and economic situation. Thus, 16 BL tumors were detected in the bled population with a time interval

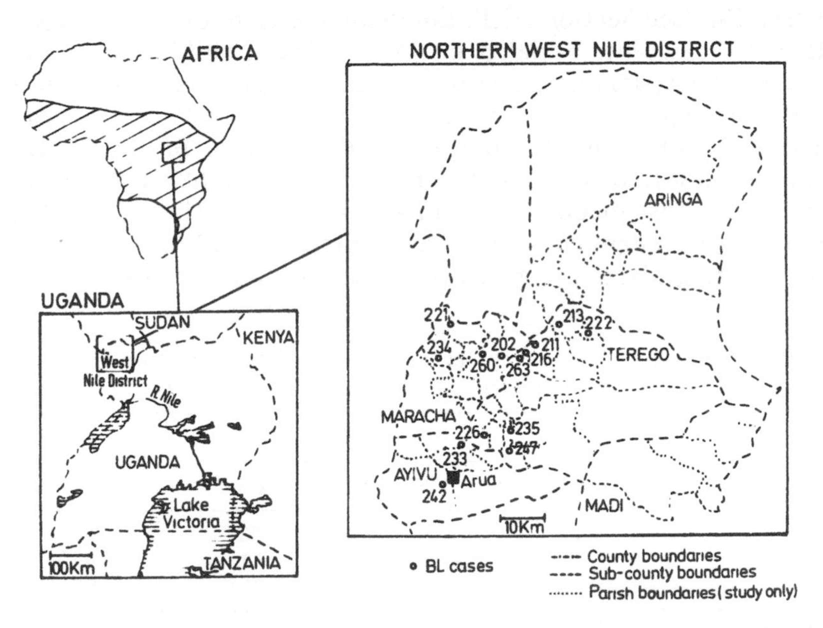

FIGURE 21. Map of Africa showing the Ugandan West Nile district where the BL prospective study was carried out (1971–1979). From the de-Thé *et al.* (1978a).

ranging from 7 months to 6 years between the main bleeding and the diagnosis of the tumor (de Thé *et al.*, 1978a; Geser *et al.*, 1982). The main results were, as can be seen in Fig. 22, that pre-BL sera exhibited higher VCA titers than age/sex/locality-matched controls. In fact, 12 of 13 EBV-associated and histologically proven BL cases had pre-BL sera with VCA titers higher than or as high as any control. The increased risk of developing BL could be estimated at approximately 30 times for children who had VCA titers 2 dilutions or more above the GMT of the corresponding normal population group standardized for age, sex, and locality. The presence of EBNA and EBV DNA sequences were established in 9 of 10 confirmed BL cases, from whom frozen biopsies were available for viral DNA investigations. The second most important finding was that VCA titers did not increase at tumor onset (see Fig. 23). The stability of VCA titers between the main bleeding and BL development was most remarkable, since it indicated that high EBV VCA titers observed in BL patients from East or West Africa did not result from disease process, but reflected a long-standing situation.

In contrast to VCA antibodies, antibodies to EA, EBNA, and herpes simple virus (HSV), cytomegalovirus (CMV), and measles virus were not elevated in pre-BL patients as compared to controls. In 7 cases, EA antibodies developed after tumor onset, but without change in VCA titers.

G. Epidemiological Evidence for a Causal Relationship

The difference in VCA titers between pre-BL sera and control sera reached a high degree of significance ($p = 0.002$) when the latest two cases described by Geser *et al.* (1982) were added to the results obtained previously (de Thé *et al.*, 1978a). As can be seen in Fig. 22, the time period between serum collection and BL occurrence did not affect the ranking of pre-BL sera vis-à-vis the control sera. The 5 cases bled 33–72 months prior to tumor development exhibited the same pattern as the 11 cases bled 7–20 months before BL onset. When clinically, histopathologically, and virologically coherent BL cases were considered, 13 of 14 had VCA titers as high as or higher than any control, 1–6 years prior to tumor development. These results, taken together with the lack of change in VCA titers after tumor onset and the lack of difference in the antibody titers to other viruses (HSV, CMV, measles) between pre-BL and control sera, made high VCA titers in pre-BL sera a unique feature bringing epidemiological evidence of causal association between EBV and BL development in Uganda. Such epidemiological evidence is consistent with the experimental data showing that EBV can immortalize human B lymphocytes *in vitro* (Pope *et al.*, 1969a) and induce lymphomas *in vivo* in certain New World primates (Epstein *et al.*, 1973; Shope *et al.*, 1973). When

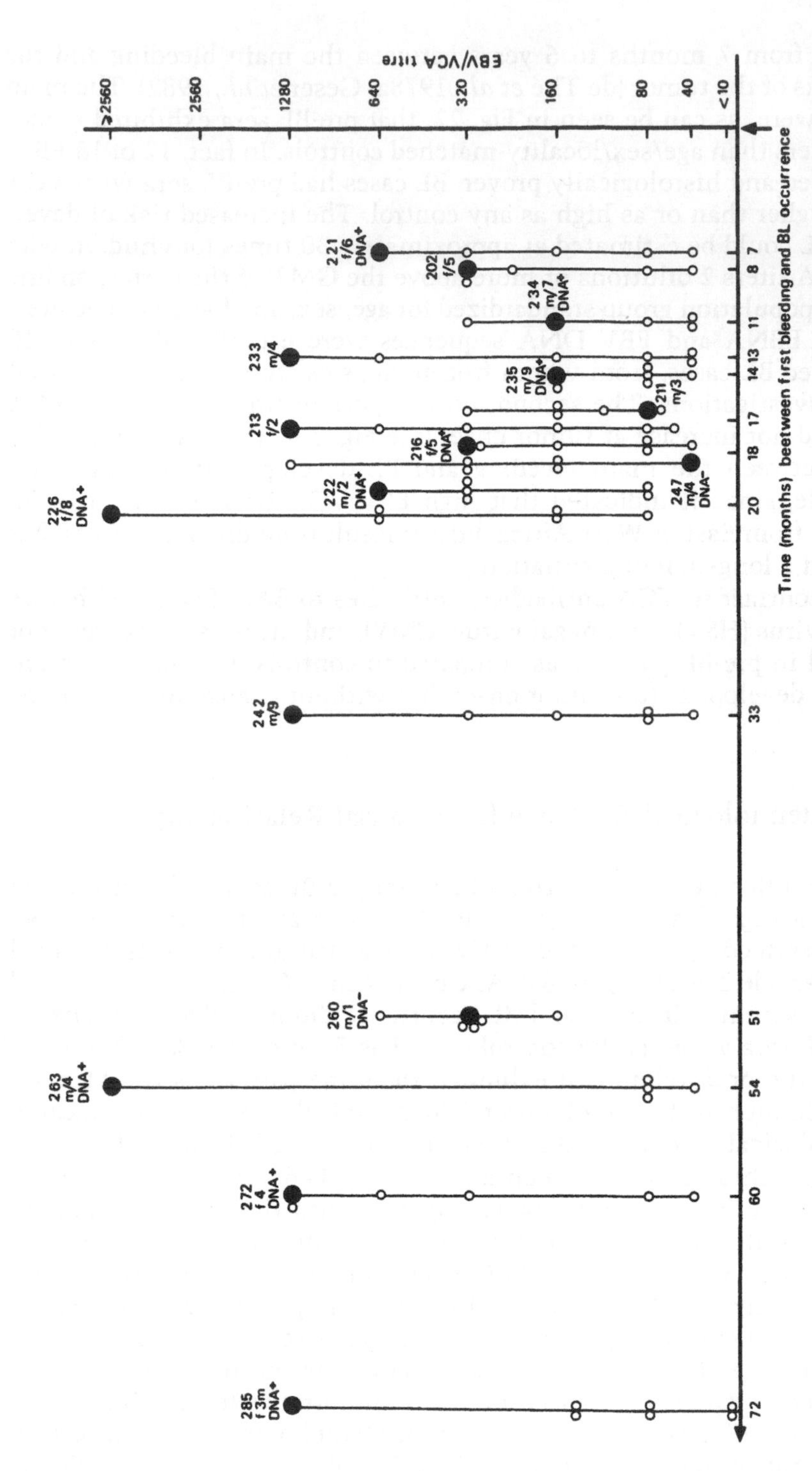

FIGURE 22. VCA antibody titers in sera collected from candidate BL cases at different periods prior to BL occurrence. (●) Future BL cases; (○) age/sex-matched controls. From de-Thé *et al.* (1978a) and Geser *et al.* (1982).

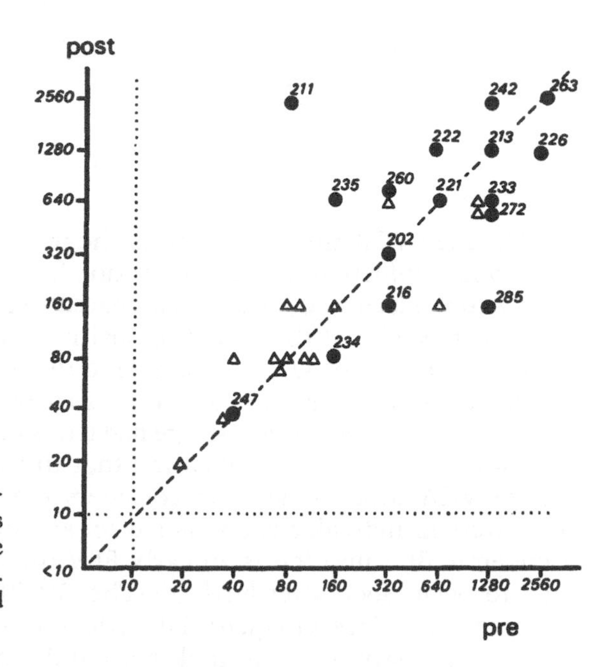

FIGURE 23. EBV VCA reactivities in BL cases (●) and controls (△) at different periods, before and after clinical onset of BL. From de-Thé *et al.* (1978a) and Geser *et al.* (1982).

epidemiological data reinforce evidence obtained through clinical and experimental studies, there is no alternative but to interpret the results of the prospective Ugandan study as offering further critical evidence of a causal relationship between EBV and BL.

Some questions remained: Why were VCA titers alone and not antibodies to EA and EBNA elevated in pre-BL sera? Why did some rare cases (such as case 247) not exhibit high VCA titers prior to or after disease onset and lack EBV markers in the tumor cells?

As discussed in Section I.B, the Pasteurian notion of a necessary agent corresponding to one disease cannot be applied to multistep degenerative diseases.

The results of the Ugandan prospective study indicate that the causal association between EBV and BL is confined to cases in which it is found both that high VCA titers precede disease onset and that EBV markers are present in the tumor cells. These "EBV-associated lymphomas" represent 96% of the endemic lymphomas described by Dennis Burkitt in children of equatorial Africa. EBV-free lymphomas, exceptional in equatorial Africa, are the rule in temperate climates (see Section IV.K). In equatorial Africa, candidates for EBV-associated lymphomas appear to be recruited among the 10% highest VCA titers as observed in the general population. If one recalls that the highest incidence in the male population between 5 and 9 years averages 20/100,000 per year (see Fig. 18), only 0.2% of the candidates will eventually develop the disease, which indicates that other cofactors are critical in BL development.

H. Epstein–Barr Virus Early Infection Acts as an Initiating Event

The high VCA titers observed in the pre-BL sera testified to a severe primary EBV infection. We saw in Section II.B.2 that a very early infection by EBV took place in Uganda as compared to South East Asia or Europe. Figure 5 showed that the GMTs of positive Ugandan sera at one year or age reached a level of 421, comparable to that observed in BL cases (GMT = 425). We saw that in the population survey in Uganda (Section II.B.2), 83% of positive sera at 1 year of age had titers of 1280 or higher, compared to 25% at 3 years of age, indicating that infection early in age resulted in high VCA titers. When one considers the extraordinary stability of VCA titers in individuals, one is led to postulate that the critical event in endemic BL causation is an early EBV infection, which we hypothesized to occur soon after birth (de-Thé, 1977). But how early can EBV infection take place in equatorial Africa? In the study by Biggar *et al.* (1978a,b), primary infection took place only after loss of maternal antibodies. In BL case 285 of the prospective study, the first serum was taken at 3 months and then at 1 year of age. Already at 3 months, this girl had an IgG VCA titer of 1280, without IgM antibodies, but with EA titer at 40–80 and EBNA titer at 640, indicating an established chronic infection by EBV, again observed at 1 year of age. Her mother had a low EBV profile at time of BL onset, suggesting poor maternal antibody protection in this case, permitting such an early infection (Geser *et al.*, 1982). As in the case of measles virus, EBV infection could take place under massive exposure in babies only partially protected by maternal antibodies. The habit of mouth kissing of certain African mothers could well favor a massive saliva transmission to babies, since 50% of the African mothers have transforming EBV in their saliva. One should recall here that in experimental viral oncology, inoculation of oncogenic RNA or DNA viruses in the few days following birth or hatching significantly increases, and sometimes is a requisite for oncogenic potential of these viruses to be expressed (Gross, 1970).

The early EBV infection probably creates a large population of EBV-infected B lymphocytes that, for a reason not yet understood, represent privileged target cells for further oncogenic events. The *in vitro* immortalization of B lymphocytes by EBV does not result in a cancerous transformation of these cells. In fact, normal B lymphocytes immortalized by EBV do not cause tumor in nude mice except when inoculated in very large amounts, in contrast to BL-derived lymphoblastoid-cell lines, which are tumorigenic in nude mice.

In a multistep carcinogenetic process, EBV early infection would act as an "initiating" event at the cellular level. There are apparently two other factors that could be etiologically associated with further steps toward tumor onset: malaria and chromosomal abnormalities.

I. Malaria

The detailed geographic distribution of BL corresponds precisely to that of holoendemic malaria. Dalldorf *et al.* (1964) and Burkitt (1969) were the first to stress this in equatorial and East Africa. Geser and colleagues, in two International Agency for Research on Cancer Annual Reports (1976, 1978), compared the level of malaria endemicity compatible with BL in the West Nile district of Uganda and in the North Mara district of Tanzania. As can be seen in Table VI, the level of parasitemia, the prevalence, and the GMTs of antibodies to malaria differed remarkably between areas of high or low incidence for BL. No significant difference was observed, however, between EBV-antibody reactivities of high- and low-altitude areas (see Table VI). It is known that individuals with sickle-cell trait and to a lesser extent with hemoglobin C trait are partially protected from the most severe forms of malaria (Allison and Clyde, 1961; Motulsky, 1964). Therefore, if malaria were involved in BL causation, children who have such genic traits should theoretically be protected against the development of BL. The first study by Pike *et al.* (1970) supported such a protective hypothesis, but failed to reach a statistically significant level, as did the study from Nigeria (Williams, 1966). A recent study by Nkrumah and Perkins (1976), in Ghana, also failed to show any protective advantage of the sickle-cell trait against BL. Their study favored, however, hemoglobin C trait to have a slight protective effect, but without reaching statistical significance.

There are many theoretical possibilities by which malaria could favor BL development. Congenital malaria is known to exist in equatorial Africa and may create conditions that favor transplacental EBV infection. On the other hand, postnatal simultaneous malaria and primary EBV infection may result in an immunopathological disorder leading to an altered cell-mediated immune control of the EBV infection, as suggested by O'Conor (1970). Greenwood and Vick (1975) and, more recently, Greenwood *et al.* (1979) reported that an antigen from *Plasmodium falciparum* was mitogenic on B cells and to a lesser extent T cells from malaria-immune but also from malaria-nonimmune individuals. A continuous antigenic stimulation of EBV-infected B lymphocytes by malaria antigen thus might represent a critical factor in BL development.

Intervention against malaria appears to be the best method of establishing whether holoendemic malaria is a causal factor for BL development. The International Agency for Research on Cancer and the Tanzanian government implemented in 1977 a malaria-intervention program in the North Mara district in which 90,000 children aged 0–10 were given chloroquine tablets twice a month. Unfortunately, these studies had dif-

TABLE VI. Malaria and Epstein–Barr Virus Burden in Populations at High or Low Incidence of Burkitt Lymphoma in East Africa[a]

			Malaria[d]			EBV antibodies[d]			
Area	Altitude (m)[b]	BL incidence[c]	Parasitemia [dry season (%)]	IFAb≥10(%)	GMT	VCA≥10(%)	GMT VCA	EA≥10(%)	GMT EA
Tanzania North Mara									
Lowlands	850–1450	15	48	99	812	96	193	11	23
High plateau	1500–2100	0	14	63	68	95	169	21	16
Uganda West Nile lowlands	600–1200	18	70	99	491	95	128	11	34

[a] From International Agency for Research on Cancer Annual Reports (1976, 1978).
[b] Above sea level. [c] Annual incidence per 100,000 children aged 5–10 years.
[d] (IFAb) Immunofluorescent antibodies; (GMT) geometric mean titers for positive sera.

ficulties: the BL incidence dropped in 1977, at the time of implementing these studies, and the treated children in 1977–1979 showed no decrease in the parasitemia level as compared to the control untreated population (International Agency for Research on Cancer, Annual Report, 1980). In the prospective Ugandan studies described in Section IV.F, no difference in malaria parasitemia or in the malaria-antibody level was found between the pre-BL specimen cases and the controls. But parasitemia can change within a very short time, and malaria-antibody titers may vary over a period of a few months. At the time of BL diagnosis, the parasitemia load of patients was lower than that of controls (International Agency for Research on Cancer, Annual Report, 1978). This seems to be due to malaria treatment that these children received prior to their consulting at hospital for BL.

In conclusion, holoendemic malaria, although a favored candidate in BL cocarcinogenesis, may not be the critical factor, and more work should be devoted to the possible role of the anopheline vector.

J. Chromosomal Anomalies

The cytogenetic abnormalities regularly observed in BL cells (Manolov and Manolova, 1972; Jarvis *et al.*, 1974; Zech *et al.*, 1976) could represent the last stage in the multistep BL carcinogenesis. The 8q– 14q+ translocation may represent a "cytogenetic convergence" of random Darwinian chromosomal changes from a specific, yet unknown, cause (Klein, 1979). A variant of the 8–14 translocation including t (2;8) and t (8;22) was recently described in African BL cell lines (Bernheim *et al.*, 1981). As will be discussed in Section IV.K, these chromosomal anomalies are not restricted to the EBV-associated BL and therefore should be considered as a separate etiological event for B lymphomas.

The suddenness of the clinical onset of BL is very intriguing and cannot be explained by the etiological factors discussed above. Lachet *et al.* (1977), analyzing the epidemiological data related to BL, concluded that a precipitating event leading to clinical onset should take place 6–8 months prior to the diagnosis of the disease.

In conclusion, the pathogenesis of BL appears to imply three or possibly four steps, each having a specific etiology. The first refers to an early EBV infection leading to a large population of EBV-infected B cells, at an intermediate state between normal and immortalized as observed *in vitro*. This "initiation" might be followed by a second step wherein a continuous mitogenic activity of malaria antigen keeps the B-cell population "alert." A third step would involve chromosomal alteration. The question as to whether chromosomal alteration reflects a selection process or results from an induction mechanism is not yet answered. The origin of the precipitating event leading to tumor may be associated with a sudden chromosomal change or could be due to another, yet unknown, factor. Thus, BL is becoming a unique model for studying multistep carcinogenesis. It also allowed the building of a bridge between chemical and viral carcinogenesis, since EBV appears to act epidemiologically and, from the viewpoint of cell biology, as an initiator of the oncogenic process.

K. Nonendemic Burkitt-Type Lymphomas and Epstein–Barr Virus

The epidemiological entity as described by Burkitt in the late 1950s corresponds to a well-characterized histopathological tumor (Berard *et al.*, 1969). That B lymphomas with similar histopathology could be seen in areas other than equatorial Africa has been established since 1967, when an issue of the *International Journal of Cancer* collected the observations of BL-type lymphomas outside Africa. Cases of children with Burkitt-type lymphomas were thus described in Europe (Finland, Sweden, England, Netherlands), in the Middle East (Irak and Turkey), South East Asia (Malaysia, Singapore, India, and New Guinea), and North America (United States and Canada). For Hoogstraten (1967), Burkitt-type lymphomas represented the most common childhood lymphoma in central Canada. However, as pointed out by Shanmugaratnam *et al.* (1967) in the same issue of the *International Journal of Cancer*, one should be very careful in applying the term Burkitt lymphoma outside the particular syndrome with clinical, histopathological, and epidemiological characteristics described by Burkitt. When only one characteristic, namely histopathology, is considered, one should use the term "nonendemic Burkitt-type lymphoma" (NEBL). As an example, BL-type lymphomas in Irak exhibit some clinical and even pathological differences from Burkitt lymphomas as observed in Africa (Al-Attar *et al.*, 1979). The frequency of gastrointestinal-tract involvement was 92% in the Iraqui patients, in contrast to African cases, in which involvement is more renal and ovarian, which is rare in Irak. CNS and bone-marrow involvement, which is frequent in patients from Uganda (Wright and Pike, 1969; Ziegler *et al.*, 1970), is an uncommon feature in the Iraqui cases. The predominance of abdominal and pelvic presentation in Irak is similar to that reported by Arsenau *et al.* (1975), but the age distribution is different, the Iraqui patients having a mean age of 7 years, the Americans a mean age of 11 years (Burchenal, 1966; Arsenau *et al.*, 1975). From the therapeutic viewpoint, both endemic and NEBLs seem to respond well to chemotherapy, with the rate of early relapse being higher in endemic areas than in nonendemic areas (Nkrumah *et al.*, 1977).

The proportion of NEBLs associated with EBV is small, usually lower than 10%. Hirshaut *et al.* (1973), studying 15 cases of American BL-type lymphomas, failed to find any significant serological differences between cases and age/sex-matched controls. By molecular hybridization techniques, absence of viral DNA in a series of American Burkitt-type lymphomas was reported by Pagano *et al.* (1973). John Ziegler *et al.* (1976), studying 20 American patients with Burkitt-type lymphoma, found 5 with a serological profile similar to that of African BL cases, but only 2 containing EBV DNA sequences in their tumor cells. The American Burkitt Lymphoma Registry (Levine *et al.*, 1972, 1975) should help to characterize clinically, epidemiologically, and serologically two groups of B lymphomas: a minority, containing EBV information that are EBV-associated (Gravell *et al.*, 1976; Goldblum *et al.*, 1977; Bornkamm *et al.*, 1976; Miyoshi *et al.*, 1978; Lenoir *et al.*, 1979) and the EBV-free majority. The EBV-positive lymphomas that develop in patients with ataxia–telangectasia (Saemundsen *et al.*, 1981) or in patients with renal allografts and treated with cyclosporin (Crawford *et al.*, 1980) provide unique situations to investigate the cocarcicogenic potential of EBV. In the cyclosporin-treated patients, it appears that a specific deficiency in cell-mediated immunity led to the development of an EBV-infected clone of B lymphocytes. The cluster of EBV-associated American Burkitt-type lymphomas described by Judson *et al.* (1977), in which 3 of 4 cases were EBV-positive and were detected within 1 year in a rural area of Pennsylvania, stresses the need for comparative epidemiological studies of EBV-associated and EBV-free B lymphomas.

Endemic and NEBLs share the same cytogenic abnormalities, i.e., the t (8;14) translocation observed in African BL. Further, NEBLs exhibited other translocations involving chromosome 8 (t 8;22 and t 8;2) (Berger *et al.*, 1979a,b; McCaw *et al.*, 1977; Van den Berghe *et al.*, 1979; Miyoshi *et al.*, 1979; Miyamoto *et al.*, 1980). Acute lymphocytic leukemias with Burkitt-type cells also present cytogenic abnormalities similar to that seen in BLs (Mitelman *et al.*, 1979; Berger *et al.*, 1979c; Slater *et al.*, 1979), the hairy-cell leukemia variant having high EBV reactivities (Sakamoto *et al.*, 1981).

The establishment of registries for leukemias and lymphomas involving epidemiological, clinical, and pathological subtyping and EBV serological data should help to determine whether EBV-associated lymphomas in temperate climates represent an etiological entity or not.

V. ROLE OF EPSTEIN–BARR VIRUS IN THE ETIOLOGY AND CONTROL OF NASOPHARYNGEAL CARCINOMA

In contrast to the situation described in Section IV for B lymphomas, the association between Epstein–Barr virus (EBV) and nasopharyngeal carcinoma (NPC) is not restricted to endemic areas, but is identical wherever the tumor is seen in the world. Suspected in Egyptian mummies dating back to 3000 years B.C. (Clifford, 1970a; Ho, 1972a), this cancer has been recognized as a major problem since the beginning of this century in south China, where it is described as the Guangdong tumor. Although one could find a fatal disease called shih-ying (described in an ancient Chinese medical document) that could have represented NPC (Ho, 1971; Wu, 1921), the reason for the lack of a good description of NPC in China before the 19th century is probably that this disease is largely confined to south China, whereas early medical writings were by physicians in northern and central China (Ho, 1970; 1972a,b). The original description of NPC in the Occident was formulated in France by Durand-Fardel (1837) for the clinical aspect and by Michaux (1944) for the histopathological aspect.

A. Pathological, Clinical, and Epidemiological Characteristics of Nasopharyngeal Carcinoma

1. Pathology

Tumors that arise in the nasopharynx (Fig. 24) are mostly epithelial in origin. Among them, benign growths represent a very small percentage and are represented by papillomas and adenomas. Among the malignant epithelial tumors, NPC represents the overwhelming majority, while adenocarcinomas and adenoid cystic carcinomas are rarities. Nonepithelial tumors include sarcomas and lymphomas, observed mostly in children and adolescents.

The long debate on the classification and histogenesis of nasopharyngeal tumors, notably the so-called lymphoepithelioma and transitional-

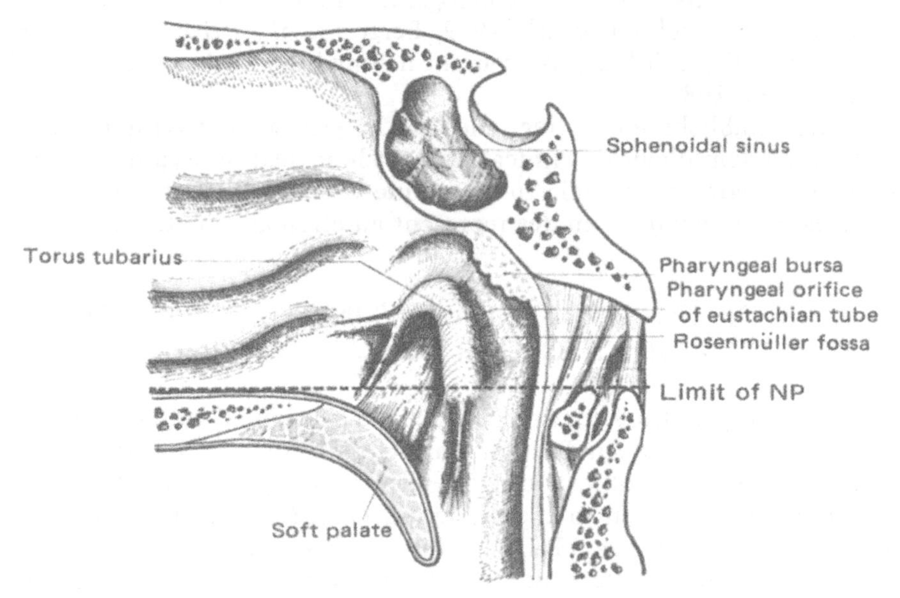

FIGURE 24. Anatomy of the nasopharyngeal area where the respiratory epithelium lies upon lymphoid submucosa.

cell carcinomas, was ended by ultrastructural studies showing that even the so-called lymphoepitheliomas contained epithelial-cell markers and were therefore epithelial in origin (Svoboda *et al.*, 1965; Gazzolo *et al.*, 1972). Therefore, irrespective of their appearance on light microscopy, the vast majority of nasopharyngeal cancers represent variants of squamous-cell carcinomas. In 1978, the World Health Organization proposed a histopathological classification in which three types of NPC were recognized (Shanmugaratnam and Sobin, 1978).

a. *Undifferentiated Carcinoma of the Nasopharyngeal Type (UCNT)*

This type tends to exhibit syncytial rather than pavement appearance of tumor cells arranged in irregular strands or loosely connected with lymphoid stroma. Tumor cells generally show vesicular nuclei with prominent nucleolei. This UCNT was formerly named lymphoepithelioma by a number of European and American pathologists.

b. *Nonkeratinizing Carcinomas*

These are tumors that show evidence of differentiation but in which the squamous maturation is not obvious. The tumor cells have well-defined cell margins and are stratified or with a pavemented arrangement, usually exhibiting a clear cell structure, sometimes referred to as clear-cell carcinomas.

c. *Keratinizing Squamous-Cell Carcinomas*

These tumors show definite evidence of differentiation with intercellular bridges and keratinization.

From the etiopathological viewpoint, it appears that one could mix types a and b and keep the well-differentiated carcinomas separate. The latter are found more commonly in older patients and in low-risk areas, are less radiosensitive, have lower EBV antibody titers than UCNTs, and are found associated with smoking and drinking, which is not the case with the UCNT type (Micheau *et al.*, 1980, 1981; Orofiamma *et al.*, submitted for publication).

2. Clinical Characteristics

The symptoms of NPC are determined by the localization of the tumor at the base of the skull, close to the cranial nerves, and by the strong lymphotropism of the tumor cells. In fact, the symptom that most frequently brings patients to consult is bilateral cervical lymph-node enlargement (Fig. 25). Less frequent symptoms are nasal obstructions, postnasal discharge, epistaxis, or partial deafness. Clinical signs reflecting invasion of the 5th, 6th, 9th, 10th, 11th, and 12th cranial nerves usually appear late during the disease.

The three modes of clinical progression of NPC described by Ho (1970) are discussed below.

a. Metastatic Type

This type is characterized by a rapid invasion of cervical lymph nodes with limited extension of the primary tumor; it is usually associated with a long survival period (5–8 years) even without adequate radiation therapy. Distant metastases involve the lung, liver, spine, and rarely the brain.

b. Invading Type

This type, which represents only 10% of cases, involves direct spread of the primary tumor into adjacent bones, muscles, cranial nerves, orbits, veins, and base of the skull, but rarely cervical lymph nodes. The prognosis of this variety is especially poor.

c. Combined Type

This type, combining direct spread of the primary tumor with involvement of regional and distant lymph nodes, represents the most common type in high-risk areas of South Asia (Fig. 25).

3. Clinical Staging

Ho (1970) first proposed clinical stages to reflect the spread of the tumor from its original site to distant metastases. Table VII describes the

FIGURE 25. Enlargement of cervical lymph nodes as the most common clinical aspect of NPC.

UICC–TNM classification, the Ho staging, that of the American Joint Committee, and the IARC–Kyoto propositions. Because the EBV antibody profile increases with the progression of the disease, it is important to reach international agreements for clinical staging to assess serology as a tool for the management of NPC and a possible marker for prognosis.

Radiation therapy remains the best treatment for both primary-site and lymph-node metastasis of NPC. The actuarial survival rate of 5 years after radiotherapy reaches 40% for stage III of the disease and 67 and 84% for stages II and I, respectively (Ho, 1978). Well-differentiated squamous-cell carcinomas have a poorer prognosis after radiation therapy, and in Europe the survival rate at 9 years after treatment is only 18% for well-differentiated carcinomas vs. 40% for undifferentiated carcinomas, all stages combined (Burkitt, 1958).

Chemotherapy has been very deceiving in NPC. Postradiotherapy, multiagent chemotherapy [bleomycin–methotrexate, velban–1-(2-chloroethyl)-3-cyclohexyl-1-nitrosourea (CCNU)] is actually under investigation (Huang, unpublished data). An interferon trial is actually being implemented in China to evaluate its usefulness in early lesions or as complement to radiotherapy (Zeng and de-Thé, unpublished).

4. Epidemiology

NPC prevalence is low in the Occidental world, where it constitutes approximately 0.25% of all cancer cases. However, around the Mediterranean Sea and North and East Africa, it accounts for approximately 7% of all cancers in males. But in large parts of South East Asia, it represents the single most prevalent tumor in males and about 20% of all cancer cases. The incidence of the tumor around the world is given in Fig. 26. To simplify, one can describe three different levels of incidence of this cancer worldwide. The highest incidence rate is observed in the southern provinces of China (Fig. 27) (Guangdong, East Guangxi, West Fuijeng), in Hong Kong and Singapore, where the general incidence is between 12 and 20/100,000 per year, reaching 98/100,000 males for Cantonese in the 45- to 50-year age group. The incidence in Vietnam, Malaysia, the Philippines, Indonesia, and Thailand is nearly as high as in south China (Muir, 1972; Hirayama, 1978; R. W. Armstrong *et al.*, 1978). Eskimos in Alaska, Greenland, and elsewhere appear to have an incidence as high as that of southern Chinese (Blot *et al.*, 1975; Nielsen *et al.*, 1977; Lanier *et al.*, 1980b).

Intermediate incidence (1.5–9/100,000 per year) is prevalent in many Mediterranean countries, including North Africa and East Africa. Muir (1972) reviewed the relative frequency of NPC in African populations and observed great variations: NPC represented between 5 and 7% of all cancers in Algeria, Tunisia, Morocco, and the Sudan, but only 0.1–1% in Egypt, Irak, Nigeria, South Africa, the Congo, and Senegal. In Israel, where this cancer is relatively rare, there are significant differences

TABLE VII. Clinical Classification and Staging of Nasopharyngeal Carcinoma (1970)

UICC (1974) classification	American Joint Committee (1976)	
	Classification	Staging
T = Primary tumor	Same	
T_{IS} Carcinoma *in situ*		I = $T_1N_0M_0$
T_0 No evidence of tumor		II = $T_2N_0M_0$
T_1 Tumor limited to one region of NP		III = T_1, T_2, T_3 with N_1M_0
T_2 Tumor extending to two regions of NP		
T_3 Tumor extending beyond NP without bone involvement		
T_4 Tumor extending beyond NP with bone involvement		
N = Regional lymph nodes		
N_0 No palpable node		
N_1 Movable homolateral node	N_1 Same less than 3 cm in diameter	
N_2 Movable contralateral or bilateral nodes	N_2 Homolateral or bilateral 3–6 cm in diameter	
N_3 Fixed contralateral or bilateral nodes	N_3 Larger than 6 cm in diameter	
M = Distant metastases		
M_0 No evidence of metastasis		IV = T_4 or any T with M_1
M_1 Metastases present		

	Ho classification	Ho (1970) staging: Stage	Ho (1970) staging: Classification	IARC–Kyoto (1978) proposed staging
T_1	Tumor confined to NP	I =	$T_1N_0M_0$	A = T_1N_0
T_2	Tumor extended to nasal fossae, oropharynx, muscles, nerves below base of skull	II =	T_2 or N_1 or both	B = N_1 irrespective of T
T_3	Tumor extending to:	III =	T_3 or N_2 or both	C = N_2 irrespective of T
T_{3a}	Below base of skull involvement	IV =	N_3	D = M present
T_{3b}	Base of skull involvement	V =	M_1	
T_{3c}	Cranial nerve involvement			
T_{3d}	Orbit and hypopharynx involvement			
N_0	No palpable lymph nodes			
N_1	Upper cervical nodes			
N_2	Lower cervical nodes			
N_3	Supraclavicular nodes			
M_0	No metastatses			
M_1	Hematogenous metastases and nodes below clavicles			

among Jews: those born in Israel, Asia, or North Africa have a higher NPC rate than Jews born in America or Europe.

In the rest of the world, including the United States and Europe, the incidence is very low, around 0.1–0.2/100,000 per year.

The situation in the southern provinces of China is epidemiologically most interesting (Fig. 27). The highest incidence for NPC appears to be associated with certain cultural patterns. In Guangdong, the main ethnic group is represented by Cantonese Chinese and boat people who are both at a very high risk, the incidence of NPC in the latter being higher than that in the former. In the Guangxi Autonomous Region, the situation is special because of the presence of many different national minorities. When one compares the map of the NPC prevalence in the Guangxi Autonomous Region with that of the geographic distribution of the minorities in the same province (Fig. 28), it becomes apparent that certain minorities, such as the Han, appear at much higher risk than others, the Yao, for example. Shanmugaratnam (1978) observed differences in the incidence of NPC among the different Chinese dialect groups living in Singapore, with Cantonese Chinese having the highest incidence of 14/100,000 per year. Ho (1972b) found a similar difference in Hong Kong between Cantonese and Hokkianese. Cantonese Chinese who had immigrated to the United States were found to have an incidence in the second generation lower than that of their original country, but still much higher than that of the surrounding Caucasian Americans (King and Haenszel, 1972; Fraumeni and Mason, 1974).

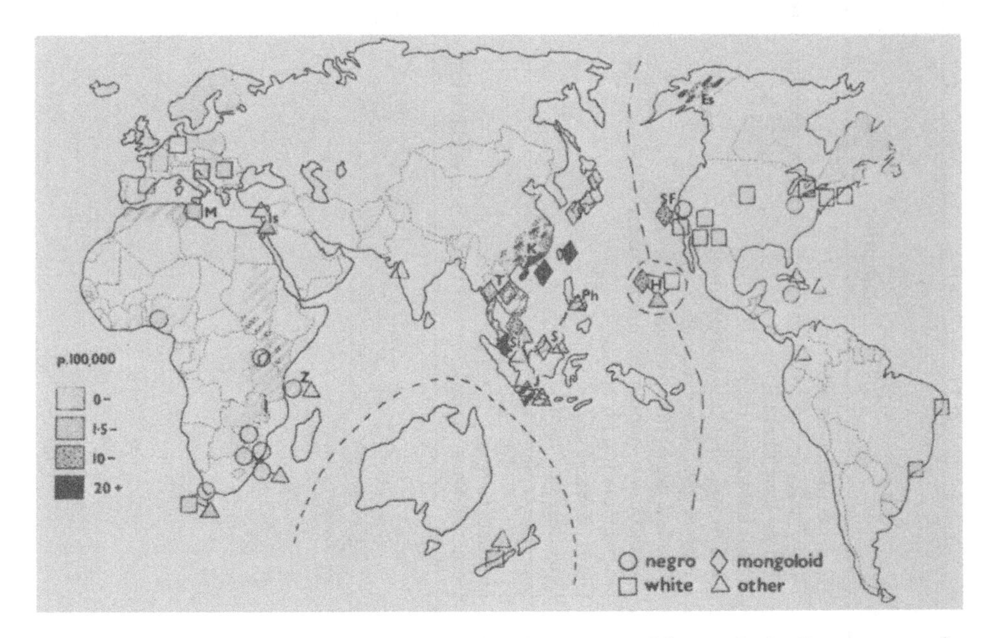

FIGURE 26. World distribution of NPC in males. Prepared by P. Cook, Department of Health and Social Security, University of Oxford.

FIGURE 27. Geographic distribution of NPC in China. The dark areas indicate regions where NPC is highest and cover the Pearl River basin. From the National Office for Cancer Control (1981).

5. Sex and Age

Male preponderance for NPC of around 2-fold was observed everywhere, but the age distribution of NPC is quite different in high-, intermediate-, and low-risk areas. Figure 29 shows a sharp rise in Hong Kong from 20–24 years of age to a peak at about 50–55 years and a fall thereafter. This fall could represent the exhaustion of a pool of susceptible individuals. It is worth noting here that NPC occurs in Swedes two decades later than in Chinese, up to the age of 70–74 years (Ho, 1972b). This is possibly due to a higher proportion of well-differentiated carcinomas in Caucasian populations than in Chinese.

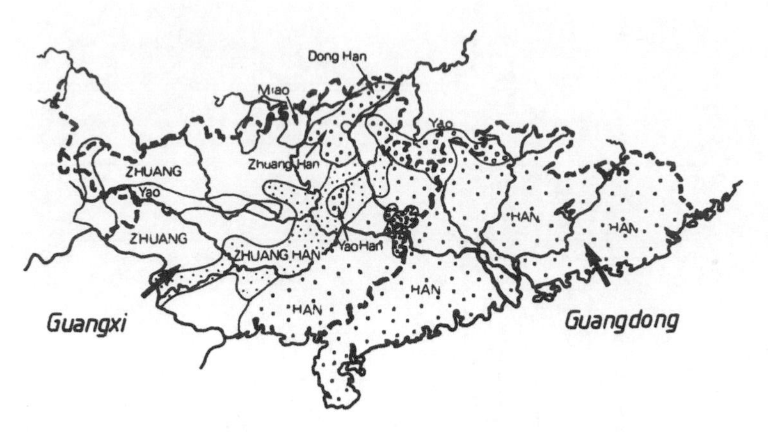

FIGURE 28. Distribution of national minorities and of NPC prevalence in the Guangxi Autonomous Region. The Han, in the east, have the highest incidence, the Zhuang appearing at intermediate risk and the Yao at lowest risk. Prepared from the Chinese Cancer Survey and a map from the National Geographic Society.

In intermediate-risk areas for NPC, the age distribution showed two peaks. The first peak observed in Tunisia (Fig. 30), Algeria, and Morocco between 10 and 24 years of age was also observed in Kuwait and the Sudan and included 15–20% of all NPC cases (Cammoun *et al.*, 1971; Saad, 1968). The second peak occurred a little later than in China. In the United States, the two peaks were also observed, the first one, under the age of 24, occurring in blacks, who were 7 times more exposed to early NPC than whites (Greene *et al.*, 1977).

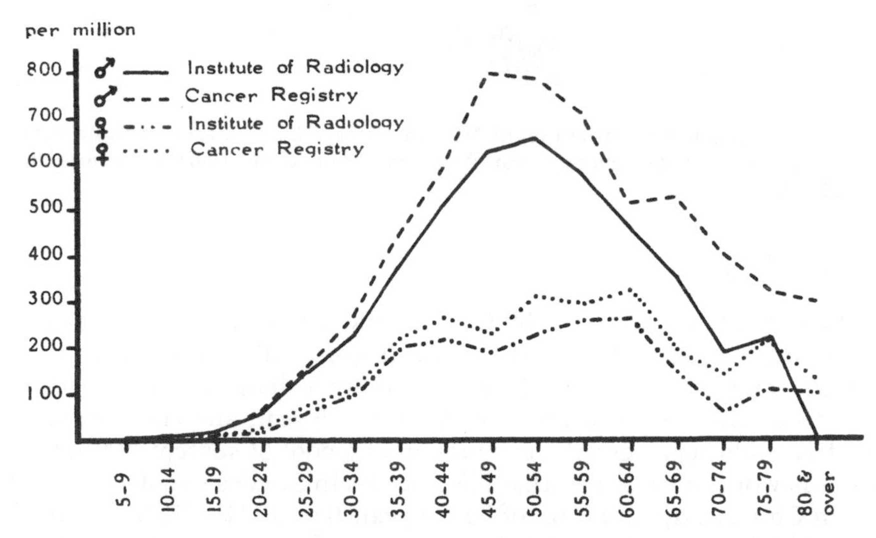

FIGURE 29. NPC age-specific incidence rates per million in Hong Kong, 1965–1968. From Ho (1970).

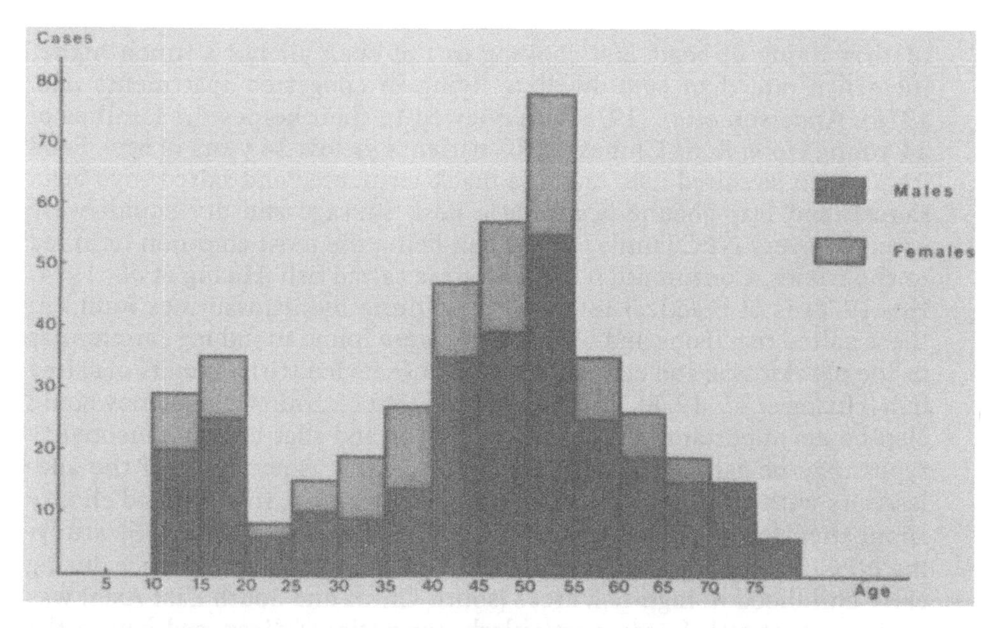

FIGURE 30. Age distribution of NPC in Tunisia (1969–1973). From Ellouz *et al.* (1978).

6. Environmental Factors and Life Style

Shanmugaratnam and Higginson (1967) were the first to conduct an in-depth case–control retrospective study for NPC, but failed to uncover any risk factor. More recently, Shanmugaratnam and Sobin (1978) interviewed nearly 400 Chinese Singaporian NPC patients and, for comparison, 600 patients with other ENT diseases and 100 other patients. They showed that NPC differed from both groups by having an association with nasal illnesses, the use of Chinese medicine, and exposure to smoke. They interpreted this latter exposure as reflecting a more traditional household than that of control patients. As pointed out by Ho (1978), critical risk factors seem to be associated with the traditional Chinese environment. R. W. Armstrong *et al.* (1978) found that Malaysian Chinese NPC patients had a lower socioeconomic status than controls. In their case–control study of NPC patients from Hong Kong, Geser *et al.* (1978) also found that risk factors involved the low occupational socioeconomic classes, history of previous illness of ear or nose after the age of 15, and salted fish given to babies after weaning. Multivariate analyses showed that the traditional life style and consumption of salted fish during weaning were independent risk factors. Ho (1971, 1972a, 1978) was the first to suggest that eating salted fish might be an important risk factor; this traditional food consumed by poor southern Chinese since early childhood contains appreciable quantities of nitrosodimethylamines (Fong and Walsh, 1971; Fong and Chan, 1973; Huang *et al.*, 1978). The factor involved is more likely to be ingested than inhaled, since the marine population living in boats and cooking in the open air has a much higher incidence rate than land dwellers living in congested apartments (Ho, 1978). Anderson *et al.* (1978) interviewed in their homes the families of 24 young Hong Kong Chinese NPC patients (below 24 years of age). Food items such as salted fish, tau-si (a black fermented and salted soya bean sauce), and laap-cheung (Cantonese pork sausage and dry squid) were eaten by every NPC family, salted fish being the most common item fed to the babies. Consumption of Cantonese salted fish (Huang *et al.*, 1974; Ho, 1978) is of practical interest, since dimethylnitrosamines found in these salted fish (Fong and Chan, 1973) were found to induce carcinomas in the nasal fossae and paranasal regions of rats fed with extracts of salted fish (Huang *et al.*, 1978). It was suggested that vitamin C deficiency could also be an important environmental factor and that certain personality types may be associated with an increased risk, since most of the adolescents with NPC were stated to have been "sickly, inactive and choosy about their food" (Anderson *et al.*, 1978). Henderson *et al.* (1976), studying NPC among Chinese Americans, found that birth or residence during early childhood in high-risk areas (south China and South East Asia) was an important risk factor, particularly for patients diagnosed before the age of 50.

The use of Euphorbiaceae plant extracts in Chinese traditional medicine could be of importance for NPC development, since croton oil extracted from *Euphorbia* is used as a laxative as well as in certain nasal balms in Chinese traditional medicine. Phorbol esters present in these extracts are, as noted in Section II.A, well-known experimental tumor promoters and are able to reactivate latent EBV infection (zur Hausen *et al.*, 1978). Ito *et al.* (1981) have recently proposed that Euphorbiaceae may play a causative role in NPC development.

7. Genetic Factors

Familial aggregation of NPC cases was observed in high-risk areas as well as in Uganda (Ho, 1972a,b; Liang, 1964; Williams and de-Thé, 1974). Studies of genetic blood-cell markers were unrewarding (Hawkins *et al.*, 1974), although studies involving red-cell enzymes and serum protein markers indicated that NPC tended to concentrate on genetically distinct subpopulations of Cantonese Chinese in Singapore (Kirk *et al.*, 1978). HLA typing has been extensively investigated among Chinese NPC patients and controls. The A2BW46 haplotype, which was found more frequently in Chinese NPC patients than in controls (Simons *et al.*, 1974, 1975), carried an increased relative risk of 1.96 (Simons *et al.*, 1976, 1977). When attention was directed to newly diagnosed cases, the haplotype AW19BW17 was observed in NPC patients more frequently than in controls; furthermore, among the long-term survivors, BW17 was found significantly decreased in frequency, as though this genetic marker were associated with a poor prognosis (Simons *et al.*, 1978). These NPC-associated haplotypes were not observed in Tunisian or Moroccan NPC patients (Betuel *et al.*, 1975; Betuel, personal communication), as though there were an NPC disease susceptibility gene close to, but outside, the *HLA A C B* area. The HLA-D and -DR typing has been started in Singapore (Chan *et al.*, 1981).

B. Association between Epstein–Barr Virus and Nasopharyngeal Carcinoma

Experimental viral oncology has shown over the past two decades that a large number of animal tumors are associated with RNA and DNA viruses. The regular presence of viral markers in tumor cells is believed to be a prerequisite for a causative association between a virus and a tumor. In contrast to the situation described for Burkitt lymphoma (Section IV), NPC, regardless of its geographic origin, is regularly associated with EBV and in an identical manner. This association is based on the presence of EBV markers in tumor cells and on an EBV-specific immune response progressing with the disease process.

1. EBV Markers and Tumor Cells

Zur Hausen *et al.* (1970) were the first to show the presence of EBV DNA sequences in NPC biopsies from Africa. Nonoyama and Pagano (1973) and Nonoyama *et al.* (1973) confirmed these results in NPC, from various origins. Desgranges *et al.* (1975a,b), comparing NPC biopsies from high-, intermediate-, and low-risk areas, showed that the virus was consistently associated with epithelial tumor cells. Because EBV was known as a lymphotropic virus, and NPC was an epithelial tumor, it was first believed that the EBV DNA detected was localized in lymphoid cells regularly present in NPC. However, the opposite was demonstrated first by Wolf *et al.* (1973) and by Desgranges *et al.* (1975a): most of the EBV DNA sequences were in epithelial cells and not in infiltrating lymphoid elements. EBV-induced nuclear antigen (EBNA) was first suspected in epithelial tumor cells by de-Thé *et al.* (1973a) and confirmed by Huang *et al.* (1974) and Klein *et al.* (1974). Simultaneous detection of EBV DNA and EBNA was obtained by Klein *et al.* (1974) when they grafted NPC tumor biopsies into nude mice, thus eliminating the lymphoid elements. They observed both EBNA and EBV DNA in now purely epithelial grafted tumors. After treatment with 5-iodo-2-deoxyuridine [(IUdR) Idoxuridine] or 5-bromodeoxyuridine or after superinfection with HR1 EBV, *in vitro* epithelial outgrowths of NPC could synthetize early antigen (EA) (Glaser *et al.*, 1976), suggesting that tumor cells have surface EBV receptors. In fact, full replication of EBV was observed in epithelial cells passaged in nude mice, and the virus was isolated (Trumper *et al.*, 1977). However, after passage in nude mice, epithelial NPC tumor cells exhibited the presence of mouse oncorna virions (de-Thé *et al.*, 1976). If the presence of EBV markers in nasopharyngeal tumor cells *in vivo* is a constant observation, the *in vitro* transformation of normal epithelial cells from nasopharyngeal mucosa was never achieved, except under exceptional circumstances with biopsies of normal nasopharyngeal mucosa of NPC-bearing patients (Desgranges and de-Thé, 1977).

2. EBV-Specific Immune Response

EBV serology has been most instrumental for establishing the association between the virus and the tumor. It is becoming increasingly useful for the management of the disease (diagnosis and prognosis). High EBV-specific antibodies have been known to be present in NPC patients since the early studies of Old *et al.* (1966), W. Henle *et al.* (1970a), de Schryver *et al.* (1969), and de-Thé *et al.* (1975a).

a. IgG Antibody Profile in the Course of NPC

As can be seen in Fig. 31, geometric mean titers (GMTs) of IgG antibodies specific for viral capsid antigen (VCA) and EA as well as complement-fixation antibodies to soluble (CF/S) or nuclear antigens (EBNA) were found greatly elevated in NPC patients from high-, intermediate-, or low-risk areas for NPC, when compared to those in patients bearing other ENT tumors or to normal individuals from the same geographic areas (de-Thé *et al.*, 1973, 1975a, 1978b).

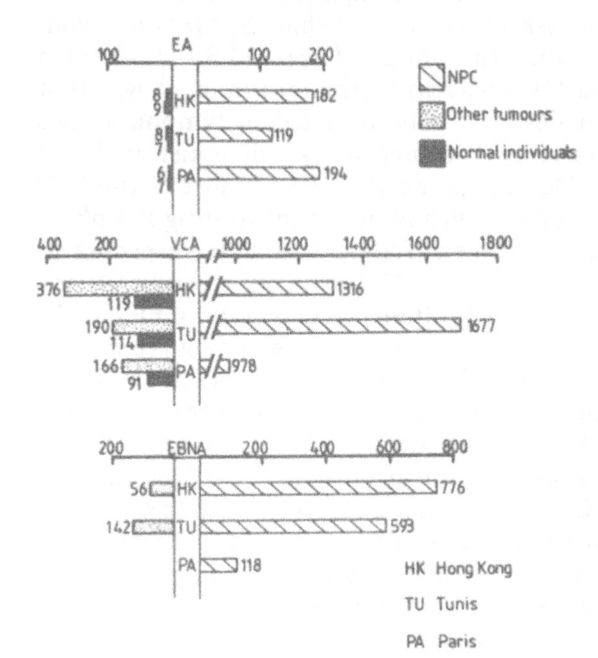

FIGURE 31. Comparison of antibody GMTs to VCA, EA, and EBNA between NPC patients and controls (patients with other ENT tumors and normal individuals) originating from various geographic areas. Prepared from data of de-Thé *et al.* (1978b).

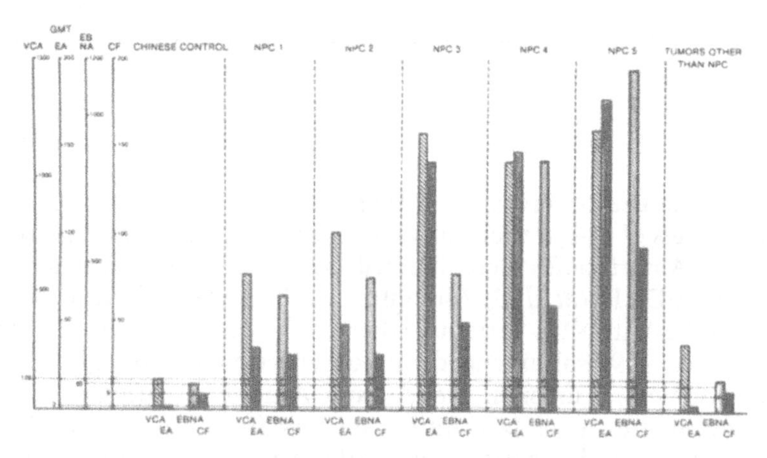

FIGURE 32. EBV reactivities according to clinical staging of Chinese NPC patients. From de-Thé *et al.* (1975a).

Antibodies to EA diffuse (D) component (W. Henle *et al.*, 1970b, 1973; de-Thé *et al.*, 1978b; Lin *et al.*, 1975) were raised by 24-fold in all NPC groups as compared with patients with other tumors and normal individuals, regardless of their geographic distribution or ethnic origin, indicating that there is an active EBV replication in these patients. The site of this replication is not yet known, and investigations aimed at searching for the presence of VCA or EA in the tumor itself have regularly failed. Antibodies to VCA do vary among geographic areas, reflecting differences in socioeconomic level. The differences among geographic areas relating to EBNA antibodies were believed to reflect genetically controlled immune response (de-Thé *et al.*, 1978b).

Figure 32 shows the GMTs of EBV-specific IgG antibodies according to the stages of the disease. As early as stage 1 of the disease, when the tumor is localized in the nasopharyngeal mucosa, the EBV profile is much higher than that of patients with other types of ENT tumors or normal controls. The EBV serological profile was found most useful in the differential diagnosis of NPC in temperate climates or in areas at intermediate risk for the tumor, where sometimes patients coming for tumorous cervical lymph nodes of unknown primary can be misdiagnosed; in such cases, the EBV serology helped by suggesting an NPC (W. Henle *et al.*, 1970a,b, 1973; de Schryver *et al.*, 1969; de-Thé *et al.*, 1975a). Because of the differences in the EBV serological profiles of NPC patients from different racial groups in Los Angeles, Henderson *et al.* (1976) questioned the strength of association between EBV and the tumor. The comparative study of NPC patients originating from Hong Kong, Tunis, and Paris (Fig. 31) showed that differences observed among ethnic groups existed, but that the ratios of the NPC patient values to those of controls from the same geographic areas were similar all around the world, indicating that a high EBV immune response was regularly associated with

NPC, the level of the response depending on the background observed in the geographic area concerned (de-Thé *et al.*, 1978b).

b. IgA Antibody Profile in Serum and Saliva in the Course of NPC

Following the observation of Wara *et al.* (1975) that NPC patients had high levels of IgA in their saliva, G. Henle and W. Henle (1976) found that such IgA antibodies were directed against EBV-determined antigens (VCA and EA-D). Ho *et al.* (1976) confirmed the presence of IgA antibodies to VCA and EA in NPC patients, whereas Desgranges *et al.* (1977a) and Pearson *et al.* (1978b) extended this observation to non-Chinese NPC patients, namely, Tunisian, French, and American patients. Figure 33 shows the evolution of IgA antibody titers from stages 1 to 5 of the disease, as compared to patients with other tumors or normal individuals (Desgranges and de-Thé, 1978). Already at stage 1, NPC patients had a GMT 10 times higher than that of patients bearing other ENT tumors or normal individuals. This has been used by Zeng and co-workers for early diagnosis of NPC in large seroepidemiological studies in China (see Section V.C.3).

Desgranges *et al.* (1977) and Desgranges and de-Thé (1978a,b) extended these serological studies to the saliva of NPC patients, which contained high levels of secretory IgA (11 S) specific for VCA and EA-D in 75 and 35% of cases, respectively. As seen in Fig. 34, secretory IgA antibodies to VCA and EA are restricted to the saliva of NPC patients (50% of cases), whereas IgG antibodies to VCA or EA were occasionally found in the saliva from patients with infectious mononucleosis (IM), Burkitt lymphoma (BL), or other tumors. Furthermore, Desgranges *et al.* (1977a) and Desgranges and de-Thé (1978b) found that plasma cells around epithelial tumor cells in NPC biopsies contained IgA molecules

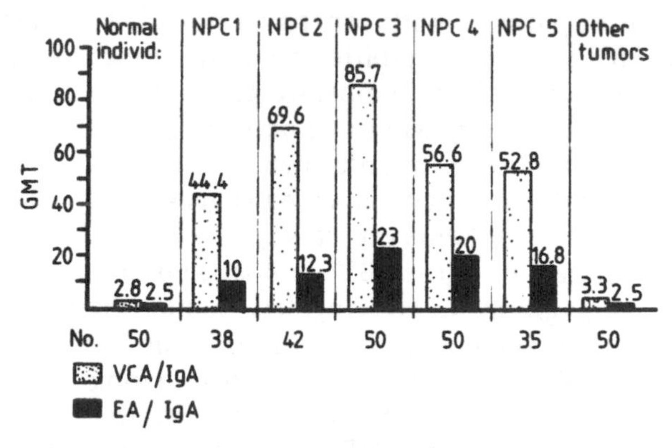

FIGURE 33. IgA antibodies to VCA and EA-D in Chinese NPC patients according to Ho's clinical stages and for comparison in patients with other ENT tumors or normal individuals. Adapted from Desgranges and de-Thé (1978b).

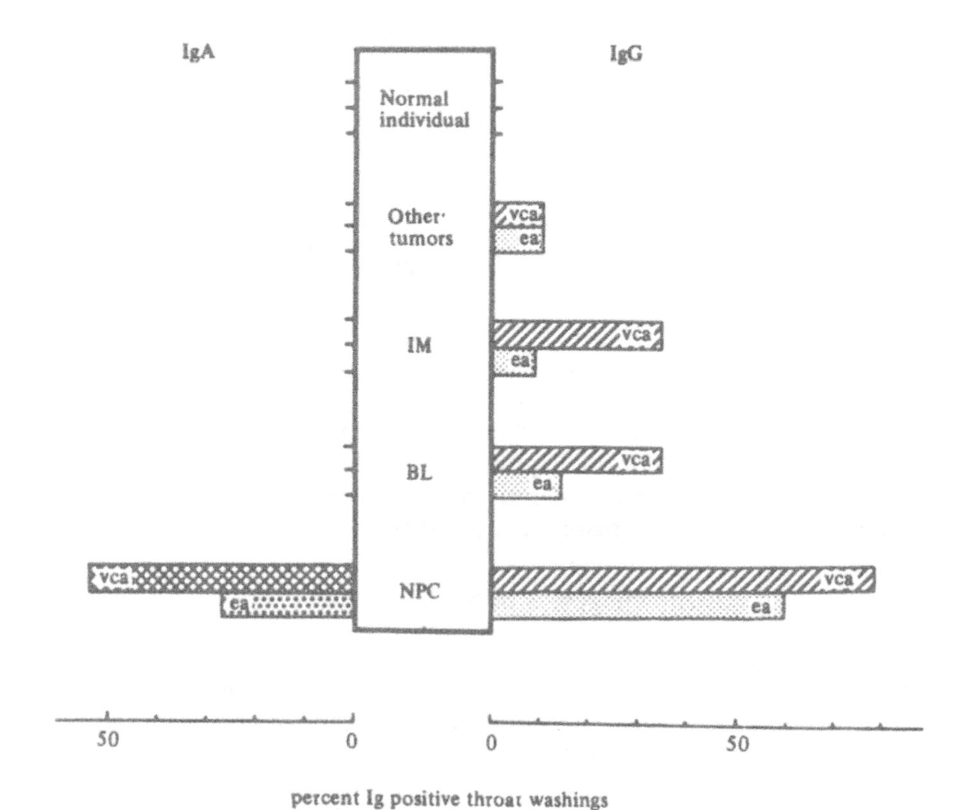

FIGURE 34. IgA and IgG antibodies to VCA and EA in throat washings from NPC patients and for comparison from patients with BL, IM, other tumors, or normal individuals. From Desgranges and de-Thé (1978a).

that were secondarily found to be EBV/VCA-specific. By immunofluorescence, epithelial tumor cells were seen to exhibit the presence of secretory pieces of their cell surface (Desgranges *et al.*, 1977b). Thus, the EBV-specific IgA present in the saliva of NPC patients originated in the tumor itself, where it could represent blocking antibodies.

c. *Antibody-Dependent Cellular Cytotoxicity*

Antibody-dependent cellular cytoxicity (ADCC), which is effective in the destruction of virus-infected cells, appears to act on virus-induced membrane antigens. Such antibodies may involve an important immune mechanism for an *in vivo* control of viral infection and in the control of virus-induced tumors (Harada *et al.*, 1975; Prather and Lausch, 1976; Pearson *et al.*, 1978a). Pearson *et al.* (1979) observed that NPC patients had high ADCC values as compared to controls. In a recent study, the same group (Mathew *et al.*, 1981) observed that ADCC antibodies were in the IgG fraction and that IgA antibodies were able to block the IgG-mediated ADCC reactions, indicating that the IgA and IgG antibodies were recognizing the same EBV-specific antigenic determinant. These results would suggest that IgA antibodies are detrimental to patients with NPC if one assumes that ADCC *in vivo* represents an active immune response against the tumor load.

C. Epstein–Barr Virus and the Management of Nasopharyngeal Carcinoma

Whatever the nature of the association between EBV and NPC is finally found to be and whenever the pathogenesis involved is discovered, EBV serology is increasingly found to be of value for the clinical management of this cancer, for its diagnosis and early detection, to assess its prognosis, and to define precancerous conditions that may lead to a new approach toward the prevention of this cancer.

1. EBV Serology for the Diagnosis of NPC

In NPC endemic areas of South East Asia, the diagnosis of NPC is evoked systematically in view of any suggesting symptom. But this is not the case in North Africa and in the Mediterranean countries, where NPC may be misdiagnosed. In countries where this cancer is exceptional, Europe and North America, NPC is often missed in cases of tumorous cervical lymph nodes without obvious primary. Furthermore, when the primary deposit in the nasopharynx is invading the submucosa without outgrowth, the observation of the nasopharynx may be clear, and only the serology should oblige the repeating of nasopharyngeal biopsies up to the point where the histopathological diagnosis is reached.

A clinically oriented international collaborative study has been set up by the National Cancer Institute, with the aim of assessing the usefulness of EBV serology in the management of NPC. Preliminary results indicate that elevated IgG EA and IgA VCA antibodies are most instrumental in speeding up NPC diagnosis (Pearson and colleagues and de Thé and colleagues, unpublished data). The development of a rapid and inexpensive serological EBV kit would permit moving the procedure from research to hospital laboratories.

EBV serology could also help to differentiate between the most common undifferentiated or poorly differentiated carcinomas (UCNT) of the nasopharynx and all other tumors that develop in this area. As can be seen in Figs. 31–33 and 35 and as discussed in Section V.B, high EBV reactivities are restricted to the undifferentiated type.

EBV serology is thus becoming a routine biological marker for this cancer, useful not only for its differential diagnostic but also for its prognostic capabilities.

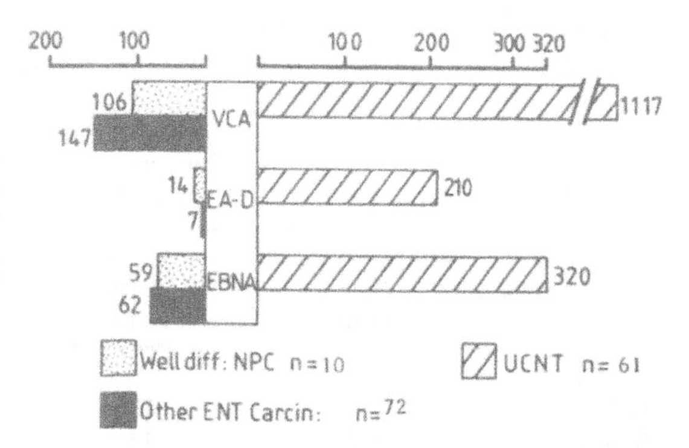

FIGURE 35. EBV serological reactivities of the undifferentiated type of NPC (UCNT) compared to the well-differentiated type and other carcinomas. The EBV profile of the well-differentiated type is similar to that of other ENT carcinomas, while the UCNT exhibits reactivities 10 times higher. From Orofiamma *et al.* (in press).

2. EBV Reactivities and Prognosis of NPC

As seen in Section V.B.2.a and Fig. 32, EBV-antibody titers increase with tumor burden from stages 1 to 5, and it could be expected that treatment that reduces or eliminates the tumor mass would be accompanied by a decline in EBV titers. In fact, declines in IgG VCA and EBNA titers were noted by de Thé *et al.* (1973b) in patients 2 years after radiotherapy. W. Henle *et al.* (1973), comparing the serological profile of NPC patients before radiotherapy and of those who survived 5 years or more, observed that the IgG VCA and IgG EA-D antibodies were substantially lower in the long-term survivors as compared to untreated patients. Table VIII gives the trend of the EBV profiles of 100 NPC patients over a 4- to 5-year period studied by W. Henle *et al.* (1977), who compared patients who survived 4–5 years to those who died within this period. They observed that EBV reactivities decreased in the group in whom radiotherapy was successful, whereas it increased in the fatal cases. When individual serological trends were studied, important variabilities were observed, but in some cases, the rise of EBV IgA VCA and IgG EA could have predicted the clinical relapse by a few months (W. Henle *et al.*, 1977).

An international collaborative study is in progress in the United States, Europe, the Far East, and North Africa to try to better assess the usefulness of EBV serology for the prognosis of NPC. Preliminary data confirm the results cited above, and after radiotherapy, it has been observed in Hong Kong (Ho *et al.*, 1981) and in Villejuif (de-Thé and colleagues, unpublished results) that the EBV profile does decrease after a year, especially the IgA VCA and IgA EA reactivities, which appear to be the most valuable markers for posttherapy follow-up of NPC.

As discussed in Section V.B.2.c, NPC sera before therapy usually exhibit high ADCC values. Pearson *et al.* (1978b) investigated two groups of NPC patients: patients who died within 2 years following the diagnosis of NPC and patients who responded well to therapy and survived more than 2 years after diagnosis. These authors found that at diagnosis, the ADCC GMT for the good survivors was significantly higher than that of

TABLE VIII. Geometric Mean Titers of Epstein–Barr Virus Reactivities of Long-Term (4–5 Year) Nasopharyngeal Carcinoma Survivors and of Fatal Nasopharyngeal Carcinoma Cases[a]

GMT	Long-term survivors (stages I/II, n = 15; stages III/IV, n = 23) First serum		Long-term survivors Last serum	Fatal cases (stages I/II, n = 32; stages III/IV, n = 33) First serum		Fatal cases Last serum
IgG VCA						
Stages I/II	182	↘	123	224	↗	314
Stages III/IV	405	↘	180	423	↗	630
IgA VCA						
Stages I/II	23	↘	9	16	↗	25
Stages III/IV	45	↘	19	30	↗	44
IgG EA						
Stages I/II	15	↘	9	25	↗	60
Stages III/IV	84	↘	22	60	↗	116
IgA EA						
Stages I/II	7	→	7	9	↗	19
Stages III/IV	22	↘	10	18	↗	32

[a] From W. Henle *et al.* (1977).

the poor-survivor group (GMT 5410 vs. 615). Furthermore, in the poor-survivor group, when ADCC titers were lower than 240, antibody titers to EA or IgA to VCA were generally high. These data suggest that ADCC titers could have a prognostic value for NPC and that low ADCC titers previously found in the sera of poor NPC survivors could be due to a blocking activity of IgA antibodies.

3. IgA and Early Detection of NPC in China

In contrast to Western countries, where a curative health system and an analytical research mind prevail, Chinese health priorities have been geared toward prevention for thousand of years.

Present Chinese priorities in cancer remain primary and secondary prevention, through the characterization of risk factors and their elimination through education and early detection through epidemiological and clinical mass surveys. Chinese consider that prevention is possible without the precise etiology or pathogenesis of a disease necessarily being known. A nationwide cancer-prevalence survey was undertaken in China between 1973 and 1976, covering more than 850,000 individuals. The results, published in an *Atlas of Cancer Mortality* by the National Office for Cancer Control (1981), showed that the main cancer sites were unevenly distributed in China.

As shown in Fig. 27, NPC prevails in southern China in the basin of the Pearl River. The strong association existing between NPC and EBV impelled the Chinese to see to what extent this association could be useful for better control of NPC. The first question was whether or not EBV serology could be used for early detection of this cancer in endemic areas. The most discriminatory and simplest EBV reactivity being the IgA VCA titer, Liu and Zeng (1979) developed an IgA VCA immunoenzymatic test, better adapted to field conditions than immunofluorescence, that permitted the surveying of very large population samples with simple equipment. In the eastern part of the Guangxi Autonomous Region, in the Zangwu county inhabited mostly by Cantonese Han, at high risk for NPC, Dr. Zeng, from the Virus Institute in Beijing, in collaboration with the Guangxi Bureau of Public Health, carried out two successive seroepidemiological surveys in 1978 and 1979, detecting 1267 IgA/VCA-positive individuals among 148,000 individuals aged 30 or more (Zeng *et al.*, 1979, 1980a). Clinical and ENT investigations of these IgA-positive individuals led to the taking of 203 biopsies and to the discovery of 46 NPC cases, half of whom were at stage 1 or 2 of the disease (Zeng *et al.*, 1980b). Careful clinical follow-up of the IgA-positive individuals as well as of biopsied individuals allowed the detection of 12 more cases of NPC within 18 months. One can calculate from these surveys that IgA-positive individuals have about 20 times the rate of NPC as compared to the general population and represent the highest-risk individuals for NPC.

D. Epstein–Barr-Virus-Related Precancerous Conditions for Nasopharyngeal Carcinoma

The Chinese seroepidemiological surveys mentioned above created an opportunity to carry out clinical examination of large populations with the aim of characterizing precancerous lesions of nasopharyngeal mucosa. In the Zangwu serological surveys organized by Zeng and coworkers, clinical examination of 1150 IgA-positive individuals allowed the detection of 335 macroscopic abnormalities of nasopharyngeal mucosa. Of the 203 biopsies taken on the most obvious abnormalities, histopathological examination showed 78 inflammatory processes, 66 lymphoid hyperplasias, and 46 NPC, as well as 13 histologically normal mucosae [Zeng *et al.*, unpublished (cited in de Thé *et al.*, 1981)]. In Guangchow, Dr. Li Chen Chuan of the Tumor Hospital organized a massive clinical survey involving more than 32,000 individuals in 1977 and 1978. On ENT examination of 494 NP mucosae, 48 nodular alterations of the mucosa were noted, 28 hyperplastic adenoids, 9 follicular hyperplasias, and 4 cases with numerous petechia and 5 atrophic mucosae, 2 of them found to be *in situ* carcinoma [C.C. Li and H. Tseung, unpublished (cited in de-Thé *et al.*, 1981)].

Biopsies from nasopharyngeal mucosae of 56 IgA-positive individuals from Zangwu, where abnormalities of the nasopharynx were observed (leukoplasia, hypertrophic mucosa, petechia), allowed the detection of 4 early cases of NPC and 14 mucosae that were noncancerous but contained EBV viral sequences or EBNA or both (de-Thé, 1981; Desgranges *et al.*, 1982). Thus, the presence of increasing IgA VCA titers in normal individuals appeared to be linked to a reactivation of EBV in the nasopharyngeal mucosa. As a working hypothesis, we consider such a reactivation as the last step preceding clinical onset of NPC. In chronological order, primary EBV infection takes place between 0 and 5 years of age in China, then a long period of latent EBV infection takes place, reactivated in the early 40s in the nasopharynx by unknown factors and reflected by an increasing level of serum IgA VCA.

How long before the clinical onset of EBV can the warning signal of IgA VCA be detected? Table IX summarizes the data obtained separately by Ho *et al.* (1978), Lanier *et al.* (1980a), and Zeng *et al.* (1982). It appears that IgA VCA could be detected 12–18 months prior to stage 1 of the clinical disease. This period is indeed critical to consider preventive intervention.

E. The Future: Causality and Prevention

NPC is a unique model among human tumors for studying the interplay between biological and chemical environmental carcinogens as well as the role of the individual's susceptibility to these oncogenic agents. How to establish the nature of the association between EBV and NPC? The fact that this association is strong, consistent, and specific in all geographic areas regardless of the level of incidence of the tumor provides strong support for, but does not prove, the causal nature of the relationship. The possibility that infection of nasopharyngeal epithelial cells by EBV is secondary to cell transformation is remote, but cannot be disproved yet. If this were the case, one should expect to observe some EBV-free NPC in populations where EBV prevalence is low. This is not the case: even in countries where EBV prevalence is low, and where up to 15% of the population is EBV-negative in their 40s and 50s, all NPC are EBV-associated. Further epidemiological evidence for a causal relationship between EBV and NPC came from the seroepidemiological surveys of Zeng showing that IgA-VCA-positive individuals were at highest risk to develop NPC. Furthermore, these preclinical events are related to EBV activities in the nasopharyngeal mucosa itself (de Thé, 1981; Desgranges *et al.*, 1982). The possibility still exists, however, that the reactivation of EBV in the nasopharynx succeeds, not precedes, the birth of the tumorous clone, itself preceding by many months or year(s) the clinical onset of the disease. Only ongoing prospective seroepidemiological studies in China with careful follow-up of the IgA-positive individuals and nasopharyngeal EBV-marker-positive individuals will separate the two alternatives. The negative-pressure apparatus developed by

TABLE IX. Epstein–Barr-Virus-Specific Immunoglobulin A Antibody before and after Diagnosis of Nasopharyngeal Carcinoma[a]

Case No.	Sex/age	Before diagnosis	After diagnosis	Interval between two bleedings (months)	Clinical stage	Histopathological diagnosis
Ho *et al.* (1978) (immunofluorescence test)						
1	M/45	20	40	30		
2	M/47	80	80	35		
3	M/37	5	80	61		
Lanier *et al.* (1980a) (immunofluorescence test)						
8		160	160	22	III	
5		<10	<10	33	I	
20		10	320	58	IV	
25		10	10	72	IV	
3		<10	40	76	II	
7		<10	<10	93	II	
29		<10	10	114	III	
Zeng *et al.* (1982) (immunoenzymatic test)						
1	M/40	10	10	10	I	Poorly differentiated carcinoma
2	M/48	80	80	15		
3	M/49	160	40	16		
4	M/35	80	160	16		
5	F/66	2.5	320	15		
6[b]	M/39	40	160	16		
7	M/49	10	20	15	II	
8	M/46	40	320	9		
9	F/56	20	1280	10		
10	F/47	80	1280	16		
11	M/50	80	1280	16	III	
12	M/67	80	40	15		

[a] From de Thé (1981). See ref. (p. 79): 1981. [b] Carcinoma *in situ*.

Zeng to collect exfoliated epithelial cells from the nasopharyngeal mucosa will allow the collection of the large numbers of specimens necessary for these studies.

Finally, antiviral interventions in these IgA-related preclinical conditions will represent a unique way to establish the role of EBV in NPC. If one could show that antiviral drugs or vaccines, or therapeutic intervention in the EBV reactivation in the nasopharyngeal mucosa decreases the incidence or the severity or even prevents the clinical onset of NPC, not only would the final proof of causal relationship be established, but also an entirely new approach for the prevention of NPC would have been discovered.

What can antiviral therapies for these pre-NPC lesions be? The first that comes to mind is interferon. Interferon is known to have not only antiviral but also antitumoral activities. Gresser and Tovey (1978) have shown that interferon acts best when inoculated prior to the development of the tumor. Its activity, once the tumor has clinically emerged, is much more difficult to evaluate. Most if not all of the interferon therapeutic clinical trials have been made on patients with late relapse of tumors resistant to radio- or chemotherapy. Thanks to Dr. Kari Cantell (Helsinki), a lymphocyte interferon clinical trial has been implemented in China, following two complementary protocols. The first incorporates early NPC cases, stages 1 and 2, in the Regional Hospital of the Guangxi Autonomous Region, in Nanning, under the responsibility of Drs. Zeng, Huang, and de-Thé; the other concentrates on late relapses in the Cancer Center of Guangchow, under the responsibility of Drs. Li Chen Chuan, Strander, and Klein. If interferon has a measurable effect on the clinical or virological level, then its intervention in IgA-related preclinical events can be envisaged.

The next critical question is whether a vaccine could be helpful in controlling NPC. Epstein (1976, 1981) repeatedly stressed the priority for an EBV vaccine. Then de-Thé (1978a) and Pearson and Scott (1978) discussed this possibility. Viral vaccines work only when inoculated prior to primary infection. Thus, for EBV-related diseases, a vaccine could eventually help to prevent IM in adolescents from high socioeconomic groups. Since BL cannot be considered a public-health priority in equatorial regions where parasitic diseases, including malaria associated with BL, kill up to 30% of children before 5 years of age, it is healthier, for the purpose of preventing BL, to propose antimalaria intervention (see Section IV.I). NPC, because of the long latent period between primary infection and development of the tumor, is a poor candidate for an EBV vaccine trial. There is, however, an avenue to explore. NPC is preceded by a reactivation of a latent EB viral infection that, like that of varicella–zoster or cytomegalovirus in renal-transplant recipients, could be under the control of a specifically impaired cell-mediated immunity. The stimulation and reinforcement of an EBV-specific cell-mediated immunity may be possible to achieve through certain vaccinal preparations, such as that proposed by Pearson and Scott (1978), made of cell-membrane vesicles.

Antiherpesvirus chemotherapy represents a third area to explore, although pessimism still prevails concerning the possibility of suppressing viral expression without hampering cellular metabolism. When viral activity is diagnosed, most of the herpesvirus-induced cell damage is already done, but not necessarily for EBV. Of the three drugs that have been tested clinically on varicella–zoster and herpes, namely Idoxuridine, adenine arabinoside (vidarabine), and acycloguanosine (acyclovir), the last may be particularly interesting because of its theoretical lack of toxicity in man. Vidarabine, on the other hand, appeared efficient in neonatal disseminated herpes simplex virus infection, but its activity against herpesvirus reactivation, such as zoster, is not yet established. New molecules such as fluoro-iodo-aracytosine (FIAC) that are still being explored in the laboratory at the cellular level may prove to be helpful for NPC patients (Lopez *et al.*, 1980).

VI. SUMMARY AND CONCLUSIONS

The epidemiological approach to the study of the Epstein–Barr virus (EBV) and of associated diseases has been very rewarding. It has enabled elucidation of how a ubiquitous virus can be the cause of geographically and ethnically restricted diseases, such as infectious mononucleosis (IM), Burkitt lymphoma (BL), and nasopharyngeal carcinoma (NPC), the critical factor for the first two diseases being the age at primary infection.

The age at primary infection varies greatly around the world and induces widely different immune responses, as described in Section II.B. Clinical syndromes associated with primary infection are clinically unnoticeable in equatorial regions where BL prevails, while IM, caused by a late EBV infection in adolescents and young adults, is well known in developed countries. When EBV infection occurs in immunodeficient children, primary infection by EBV may result in deadly syndromes, including lymphomatous multiclonal growths.

The causal association between endemic BL and the EBV is now epidemiologically established thanks to the long-term prospective seroepidemiological study in Uganda. High viral capsid antigen (VCA) titers precede BL clinical development by many years and appear to be related to a very early EBV infection, possibly in the first few months of life. Such a very early EBV infection could be due to mediocre maternal antibody protection and represents an "initiating event" for later development of BL. A large population of B lymphocytes infected by EBV and intermediate between the normal and the cancerous state appears to be specifically susceptible to further oncogenic steps. The exact role of malaria is not yet known. The chromosomal anomalies observed in endemic BL (translocations involving choromosome 8) are also found in nonendemic lymphomas observed in temperate climates and having a similar histopathology. In these temperate-climate Burkitt-type lymphomas, EBV markers are rarely present and malaria is indeed not involved. Thus, endemic BL emerges as a unique model in which the herpes EBV plays the role of an environmental biological carcinogen, where early infection represents a critical risk factor, responsible for the initiation of the endemic tumor.

NPC is an epithelial tumor, endemic in certain ethnic groups and geographic areas. NPC is strongly associated with EBV wherever it is observed in the world, in sharp contrast to the situation observed for Burkitt-type lymphomas. How a lymphotropic herpesvirus could infect epithelial cells that lack EBV receptors is not yet known. The age of primary infection by EBV does not appear critical for NPC genesis, but a reactivation of a latent EBV infection in the nasopharynx precedes clinical onset of the tumor by a period long enough (1–2 years) that IgA VCA serological markers of reactivation can be used as a tool for early detection of this cancer in endemic areas, such as China. Precancerous conditions are being actively studied, since they may offer a unique opportunity to try antiviral interventions as a way to stop EBV reactivation and possibly to prevent clinical onset of the disease. If this were achieved, both the proof of the causal role of the EBV in NPC development and a new approach for its prevention would be in sight.

The EBV thus appears as a unique virus to investigate latency, herpes-associated multistep oncogenesis in man, and the boundaries between chemical and biological carcinogens of the environment. It may prove to represent the first example of an oncogenic virus causally involved in two human tumors, without representing a necessary or a sufficient, factor. EBV serological markers can now be used for better management of the tumor (early diagnosis and surveillance). Whether anti-EBV intervention can lead to better control (treatment or prevention or both) of NPC represents the main question to be answered in coming years.

ACKNOWLEDGMENTS. The research described herein was supported by the Centre National de la Recherche Scientifique (GIS 122003), the National Cancer Institute (Contract No. 1 CP 91035), the Cancer Research Institute, New York, and the Association pour le Développement de la Recherche sur le Cancer. I deeply appreciate the secretarial assistance of B. Maret.

REFERENCES

Adatia, A.K., 1970, Dental aspects, in: *Burkitt's Lymphoma* (D.P. Burkitt and D.H. Wright, eds.), pp. 34–42, Livingstone, Edinburgh and London.

Addy, P.A., Ahiabor, T., Amane, C., and Pappoe, M.A., 1978, Jaundice and infectious mononucleosis in Ghana, *Afr. J. Med. Sci.* 7(2):85–92.

Al-Attar, A., Al-Mondhiry, H., Al-Bahrani, Z., and Al-Saleem, T., 1979, Burkitt's lymphoma in Iraq: Clinical and pathological study of 47 patients, *Int. J. Cancer* **23**:14–17.

Allison, A.C., and Clyde, D.F., 1961, Malaria in African children with deficient erythrocyte glucose-6-phosphate dehydrogenase, *Br. Med. J.* **1**:1346–1349.

American Joint Committee for Cancer Staging and End Results Reporting, 1976, Staging of cancer at head and neck sites, in: *Preliminary Handbook on Classification and Staging* (Task Force on Head and Neck Sites, eds.), pp. 61–72, American Joint Committee, Chicago.

Anderson, E.N., Jr., Anderson, M.L., and Ho, J.H.C., 1978, A study of the environmental backgrounds of young Chinese nasopharyngeal carcinoma patients, 1978, in: *Nasopharyngeal Carcinoma: Etiology and Control* (G. de-Thé and Y. Ito, eds.), pp. 231–239, IARC Scientific Publication No. 20, World Health Organization, Geneva.

Andiman, W.A., 1979, The Epstein–Barr virus and EBV infection in childhood, *J. Pediatr.* **95**:171–182.

Armstrong, D., Henle, G., and Henle, W., 1966, Complement-fixation tests with cell lines derived from Burkitt's lymphoma and acute leukemias, *J. Bacteriol.* **91**:1257–1262.

Armstrong, R.W., Kannan-Kutty, M., and Armstrong, M.J., 1978, Self-specific environments associated with nasopharyngeal carcinoma in Selangor, Malaysia, *Soc. Sci. Med.* **12**:149–156.

Arsenau, J.C., Canellos, G.P., Banks, P.M., Berard, C.W., Granlick, H.R., and de-Vita, V.T., Jr., 1975, American Burkitt's lymphoma: Clinicopathological study of thirty cases. I. Clinical factors relating to prolonged survival, *Am. J. Med.* **58**:314–321.

Bar, R.S., Delor, C.J., Clausen, K.P., Hurtubise, P., Henle, W., and Hewetson, J.F., 1974, Fatal infectious mononucleosis in a family, *N. Engl. J. Med.* **290**:363–367.

Baron, D., Benz, W.C., Carmichael, G., Yocum, R.R., and Strominger, J.L., 1975, Assay and partial purification of Epstein–Barr nuclear antigen, in: *Proceedings of the International Workshop on the Production, Concentration and Purification of Epstein–Barr Virus*, pp. 257–262, IARC Press, Lyon.

Bender, C.E., 1962, Recurrent mononucleosis, *J. Am. Med. Assoc.* **182**:954–956.

Berard, C., O'Conor, G.T., Thomas, L.B., and Torloni, H., 1969, Histopathological definition of Burkitt's tumor, *Bull. WHO* **40**:601–607.

Beremblum, I., 1940, The mechanisms of carcinogenesis: A study of the significance of cocarcinogenic action and related phenomena, *Cancer Res.* **1**:807.

Beremblum, I., 1978, Historical perspectives, in: *Carcinogenesis—A Comprehensive Survey*, Vol. 2 (T.J. Slaga, A. Sivak, and R.K. Boutwell, eds.), p. 1, Raven Press, New York.

Berger, R., Bernheim, A., Fellous, M., and Brouet, J.C., 1979a, Cytogenetic study of a European Burkitt's lymphoma cell line, *J. Natl. Cancer Inst.* **62**:1187–1192.

Berger, R., Bernheim, A., Weh, H.J., Flandrin G., Daniel, M.T., Brouet, J.C., and Colbert, N., 1979b, A new translocation in Burkitt tumor cells, *Hum. Genet.* **53**:111–112.

Berger, T., Bernheim, A., Flandrin, G., Daniel, M.T., Schaison, G., Brouet, J.C., and Bernard, J., 1979c, Translocation t(8;14) dans la leucémie lymphoblastique de type Burkitt, *Nouv. Presse Med.* **8**(3):181–183.

Bernheim, A., Berger, R., and Lenoir, G., 1981, Cytogenetic studies on African Burkitt's lymphoma cell lines: t(8;14), t(2;8) and t(8;22) translocations, *Cancer Genet. Cytogenet.* **3**:307–315.

Betuel, H., Cammoun, M., Colombani, J., Day, N., Ellouz, R., and de Thé, G., 1975, The relationship between nasopharyngeal carcinoma and the HL-A system among Tunisians, *Int. J. Cancer* **16**:249–254.

Biggar, R.J., and Nkrumah, F.K., 1979, Burkitt's lymphoma in Ghana: Urban rural distribution, time–space clustering and seasonality, *Int. J. Cancer* **23**:330–336.

Biggar, R.J., Henle, W., Fleisher, G., Bocker, J., Lennette, E.T., and Henle, G., 1978a, Primary Epstein–Barr virus infection in African infants. I. Decline of maternal antibodies and time of infection, *Int. J. Cancer* **22**:239–243.

Biggar, R.J., Henle, G., Bocker, S., Lennette, E.T., Fleisher, G., and Henle, W., 1978b, Primary Epstein–Barr virus infection in African infants. II. Clinical and serological observations during seroconversion, *Int. J. Cancer* **22**:244–250.

Black, F.L., Woodall, J.P., Evans, A.S., Liebhaber, H., and Henle, G., 1970, Prevalence of antibody against viruses in Tiriyo, an isolated Amazon tribe, *Am. J. Epidemiol.* **91**:430–438.

Blot, W.J., Lanier, A., and Fraumeni, J.F., Jr., 1975, Cancer mortality among Alaskan natives, 1960–1969, *J. Natl. Cancer Inst.* **55**:547–554.

Bornkamm, G.W., Stein, H., Lennert, K., Rüggeberg, F., Bartels, H., and Zur Hausen, H., 1976, Attempts to demonstrate virus-specific sequences in human tumors. IV. EB viral DNA in European Burkitt's lymphoma and immunoblastic lymphadenopathy with excessive plasmocytosis, *Int. J. Cancer* **17**:177–181.

Britton, S., Andersson-Anvret, M., Gergely, P., Henle, W., Jonal, M., Klein, G., Sandstedt, B., and Svedmyr, E., 1978, Epstein–Barr virus immunity and tissue distribution in a fatal case of infectious mononucleosis, *N. Engl. J. Med.* **298**:89–92.

Brown, T.D.K., Ernberg, I., Lamon, E.W., and Klein, G., 1974, Detection of Epstein–Barr virus (EBV)-associated nuclear antigen in human lymphoblastoid cell lines by means of an I^{125}-IgG-binding assay, *Int. J. Cancer* **13**:784–794.

Brubaker, G., Geser, A., and Pike, M.C., 1973, Burkitt's lymphoma in the North Mara district of Tanzania, 1964–1970: Failure to find evidence of space–time clustering in a high risk isolated rural area, *Br. J. Cancer* **28**:469–472.

Burchenal, J.H., 1966, Geographical chemotherapy, Burkitt's tumor as stalking horse for leukemia: Presidential address, *Cancer Res.* **26**:2393–2405.

Burkitt, D., 1958, A sarcoma involving the jaws in African children, *Br. J. Surg.* **46**:218.

Burkitt, D.P., 1962a, A children's cancer dependent on climatic factors, *Nature (London)* **194**:232–234.

Burkitt, D.P., 1962b, A "tumor safari" in East and Central Africa, *Br. J. Cancer* **16**:379–386.

Burkitt, D.P., 1967, Burkitt's lymphoma outside the known endemic area of Africa and New Guinea, *Int. J. Cancer* **2**:562–565.

Burkitt, D.P., 1969, Etiology of Burkitt's lymphoma: An alternative to a vectored virus, *J. Natl. Cancer. Inst.* **42**:19–28.

Burkitt, D.P., 1970, Geographical distribution, in: *Burkitt's Lymphoma* (D.P. Burkitt and D.H. Wright, eds.), pp. 186–197, Livingstone, Edinburgh and London.

Burkitt, D.P., and Wright, D.H., 1966, Geographical and tribal distribution of the African lymphoma in Uganda, *Br. Med. J.* **5487**:569–573.

Cammoun, M., Vogt-Hoerner, G., and Mourali, N., 1971, Les tumeurs du naso-pharynx en Tunisie: Etude anatomo-clinique de 143 observations, *Tunis. Med.* **49**(3):131–141.

Carter, R.L., and Penman, H.G., 1969, *Infectious Mononucleosis*, Blackwell, Oxford.

Carvalho, R.P.S., Evans, A.D., Forst, P., Dalldorf, G., Camargo, M.F., and Jamra, M., 1973, EBV infections in Brazil. I. Occurrence in normal persons, in lymphomas and in leukemias, *Int. J. Cancer* **11**:191–201.

Chan, S.H., Day, N.E., Khor, T.H., Kunaratnam, N., and Chia, K.B., HLA markers in the development and prognosis of NPC in Chinese, in: *Cancer Campaign, Vol. 5, Nasopharyngeal Carcinoma* (E. Grundmann, G.R.F. Krueger, and D.V. Ablashi, eds.) pp. 205–211, Gustav Fisher Verlag, Berlin.

Chang, R.S., Lewis, J.S., Reynolds, R.D., Sullivan, M.J., and Neuman, J., 1978, Oropharyngeal excretion of Epstein–Barr virus by patients with lymphoproliferative disorders and by recipients of renal homografts, *Ann. Intern. Med.* **88**:111–123.

Chang, R.S., Char, D.F., Jones, J.H., and Halstead, S.B., 1979, Incidence of infectious mononucleosis at the Universities of California and Hawaii, *J. Infect. Dis.* **140**(4):479–86.

Cheeseman, S.H., Henle, W., Rubin, R.H., Tolkoff-Rubin, N.E., Cosimi, B., Cantell, K., Winkle, S., Herrin, J.T., Black, P.H., Russell, P.S., and Hirsch, M.S., 1980, Epstein–Barr virus infection in renal transplant recipients: Effects of antithymocyte globulin and interferon, *Ann. Intern. Med.* **93**:39–42.

Clifford, P., 1970a, A review on the epidemiology of nasopharyngeal carcinoma, *Int. J. Cancer* **5**:287–309.

Clifford, P., 1970b, Treatment, in: *Burkitt's lymphoma* (D.P. Burkitt and D.H. Wright, eds.), pp. 52–63, Livingstone, Edinburg and London.

Collected reports of cases of Burkitt's lymphoma from countries outside the endemic areas, 1967, *Int. J. Cancer* **2**:559–609.

Connelly, R.R., and Christine, B.W., 1974, A cohort study of cancer following infectious mononucleosis, *Cancer Res.* **34**:1172–1178.

Crawford, D.H., Epstein, M.A., Achong, B.G., Finerty, S., Newman, J., Liversedge, S., Tedder, R.S., and Steward, J.W., 1979, Virological and immunological studies on a fatal case of infectious mononucleosis, *J. Infection* **1**:37–48.

Crawford, D.H., Thomas, J.A., Janossy, G., Sweny, P., Fernando, O.N., Moorhead, J.F., and Thompson, J.H., 1980, Epstein–Barr virus nuclear antigen positive lymphoma after cyclosporin-A treatment in patient with renal allograft, *Lancet* :1355–1356.

Dalldorf, G., Linsell, C.A., Barnhart, F.E., and Martyn, R., 1964, An epidemiological approach to the lymphomas of African children and Burkitt's sarcoma of the jaws, *Perspect. Biol. Med.* **7**:435–449.

Dambaugh, T., Nkrumah, F.K., Biggar, R.J., and Kieff, E., 1979, Epstein–Barr virus RNA in Burkitt's tumor tissue, *Cell* **16**:313–322.
Dameshek, W., 1969, Speculations on the nature of infectious mononucleosis, in: *Infectious Mononucleosis* (R.L. Carter and H.G. Penman, eds.), pp. 225–240, Blackwell, London.
Davidsohn, I., and Nelson, D.A., 1969, The blood, in: *Todd–Sanford's Clinical Diagnosis by Laboratory Methods*, 14th ed. (I. Davidsohn and J.B. Henry, eds.), Chapt. 5, W.B. Saunders, Philadelphia.
De Schryver, A., Friberg, S., Klein, G., Henle, W., Henle, G., de Thé, G., Clifford, P., and Ho, J.H.C., 1969, Epstein–Barr virus-associated antibody patterns in carcinoma of the post-nasal space, *Clin. Exp. Immunol.* **5**:443–459.
Desgranges, C., and de-Thé, G., 1977, Epstein–Barr virus induces viral nuclear antigen in nasopharyngeal epithelial cells, *Lancet* **2**:1286–1287.
Desgranges, C., and de-Thé, G., 1978a, Presence of EBV-specific IgA in saliva of nasopharyngeal carcinoma (NPC) patients: Their activity, origin and possible clinical value, in: *Nasopharyngeal Carcinoma* (G. de Thé and Y. Ito, eds.), pp. 459–469, IARC Scientific Publication No. 20, World Health Organization, Geneva.
Desgranges, C., and de-Thé, G., 1978b, IgA and nasopharyngeal carcinoma, in: *Oncogenesis and Herpesviruses III* (G. de Thé, Y. Ito, and F. Rapp, eds.), pp. 883–891, IARC Scientific Publication No. 25, Lyon.
Desgranges, C., Wolf, H., de-Thé, G., Shanmugaratnam, K., Ellouz, R., Cammoun, N., Klein, G., and Zur Hausen, H., 1975a, Nasopharyngeal carcinoma. X. Presence of Epstein–Barr genomes in epithelial cells of tumors from high and medium risk areas, *Int. J. Cancer* **16**:7–15.
Desgranges, C., Wolf, H., Zur Hausen, H., and de-Thé, G., 1975b, Further studies on the detection of the Epstein–Barr viral DNA on nasopharyngeal carcinoma biopsies from different parts of the world, in: *Oncogenesis and Herpesviruses II*, Vol. 2 (G. de Thé, M.A. Epstein, and H. Zur Hausen, eds.), pp. 191–193, IARC Scientific Publication No 11, Lyon.
Desgranges, C., de-Thé, G., Ho, J.H.C., and Ellouz, R., 1977a, Neutralizing EBV-specific IgA in throat washings of nasopharyngeal carcinoma (NPC) patients, *Int. J. Cancer* **19**:627–633.
Desgranges, C., Li, J.Y., and de-Thé, G., 1977b, EBV specific secretory IgA in saliva of NPC patients: Presence of secretory piece in epithelial malignant cells, *Int. J. Cancer* **20**:881–886.
Desgranges, C., Bornhamm, G.W., Zeng, Y., Wang, P.C., Zhu, J.S., Shang, H., and de-Thé, G., 1982, Detection of Epstein–Barr viral DNA in the nasopharyngeal mucosa of Chinese with IgA/EBV-specific antibodies, *Int. J. Cancer* **29**:187–191.
de-Thé, G., 1977, Is Burkitt's lymphoma related to a perinatal infection by Epstein–Barr virus?, *Lancet* **1**:335–338.
de-Thé, G., 1978a, Epstein–Barr virus: Is it timely to discuss a vaccine?, *Biomedicine* **28**:15–17.
de-Thé, G., 1978b, Events early in life and risk to develop Burkitt's lymphoma—Role of neonatal Epstein–Barr virus infection and heavy malaria burden, in: *Tumors of Early Life in Man and Animals* (L. Severi, ed.), pp. 225–231, Institute of Pathological Anatomy and Histology, Perugia.
de-Thé, G., 1979, The epidemiology of Burkitt's lymphoma: Evidence for a causal relationship with Epstein–Barr virus, *Am. J. Epidemiol.* **1**:32–54.
de-Thé, G., 1980a, The role of the Epstein–Barr virus in human diseases: Infectious mononucleosis (IM), Burkitt's lymphoma (BL) and nasopharyngeal carcinoma (NPC), in: *Viral Oncology* (G. Klein, ed.), pp. 769–797, Raven Press, New York.
de-Thé, G., 1980b, Multistep carcinogenesis, Epstein–Barr virus and human malignancies: Viruses in naturally occurring cancers, *Cold Spring Harbor Conf. Cell Prolif.* **7**:11–21.
de-Thé, G., 1981, The Chinese epidemiological approach of nasopharyngeal carcinoma research and control in China, *Yale J. Biol. Med.* **54**:33–39.
de-Thé, G., and Lenoir, G., 1977, Comparative diagnosis of Epstein–Barr virus related di-

seases: Infectious mononucleosis, Burkitt's lymphoma and nasopharyngeal carcinoma, in: *Comparative Diagnosis of Viral Diseases* (E. Kurstak, ed.), pp. 196–241, Academic Press, New York.

de-Thé, G., and Zeng, Y., 1982, Epidemiology of EBV: Recent results on endemic versus nonendemic BL and EBV related pre-NPC conditions, in *Advances in Comparative Leukemia Research* (D. Yohn and J.R. Blakeslee, eds.), pp. 3–15, Elsevier/North Holland, New York.

de-Thé, G., Ablashi, D.V., Liabeuf, A., and Mourali, N., 1973a, Nasopharyngeal carcinoma (NPC). VI. Presence of an EBV nuclear antigen in fresh tumor biopsies: Preliminary results, *Biomedicine* **19**:349–352.

de-Thé, G., Sohier, R., Ho, J.H.C., and Freund, R.J., 1973b, Nasopharyngeal carcinoma. IV. Evolution of complement-fixing antibodies during the course of the disease, *Int. J. Cancer* **12**:368–377.

de-Thé, G., Ho, J.H.C., Ablashi, D.V., Day, N.E., Macario, A.J.L., Martin-Bertholon, M.C., Pearson, G., and Sohier, R., 1975a, Nasopharyngeal carcinoma. IX. Antibodies to EBNA and correlation with response to other EBV antigens in Chinese patients, *Int. J. Cancer* **16**:713–721.

de-Thé, G., Day, N.E., Geser, A., Lavoué, M.F., Ho, J.H.C., Simons, M.J., Sohier, R., Tukei, P., Vonka, V., and Zavadova, H., 1975b, Sero-epidemiology of the Epstein–Barr virus—Preliminary analysis of an international study, in: *Oncogenesis and Herpesviruses II*, Vol. 2 (G. de Thé, M.A. Epstein, and H. Zur Hausen, eds.), pp. 3–16, IARC Scientific Publication No. 11, Lyon.

de-Thé, G., Vuillaume, M., Giovanella, B.C., and Klein, G., 1976, Epithelial characteristics of tumor cells in nasopharyngeal carcinoma (NPC) passaged in nude mice—an ultrastructural study, *J. Natl. Cancer Inst.* **57**:1101–1105.

de-Thé, G., Geser, A., Day, N.E., Tukei, P.M., Williams, E.H., Beri, D.P., Smith, P.G., Dean, A.G., Bornkamm, G.W., Feorino, P., and Henle, W., 1978a, Epidemiological evidence for a causal relationship between Epstein–Barr virus and Burkitt's lymphoma: Results of the Ugandan prospective study, *Nature (London)* **274**:756–761.

de-Thé, G., Lavoué, M.F., and Muenz, L., 1978b, Differences in EBV antibody titers of patients with nasopharyngeal carcinoma originating from high, intermediate and low incidence areas, in: *Nasopharyngeal Carcinoma: Etiology and Control* (G. De Thé and Y. Ito, eds.), pp. 471–481, IARC Press, Lyon.

de-Thé, G., Desgranges, C., Zeng, Y., Wang, P.C., Bornkamm, G.W., Zhu, J.S., and Shang, M., 1981, Search for pre-cancerous lesions and EBV markers in the nasopharynx of IgA positive individuals, in: *Cancer Campaign*, Vol. 5, *Nasopharyngeal Carcinoma* (E. Grundmann, G.R.F. Krueger and D.V. Ablashi eds.), pp. 111–117, Gustav Fischer, Stuttgart.

Diehl, V., Henle, W., and Kohn, G., 1968, Demonstration of a herpes group virus in cultures of peripheral leukocytes from patients with infectious mononucleosis, *J. Virol.* **2**:663–669.

Durand-Fardel, 1837, Cancer du pharynx—ossification dans la substance musculaire du coeur, *Bull. Soc. Anat. (Paris)* **12**:73–80.

Ellouz, R., Cammoun, M., Ben-Altia, R., and Bahi, J., 1978, NPC in children and adolescents in Tunisia: Clinical aspects and paraneoplastic syndrome, in: *Etiology and Control* (G. de-Thé and Y. Ito, eds.), pp. 115–129, IARC Scientific Publication No. 20, Lyon.

Epstein, M.A., 1976, Implications of a vaccine for the prevention of Epstein–Barr virus infection: Ethical and logistic considerations, *Cancer Res.* **36**:711–714.

Epstein, M.A., 1981, What we could do with an EBV vaccine, *Lancet* **1**:759–761.

Epstein, M.A., Achong, B.G., and Barr, Y.M., 1964, Virus particles in cultured lymphoblasts from Burkitt's lymphoma, *Lancet* **1**:702–703.

Epstein, M.A., Hunt, R.D., and Rabin, H., 1973, Pilot experiments with EB virus in owl monkeys (*Aotus trivirgatus*). I. Reticuloproliferative disease in an inoculated animal, *Int. J. Cancer* **12**:309–318.

Evans, A.S., 1974, New discoveries in infectious mononucleosis, *Mod. Med.* **1**:18–24.
Evans, A.S., 1978, Infectious mononucleosis and related syndromes, *IARC (Int. Agency Res. Cancer) Sci. Publ.* **20**:591–606.
Evans, A.S., and Campos, L.E., 1971, Acute respiratory diseases in students at the University of the Philippines, *Bull. WHO* **45**:103–112.
Evans, A.S., and Niederman, J.C., 1976, Epstein–Barr virus, in: *Viral Infections of Humans: Epidemiology and Control* (A.S. Evans, ed.), pp. 209–234, Plenum Press, New York.
Evans, A.S., Niederman, J.C., and McCollum, R.W., 1968, Seroepidemiological studies of infectious mononucleosis with EB virus, *N. Engl. J. Med.* **297**:1121–1127.
Fialkow, P.J., Klein, G., Gartler, S.M., and Clifford, P., 1970, Clonal origin for individual Burkitt tumors, *Lancet* **1**:384–386.
Fong, Y.Y., and Chan, W.C., 1973, Bacterial production of di-methyl nitrosamine in salted fish, *Nature (London)* **243**:421–423.
Fong, Y.Y., and Walsh, E.O., 1971, Carcinogenic nitrosamines in Cantonese salt-dried fish, *Lancet* **1**:1032.
Fraumeni, J.F., Jr., and Mason, T.J., 1974, Cancer mortality among Chinese Americans, 1950–1969, *J. Natl. Cancer Inst.* **52**:659–665.
Gazzolo, L., de-Thé, G., and Vuillaume, M., 1972, Nasopharyngeal carcinoma. II. Ultrastructure of normal mucosa tumor biopsies and subsequent epithelial growth *in vitro*, *J. Natl. Cancer Inst.* **48**:73–86.
Gerber, P., 1972, Activation of Epstein–Barr virus by 5-bromo-deoxyuridine in virus free human cells, *Proc. Natl. Acad. Sci. U.S.A.* **69**:83–85.
Gerber, P., and Deal, D.R., 1970, Epstein–Barr virus-induced viral and soluble complement-fixing antigens in Burkitt's lymphoma cell cultures, *Proc. Soc. Exp. Biol. Med.* **134**:748–751.
Gerber, P., Purcell, R.H., Rosenblum, E.N., and Walsh, J.H., 1969, Association of EB-virus infection with the post-perfusion syndrome, *Lancet* **1**:593–596.
Gerber, P., Goldstein, L.L., Lucas, S., Nonoyama, M., and Perlin, E., 1972, Oral excretion of Epstein–Barr viruses by healthy subjects and patients with infectious mononucleosis, *Lancet* **2**:988–989.
Geser, A., and de-Thé, G., 1972, Does the Epstein–Barr virus play an etiological role in Burkitt's lymphoma?, in: *Oncogenesis and Herpesviruses* (P.M. Biggs, G. de Thé, and L.N. Payne, eds.), pp. 372–375, IARC Scientific Publication No. 2, Lyon.
Geser, A., Charnay, N., Day, N.E., de-Thé, G., and Ho, J.H.C., 1978, Environmental factors in the etiology of nasopharyngeal carcinoma: Report on a case–control study in Hongkong, in: *Nasopharyngeal Carcinoma: Etiology and Control* (G. de Thé and Y. Ito, eds.), pp. 213–229, IARC Scientific Publication No. 20, Lyon.
Geser, A., de-Thé, G., Lenoir, G., Day, N.E., and Williams, E.H., 1982, Final case reporting from the Ugandan prospective study of the relationship between EBV and Burkitt's lymphoma. *Int. J. Cancer* **29**:397–400.
Glaser, R., and Nonoyama, M., 1974, Host cell regulation of induction of Epstein–Barr virus, *J. Virol.* **14**:174–176.
Glaser, R., and Rapp, F., 1972, Rescue of Epstein–Barr virus from somatic cell hybrids of Burkitt lymphoblastoid cells, *J. Virol.* **10**:288–296.
Glaser, R., de-Thé, G., Lenoir, G., and Ho, J.H.C., 1976, Superinfection of nasopharyngeal carcinoma epithelial tumor cells with Epstein–Barr virus, *Proc. Natl. Acad. Sci. U.S.A.* **73**:960–963.
Goldblum, N., Ben-Bassat, H., Mitrani, S., Andersson-Anvret, M., Goldblum, T., Aghai, E., Ramot, B., and Klein, G., 1977, A case of an Epstein Barr virus genome-carrying lymphoma in an Israeli-Arab child, *Eur. J. Cancer* **13**:693–698.
Goldstein, G., Klein, G., Pearson, G., and Clifford, P., 1969, Direct membrane immunofluorescence reaction of Burkitt's lymphoma cells in culture, *Cancer Res.* **29**:749–752.
Granlund, D.J., and Andrese, A.P., 1977, Detection of Epstein–Barr virus antigen with enzyme-conjugated antibody, *Int. J. Cancer* **20**:495–499.

Gravell, M., Levine, P.H., McIntyre, R.F., Land, V.J., and Pagano, J.S., 1976, EBV in an American patient with Burkitt's lymphoma: Detection of viral genome in tumor tissue and establishment of a tumor derived cell line, *J. Natl. Cancer Inst.* **56**:701.

Greene, M.H., Fraumeni, J.F., Jr., and Hoover, R., 1977, Nasopharyngeal cancer among youths in United States: Racial variations by cell type, *J. Natl. Cancer Inst.* **58**:1267–1270.

Greenwood, B.M., and Vick, R., 1975, Evidence for a malaria mitogen in human malaria, *Nature (London)* **257**:592.

Greenwood, B.M., Oduloju, A.J., and Platts-Mill, T.A.E., 1979, Partial characterization of a malaria mitogen, *Trans. R. Soc. Trop. Med. Hyg.* **73**:178.

Gresser, I., and Tovey, M., 1978, Anti-tumor effects of interferon, *Biochim. Biophys. Acta* **516**:231–247.

Grose, C., and Feorino, P.M., 1972, Epstein–Barr virus and Guillain–Barré syndrome, *Lancet* **1**:1285–1287.

Gross, L., 1970, *Oncogenic Viruses*, Pergamon Press, Oxford,

Gu, S.Y., and Zeng, Y., 1978, Complement fixing antibody levels in sera from population groups in Kwangchow and in Peking, *Chin. J. ENT*, **1**:23–25.

Hallee, T.J., Evans, A.S., Niederman, J.C., Brooks, C.M., and Voegtly, J.H., 1974, Infectious mononucleosis as the U.S. Military Academy: A prospective study of a single class over four years, *Yale J. Biol. Med.* **47**:182–195.

Hampar, B., Derge, J.G., Martos, L.M., and Walker, J.L., 1972, Synthesis of Epstein–Barr virus after the activation of the viral genome in a virus negative human lymphoblastoid cell (Raji) made resistant to five bromodeoxyuridine, *Proc. Natl. Acad. Sci. U.S.A.* **69**:78–82.

Harada, M., Pearson, G., Redmon, L., Winters, E., and Kasuga, S., 1975, Antibody production and interaction with lymphoid cells in relation to tumor immunity in the Moloney sarcoma virus system, *J. Immunol.* **114**:1318–1322.

Hawkins, B.R., Simons, M.J., Goh, E.H., Chia, K.B., and Shanmugaratnam, K., 1974, Immunogenetic aspects of nasopharyngeal carcinoma. II. Analysis of ABO, rhesus and MNS's red cell systems, *Int. J. Cancer* **13**:116–121.

Heath, C.W., Brodsky, A.L., and Potolsky, A.I., 1972, Infectious mononucleosis in general population, *Am. J. Epidemiol.* **95**:46–52.

Hehlmann, R., Walther, B., Zöllner, N., Wolf, H., and Deinhardt, F., 1980, Infectious mononucleosis and acute monocytic leukaemia, *Lancet* :652–653.

Henderson, B.E., Louie, E., Jing, J.S.H., Buell, P., and Gardner, M.B., 1976, Risk factors associated with nasopharyngeal carcinoma, *N. Engl. J. Med.* **295**:1101–1106.

Henle, G., and Henle, W., 1976, Epstein–Barr virus-specific IgA serum antibodies as an outstanding feature of nasopharyngeal carcinoma, *Int. J. Cancer* **17**:1–7.

Henle, G., Henle, W., and Diehl, V., 1968, Relation of Burkitt tumor associated herpes-type virus to infectious mononucleosis, *Proc. Natl. Acad. Sci. U.S.A.* **59**:94–101.

Henle, G., Henle, W., Clifford, P., Diehl, V. Kafuko, G.W., Kirya, B.G., Klein, G., Morrow, R.H., Munube, G.M.R., Pike, P., Tukei, P.M., and Ziegler, J.L., 1969, Antibodies to Epstein–Barr virus in Burkitt's lymphoma and control groups, *J. Natl. Cancer Inst.* **43**:1147–1157.

Henle, G., Henle, W., Klein, G., Gunren, P., Clifford, P., Morrow, R., and Ziegler, J., 1971, Antibodies to early Epstein–Barr virus induced antigens in Burktt's lymphoma, *J. Natl. Cancer Inst.* **46**:861–871.

Henle, G., Henle, W., and Horwitz, C.A., 1974, Antibodies to Epstein–Barr virus-associated nuclear antigen in infectious mononucleosis, *J. Infect. Dis.* **130**:231–239.

Henle, W., Henle, G., Harrison, F.S., Joyner, C.R., Klemola, E., Paloheimo, J., Scriba, M., and Von Essen, F., 1968, Antibody responses to the Epstein–Barr virus and cytomegaloviruses after open-heart and other surgery, *N. Engl. J. Med.* **282**:1068–1074.

Henle, W., Henle, G., Burtin, P., Cachin, Y., Clifford, P., de Schryver, A., de Thé, G., Diehl, V., Ho, J.H.C., and Klein, G., 1970a, Antibodies to Epstein–Barr virus in nasopharyngeal carcinoma, other head and neck neoplasms and control groups, *J. Natl. Cancer. Inst.* **44**:225–231.

Henle, W., Henle, G., Zajac, B.A., Pearson, G., Waubke, R., and Scriba, M., 1970b, Differential reactivity of human serums with early antigens induced by Epstein–Barr virus, *Science* **169**:188–190.

Henle, W., Ho, J.H.C., Henle, G., and Kwan, H.C., 1973, Antibodies to Epstein–Barr virus related antigens in nasopharyngeal carcinoma: Comparison of active cases and long term survivors, *J. Natl. Cancer Inst.* **51**:361–369.

Henle, W., Ho, J.H.C., Henle, G., Chan, J.C.W., and Kwan, H.C., 1977, Nasopharyngeal carcinoma: Significance of changes in Epstein–Barr virus related antibody patterns following therapy, *Int. J. Cancer* **20**:663–672.

Hewetson, J.F., Gothoskar, B., and Klein, G., 1972, Radioiodine-labelled antibody test for the detection of membrane antigens associated with Epstein–Barr virus, *J. Natl. Cancer Inst.* **48**:87–94.

Hirayama, T., 1978, Descriptive and analytical epidemiology of nasopharyngeal cancer, in: *Nasopharyngeal Carcinoma: Etiology and Control* (G. de Thé and Y. Ito, eds.), pp. 167–190, IARC, Lyon.

Hirshaut, Y., Cohen, M.H., and Stevens, D.A., 1973, Epstein–Barr virus antibodies in African and American Burkitt's lymphomas, *Lancet* **1**:114–116.

Ho, J.H.C., 1970, The natural history and treatment of nasopharyngeal carcinoma (NPC), in: *Oncology*, Vol. 4 (R. Lee Clark, R. Cumley, J.E. McCay, and M. Copeland, eds.), pp. 1–14, Medical Publishers, Year Book, Chicago.

Ho, J.H.C., 1971, Incidence of nasopharyngeal cancer in Hongkong, UICC Bulletin, *Cancer* **9**(2):5.

Ho, J.H.C., 1972a, Current knowledge of the epidemiology of nasopharyngeal carcinoma: A review, in: *Oncogenesis and Herpesviruses* (P. Biggs, de Thé, and L. Payne, eds.), pp. 357–366, IARC, Lyon.

Ho, J.H.C., 1972b, Nasopharyngeal carcinoma (NPC), in: *Advances in Cancer Research* (G. Klein, S. Weinhouse, and A. Haddow, eds.), pp. 59–92, Academic Press, New York.

Ho, J.H.C., 1978, An epidemiologic and clinical study of nasopharyngeal carcinoma, *Int. J. Radiol. Oncol. Biol. Phys.* **4**:181–198.

Ho, J.H.C., Ng, M.H., Kwan, H.C., and Chan, J.C.W., 1976, Epstein–Barr virus specific IgA and IgG serum antibodies in nasopharyngeal carcinoma, *Br. J. Cancer* **34**:655–660.

Ho, J.H.C., Kwan, H.C., Ng, M.H., and de-Thé, G., 1978, Serum IgA antibodies to EBV capsid antigens preceeding symptoms of nasopharyngeal carcinoma, *Lancet* **1**:436–437.

Ho, J.H.C., Lau, W.H., Kwan, H.C., Chan, C.L., Au, G.K.H., Saw, D., and De Thé, G., 1981, Diagnostic and prognostic serological markers in nasopharyngeal carcinoma, in: *Cancer Campaign*, Vol. 5, *Nasopharyngeal Carcinoma*, (E. Grundmann, G.R.F. Krueger, and D.V. Ablashi, eds.), pp. 219–224, Gustav Fischer, Stuttgart.

Hoogstraten, J., 1967, Observations on Burkitt's tumor in central and northern Canada, *Int. J. Cancer* **2**:566–575.

Huang, D.P., Ho, J.H.C., Henle, W., and Henle, G., 1974, Demonstration of Epstein--Barr virus associated nuclear antigen in nasopharyngeal carcinoma cells from fresh biopsies, *Int. J. Cancer* **14**:580–588.

Huang, D.P., Ho, J.H.C., Saw, D., and Teoh, T.B., 1978, Carcinoma of the nasal and paranasal regions in rats fed Cantonese salted marine fish, in: *Nasopharyngeal Carcinoma: Etiology and Control* (G. de Thé and Y. Ito, eds.), pp. 315–328, IARC, Lyon.

Huebner, R.J., 1957, The virologist's dilemma, *Ann. N. Y. Acad. Sci.* **67**:430.

Huebner, R.J., and Todaro, G.J., 1969, Oncogenes of RNA tumor viruses as determinants of cancer, *Proc. Natl. Acad. Sci. U.S.A.* **64**:1087.

International Agency for Research on Cancer (IARC), Annual Report, 1968, Internal Technical Report Series 68/004, Planning conference for epidemiological studies on Burkitt's lymphoma and infectious mononucleosis, Nairobi, Kenya, p. 35.

International Agency for Research on Cancer, Annual Report, 1976, WHO Publications, Geneva.

International Agency for Research on Cancer, Annual Report, 1978, WHO Publications, Geneva.

International Agency for Research on Cancer, Annual Report, 1980, WHO Publications, Geneva.

Ito, Y., Kawanishi, M., Hirayama, T., and Takabayashi, S., 1981, Combined effect of the extracts from *Croton Tiglium, Euphorbia lathyris* or *Euphorbia tizucalli* and N-butyrate on Epstein–Barr virus expression in human lymphoblastoid P3HR-1 and Raji cells, *Cancer Lett.* **12**:175–180.

Jarrett, W.F.H., McNeil, P.E., Laird, H.M., O'Neil, B.W., Murphy, J., Campo, M.S., and Moar, M.H., 1980, Papilloma viruses in benign and malignant tumors of cattle, in: *Viruses in Naturally Occurring Cancers* (M. Essex, G. Todaro, H. Zur Hausen, eds.), *Cold Spring Harbor Conf. Cell Prolif.* **7**:215–222.

Jarvis, J.E., Ball, G., and Rickinson, A.B., 1974, Cytogenetic studies on human lymphoblastoid cell lines from Burkitt's lymphoma and other sources, *Int. J. Cancer* **14**:716.

Jennings, E., M.D. thesis, 1973, Yale School of Medicine.

Jondal, M., 1976, Antibody-dependent cellular cytotoxicity (ADCC) against Epstein–Barr virus-determined membrane antigens. I. Reactivity in sera from normal persons and from patients with acute infectious mononucleosis, *Clin. Exp. Immunol.* **25**:1–5.

Judson, S.C., Henle, W., and Henle, G., 1977, A cluster of Epstein–Barr virus associated American Burkitt's lymphoma, *N. Engl. J. Med.* **297**:464–468.

Kafuko, G.W., Henderson, B.E., Kirya, G.G., Manube, G.M., Tukei, P., Day, N.E., Henle, G., Henle, W., Morrow, R.H., Pike, M.C., Smith, P.G., and Williams, E. H., 1972, Epstein–Barr virus antibody levels in children from the West Nile district of Uganda: Report of a field study, *Lancet* **1**:706–709.

King, H., and Haenszel, 1972, Cancer mortality among foreign and native-born Chinese in the United States, *J. Chron. Dis.* **26**:623–646.

Kirk, R.L., Blake, N.M., Serjeantson, S., Simons, M.J., and Chan, S.H., 1978, Genetic components in susceptibility to nasopharyngeal carcinoma, in: *Nasopharyngeal Carcinoma: Etiology and Control* (G. de Thé and Y. Ito, eds.), pp. 283–297, IARC Scientific Publication No. 20, Lyon.

Klein, G., 1979, Lymphoma development in mice and humans: Diversity of initiation is followed by a convergent cytogenetic evolution, *Proc. Natl. Acad. Sci. U.S.A.* **76**:2442.

Klein, G., Clifford, P., Klein, E., Smith, R.T., Minowada, J., Kourilski, F.M., and Burchenal, J.H., 1967, Membrane immunofluorescence reactions of Burkitt's lymphoma cells from biopsy specimens and tissue culture, *J. Natl. Cancer Inst.* **39**:1027–1044.

Klein, G., Pearson, G., Nadkarni, J.S., Nadkarni, J.J., Klein, E., Henle, G., Henle, W., and Clifford, P., 1968, Relation between Epstein–Barr viral and cell membrane immunofluorescence of Burkitt tumor cells. I. Dependence of cell membrane immunofluorescence on presence of EB virus, *J. Exp. Med.* **128**:1011–1020.

Klein, G., Pearson, G., Henle, G., Henle, W., Goldstein, G., and Clifford, P., 1969, Relation between Epstein–Barr viral and cell membrane immunofluorescence in Burkitt tumor cells. III. Comparison of blocking of direct membrane immunoflurescence and anti-EBV reactivities of different sera, *J. Exp. Med.* **129**:697–705.

Klein, G., Giovanella, B.C., Lindahl, T., Fialkow, P.J., Singh, S., and Stehlin, J., 1974, Direct evidence for the presence of Epstein–Barr virus DNA and nuclear antigen in malignant epithelial cells from patients with anaplastic carcinoma of the nasopharynx, *Proc. Natl. Acad. Sci. U.S.A.* **71**:4737–4741.

Koch, R., 1980, Uber bakteriologische Forschung, in: Verh. X. Int. Med. Congr. Berlin, 35.

Kvale, G., Holby, E.A., and Pedersen, E., 1979, Hodgkin's disease in patients with previous infectious mononucleosis, *Int. J. Cancer* **23**(5):593–597.

Lachet, B., Day, N.E., de-Thé, G., and Dufour, J., 1977, Temps de latence du lymphôme de Burkitt en Ouganda, *Med. Biol. Environ.* **5**:60–67.

Lang, D.J., Garruto, R.M., and Gajdusek, D.C., 1977, Early acquisition of cytomegalovirus and Epstein–Barr virus antibody in several isolated Melanesian populations, *Am. J. Epidemiol.* **105**:480–487.

Lange, B., Arbeter, A., Hewetson, J., and Henle, W., 1978, Longitudinal study of Epstein–Barr virus antibody titers and excretion in pediatric patients with Hodgkin's disease, *Int. J. Cancer* **22**(5):521–527.

Lanier, A.P., Henle, W., Bender, T.R., Henle, G., and Talbot, M.L., 1980a, Epstein–Barr virus-specific antibody titers in seven Alaskan natives before and after diagnosis of nasopharyngeal carcinoma, *Int. J. Cancer* **26**:133–137.

Lanier, A.P., Bender, T., Talbot, M., Wilmeth, S., Tschopp, C., Henle, W., Henle, G., Ritter, D., and Terasaki, P., 1980b, Nasopharyngeal carcinoma in Alaskan Eskimos, Indians and Aleuts: A review of cases and study of Epstein–Barr virus, HLA and environmental risk factors, *Cancer* **46**(9):2100–2006.

Lenoir, G., Philip, T., Bornkamm, G.W., Gillet, P., Bryon, P.A., Bouvier, R., Dodat, H., Dore, J.F., Burnat-Mentigny, M., and Hermier, M., 1979, Lymphôme de Burkitt associé au virus d'Epstein–Barr chez un enfant français, *Nouv. Presse Med.* **49**:4031–4034.

Levine, P.H., O'Conor, G.T., and Berard, C.W., 1972, Antibodies to Epstein–Barr virus in American patients with Burkitt's lymphoma, *Cancer* **30**:610–615.

Levine, P.H., Cho, B.R., Connelly, R.R., Berard, C.W., O'Conor, C.T., Dorfman, R.F., Easton, J.M., and de Vita, V.T., 1975, The American Burkitt's Lymphoma Registry: A progress report, *Ann. Intern. Med.* **83**:31–36.

Liang, P.C., 1964, Studies on nasopharyngeal carcinoma in Chinese: Statistical and laboratory investigations, *Chin. Med. J.* **83**:373–390.

Lin, T.M., Chen, K.P., Lin, C.C., Hsu, M.M., Tu, S.M., Chiang, T.C., Jung, P.F., and Hirayama, T., 1975, Retrospective study on nasopharyngeal carcinoma, *J. Natl. Cancer Inst.* **51**:1403–1408.

Lindahl, T., Klein, G., Reedman, B.M., Johansson, B., and Singh, S., 1974, Relationship between Epstein–Barr virus DNA and the EBV-determined nuclear antigen (EBNA) in Burkitt's lymphoma biopsies and other lympho-proliferative malignancies, *Int. J. Cancer* **13**:764–772.

Liu, Y.X., and Zeng, Y., 1979, Detection of EB-virus specific IgA antibody from patients with NPC by immunoenzymatic method, *Chin. J. Oncol.* **1**:8.

Lopez, C., Watanabe, K.A., and Fox, J.J., 1980, 2'-Fluoro-5-iodo-aracytosine, a potential and selective anti-herpesvirus agent, *Antimicrob. Agents Chemother.* **17**:803–806.

Manolov, G., and Manolova, Y., 1972, Marker bank in one chromosome 14 from Burkitt's lymphomas, *Nature (London)* **237**:33.

Marker, S.C., Ascher, N.L., Kalis, J.M., Simmons, R.L., Najarian, J.S., and Balfour, H.H., Jr., 1979, Epstein–Barr virus antibody responses and clinical illness in renal transplant recipients, *Surgery* **85**(4):433–440.

Mathew, G.D., Qualtière, L.F., Neel, H.B., III, and Pearson, G.R., 1981, IgA antibody, antibody-dependent-cellular-cytoxicity and prognosis in patients with nasopharyngeal carcinoma, *Int. J. Cancer* **27**:175–180.

Matsuo, T., Nishi, S., Hirasi, H., and Osato, T., 1976, Studies on Epstein–Barr virus-related antigens. I. Indirect single-radial immunodiffusion as a useful method for detection and assay of soluble antigen, *Int. J. Cancer* **18**:453–457.

McCaw, B.K., Epstein, A.L., Kaplan, H.S., and Hecht, F., 1977, Chromosome 14 translocation in African and North American Burkitt's lymphoma, *Int. J. Cancer* **19**:482–486.

Michaux, L., 1944, Carcinome de base du crâne, cited in Godtfredson (19). Ophthalmologic and neurologic symptoms at malignant nasopharyngeal tumors: Clinical study comprising 454 cases, with special reference to histopathology and possibility of earlier recognition, *Acta Psychiatr. Scand. Suppl.* **34**:1–323.

Micheau, C., de-Thé, G., Orofiamma, B., Schwaab, G., Brugère J., Tursz, T., Sancho-Garnier, H., and Cachin, Y., 1980, Carcinomas of the nasopharynx: Relationship between histological type and anti-Epstein–Barr virus serology, *Nouv. Presse Med.* **9**(1):21–24.

Micheau, C., de-Thé, G., Orofiamma, B., Schwaab, G., Brugère, J., Tursz, T., Sancho-Garnier, H., and Cachin, Y., 1981, Practical value of classifying nasopharyngeal carcinoma in two major microscopical types, in: *Cancer Campaign*, Vol. 5, *Nasopharyngeal Car-*

cinoma (E. Grundmann, G.R.F. Krueger, and D.V. Ablashi, eds.), pp. 51–57, Gustav Fischer, Stuttgart.

Miller, G., and Beebe, G.W., 1973, Infectious mononucleosis and the empirical risk of cancer, *J. Natl. Cancer Inst.* **50**:315–321.

Miller, G., Niederman, J.C., and Stitt, D.A., 1972a, Infectious mononucleosis: Appearance of neutralizing antibody to Epstein–Barr virus measured by inhibition of formation of lymphoblastoid cell lines, *J. Infect. Dis.* **125**:403–406.

Miller, G., Shope, T., Lisco, H., Stitt, D., and Lipman, M., 1972b, Epstein–Barr virus: Transformation, cytopathic changes and viral antigens in squirrel monkey and marmoset leukocytes, *Proc. Natl. Acad. Sci. U.S.A.* **69**:383–387.

Miller, G., Niederman, J.C., and Andrews, L.L., 1973, Prolonged oropharyngeal excretion of Epstein–Barr virus after infectious mononucleosis, *N. Engl. J. Med.* **188**:229–232.

Miller, G., Shope, T., and Coope, D., 1974, Aspects of the pathogenesis of lymphoma in cotton-top marmosets following inoculation of EB virus, in: *Mechanisms of Virus Diseases*, Vol. 1 (W.S. Robinson and F.C. Fox, eds.), pp. 429–454, W.A. Benjamin, Menlo Park, California.

Mitelman, F., Andersson-Anvret, M., Brandt, L., Catovsky, D., Klein, G., Manolov, G., Manolova, Y., Mark-Vendel, E., and Nilsson, P.G., 1979, Reciprocal 8;14 translocation in EBV-negative B-cell acute lymphocytic leukemia with Burkitt type cells, *Int. J. Cancer* **24**:27–33.

Miyamoto, K., Miyano, K., Miyoshi, I., Hamasaki, K., Nishihara, R., Terao, S., Kimura, I, Maeda, K., Matsumara, K., Nishijima, K., and Tanaka, T., 1980, Chromosome 14q+ in a Japanese patient with Burkitt's lymphoma, *Acta Med. Okayama* **34**:61–65.

Miyoshi, I., Hiraki, S., Tsubota, T., Uno, J., Nakamura, K., Ota, T., Hikita, T., Hayashi, K., Kataoka, M., Tanaka, T., Kimura, I., and Sairenui Hinuma, Y., 1978, Epstein–Barr virus positive Japanese Burkitt lymphoma, *Gann* **69**:449–450.

Miyoshi, I., Hiraki, S., Kimura, I., Miyamoto, K., and Sato, J., 1979, 2/8 translocation in a Japanese Burkitt's lymphoma, *Experientia* **35**:742–743.

Moar, M.H., Siegert, W., and Klein, G., 1977, Detection and localization of Epstein–Barr virus associated early antigens in single cells by autoradiography using I^{125} labelled antibodies, *Intervirology* **8**:226–239.

Morgan, D.G., Miller, G., Niederman, J.C., Smith, H.W., and Dowaliby, J.M., 1979, Site of Epstein–Barr virus replication in the oropharynx, *Lancet* :1154–1157.

Morrow, R.H., Kisuule, A., Pike, M.C., and Smith, P., 1976, Burkitt's lymphoma in the Mengo districts of Uganda: Epidemiologic features and their relationship to malaria, *J. Natl. Cancer Inst.* **56**:479–483.

Morrow, R.H., Pike, M.C., and Smith, P.G., 1977, Further studies of space–time clustering of Burkitt's lymphoma in Uganda, *Br. J. Cancer* **35**:668–673.

Motulsky, A.G., 1964, Hereditary red cell traits and malaria, *Am. J. Trop. Med. Hyg.* **13**:147–158.

Muir, C.S., 1972, Nasopharyngeal carcinoma in non-Chinese populations, in: *Oncogenesis and Herpeviruses* (P.M. Biggs, G. de Thé, and L.N. Payne, eds.), pp. 367–371, IARC, Lyon.

Munoz, N., Davidson, R.J., Witthoff, B., Ericsson, J.E., and de-Thé, G., 1978, Infectious mononucleosis and Hodgkin's disease, *Int. J. Cancer* **22**:10–13.

National Office for Cancer Control, China's Ministry of Health, and the Nanjing Institute of Geography, Chinese Academy of Medical Sciences, 1981, *Atlas of Cancer Mortality in the People's Republic of China* (Dr. LiPing) (in press).

Niederman, J.C., McCollum, R.W., Henle, G., and Henle, W., 1968, Infectious mononucleosis: Clinical manifestations in relation to Epstein–Barr virus antibodies, *J. Am. Med. Assoc.* **203**:205–209.

Nielsen, N.H., Mikkelsen, F., and Hansen, J.P., 1977, Nasopharyngeal cancer in Greenland, *Acta Pathol. Microbiol. Scand.* **85**(6):850–858.

Nkrumah, F.K., and Perkins, I.V., 1976, Sickle cell trait, hemoglobin C trait and Burkitt's lymphoma, *Am. J. Trop. Med. Hyg.* **25**:633–636.

Nkrumah, F.K., Perkins, I.V., and Biggar, R.J., 1977, Combination chemotherapy in abdominal Burkitt's lymphoma, *Cancer* **40:**1410–1416.
Nonoyama, M., and Pagano, J.S. 1973, Homology between Epstein–Barr viruses DNA and viral DNA from Burkitt's lymphoma and nasopharyngeal carcinoma determined by DNA–DNA reassociation kinetics, *Nature* (*London*) **242:**44–47.
Nonoyama, M., Huang, D.P., Pagano, J.S., Klein, G., and Singh, S., 1973, DNA of Epstein–Barr virus detected in tissue of Burkitt's lymphoma and nasopharyngeal carcinoma, *Proc. Natl. Acad. Sci. U.S.A.* **70:**3265–3268.
O'Conor, G.T., 1961, Malignant lymphoma in African children. II. A pathological entity, *Cancer* **12:**270–283.
O'Conor, G.T., 1970, Persistent immunologic stimulation as a factor in oncogenesis with special reference to Burkitt's tumor, *Am. J. Med.* **48:**279–285.
O'Conor, G.T., and Davies, J.N.P., 1960, Malignant tumors in African children with special reference to malignant lymphoma, *J. Pediatr.* **56:**526–535.
Old, L.J., Clifford, P., Boyse, E.A., de Harven, E., Geering, G., Oettgen, H.F., and Williamson, B., 1966, Precipitating antibody in human serum to an antigen present in cultured Burkitt lymphoma cells, *Proc. Natl. Acad. Sci. U.S.A.* **56:**1699–1704.
Olweny, C.L.M., Atine, I., Kaddu-Mukasa, A., Owor, R., Anderson, M., Klein, G., Henle, W., and de Thé, G., 1977, Epstein–Barr virus genome studies in Burkitt and non-Burkitt lymphomas in Uganda, *J. Natl. Cancer Inst.* **58:**1191–1196.
Olweny, C.L.M., Katongole-Mbidde, E., Otim, D., Lwanga, S.K., Magrath, I.T., and Ziegler, J.L., 1980, Long term experience with Burkitt's lymphoma in Uganda, *Int. J. Cancer* **26:**261–266.
Orofiamma, B., Micheau, C., Schwaab, G., Sancho-Garnier, H., de-Thé, G., and Brugère, J., 1982, Carcinomes du rhinopharynx: Relations entre types histologiques et caractéristiques étio-épidémiologiques, *Nouv. Presse Med.* (submitted).
Pagano, J.S., Huang, D.H., and Levine, P., 1973, Absence of Epstein–Barr viral DNA in American Burkitt's lymphoma, *N. Engl. J. Med.* **289:**1395–1399.
Paul, J.R., and Bunnell, W.W., 1932, The presence of heterophile antibodies in infectious mononucleosis, *Am. J. Med. Sci.* **183:**91–104.
Pearson, G.R., and Orr, T.W., 1976, Antibody-dependent lymphocyte-cytotoxicity against cells expressing Epstein–Barr virus antigens, *J. Natl. Cancer Inst.* **56:**485–488.
Pearson, G.R., and Qualtière, L.F., 1978, Papain solubilization of the Epstein–Barr virus-induced membrane antigen, *J. Virol.* **28:**344–351.
Pearson, G.R., and Scott, R.E., 1978, Potential for immunization against herpesvirus infection with plasma membrane vesicles, in: *Oncogenesis and Herpesviruses III* (G. de Thé, W. Henle, and F. Rapp, eds.), pp. 1019–1025, IARC Scientific Publication No. 24, Lyon.
Pearson, G.R., Dewey, F., Klein, G., Henle, G., and Henle, W., 1970, Relation between neutralization of Epstein–Barr virus and antibodies to cell membrane antigens induced by the virus, *J. Natl. Cancer Inst.* **45:**989–995.
Pearson, G.R., Prevost, J.M., and Orr, T.W., 1975, Detection of virus-associated antigens with the antibody-dependent lymphocyte cytotoxicity test, in: *Proceedings of the International Workshop on the Production, Concentration and Purification of Epstein–Barr Virus*, pp. 263–275, IARC, Lyon.
Pearson, G.R., Johansson, B., and Klein, G., 1978a, Antibody-dependent cellular cytotoxicity against Epstein–Barr virus associated antigens in African patients with nasopharyngeal carcinoma, *Int. J. Cancer* **22:**120–125.
Pearson, G.R., Coates, H.L., Needl, H.B., Levine, P., Ablashi, D., and Easton, J., 1978b, Clinical evaluation of EBV serology in American patients with nasopharyngeal carcinoma, in: *Nasopharyngeal Carcinoma: Etiology and Control* (G. de Thé and Y. Ito, eds.), pp. 439–448, IARC, Lyon.
Pearson, G.R., Qualtière, L.F., Klein, G., Norin, T., and Bal, I.S., 1979, Epstein–Barr virus-specific antibody-dependent cellular cytotoxicity in patients with Burkitt's lymphoma, *Int. J. Cancer* **24:**402–406.

Peto, R., 1977, Epidemiology, multistage models and short-term mutagenicity tests, *Cold Spring Harbor Conf. Cell Prolif.* **4**:1403.

Pfeiffer, E., 1889, Drüsenfieber, *Jahrb. Kinderheilkd.* **29**:257–264.

Pickens, S., and Murdoch, J.M., 1979, Infectious mononucleosis in the elderly, *Age Ageing* **8**(2):93–95.

Pike, M.C., Williams, E.H., and Wright, B., 1967, Burkitt's tumor in the West Nile district of Uganda, 1961–1965, *Br. Med. J.* **2**:395–399.

Pike, M.C., Morrow, M.D., Kisuule, A., and Mafigiri, J., 1970, Burkitt's lymphoma and sickle cell trait in Uganda, *Br. J. Prev. Soc. Med.* **24**:39–41.

Pope, J.H., Horne, M.K., and Scott, W., 1969a, Identification of the filtrable leucocyte transforming factor of QIMR-WILL cells as herpesvirus, *Int. J. Cancer* **4**:255–260.

Pope, J.H., Horne, M.K., and Wetters, E.J., 1969b, Significance of a complement-fixing antigen associated with herpes-like virus and detected in the Raji cell line, *Nature (London)* **222**:166–167.

Prather, S.O., and Lausch, R.N., 1976, Kinetics of serum factors mediating blocking, unblocking and antibody-dependent cellular cytotoxicity in hamsters given isografts of Para-7 tumor cells, *Int. J. Cancer* **17**:380–388.

Purtilo, D.T., 1980, Registry of fatal infectious mononucleosis and Epstein–Barr virus infections, *J. Am. Med. Assoc.* **243**(18):1806.

Purtilo, D.T., Cassel, C.K., Yang, J.P.S., Harper, R., Stephenson, S.R., Landing, B.H., and Vawter, G.F., 1975, X-linked recessive progressive combined variable immunodeficiency (Duncan's disease), *Lancet* **1**:937–941.

Purtilo, D.T., Paquin, L., DeFlorio, D., Virzi, F., and Sakhuia, R., 1979, Immunodiagnosis and immunopathogenesis of the X-linked recessive lymphoproliferative syndrome, *Semin. Hematol.* **16**:309–343.

Reedman, B.M., and Klein, G., 1973, Cellular localization of an Epstein–Barr virus (EBV) associated complement-fixing antigen in producer and non-producer lymphoblastoid cell lines, *Int. J. Cancer* **2**:499–520.

Reedman, B.M., Klein, G., Pope, J.H., Walters, M.K., Hilgers, J., Singh, S., and Johansson, B., 1974, Epstein–Barr virus associated complement fixing and nuclear antigens in Burkitt's lymphoma biopsies, *Int. J. Cancer* **13**:755–763.

Rickinson, A.B., Jarvis, J.E., Crawford, D.H., and Epstein, M.A., 1974, Observations on the type of infection by Epstein–Barr virus in peripheral lymphoid cells of patients with infectious mononucleosis, *Int. J. Cancer* **14**:704–715.

Rivers, T., 1937, Viruses and Koch's postulates, *J. Bacteriol.* **33**:1–12.

Robinson, J.E., Brown, N., Andiman, W., Halliday, K., Francke, U., Robert, M.F., Andersson-Anvret, M., Horstmann, D., and Miller, G., 1980, Diffuse polyclonal B-cell lymphoma during primary infection with Epstein–Barr virus, *N. Engl. J. Med.* **302**:1293–1297.

Rocchi, G., Hewetson, J., and Henle, W., 1973a, Specific neutralizing antibodies in Epstein–Barr virus-associated diseases, *Int. J. Cancer* **11**:637–647.

Rocchi, G., Hewetson, J., Henle, W., and Henle, G., 1973b, Antigen expression and colony formation of lymphoblastoid cell lines after superinfection with Epstein–Barr virus, *J. Natl. Cancer Inst.* **50**:307–314.

Rosdahl, N., Larsen, S.O., and Clemmesen, J., 1974, Hodgkin's disease in patients with previous infectious mononucleosis: 30 years' experience, *Br. Med. J.* **2**:253–256.

Saad, A., 1968, Observations on nasopharyngeal carcinoma in the Sudan, in: *Cancer in Africa* (C.A. Linsell and G.L. Limms, eds.), pp. 281–285, East African Publishing House, Nairobi.

Saemundsen, E., Berkel, A.I., Henle, W., Henle, G., Anvret, M., Sanal, Ö., Ersoy, F., Caglar, M., Klein, G., 1981, Epstein–Barr virus carrying lymphoma in a patient with ataxia–telangiectasia, *Br. Med. J.* **282**:425–427.

Sakamoto, K., Aiba, M., Katayama, I., Sullivan, J., Humphreys, R., and Purtilo, D.T., 1981, Antibodies to Epstein–Barr virus specific antigens in patients with hairy-cell leukemia, *Int. J. Cancer* **27**:453–458.

Sawyer, R.N., Evans, A.S., Niederman, J.C., and McCollum, R.W., 1971, Prospective study of a group of Yale University freshmen. I. Occurrence of infectious mononucleosis, *J. Infect. Dis.* **123**:263–269.

Shanmugaratnam, K., 1978, Variations in nasopharyngeal carcinoma incidence among specific Chinese communities (dialect groups) in Singapore, in: *Nasopharyngeal Carcinoma: Etiology and Control* (G. de-Thé and Y. Ito, eds.), pp. 191–198, IARC, Lyon.

Shanmugaratnam, K., and Higginson, J., 1967, Aetiology of nasopharyngeal carcinoma: Report on a retrospective survey in Singapore, in: *Cancer of the Nasopharynx* (C.S. Muir and K. Shanmugaratnam, eds.), UICC Monograph Series, Vol. 1, pp. 130–137, Munksgaard, Copenhagen.

Shanmugaratnam, K., and Sobin, L.H., 1978b, Histological typing of upper respiratory tract tumors, in: *International Histological Classification of Tumors,* No. 19, World Health Organization, Geneva.

Shanmugaratnam, K., Tan, K.K., and Lee, K.W., 1967, Lymphoma of the Burkitt type in Singapore, *Int. J. Cancer* **2**:576–580.

Shope, T., Dechairo, D., and Miller, G., 1973, Malignant lymphoma in cotton-top marmosets following inoculation of Epstein–Barr virus, *Proc. Natl. Acad. Sci. U.S.A.* **70**:2487–2491.

Simons, M.J., Wee, G.B., Day, N.E., de-Thé, G., Morris, P.J., and Shanmugaratnam, K., 1974, Immuno-genetic aspects of nasopharyngeal carcinoma. I. Difference in HL-A antigen profiles between patients and comparison groups, *Int. J. Cancer* **13**:122–134.

Simons, M.J., Wee, G.B., Day, N.E., Chan, S.H., Shanmugaratnam, K., and de Thé, G., 1975, Immunogenetic aspects of nasopharyngeal carcinoma (NPC). IV. Probable identification of an HL-A second antigen associated with a high risk for NPC, *Lancet* **1**:142–143.

Simons, M.J., Wee, G.B., Goh, E.H., Chan, S.H., Shanmugaratnam, K., Day, N.E., and de Thé, G., 1976, Immunogenetic aspects of nasopharyngeal carcinoma. IV. Increased risk in Chinese of nasopharyngeal carcinoma associated with a Chinese-related HLA profile (A2, Singapore 2), *J. Natl. Cancer Inst.* **57**:977–980.

Simons, M.J., Wee, G.B., Singh, D., Dharmalingham, S., Yong, N.K., Chau, J.C.W., Ho, J.H.C., Day, N.E., and De Thé, G., 1977, Immunogenetic aspects of nasopharyngeal carcinoma. V. Confirmation of a Chinese related HLA profile (A2, Singapore 2) associated with an increased risk for nasopharyngeal carcinoma, in: *Epidemiology and Cancer Registries in the Pacific, Natl. Cancer Inst. Monogr.* **47**:147–151.

Simons, M.J., Chan, S.H., Wee, G.B., Shanmugaratnam, K., Goh, E.H., Ho, J.H.C., Chau, J.C.W., Darmalingham, S., Prasad, U., Betuel, H., Day, N.E., and de-Thé, G., 1978, Nasopharyngeal carcinoma and histocompatibility antigens, in: *Nasopharyngeal Carcinoma: Etiology and Control* (G. de-Thé and Y. Ito, eds.), pp. 271–282, Scientific Publication No. 20, IARC, Lyon.

Slater, R.M., Philip, P., Badsberg, E., Behrendt, H., Hansen, N.E., and Van Heerde, P., 1979, A 14q+ chromosome in a B-cell acute lymphocytic leukemia and in a leukemic non-endemic Burkitt's lymphoma, *Int. J. Cancer* **23**:639–647.

Sohier, R., 1971, Discordance au cours de la mononucléose infectieuse entre l'élaboration des anticorps fixant le complément avec un antigène soluble et des anticorps révélés par immunofluorescence, *C. R. Acad. Sci. Ser.* **272**:3098–3101.

Sohier, R., and de-Thé, G., 1978, EBNA antibodies are more useful than complement-fixing antibody in monitoring infectious mononucleosis patients, *Biomedicine* **29**:170–173.

Sohier, R., Lenoir, G.M., and Lamelin, J.P.L., 1981, Les formes mortelles de l'infection primaire à virus d'Epstein–Barr avec ou sans mononucléose infectieuse, *Ann. Med. Intern.* **132**:48–57.

Sprunt, T.P., and Evans, F.A., 1920, Mononuclear leucocytosis in reaction to acute infections (in infectious mononucleosis), *Bull. Johns Hopkins Hosp.* **31**:410–417.

Stancek, D., and Robensky, J., 1979, Enhancement of Epstein–Barr virus antibody production in systemic lupus erythematosus patients, *Acta. Virol. (Prague)* **23**(2):168–169.

Stephens, R., Traul, K., Gaudreau, P., Teh, J., Fisher, L., and Mayyasi, S.A., 1977, Comparative studies on EBV antigens by immunofluorescence and immunoperixodase techniques, *Int. J. Cancer* **19**:305–316.

Sullivan, J.L., Wallen, W.C., and Johnson, F.L., 1978, Epstein–Barr virus infection following bone-marrow transplantation, *Int. J. Cancer* **22**(2):132–135.
Sumaya, C.V., 1977, Primary Epstein–Barr virus infections in children, *Pediatrics* **59**(1):16–21.
Svedmyr, E., and Jondal, M., 1975, Cytotoxic effector cells specific for B cell lines transformed by Epstein–Barr virus are present in patients with infectious mononucleosis, *Proc. Natl. Acad. Sci. U.S.A.* **7**:1622.
Svedmyr, A., Demissie, A., Klein, G., and Clifford, P., 1970, Antibody patterns in different human sera against intracellular and membrane antigen complexes associated with Epstein–Barr virus, *J. Natl. Cancer Inst.* **44**:595–610.
Svoboda, D., Kirchner, K.R., and Shanmugaratnam, K., 1965, Ultrastructure of nasopharyngeal carcinomas in American and Chinese patients: An application of electron microscopy to geographic pathology, *Exp. Mol. Pathol.* **4**:189–204.
Thestrup-Pedersen, K., Esmann, V., Bisballe, S., Jensen, J.R., Pallesen, G., Hastrup, J., Madsen, M., Thorling, K., Masucci, M.G., Saedmundsen, A.K., and Ernberg, I., 1980, Epstein–Barr virus induced lymphoproliferative disorder converting to fatal Burkitt-like lymphoma in a boy with interferon-inducible chromosomal defect, *Lancet* :997–1002.
Tischendorf, P., Shramek, G.J., Balagdas, R.C., Deinhardt, F., Knospe, W.H., Noble, G.R., and Maynard, J.E., 1970, Development and persistence of immunity to Epstein–Barr virus in man, *J. Infect. Dis.* **122**:401–409.
Tovey, M.G., Lenoir, G., and Begon-Lours, J., 1978, Activation of latent Epstein–Barr virus by antibody to human IgM, *Nature (London)* **276**:270–272.
Trumper, P.A., Epstein, M.A., Giovanella, G.C., and Finerty, S., 1977, Isolation of infectious EB virus from epithelial tumor cells of nasopharyngeal carcinoma, *Int. J. Cancer* **20**:655–662.
UICC, 1974, *TNM Classification of Malignant Tumors*, 2nd ed., Geneva.
University Health Physicians and PHS Laboratories, 1971, A joint investigation: Infectious mononucleosis and its relationship to EB virus antibody, *Br. Med. J.* **4**:643–646.
USPHS Advisory Committee on Smoking and Health, 1964, PHS Publ. No. 1103, USPHS, Washington, D.C.
Van den Berghe, H., Parloir, C., Gosseye, S., Englebienne, V., Cornu, G., and Sokal, G., 1979, Variant traslocation in Burkitt's lymphoma, *Cancer Genet. Cytogenet.* **1**:9–14.
Virelizier, J.L., Lenoir, G., and Griscelli, C., 1978, Persistent Epstein–Barr virus infection in a child with hypergammaglobulinaemia and immunoblastic proliferation associated with a selective defect in immune interferon secretion, *Lancet* **2**:231–234.
Vonka, V., and Kutinova, L., 1973, Increase of EB virus positive cells in cultured Burkitt lymphoma cells after short-time exposure to heat, *Neoplasma* **30**:349–352.
Vonka, V., Benyesh-Melnick, M., and McCombs, R.J., 1970, Antibodies in human sera to soluble and viral antigens found in Burkitt's lymphoma and other lymphoblastoid cell lines, *J. Natl. Cancer Inst.* **44**:865–872.
Vonka, V., Kutinova, L., Drobnik, J., and Bräuerova, J., 1972, Increase of Epstein–Barr virus positive cells in EB 3 cultures after treatment with cisdichloro-diammine-platinum (11), *J. Natl. Cancer Inst.* **48**:1277–1281.
Wara, W.M., Wara, D.W., Phillips, T.L., and Ammahh, A., 1975, Elevated IgA in carcinoma of the nasopharynx, *Cancer* **35**:1313–1315.
Williams, A.O., 1966, Hemoglobin genotypes ABO blood groups and Burkitt's tumor, *J. Med. Genet.* **3**:177–179.
Williams, E.H., and de-Thé, G., 1974, Familial aggregation in nasopharyngeal carcinoma, *Lancet*, **2**:295.
Williams, E.H., Spit, P., and Pike, M.C. 1969, Further evidence of space–time clustering of Burkitt's lymphoma patients in the West Nile district of Uganda, *Br. J. Cancer*, **23**:235–246.
Williams, E.H., Day, N.E., and Geser, A., 1974, Seasonal variation in onset of Burkitt's lymphoma in the West Nile district of Uganda, *Lancet* **1**:19–22.

Williams, E.H., Smith, P.G., Day, N.E., Geser, A., Ellice, J., and Tukei, P., 1978, Space–time clustering of Burkitt's lymphoma in the West Nile district of Uganda, 1961–1975, *Br. J. Cancer* **37**:109–122.

Wolf, H., Zur Hausen, H., and Becker, V., 1973, EB viral genomes in epithelial nasopharyngeal carcinoma cells, *Nature (London) New Biol.* **244**:245–257.

World Health Organization, 1970, Mortality from malignant neoplasms, 1955–1965, Geneva.

Wright, D.H., and Pike, P.A., 1969, Bone-marrow involvement in Burkitt's tumor, *Br. J. Haematol.* **15**:409–416.

Wright, D.H., Schumacher, H.R., Thomas, W.J., Duval-Arnould, B.J., Creegan, W.J., Forman, D.S., Strong, D.M., Robinson, J.E., and Miller, G., 1980, Polyclonal B cell lymphoma during infection with Epstein–Barr virus (letter to editor), *N. Engl. J. Med.* **303**(18):1064–1065.

Wu, C.H., 1921, in: *The Encyclopaedia of Chinese Medical Terms*, Vol. 1, p. 756, Commercial Press, Shanghai (in Chinese).

Zech, L., Haglund, U., and Nilsson, K., 1976, Characteristic chromosomal abnormalities in biopsies and lymphoid cell lines from patients with Burkitt and non-Burkitt lymphomas, *Int. J. Cancer* **17**:47.

Zeng, Y., Shang, M., Liu, C.R., Cheng, Y.H., Ou, R.S., Li, X.H., Gan, B.W., Hu, M.G., Chen, M., He, S.Q., and Mu, G.P., 1979, Detection of IgA antibody to EB virus VCA from patients with nasopharyngeal carcinoma by immunofluorescent test, *Chin. J. Oncol.* **1**:81–83.

Zeng, Y., Liu, Y.X., Liu, C.R., Chen, S.W., Wei, J.N., Zhu, J.S., and Zai, H.J., 1980a, Application of an immunoenzymatic method and an immunoautoradiographic method for a mass survey of nasopharyngeal carcinoma, *Intervirology* **13**:162–168.

Zeng, Y., Pi, G.H., and Zhao, W.P., 1980b, Detection of EBNA by anticomplement immunoenzymatic method, *Acta Acad. Med. Sin.* **2**:134–135.

Ziegler, J.L., 1972, Chemotherapy of Burkitt's lymphoma, *Cancer* **30**:1534–1540.

Ziegler, J.L., 1977, Treatment results of 54 American patients with Burkitt's lymphoma are similar to the African experience, *N. Engl. J. Med.* **297**:75–80.

Ziegler, J.L., Bluming, A.Z., Morrow, R.H., Fass, L., and Carbone, P.P., 1970, Central nervous system involvement in Burkitt's lymphoma, *Blood* **36**:718–728.

Ziegler, J.L., Andersson, M., Klein, G., and Henle, W., 1976, Detection of Epstein–Barr virus DNA in American Burkitt's lymphoma, *Int. J. Cancer* **17**:701–706.

Zur Hausen, H., Schulte-Holthausen, H., Klein, G., Henle, W., Henle, G., Clifford, P., and Santesson, L., 1970, EBV DNA in biopsies of Burkitt's tumors and anaplastic carcinomas of the nasopharynx, *Nature (London)* **228**:1056–1058.

Zur Hausen, H., O'Neill, F.J., Freese, U.K., and Hecher, E., 1978, Persisting oncogenic herpevirus induced by tumor promotor TPA, *Nature (London)* **272**:373–375.

CHAPTER 3

Biochemistry of Epstein–Barr Virus

Elliott Kieff, M.D., Ph.D., Timothy Dambaugh, Ph.D., Walter King, Ph.D., Mark Heller, Ph.D., Andrew Cheung, Ph.D., Vicky van Santen, Ph.D., Mary Hummel, Ph.D., Christopher Beisel, and Susan Fennewald

I. INTRODUCTION

The objective of this introduction is to develop concisely the context in which biochemical studies of Epstein–Barr virus (EBV) have been undertaken. Elements of history and biology are relevant to an understanding of the somewhat unusual course of biochemical research with EBV.

The history of EBV begins with two discoveries: the description of an endemic lymphoma among children in equatorial Africa by Dennis Burkitt (1962) and the subsequent identification of a new virus in culture of the lymphoma cells by Epstein *et al.* (1964). Since some viruses were known to cause malignancy after experimental infection of animals and since some naturally occurring animal cancers were known to be transmitted by virus infection, the finding of a new virus in cells of an endemic human tumor caught the attention of the scientific community. From the moment of its discovery in cultures of tumor-biopsy cells, an important question has been whether EBV is the cause of African Burkitt lymphoma.

EBV was the fifth human herpesvirus to be discovered. Prior to its discovery, a great deal of research had been done with the other human

ELLIOTT KIEFF, M.D., Ph.D., TIMOTHY DAMBAUGH, Ph.D., WALTER KING, Ph.D., MARK HELLER, Ph.D., ANDREW CHEUNG, Ph.D., VICKY VAN SANTEN, Ph.D., MARY HUMMEL, Ph.D., CHRISTOPHER BEISEL, and SUSAN FENNEWALD • Division of Biological Sciences, The University of Chicago, 910 East 58th Street, Chicago, Illinois 60637.

herpesviruses: herpes simplex virus type 1 (HSV-1), herpes simplex virus type 2 (HSV-2), cytomegalovirus (CMV), and varicella–zoster virus (VZV). There was substantial knowledge of the biology and morphology of HSV, CMV, and VZV and of the biochemistry of HSV. Although morphologically indistinguishable from these other herpesviruses, EBV has little if any antigenic relatedness to them. EBV also differs from these and other viruses in its narrowly restricted host range *in vitro*. *In vitro*, EBV will infect only primate lymphocytes of the immunoglobulin-producing or B-lymphocyte series (W. Henle *et al.*, 1967; Pope *et al.*, 1968b; Jondal and Klein, 1973).

After infection with EBV, lymphocytes express at least one new antigen specified by the virus, and the infected cells begin to proliferate (Pope *et al.*, 1971; Gerber and Hoyer, 1971; Reedman and Klein, 1973; Moss and Pope, 1975). Some of the proliferating cells have the potential to be grown as continuous cell lines, a property that normal, uninfected lymphocytes lack (W. Henle *et al.*, 1967; Pope *et al.*, 1968a,b; Nilsson *et al.*, 1971). Studies with other viruses and with EBV have shown a link between growth transformation *in vitro* and oncogenic potential *in vivo* (Shope *et al.*, 1973; Epstein *et al.*, 1973; Deinhardt *et al.*, 1975). Therefore, the mechanism by which EBV enhances the replication of cells *in vitro* has also been a focal point of EBV research.

Cultures of EBV-infected lymphocytes vary in their permissiveness for viral replication. While most cultures are nonpermissive, replication does occur in a small fraction of cells in some cultures. There are four important consequences of the largely nonpermissive outcome of EBV infection *in vitro*: First, the maintenance of latency in continuous cell cultures makes it possible to study the biochemistry of latent infection in a homogeneous population of cells, *in vitro*. Since lymphocytes are an important site of latent EBV infection in humans, latent lymphocyte infection *in vitro* is relevant to natural human infection. Second, the limited permissiveness of lymphoblastoid-cell lines for EBV replication has made it difficult to study virus replication and the components of mature virus. In recent years, studies requiring viral replication have been somewhat facilitated by the selection of clones of infected lymphocytes that are more permissive of virus replication (Hinuma *et al.*, 1967; Miller and Lipman, 1973a,b; Pizzo *et al.*, 1978), by the recognition that inducers of cellular differentiation can induce virus replication in some nonpermissively infected cells (Hampar *et al.*, 1972; Gerber, 1972; Kallin *et al.*, 1979; zur Hausen *et al.*, 1978), by the discovery that EBV produced by P3HR-1 cultures can induce productive infection in some latently infected cells (W. Henle *et al.*, 1970; Yajima and Nonoyama, 1976), and by the use of recombinant DNA technology to produce large quantities of viral DNA in bacterial cultures (Dambaugh *et al.*, 1980a). Third, most biochemical studies of EBV have utilized chronically infected lymphocyte cultures that, though partially permissive of virus replication, can be induced to be more permissive. These cultures must be passed for months to obtain adequate quantities of virus for purification and biochemical analysis. Several of these cell lines have been passaged continuously for years. The stability of the viral genome under these conditions is unknown. EBV-infected cells usually contain multiple copies of the viral genome (zur Hausen *et al.*, 1972; Kawai *et al.*, 1973; Pritchett *et al.*, 1976). It is generally assumed that the presence of multiple copies promotes stability of the viral genome in the infected cells. This has not been rigorously examined. Fourth, genetic approaches are important for analysis of the significance of biochemical observations. Although EBV can be cloned by transformation of lymphocytes in semisolid media (Moss and Pope, 1972; Yamamoto and Hinuma, 1976; Sugden and Mark, 1977; Henderson *et al.*, 1977a) or by limiting virus dilution, the delayed and uncertain yield of virus from such experiments has frustrated attempts to derive and characterize EBV mutants. One experiment that did yield interesting results was the cloning of a culture of partially permissive cells, Jijoye, to obtain a clone of more permissive cells (Hinuma *et al.*, 1967). In this instance, the virus produced by the more permissive clone, P3HR-1, differs from the virus produced by the parent culture, Jijoye, and from virus produced by other cultures in that the P3HR-1 virus lacks the ability to growth-transform noninfected B lymphocytes (Miller *et al.*, 1974; Menezes *et al.*, 1975; Ragona *et al.*, 1980).

Early attempts to extend the host range of EBV led to the observation that B lymphocytes of other primates were infectable (Miller *et al.*, 1972). Although similar to human B lymphocytes in their response to EBV infection, lymphocytes of New World primates, especially marmosets, tend to be more permissive for virus replication than human lymphocytes (Miller and Lipman, 1973a,b). Several Old World primates were discovered to be infected with herpesviruses that are closely related to EBV in structure, antigenicity, and biological properties (Falk, *et al.*, 1976; Gerber *et al.*, 1976b). These viruses are useful for experimental studies in their natural hosts and for biochemical comparison with human EBV.

II. STRUCTURAL COMPONENTS OF THE VIRUS

A. Morphology of the Virus

The few detailed studies of the structure of Epstein–Barr virus (EBV) are of thin sections of lymphoblastoid-cell cultures in which a small percentage of the cells are productively infected or are of virus purified from such partially permissive cultures (Fig. 1) (Epstein *et al.*, 1965, 1967; O'Conor and Rabson, 1965; Stewart *et al.*, 1965; Hummeler *et al.*, 1966; Toplin and Schidlovsky, 1966; Epstein and Achong, 1968, 1970, 1979; Pope *et al.*, 1968a; Dolyniuk *et al.*, 1976a,b). The innermost component of the virus particle, the core, varies in appearance with the plane

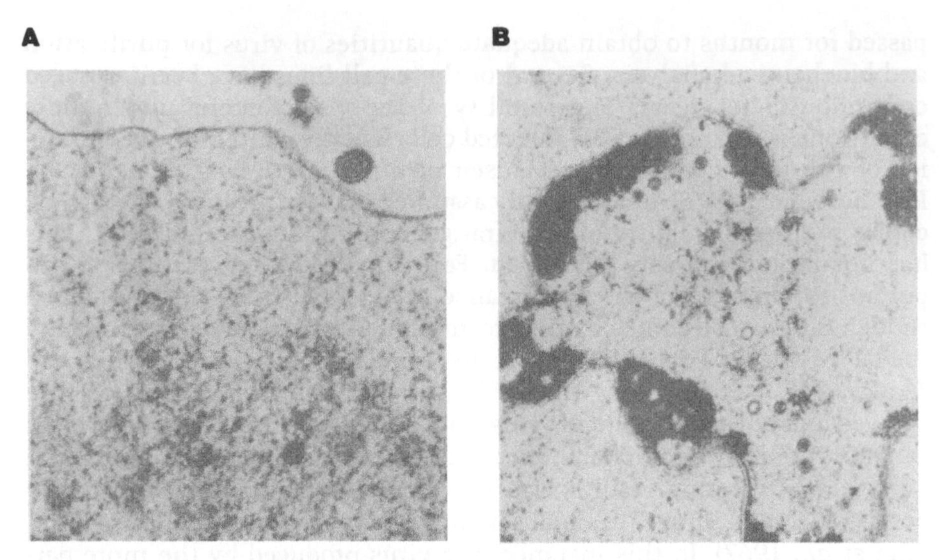

FIGURE 1. Electron photomicrographs of thin sections of (A) an enveloped extracellular virus located near the cytoplasmic membrane of an infected lymphoblast and (B) an infected lymphoblast nucleus containing both empty and full nucleocapsids.

of sectioning. The core appears similar to the toroid of herpes simplex virus (HSV) (Furlong *et al.*, 1972) and therefore probably consists of DNA coiled around protein. The core is surrounded by the icosahedral capsid. Hexameric capsomeres are commonly seen in negatively stained preparations of partially enveloped or maturing virus and in negatively stained preparations of virus treated with nonionic detergent. Pentameric capsomeres should also be present to facilitate puckering of the capsid. The capsid is enveloped by a trilaminar membrane with spikes on its outer surface. The space between the capsid shell and the envelope is filled with amorphous tegument.

B. Polypeptides of the Virus

Enveloped virus is generally purified from extracellular fluids from partially permissive cultures by velocity sedimentation in dextran gradients (Dolyniuk *et al.*, 1976a). Two isolates of viruses have been most extensively studied (Dolyniuk *et al.*, 1976a,b; Mueller-Lantzsch *et al.*, 1979, 1980a,b); virus produced by B95-8, a culture of a clone of marmoset lymphocytes infected with virus from a culture of lymphocytes from a patient with infectious mononucleosis (Miller and Lipman, 1973a), and virus produced by P3HR-1, a clone of the Jijoye Burkitt-tumor-cell line (Hinuma *et al.*, 1967). At a level of purity of B95-8 virus at which at least 98% of the labeled protein is viral, 33 component polypeptides could be resolved in denaturing polyacrylamide gels (Fig. 2) (Dolyniuk *et al.*, 1976a). On the basis of experience with the more thoroughly studied polypeptides of HSV (Spear, 1980), some of the 33 EBV polypeptides are likely to share common polypeptide components and differ in molecular weight as a consequence of varying degrees of posttranslational modifi-

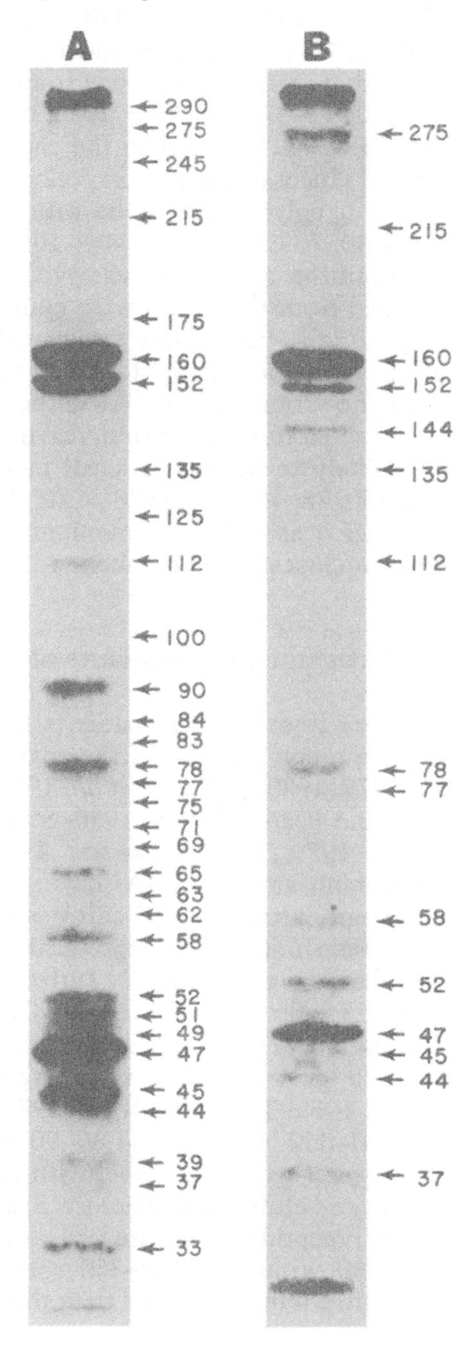

FIGURE 2. Fluorograms of ^{35}S-labeled polypeptides of enveloped virus (A) and nucleocapsids (B) separated on sodium dodecyl sulfate–polyacrylamide gels. The numbers indicate the estimated molecular weights in kilodaltons (Dolyniuk *et al.*, 1976a,b).

cation, principally glycosylation or phosphorylation. Some of the minor components may be products of partial proteolysis of major components. Of the 33 polypeptides, 18 are more abundant than the others and differ in molar abundance among themselves by 20- to 30-fold. The apparent sizes of the most abundant polypeptides are 290, 160, 152, 90, 78, 47, 42–44, 33, and 28 × 10^3 daltons (d), based on their electrophoretic mobility in acrylamide gels under denaturing conditions. Nonionic detergents that strip the envelope from the nucleocapsid solubilize the 290, 152, 90, 42–44, and 33 × 10^3 d components, suggesting that these polypeptides are external to the capsid. On the basis of incorporation of labeled glucosamine and on periodic acid–Schiff (PAS), staining, the 290 × 10^3 d polypeptide is the major glycosylated envelope protein. The 120–130, 70–90, and 42–44 × 10^3 d polypeptides label less heavily with glucosamine and stain less with PAS. Since these polypeptides are stripped from the virion with nonionic detergents and are glycosylated, they are also tentatively designated as envelope components. The 150 × 10^3 d polypeptide is the major nonglycosylated polypeptide removed from the virion by nonionic detergent. Since this abundant polypeptide is extracted from the virion, leaving the nucleocapsid apparently intact, it is likely to be the principal membrane matrix or tegument protein. The 160, 78, 44–47, and 28 × 10^3 d polypeptides remain associated with the nucleocapsid through nonionic-detergent extraction and are the principal nucleocapsid components.

C. Antigenic Determinants of the Virus

Sera from infected patients have been used to identify the major antigenic polypeptides of the envelope and capsid (Mueller-Lantzsch *et al.*, 1979, 1980a,b; Kallin *et al.*, 1979; Pearson and Qualtière, 1978; Qualtière and Pearson, 1979, 1980; Strnad *et al.*, 1979; Thorley-Lawson and Edson, 1979; Thorley-Lawson, 1979). Most studies of the envelope or membrane antigens (MA) have prepared the antigens from nonionic-detergent extracts of cell cultures in which as many as 50% of the cells have been induced to express MA. In some studies, the outer plasma membrane was specifically surface-labeled by iodination or borohydride reduction (Strnad *et al.*, 1979; Qualtière and Pearson, 1980; Mueller-Lantzsch *et al.*, 1980b). In other studies, plasma membranes or virus were purified prior to nonionic-detergent extraction (Thorley-Lawson and Edson, 1979; Thorley-Lawson, 1979). The major surface-labeled antigens are 220–350 × 10^3 d and 80–90 × 10^3 d. Both groups of antigens are heterogeneous in size, label with radioactive glucosamine, bind to ricin lectin, and elute with galactose. Either or both could consist of more than one polypeptide. There are two rather distinct components, 220–250 and 320–350, within the 220–350 × 10^3 d complex. The 220–250 component is slightly more abundant than the 320–350 in extracts of P3HR-1 virus or P3HR-1 cells, while the 320–350 component is much more abundant than the 220–250 on the surface of B95-8 virus or B95-8 cells. Two sets of monoclonal antibodies recognize both the 220–250 and 320–350 components (Hoffman *et al.*, 1980; Thorley-Lawson and Geilinger, 1980), indicating shared antigenicity through either common polypeptide or oligosaccharide components. These monoclonal antibodies also neutralize virus, confirming that the 220–350 × 10^3 d complex is on the outer surface of the virus. With the reactivity demonstrated by human sera to these antigens, it is likely that the 220–350 complex is an important neutralizing antigen in natural human infection. The 80–90 × 10^3 d glycoprotein complex may also be an important neutralizing antigen. Surface labeling indicates that it is also on the outer surface of the virus envelope. Convalescent human sera react with antigenic determinents of the 80–90 × 10^3 d complex.

Superinfection of latently infected Raji cells with the P3HR-1 virus induces the synthesis of both the 220–350 and 80–90 × 10^3 d complexes (Qualtière and Pearson, 1980; Mueller-Lantzsch *et al.*, 1980a,b). A 70 × 10^3 d polypeptide appears on the cell surface within 2 hr of P3HR-1 superinfection and before the 80–90 × 10^3 d polypeptide can be detected. Over the ensuing 24–48 hr, the relative intensity of heterogeneously migrating polypeptide species in the size range 80–90 × 10^3 d increases, while the amount of 70 × 10^3 d polypeptide remains constant. The 70 × 10^3 d polypeptide is not found on the surface of the virus envelope. These data suggest that the 70 × 10^3 d protein is either a precursor to the 80–90 × 10^3 d glycoprotein or an earlier virus polypeptide that binds to the cellular plasma membrane and is not incorporated into the viral envelope. In one study, the synthesis of the 70 and 80–90 × 10^3 d polypeptides was unaffected by the use of cytosine arabinoside to inhibit viral DNA synthesis. In other studies, the 80–90 × 10^3 d complex was reduced by treatment with phosphonoacetic acid, a better-characterized inhibitor of EBV DNA synthesis.

The 43 × 10^3 d envelope polypeptide is labeled less heavily by surface iodination and is less evident in immunoprecipitates with human sera than the 220–350 and 80–90 × 10^3 d complexes (Mueller-Lantzsch *et al.*, 1980a,b; Thorley-Lawson, 1979; Thorley-Lawson and Edson, 1979; Strnad *et al.*, 1979; Qualtière and Pearson, 1980; Kallin *et al.*, 1979). A 130 × 10^3 d polypeptide has also been detected as a minor component in immunoprecipates of glucosamine-labeled cell extracts. The 140–150 × 10^3 d membrane matrix or tegumentary protein is usually not evident in immunoprecipitates of surface-labeled proteins, but can be clearly identified in immunoprecipitates of ^{35}S-labeled polypeptides with human antisera. Trypsin treatment of infected cells before surface iodination exposes the 140–150 × 10^3 d polypeptide to iodination. The exposure of this protein after trypsin treatment suggests that it is probably a membrane matrix protein rather than an exterior-membrane or tegument protein.

The 150–160 × 10^3 d polypeptide, the major component of the nucleocapsid, is also the major capsid component precipitated by immune human sera (Mueller-Lantzsch *et al.*, 1980a,b; Kallin *et al.*, 1979). This polypeptide is probably an important component of the viral capsid antigen (VCA) complex detected by immunofluorescence in the nucleus and cytoplasm of permissively infected cells (G. Henle and W. Henle, 1966; Mayyasi *et al.*, 1967).

Little is known about the biochemistry of the antigenic determinants or structural polypeptides of the primate EBV-related viruses herpesvirus pan (HVPan) and herpesvirus papio (HVPapio). Human and primate sera that have activity in homologous assays cross-react extensively in heterologous VCA, MA, and neutralization tests, suggesting that EBV, HVPan, and HVPapio have similar antigenic determinants on their capsids and envelopes (Gerber *et al.*, 1976b, 1977; Neubauer *et al.*, 1979; Falk *et al.*, 1976).

D. Viral Deoxyribonucleic Acid

Comparative analysis of EBV DNAs is a particularly important issue in EBV research. Because of the nonpermissive outcome of EBV infection *in vitro*, more of the phenotypic properties of the genome are cryptic in culture than is the case with other herpesviruses. Therefore, elucidation and comparison of the phenotypic properties of EBV isolates are more difficult and less precise. Further, although seroepidemiological studies indicate that infection with EBV is prevalent in all human populations, the virus has a unique disease association with Burkitt lymphoma and nasopharyngeal cancer. Burkitt lymphoma is a distinctive endemic disease in equatorial Africa. EBV is almost invariably found in every African Burkitt-tumor cell (Reedman *et al.*, 1974), while elsewhere EBV is not a regular passenger in malignant B cells. EBV is also almost invariably present in anaplastic nasopharyngeal-cancer cells (zur Hausen *et al.*, 1970; Wolf *et al.*, 1975; Klein *et al.*, 1974). This disease occurs worldwide in association with EBV (Pagano *et al.*, 1975; Desgranges *et al.*, 1975). However, there is endemic clustering of cases in southern China (de Thé, 1979). One hypothesis that could explain the restricted geography of Burkitt lymphoma and the high incidence of nasopharyngeal cancer in southern China is geographic strain variation among EBV isolates (Kieff and Levine, 1974). Other hypotheses that recognize other infectious, nutritional, or human genetic or immune cofactors could account for some of the epidemiological findings. However, the possibility that there may be biologically significant differences among EBV strains is still an important issue in studies of EBV DNA.

The DNA of the B95-8 isolate of EBV has been the prototype for detailed restriction-endonuclease mapping, for cloning into *Escherichia coli* plasmids and into bacteriophase lambda, and for nucleotide sequencing (Given and Kieff, 1978, 1979; Given *et al.*, 1979; Kieff *et al.*, 1979;

Dambaugh, 1980; Dambaugh *et al.*, 1980a,b; Skare and Strominger, 1980). Comparative restriction-endonuclease maps of eight other EBV isolates have been derived (Raab-Traub *et al.*, 1980; Bornkamm *et al.*, 1980; Heller *et al.*, 1981a). DNA fragments from two other EBV isolates, AG876 and W91, have been partially cloned into plasmid vectors (Raab-Traub *et al.*, 1980; Dambaugh, 1980). AG876 is a culture of an African Burkitt-lymphoma biopsy (Pizzo *et al.*, 1978). W91 is a clone of marmoset cells infected with an African Burkitt-lymphoma EBV isolate (Miller *et al.*, 1976). These studies of EBV DNAs provide most of our current knowledge of the similarity and variation among EBV isolates.

1. General Features

The DNA of EBV is a linear, double-stranded molecule of approximately 170×10^3 nucleotide base pairs (bp) (Pritchett *et al.*, 1975a,b; Hayward and Kieff, 1977). The density of EBV DNA in neutral CsCl is $1.718 \text{ g} \cdot \text{cm}^{-3}$ (Wagner *et al.*, 1970; Jehn *et al.*, 1972; Pritchett *et al.*, 1975a). Assuming that the DNA consists entirely of unmodified bases, the overall base composition that corresponds to this density is 57% guanine plus cytosine (G+C) and 43% adenine plus thymine. The nucleotide composition of EBV DNA has not been directly determined. Comparison of the restriction endonuclease fragments of virion DNA digested with the isoschizomers *Msp*I and *Hpa*II indicates that CG residues are not methylated in virion DNA. The DNA in most virus preparations from chronically infected cultures is extensively and randomly nicked by endonucleases (Given *et al.*, 1979). As with HSV, as much as 50% of the single strands of nascent virion DNA is intact, as shown by alkaline sucrose sedimentation (Pritchett *et al.*, 1975a).

The general structure of EBV DNAs is shown in Fig. 3 (Dambaugh *et al.*, 1980b). There is a terminal repeat (TR) at both ends of the DNA (Given and Kieff, 1978; Given *et al.*, 1979; Kintner and Sugden, 1979). TR is 500 bp long. The repetitions are tandem and direct and vary in number at each end. There are usually 4–8 copies per end, although some ends have as many as 12 copies. TR is in the same orientation at both ends of the DNA. There is no homology between TR and internal DNA. Although the functions of TR are not fully known, one recognized role is to facilitate circularization of EBV DNA following infection (Lindahl *et al.*, 1976). Inside the infected cells, TRs at opposite ends of the DNA become covalently linked and can be cloned as a covalently joined fragment (Dambaugh *et al.*, 1980a,b). TR could facilitate circularization by base-pairing between the complementary sequences at both ends or by repeating a sequence that could bind a bivalent protein or protein complex. It is not known whether a protein is involved in circularization of the EBV genome. The 5′ ends of virion DNA are susceptible to lambda-exonuclease digestion, indicating that there is no protein covalently linked to the 5′ terminus (Hayward and Kieff, 1977).

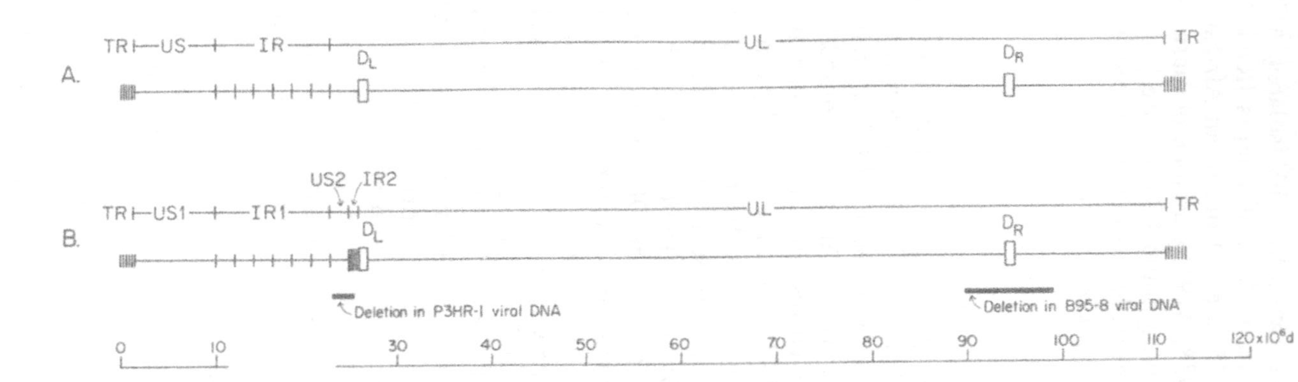

FIGURE 3. Schematic map of the structural features of prototype EBV virion DNA. (A) An early version of a schematic map utilized in Figs. 4C, 5A, 6, and 7. (B) An updated structural map that includes newly recognized features of EBV DNA structure. (TR) Variable number of direct, tandem repeats of a 500-bp sequence present at both ends of the molecule. (US1) [US in (A)] A 14 × 10^3 bp short unique sequence region that lies between TR and IR1. (IR1) [IR in (A)] The 3071-bp sequence present in variable of nonintegral numbers of direct, tandem repeats in different EBV isolates. Because of this variability, the map scale in this region is interrupted in (B). (IR2) A newly discovered internal repetition of a 123-bp sequence found in 11 direct, tandem copies in B95-8 DNA. (US2) A 3000-hp ultrashort unique sequence region that lies between IR1 and IR2. As indicated by the dark horizontal bar, most of US2 and all of IR2 are deleted from P3HR-1 viral DNA. D_L and D_R define a sequence of approximately 2000 bp that is duplicated near the left and right ends of UL, the 135 × 10^3 bp long unique sequence region. It is not known at present whether IR2 is represented in D_R. The long horizontal bar indicates a deletion from B95-8 viral DNA of a 13 × 10^3 bp DNA segment that includes D_R.

In addition to TR, EBV DNA contains two nonhomologous internal repeats, a long internal repeat designated IR1 (previously designated IR) and IR2, a short internal repeat (Kieff *et al.*, 1979; Given and Kieff, 1978; Given *et al.*, 1979; Rymo and Forsblum, 1978; Hayward *et al.*, 1980; Cheung and Kieff, 1981, 1982; Dambaugh and Kieff, 1982). IR1 and IR2 separate the unique sequences of EBV DNA into a 14×10^3 bp short unique region, US1, and a 135×10^3 bp long unique region, UL. An ultrashort unique region of approximately 3000 bp, US2, separates IR1 and IR2 (Dambaugh and Kieff, 1982). The location of these unique and repetitive sequences is indicated in Fig. 3. IR1 is 3071 bp and lies between US1 and US2, while IR2 lies between US2 and UL. The number of direct tandem repeats of IR1 varies among molecules of the same EBV isolate and among EBV isolates (Given and Kieff, 1979). Detailed restriction-endonuclease maps have been derived for IR1. The complete nucleotide sequences of IR1 and of the junctures between IR1 and US1 and US2 have recently been determined (Cheung and Kieff, 1981; Cheung and Kieff, 1982). IR1 is 67% G+C. IR1 has a single *Bam*HI site. Within the 250 nucleotides to the left of the *Bam*HI site in IR1 are sequences of 14 and 35 nucleotides that are directly repeated within the 250 nucleotides to the right of the *Bam*HI site. The transition from IR1 to US2 occurs at a point several hundred nucleotides short of completion of a full copy of IR1. The 14 and 35 bp direct repeats within IR1 are not deleted from the incomplete copy of IR1 adjacent to US2.

As described in Section III.B, IR1 encodes an abundant RNA in latently infected cells growth-transformed by EBV. There is a "TATAA" box within IR1 located 51 nucleotides to the 5′ end of an AUG. The repetition of the IR1 sequence in EBV DNA could reinforce the action of this "promoter." The transcription of RNAs from this repeated promoter would then result in a family of RNAs increasing in size by fixed increments. The expected family of RNAs would have a common 3′ end.

The second internal repetition in EBV DNA, IR2, is 123 bp long (Dambaugh and Kieff, 1982). IR2 is present as 11 direct, tandem repeats in the B95-8 EBV isolate, but may be present in as many as 13–15 copies in the W91 and AG876 isolates. The 1300 bp of US2 nearest to IR1 and the 600 bp of US2 closest to IR2 seem to be constant among most EBV isolates. The remaining 1100 bp of US2, located to the right of the single *Bam*HI site in US2, may be partially or completely deleted in some viral isolates. Nonetheless, there are no apparent differences in the biological properties of these isolates *in vitro*. Furthermore, some isolates from Africa and the United States have a full complement of US2 sequences while others do not. Thus, the presence of the sequences is not associated with a specific geography (King *et al.*, 1982).

The UL region is 135×10^3 bp. DNA mapping at $26–28 \times 10^6$ d at the left end of UL near the juncture with IR2 has homology to DNA mapping at $93–95 \times 10^6$ d near the right end of UL (Raab-Traub *et al.*, 1980; Heller *et al.*, 1981a). Duplication is present in all EBV isolates except B95-8, which has a deletion of 13 × 10^3 bp around the rightward duplication in UL (Raab-Traub *et al.*, 1978, 1980). Deletion of this region does not interfere with the ability of the B95-8 isolate to growth-transform cells in culture, to cause tumors in primates, or to replicate in some of the transformed cells *in vitro* (Miller *et al.*, 1976). Near the center of UL, at 63–66 × 10^6 d, the 1200 bp *Hinf*I fragment of *Bam*HI K has homology to a repeated sequence in human and mouse cell DNA (Heller *et al.*, 1982b,c). This sequence does not have homology to the Simian virus 40 origin of replication. The nucleotide sequence of this fragment is being determined in order to compare it with known repeated sequences in human and mouse DNAs.

2. Range of Variability among EBV DNAs

Comparison of the restriction-endonuclease fragments and the linkage maps of nine EBV DNAs reveals sufficient variability to specifically identify each of the DNAs (Given and Kieff, 1978; Raab-Traub *et al.*, 1980; Dambaugh *et al.*, 1980b; Heller *et al.*, 1981a; Bornkamm *et al.*, 1980; Fischer *et al.*, 1981). Although four of the DNAs are from Africans with Burkitt lymphoma, three from Americans with infectious mononucleosis, one from an Australian with myeloblastic leukemia, and one from a patient with nasopharyngeal cancer, no features have been identified that correlate with a specific geographic or pathological origin. Most of the variations observed among the DNAs are in the numbers of copies of repeat elements or could be the consequence of mutations at existing or potential restriction-endonuclease sites, since there is no discernible change in the order or size of the surrounding DNA sequences (Heller *et al.*, 1981a; Dambaugh *et al.*, 1980b). Similar kinds of differences have been observed with passage of the B95-8 virus in cord-blood cells (Raab-Traub *et al.*, 1980; King *et al.*, 1980; Dambaugh *et al.*, 1980b). While it is clear from these studies that the isolates do not segregate into discernible subtypes, it is premature to conclude that sufficient variability could not exist to account for the differences in the epidemiology of the diseases associated with EBV infection. Relatively minor changes in nucleotide sequence could have profound effects on expression of one or several genes of pathophysiological significance. The level of resolution of most of the comparative restriction-endonuclease studies of different EBV isolates is of the order of several hundred nucleotides. This level of investigation is barely adequate to detect the variability in the US2 region described in Section II.D.1. Similar variability has also been detected near the rightward duplication of DNA in UL (Raab-Traub *et al.*, 1980). The latter has not been investigated.

3. DNA of the P3HR-1 Isolate

P3HR-1 cells are a high-level virus-producing clone of cells derived from the Jijoye cell line (Hinuma *et al.*, 1967). The Jijoye cell line was established from an African Burkitt-tumor biopsy. As described in Section I, EBV from P3HR-1 cultures differs from the virus produced by Jijoye and from all other EBV isolates in its inability to growth-transform lymphocytes *in vitro*. This inability to induce growth transformation is not abolished by end-point dilution of P3HR-1 virus, suggesting that it is not a consequence of defective interfering particles (Miller *et al.*, 1974). Furthermore, the P3HR-1 virus can infect B lymphocytes and induce abortive or productive replication depending in part on the presence of a latent EBV genome in the infected cell (W. Henle *et al.*, 1970; Yajima and Nonoyama, 1976). The P3HR-1 line has been passaged for over a decade in continuous culture. Analysis of the DNA of the P3HR-1 virus indicates that it is heterogeneous (Hayward *et al.*, 1976; Hayward and Kieff, 1977; Sugden *et al.*, 1976; Delius and Bornkamm, 1978). Approximately half the molecules are similar in structure to standard EBV DNAs except for the deletion of almost the entire US2 region (Kieff *et al.*, 1979; King *et al.*, 1980, 1982; Heller *et al.*, 1981a). In contrast, Jijoye has more of the US2 region (King *et al.*, 1982). Since Jijoye virus can transform cells and P3HR-1 cannot, the data suggest that sequences within the US2 region are important in the initiation of transformation. One attempt to approach this issue has been to superinfect the latently infected, Burkitt-tumor-biopsy-derived Raji cell line with P3HR-1 virus in an attempt to obtain recombinants between P3HR-1 and the endogenous Raji EBV DNA (Fresen and zur Hausen, 1976; Fresen *et al.*, 1979; Yajima *et al.*, 1978). Small amounts of transforming virus are released following superinfection, and this DNA can be amplified by cellular growth transformation. The P3HR-1 and endogenous Raji genomes can be distinguished at several sites with two restriction endonucleases and presumably at other sites with additional enzymes (Heller *et al.*, 1981a).

At least two minor populations of P3HR-1 molecules can be differentiated by analysis of electron micrographs of partially denatured molecules (Delius and Bornkamm, 1978). The most common minor population appears to consist of a repeating unit of 35–40 × 10^6 d. Restriction-endonuclease analysis of P3HR-1 DNA reveals the presence of defective molecules in which noncontinguous parts of UL, US, and TR have recombined in new linkage arrangements (Heller *et al.*, 1981a). The defective molecules may be important in the induction of the permissive infection observed on superinfection of Raji cells with P3HR-1 virus. The defective molecules could achieve this effect by turning on late viral functions in a manner analagous to alpha or beta genes of HSV (Honess and Roizman, 1974) or by titrating a putative repressor of lytic infection.

4. Primate EBV-Related Agents

Many Old World primate species, including chimpanzees, baboons, gorillas, and orangutans, are known to be latently infected with herpesviruses that are antigenically cross-reactive with EBV (Gerber *et al.*, 1976b, 1977; Falk *et al.*, 1976; Goldman *et al.*, 1968; Gerber and Birch,

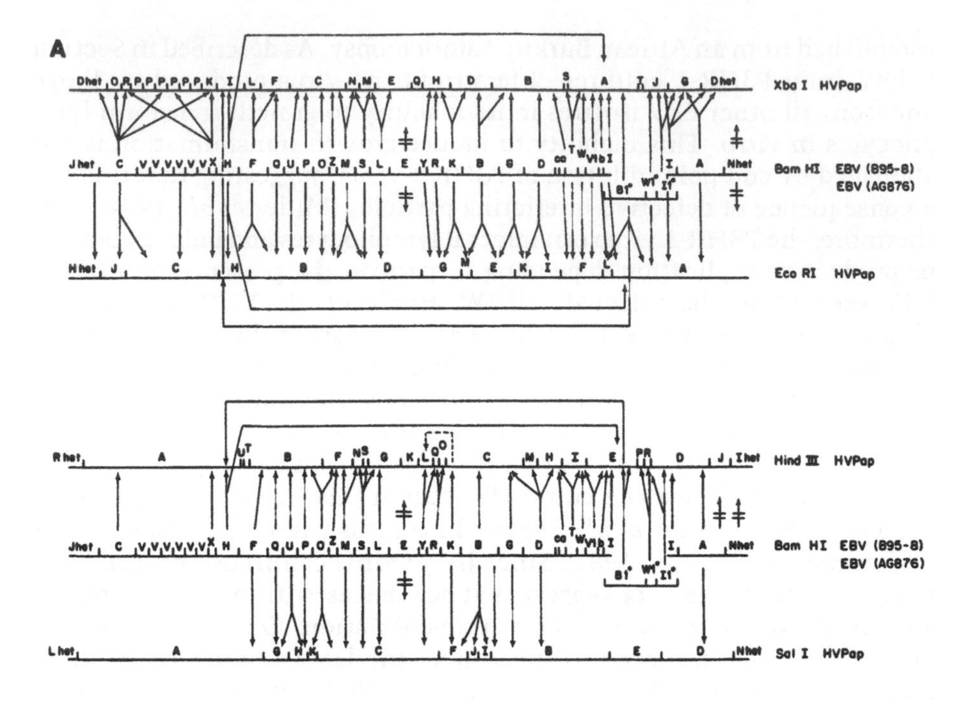

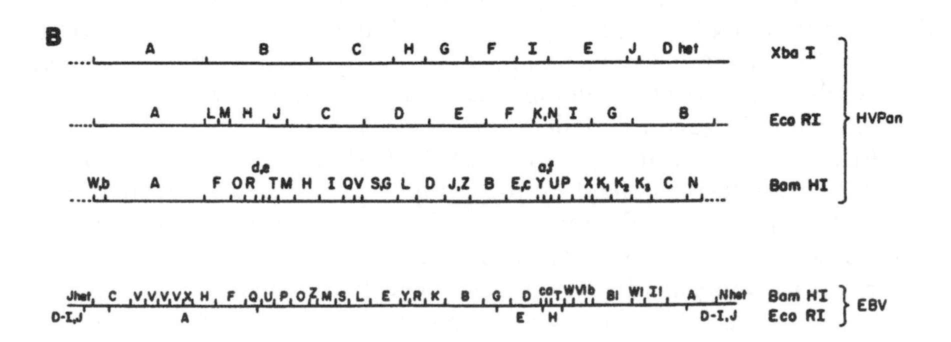

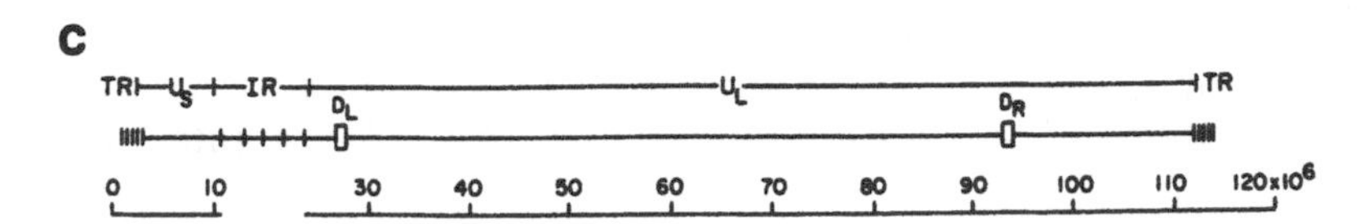

FIGURE 4. Colinearity of HVPapio, HVPan, and EBV DNAs. (A) Summary of homology between HVPapio and EBV DNAs. The *Bam*HI linkage map representative of a prototypical EBV genome is shown aligned with the *Xba*I, *Eco*RI, *Hind*III, and *Sal*I linkage maps of HVPapio DNA. Clones of the indicated *Bam*HI EBV DNA fragments from B95-8 and AG876 viruses were labeled *in vitro* and cross-hybridized to Southern blots of HVPapio DNA. The

1967; Stevens *et al.*, 1970; Landon *et al.*, 1968; Levy *et al.*, 1971; Rabin *et al.*, 1978, 1980). The viruses of chimpanzees and baboons, HVPan and HVPapio, respectively, have been studied most extensively. HVPan and HVPapio are endemic in their respective primate species. Like that of EBV, the host ranges of HVPan and HVPapio are restricted to primate B lymphocytes, in which these viruses induce growth transformation and intranuclear antigen (Gerber *et al.*, 1976b, 1977; Ohno *et al.*, 1979; Rabin *et al.*, 1980). Although limited cross-reactions have been detected, the intranuclear antigens specified by each virus have major non-cross-reactive determinants. The VCA, MA, and early antigens specified by each of these viruses cross-react extensively with antisera from the heterologous primate species.

Analyses of HVPapio and HVPan DNAs were undertaken to define the relationship of these viruses to EBV and to each other. EBV, HVPapio, and HVPan DNAs have 40% homology in each pairwise comparison (Gerber *et al.*, 1976b; Falk *et al.*, 1976; Heller *et al.*, 1982a). The size and organization of HVPapio and HVPan DNA are similar to those of EBV DNA (Heller *et al.*, 1981b, 1982a; Heller and Kieff, 1981; Lee *et al.*, 1981). Like EBV, DNA, HVPan and HVPapio DNAs are also 170 $\times$ 10^3 bp. Tandem direct repeats of a 3100 bp sequence, IR1, separates HVPapio and HVPan DNAs into a short, 14 $\times$ 10^3 bp, unique DNA segment, US, and a long, 135 $\times$ 10^3 bp, unique DNA segment, UL. In HVPapio and HVPan DNAs, as in EBV DNA, there is homology between DNAs at 26–28 and 93–95 $\times$ 10^6 d in UL. Restriction-endonuclease fragments from both ends of HVPapio DNA vary in the number of repeats of a 550 bp sequence, TR, which hybridize to each other. The data identifying the TR in HVPan DNA are much less complete. The terminal fragments are heterogeneous in size, suggesting that there are varying numbers of copies of the TR. Thus, both HVPan and HVPapio probably have a variable number of copies of direct tandem TRs similar to EBV DNA. The common features of EBV, HVPapio, and HVPan DNAs are shown in Fig. 4.

The extent of physical colinearity between EBV and HVPapio and HVPan DNAs was investigated by hybridizing labeled cloned fragments

arrows drawn between the EBV and HVPapio maps indicate the presence of homologous sequences in the two DNAs and the conservation in HVPapio DNA of structural features characteristic of EBV DNA. (↟) Indicates that a labeled EBV probe did not hybridize to HVPapio DNA. (B) Summary of homology between HVPan and EBV DNAs. An analysis similar to that shown in (A) was conducted for HVPan DNA. The *Xba*I, *Eco*RI, and *Bam*HI linkage maps of HVPan DNA are aligned above the *Bam*HI and *Eco*RI linkage maps of EBV DNA to indicate the colinear arrangement of homologous sequences and structural features in the two DNAs. The dashed lines in the HVPan maps indicate a lack of homology between HVPan DNA and the terminal repeat (TR) and unique sequences to the left of 6 $\times$ 10^6 d and to the right of 108 $\times$ 10^6 d in EBV. The actual numbers and sizes of HVPan DNA fragments near the termini that lack homology to EBV DNA are unknown. (C) Structural map of EBV, HVPapio, and HVPan DNAs. The various structural freatures shown are described in the Fig. 3 caption and in the text. (Heller *et al.*, 1981b, 1982a; Heller and Kieff, 1981).

of EBV DNA, ranging in complexity from 0.8–6.4 × 10^6 d and representing the entire EBV genome, to Southern blots of fragments of HVPapio or HVPan DNA (Heller and Kieff, 1981; Heller *et al.* 1982a). The labeled EBV DNA fragments hybridize to HVPapio and HVPan DNA fragments that correspond to the same map positions as the EBV DNA fragments (Fig. 4). EBV DNA fragments containing the duplicated region in UL of EBV DNA hybridized to both regions in UL of HVPapio and HVPan DNAs, indicating that the duplicated sequences are conserved among these DNAs. One region of nonhomology was detected in the UL of EBV and HVPapio DNA at 54–59 × 10^6 d. EBV and HVPan DNA have homology in this region. There was no homology detected between the TR of EBV and the TRs of HVPapio or HVPan. There was also no homology between the TR of HVPapio and the TR of HVPan. The lack of homology among the TRs of EBV, HVPan, and HVPapio and the difference in size of the TRs of EBV and HVPapio suggest that the specific nucleotide sequence of the TRs is not highly conserved among the primate EBV-related agents. Two inferences could be drawn from the lack of TR homology: First, the greater sequence divergence of TR favors the hypothesis, considered in Section II.D.1, that TR may function only to facilitate circularization of the DNA. If TR does not encode an essential protein, its sequences may be subject to greater variation than other regions of the genome. Second, divergence of TR might also be expected if TR is a region of the viral genome that may be involved in interaction with cellular components that differ among human, chimpanzee, and baboon B lymphocytes.

III. TRANSFORMATION AND LATENCY WITHIN B LYMPHOCYTES

After infection with Epstein–Barr virus (EBV), some and probably most B lymphocytes replicate perpetually in culture (W. Henle *et al.*, 1967; Pope *et al.*, 1968b, 1971; Gerber *et al.*, 1969; Sugden and Mark, 1977; Henderson *et al.*, 1977a,b, 1978). The continued presence of the EBV intranuclear antigen (EBNA) and the EBV genome indicates that the cells are infected. The infection is latent in most or all of the cells. The frequency of spontaneous permissive infection varies with cell type and ranges from 0 to 1% among cell lines derived by infection of neonatal human cells, from 0 to 3% among adult human cell lines, and from 0 to 10% among marmoset cell lines (Nilsson, 1971; Nilsson *et al.*, 1972; Gerber *et al.*, 1976a; Miller and Lipman, 1973a,b). Thus, on a statistical basis, marmoset lymphocytes tend to be more permissive than adult human lymphocytes, which are in turn more permissive than neonatal human lymphocytes. Within each category, some cultures are completely nonpermissive of viral replication. Maintenance of partially permissive cultures in rich media with frequent feedings usually results in rapid cell growth and selection of cells that are nonpermissive for viral replication.

In almost all latently infected cultures, there are two classes of new antigenic activity. One antigen is detectable on the cell surface using immune T lymphocytes from patients recovering from primary EBV infection. This antigen, lymphocyte-detected membrane antigen (LYDMA), cannot be detected with immune sera and has not been further characterized (Svedmyr and Jondal, 1975). The second antigenic activity is the intranuclear antigen, EBNA (Reedman and Klein, 1973). EBNA, and presumably also Lydma are present in latently infected cells *in vivo*, including nonmalignant B lymphocytes, Burkitt tumor lymphoblasts, and nasopharyngeal cancer cells. In contrast, none of the antigens associated with abortive or productive viral infection has been identified in latently infected cells in human tissues, including latently infected normal peripheral B lymphocytes, malignant Burkitt-tumor cells, and nasopharyngeal-cancer cells (Reedman *et al.*, 1974; Klein *et al.*, 1974, 1976; Robinson *et al.*, 1981). EBNA continues to be expressed when these cells are explanted and maintained *in vitro* or grown in nude mice. Increasing evidence suggests that EBNA is a complex of polypeptides. It is not known whether EBNA and LYDMA share components.

Phosphonoacetic acid (PAA), phosphonoformic acid (PFA), and acycloguanosine (ACG) have been shown to inhibit EBV DNA replication in partially permissive cultures (Summers and Klein, 1976; Thorley-Lawson and Strominger, 1976; Colby *et al.*, 1980). Under the conditions used, these drugs preferentially inhibit herpesvirus DNA polymerases, suggesting that EBV DNA is replicated by a viral polymerase during productive infection (Datta *et al.*, 1980). Although these drugs have no effect on the number of copies of EBV DNA in latently infected, growth-transformed cells, they can inhibit the establishment of latent infection and growth transformation. These latter data suggest that EBV-mediated DNA replication is a necessary condition to achieve latent infection and growth transformation and that once this state is achieved, the viral DNA is replicated by cellular polymerase that is less sensitive to PAA, PFA, and ACG. The relative amount of viral DNA in synchronized cultures remains constant over the cell cycle. A slight increase during early S phase suggests that viral DNA is replicated early during the synthesis of latently infected cell DNA (Hampar *et al.*, 1974c).

The term "restringent" has been used to denote the latent state of EBV infection in the growth-transformed B lymphocytes (Orellana and Kieff, 1977; Powell *et al.*, 1979; King *et al.*, 1980, 1981). The rationale for coining a new term was that no existing term adequately described this state. The infection is nonpermissive for viral replication, but it is not entirely latent in that the viral genome is actively transcribed, viral RNA is processed and transported, and viral protein or proteins are produced. Some viral gene product is presumed to be necessary for initiating and maintaining cellular growth transformation. Other possibilities, including a *cis*-acting effect of the viral genome, have not been rigorously excluded but seem unlikely, since transformation is very sensitive to UV inactivation of the virus (Henderson *et al.*, 1978). To the extent that maintenance of growth transformation requires active expression of the EBV genome, viral gene expression during primary infection and subsequent growth transformation may be different from expression in latent infections of B lymphocytes transformed prior to infection, in hybrids between EBV-infected lymphocytes and nonlymphoid cells, and in nonlymphoid cells infected by microinjection or by receptor transplantation (Glaser and O'Neill, 1972; Fresen and zur Hausen, 1977; Graessman *et al.*, 1980; Volsky *et al.*, 1980). Whether the term restringent should be used for describing a latently infected African Burkitt-tumor-cell culture is therefore an open issue, since it is uncertain whether the typical infected tumor cell was initially growth-transformed by EBV. Furthermore, the chromosomal change that accompanies malignant transformation (Manolov and Manolova, 1972) may be associated with a change in the state of EBV infection.

A. Viral Deoxyribonucleic Acid in Latently Infected Cells

Cultures of latently infected lymphocytes explanted from healthy seropositive donors, cultures of latently infected lymphocytes that were infected with EBV *in vitro*, and cultures of Burkitt-tumor biopsies have been examined for the persistence of EBV DNA. In most instances, the latently infected cells contain multiple copies of the EBV genome (zur Hausen and Schulte-Holthausen, 1970; Nonoyama and Pagano, 1971, 1973; Nonoyama *et al.*, 1973; Kawai *et al.*, 1973; Kieff and Levine, 1974; Pritchett *et al.*, 1976; Sugden *et al.*, 1979). DNA–DNA reassociation kinetics indicate that within the limits of accuracy of these analyses (5–10%), the entire EBV genome is present. In only two instances have studies indicated that an incomplete EBV genome is present in latently infected cells. The kinetics of DNA–DNA reassociation between labeled viral DNA and Namalwa-cell DNA indicate that the Namalwa-cell line contains one or two copies of an incomplete EBV genome (Pritchett *et al.*, 1976). However, labeled Namalwa-cell DNA hybridizes to all the larger *Eco*RI fragments of EBV DNA, indicating that Namalwa contains sequences spread over at least 80% of the EBV genome (Heller and Kieff, unpublished). In a second series of studies in which labeled viral DNA was hybridized to DNA from nasopharyngeal-carcinoma biopsies, the reassociation kinetics suggested that these tissues also contain less than the complete viral genome (Pagano *et al.*, 1975; Pagano and Huang, 1976). Since nasopharyngeal carcinoma is an infiltrative malignant process, the tumor cells are mixed with normal stromal cells and the number of EBV DNA copies per cell is lower than in Burkitt-tumor tissue, which is a more uniform tumor composed primarily of infected lymphocytes. As with Namalwa cells, observation of a change in reassociation kinetics could be expected as a consequence of degradation of the probe DNA with long intervals of hybridization. It was not demonstrated in either study that the residual labeled single-stranded viral DNA, which did not hybridize to Namalwa or nasopharyngeal-carcinoma DNA, was still able to hybridize to viral DNA. Neither was it demonstrated that the unhybridized viral DNA mapped to specific "lost" region(s) of the viral genome.

The Raji cell line, a culture of nonpermissively infected Burkitt-tumor cells, has been the prototype for many studies of the physical state of viral DNA in nonpermissively infected cells. By complementary RNA (cRNA) filter hybridization and by DNA–DNA reassociation kinetics, Raji cells are estimated to carry 50 copies of viral DNA. The viral DNA in Raji cells is not covalently linked to cell DNA, although it may be loosely associated with cellular chromatin (Nonoyama and Pagano, 1972). Some of the episomal EBV DNA in Raji cells sediments at 65 S and 100 S in neutral glycerol velocity gradients, the expected values for open and supercoiled circles equal in length to EBV DNA (Lindahl *et al.*, 1976). The 100 S DNA sediments at 300 S in alkaline glycerol gradients, indicating that it contains no nicks, gaps, or ribonucleotides (Lindahl *et al.*, 1976). X-irradiation converts the 100 S DNA to an open 65 S form that is identical in length to viral DNA (Lindahl *et al.*, 1976). The Raji DNA circles are formed by covalent joining of the termini. The joining of the terminal restriction-endonuclease fragments of Raji DNA into a new juncture fragment has been confirmed in Southern blots of Raji DNA (Heller *et al.*, 1981a). Viral DNA in Raji cells is partially resistant to cleavage by *Hpa*II, but is sensitive to the isoschizomer *Msp*I, while virion DNA is sensitive to both enzymes (Desrosiers *et al.*, 1979; Kinter and Sugden, 1981; Fennewald and Kieff, unpublished observations). This strongly suggests that Raji DNA is partially methylated at CG sites and raises the possibility that methylation may play a role in control of transcription of intracellular Raji DNA. An attempt has been made to evaluate this hypothesis by examining the differential sensitivity of Raji EBV long-internal-repeat (IR1) and *Bam*HI T DNA to *Hpa*II and *Msp*I (Fennewald and Kieff, unpublished observations). IR1 *Bam*HI T, which does not encode messenger RNA in Raji cells. Both regions of the endogenous Raji DNA were partially resistant to *Hpa*II, indicating partial methylation. The effects of chemical induction of EBV expression in Raji cells include enhanced messenger RNA (mRNA) transcription (King *et al.*, 1981) and a change in the pattern of methylation of *Bam*HI T DNA.

Circular episomal DNA the length of linear EBV DNA has been found in other nonpermissively infected cells and in Burkitt-tumor tissue (Adams *et al.*, 1977; Kaschka-Dierich *et al.*, 1976, 1977). The restriction-endonuclease maps of intracellular EBV DNAs from a variety of sources, including Burkitt-tumor-cell cultures and cultures of latently infected cells from seropositive patients, are similar to those of virion DNA except that the two terminal fragments of linear virion DNA are joined into a single fragment in intracellular DNA (Heller *et al.*, 1981a). Changes in EBV DNA during establishment of latent infection were investigated by comparing B95-8 virion DNA with intracellular EBV DNA from clones of normal neonatal lymphocytes latently infected and growth-transformed by the B95-8 virus (King *et al.*, 1980; Raab-Traub *et al.*, 1980; Dambaugh *et al.*, 1980b). Several differences were found between the restriction-endonuclease profiles of virion DNA and the intracellular DNA from the transformed cells: First, the two terminal *Eco*RI fragments of viral DNA migrate as a single larger fragment in intracellular DNA. The larger size of the terminal fragments in intracellular DNA is due in part to the covalent joining of both ends of virion DNA into a single fragment in intracellular DNA. The putative joined ends from an *in vitro* growth-transformed cell line were cloned into Charon 4A and shown to consist of sequences from both ends of virion DNA (Dambaugh *et al.*, 1980a). Second, there is variation in the size of the joined end fragments from several independently derived latently infected, growth-transformed neonatal lymphocytes (Raab-Traub *et al.*, 1980). This variation is in part attributable to fluctuation in the number of terminal repeats (TRs) in the joined end fragments. Some of the variation may be due to the loss of copies of TR during cloning. The latter phenomenon occurs commonly during propagation in *E. coli* of recombinant clones of the joined *Eco*RI termini. Digestion of the DNA of a single such clone of Charon 4A yields molecules varying in the number of repeats of TR (Dambaugh *et al.*, 1980a). Third, in one cell line, the fragments containing the sequences from the right terminus of EBV DNA are $12–25 \times 10^3$ bp larger than the joined terminal fragment in other cell lines. This larger size could be a consequence of 24–50 more copies of TR, but could also be due to rearrangement or duplication of other viral DNA. Fourth, the size of the *Eco*RI A fragment varies among different cell lines from 23 to 37×10^6 d, probably because of differing numbers of repeats of IR1. Fifth, in one of the cell lines, the *Eco*R1 I and J fragments lack the *Eco*R1 site between them. The size of the new *Eco*R1 fragment is equal to the combined size of *Eco*R1 I and J, suggesting that there is no major change in the DNA.

The restriction-endonuclease maps of EBV DNA were investigated in three nonpermissively infected cell lines that have been passaged for several years in culture and rapidly grown for several months prior to harvest of the intracellular viral DNA (Heller *et al.*, 1981a). Defective DNA molecules were found to accumulate as a minor population in each of the DNAs. The defective molecules consisted of linked sequences from various noncontiguous regions of the EBV genome. Many of these defective molecules resemble the defective populations in P3HR-1 virion DNA, suggesting a common mechanism in the origin of both types of defectives. In particular, DNA in the *Eco*R1 B fragment was frequently linked to DNA that maps elsewhere in standard EBV DNA, suggesting that there may be an origin of DNA replication in this region. As described in Section II.D.1, one fragment from the region is homologous to a repeated sequence in cell DNA but is not homologous to the Simian virus 40 (SV40) origin of replication. Other atypical linkages included unusual juncture fragments between the left and right ends of the DNA. None of the unusual fragments has been cloned for precise mapping and sequencing.

In nonpermissively infected cells, the EBV genome is replicated by a cellular polymerase. The generation of defective viral DNA molecules could be a consequence of replication of the viral DNA by the cellular polymerase. Rapid cell growth could enhance the generation of defective viral molecules by further stressing the cellular enzymes that are involved in viral DNA replication.

There have been a number of attempts to determine whether EBV DNA is integrated into cell DNA of Raji cells (Nonoyama and Pagano, 1972; Adams *et al.*, 1973). Raji cells were chosen because they have been the prototype for studies of latently infected Burkitt-tumor-cell lines. However, Raji cells contain numerous circular episomal copies of EBV DNA. A small fraction of EBV DNA from Raji cells bands at a lower density than virion DNA, suggesting linkage to cell DNA (Adams *et al.*, 1973). On shearing, the less dense EBV DNA rebands at a higher density, further suggesting previous linkage to cell DNA. The data do not exclude the possibility that the small fraction of viral DNA initially banded at lower density as a consequence of tightly adherent protein or trapping by cell DNA. Studies of a malignant human B-cell line infected with EBV *in vitro* suggest that at least part of the EBV genome can integrate (Andersson-Anvret and Lindahl, 1978). In these experiments, the cells contained one or a few EBV DNA copies per cell, and the density of EBV DNA was near the density of cell DNA. The possibility that cytosine methylation or recombinant forms of viral DNA could lower the density of intracellular EBV DNA was not excluded in these studies. Whether integration of EBV DNA into cellular DNA is necessary or even associated with cellular growth transformation remains an open issue. One approach would be to demonstrate that a new restriction fragment of EBV DNA appears after growth transformation and that the fragment consists of both cellular and viral DNA sequences. Likely sites within the viral genome for covalent linkage to cellular DNA include TR and the site in *Bam*HI K that has homology to cellular DNA. The latter could provide a site for homologous recombination into repetitive cell sequences. If a new restriction-endonuclease fragment could be identified, recovery of the fragment by molecular cloning and sequencing of the region(s) of linkage between cellular and viral DNA should yield insight into the molecular mechanisms of recombination between EBV and cell DNA. A second, complementary approach would be to demonstrate incorporation of viral DNA into a specific cellular chromosome by *in situ* hybridization. Identification of a specific chromosomal site might give some indication of the mechanism of a *cis*-acting effect. Finally, it is possible that EBV DNA may not integrate or may integrate on an irregular basis and that cellular growth transformation and synthesis of progeny EBV DNA in latently infected cells could be accomplished by episomal EBV DNA alone.

B. Viral Ribonucleic Acids in Latently Infected, Growth-Transformed Cells and in Burkitt-Tumor Cells *in Vitro*

Analyses of viral RNA from latently infected, growth-transformed B lymphocytes or from latently infected African Burkitt-tumor lymphoblasts grown in culture indicate that there is extensive transcription of the viral genome (Hayward and Kieff, 1976; Pritchett *et al.*, 1975b; Orellana and Kieff, 1977; Powell *et al.*, 1979; King *et al.*, 1980, 1981). RNA from either class of latently infected B lymphocytes hybridizes to as much as 30% of randomly labeled EBV DNA. Only stable RNAs are detected in this kind of analysis. If these stable RNAs are encoded by one DNA strand or the other, but never by both complementary strands, the data would indicate that stable RNA is encoded by 60% of the single-stranded length of EBV DNA. To evaluate the possible significance of transcripts from complementary strands, RNA was permitted to anneal to itself under the same conditions used in RNA–DNA hybridizations. The RNA was then treated with pancreatic ribonuclease. The putative double-stranded RNA was then denatured and hybridized to labeled viral DNA. The RNA hybridized to less than 5% of EBV DNA, indicating that only a small part of the RNA that hybridizes to EBV DNA is encoded by complementary DNA strands (Powell, Dambaugh, and Kieff, unpublished).

The polyadenylated RNA fraction from latently infected, growth-transformed cells or from cultures of Burkitt-tumor lymphoblasts hybridizes to 20% of EBV DNA (Orellana and Kieff, 1977; Powell *et al.*, 1979; King *et al.*, 1980, 1981). The hybridization is initially rapid and reaches an early plateau, indicating that the polyadenylated RNA fraction is enriched for RNA encoded by this 20% of EBV DNA. Polyadenylated RNA does not hybridize further to viral DNA, indicating specific enrichment for those RNAs that selectively accumulate in the polyadenylated RNA fraction.

Polyribosomal RNA or cytoplasmic polyadenylated RNA from latently infected, growth-transformed cells or from cultures of Burkitt lymphocytes hybridizes to 10% of EBV DNA, indicating that a subset of the polyadenylated RNA selectively accumulates on polyribosomes (Hayward and Kieff, 1976; Pritchett *et al.*, 1975b; Orellana and Kieff, 1977; Powell *et al.*, 1979; King *et al.*, 1980, 1981). Again, additional RNAs are not detected with longer hybridizations. From the kinetics of hybridization of latently infected cellular RNAs to viral DNA, viral RNA is estimated to be 0.06% of the cellular nuclear RNAs, 0.06% of the polyadenylated RNA, and 0.003% of the polyribosomal RNA. Earlier lower estimates of the abundance of viral RNA were due to degradation of the RNA during extraction and failure to correct hybridization kinetics for the effect of DNA–DNA and RNA–DNA hybridization involving the IR1 sequence in EBV DNA.

The results obtained from the kinetics of hybridization of cellular

RNAs to viral DNA indicate that nuclear RNAs should map more broadly in EBV DNA than should polyadenylated RNA and that polyribosomal RNA should be restricted to about 20% of the length of EBV DNA. The studies that have been done are consistent with these expectations (Powell *et al.*, 1979; King *et al.*, 1980, 1981). The most precise mapping data available are for RNA from two cell lines; Raji, a Burkitt-tumor-cell line, and IB-4, a clone of neonatal cells transformed *in vitro* by B95-8 virus (King *et al.*, 1980, 1981; van Santen *et al.*, 1981). The results are remarkably similar and consistent with earlier, less precise, data obtained with RNA from the Namalwa Burkitt-tumor-cell line (Powell *et al.*, 1979; Kieff *et al.*, 1978). Much of the increased precision of the more recent analysis is due to the extensive use of cloned restriction-endonuclease fragments of EBV DNA (Dambaugh *et al.*, 1980a). Polyribosomal RNA is encoded by 40% of IR1 and by at least 25% of the adjacent ultrashort unique region (US2)–short internal repeat (IR2) region (*Bam*HI X and H fragments) and by DNA that maps at 63–66 and 110–115 $\times$ 10^6 d in EBV DNA (King *et al.*, 1980, 1981; van Santen *et al.*, 1981). These results are summarized in Fig. 5. Nuclear RNA is encoded by larger fractions of those viral DNA fragments that encode polyribosomal RNA and by other fragments of EBV DNA, including the *Eco*RI F and G fragments. The finding that nuclear RNAs are encoded by larger fractions of the fragments that encode polyribosomal RNA suggests that the RNAs are derived from larger nuclear precursors by cleavage or splicing. However, the possibility that the additional complexity of nuclear RNA may be due to transcripts that are not precursors for messenger RNA has not been formally excluded. The sizes of the nuclear RNAs encoded by the IR1–US2–IR2 region are larger than the RNAs, as would be expected for precursor molecules (van Santen and Kieff, in preparation). Some of the nuclear RNAs are smaller than mRNA and are therefore not precursors of mRNA. The latter may be stable splicing or cleavage products of maturing RNAs. Further, the nuclear RNA encoded by *Eco*RI F and G is unlikely to be part of a giant precursor RNA, since *Eco*RI F and G map approximately 30,000 bp away from the other fragments that encode polyribosomal RNA. Nuclear RNAs of sufficient size to extend from the IR1–US2 or 63–66 $\times$ 10^3 d regions through *Eco*RI F or G have not been detected. In an attempt to detect introns, polyribosomal RNA was hybridized to Southern blots of fragments of recombinant clones of EBV DNA from the IR1–US2–IR2 region. If there are large introns in EBV transcripts, an internal DNA fragment might not hybridize to polyribosomal RNA. Most of the DNA fragments tested were larger than 1000 bp, and all hybridized to polyribosomal RNA.

The size of the cytoplasmic polyadenylated RNAs encoded by the short unique region (US1), by IR1–US2 by the 63–66 $\times$ 10^6 d region of the long unique region (UL), and by the 110–115 $\times$ 10^6 d region of UL was determined by Northern-blot analysis (van Santen *et al.*, 1981). The region of US1 from 5 to 7 $\times$ 10^6 d encodes two low-abundance RNAs of

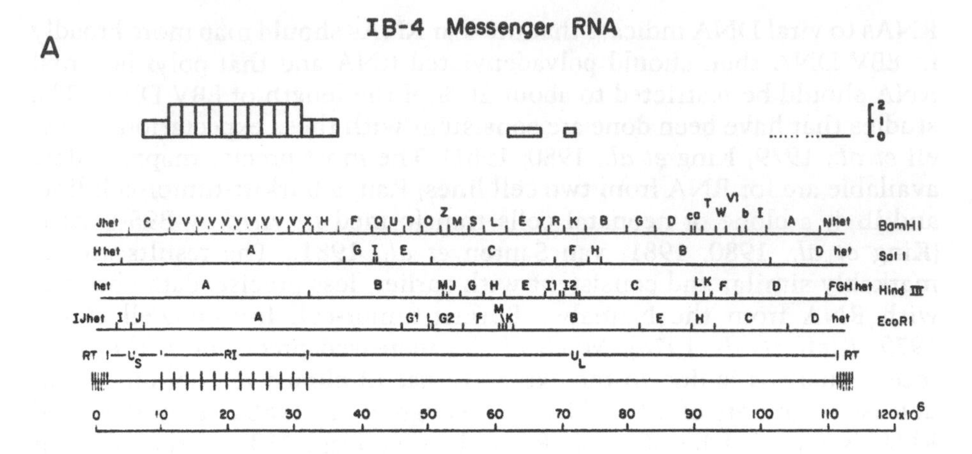

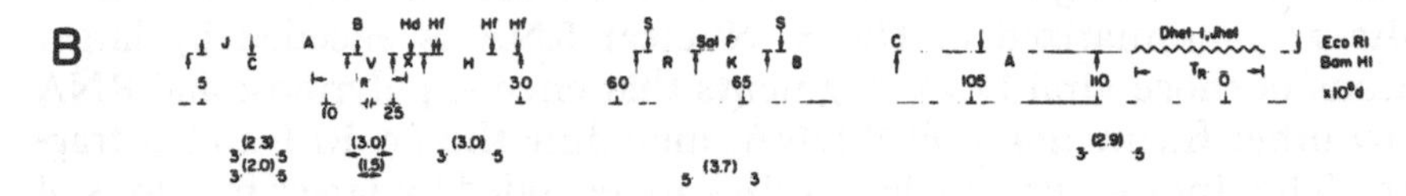

FIGURE 5. (A) Physical map of *Bam*HI, *Sal*I, *Hind*III, and *Eco*RI restriction-endonuclease fragments of EBV (B95-8) DNA with the location of DNA sequences that encode stable polyadenylated polyribosomal RNA in the latently infected, growth transformed IB4 cell line. The height of the bar is proportional to the extent of hybridization, to blots of viral DNA, of ^{32}P-labeled cDNA complementary to RNA. (B) Map of EBV mRNAs in IB-4 cells, showing size (in kb) and orientation to restriction-enzyme map of three regions of the EBV genome with map coordinates (in Md). Unlabeled arrows above the line (↓) indicate *Eco*RI restriction sites; arrows below the line (↑) indicate *Bam*HI restriction sites. Other restriction sites identified are: *Bgl*II (B), *Hind*III (Hd), *Hin*fI (Hf), and *Sal*I (S). For simplicity, only one copy of the IR1 (I_R) is shown.

2.3 and 2.0 kb; IR1 and US2 encode one or two abundant 3-kb RNAs and a less abundant 1.5-kb RNA. The 63–66 and 110–115 $\times$ 10^6 d regions encode RNAs of 3.7 and 2.9 kb, respectively.

To determine the 5′ to 3′ orientation of these cytoplasmic polyadenylated RNAs, short-length, oligodeoxythymidylic acid [oligo(dT)]-primed cDNA was synthesized as a probe for the 3′ end of the RNA. The hybridization of the oligo(dT)-primed cDNA to Southern blots of recombinant EBV DNA fragments was compared with the hybridization of oligodeoxynucleotide-primed cDNA, a probe representative of full-length RNA (Fig. 5). The most easily interpretable data were those obtained with the 63–66 $\times$ 10^6 d region. The 3′-specific cDNA hybridized to the 63–66 $\times$ 10^6 d DNA fragment and not to the 61–63 $\times$ 10^6 d DNA fragment, while randomly primed cDNA hybridized to both fragments. Thus, the 3′ end of the RNA encoded by this region maps to the right at 63–66 $\times$ 10^6 d and the 5′ end of the RNA maps to the left at 61–63 $\times$ 10^6 d.

Both 3′-specific and randomly primed cDNA probes hybridized to the 105–110 and 110–115 $\times$ 10^6 d fragments. However, randomly primed cDNA hybridized more to the 110–115 $\times$ 10^6 d fragment, suggesting that the 3′ end was in the 105–110 $\times$ 10^6 d fragment and the 5′ end in the 110–115 $\times$ 10^6 d fragment. Determination of orientation of the RNAs encoded by the US1–IR1–US2–IR2 region is not possible, since randomly primed cDNA hybridizes to all the fragments tested from this region and the 3′-specific cDNA hybridizes to the 5–7 $\times$ 10^6 d region of US1, to the large *Bam*HI–*Bgl*II component of IR1 and to most of the left part of US2. There are more 3′ ends than there are RNAs of different sizes identified by DNA from this region on Southern blots. One resolution of the available data postulates the existence of a second 3-kbp RNA encoded by IR2 and the region just to the right of IR2 in UL. Direct evidence for this putative second 3-kbp RNA has not been obtained. In an attempt to resolve whether IR1 and IR2 encode different RNAs, Northern-blot analyses were carried out with nuclear RNAs using labeled probes including IR1 and IR2 and part of US2 (van Santen and Kieff, in preparation). The results are that all three probes hybridize to a family of large nuclear RNAs differing in size from each other in fixed increments. This suggests that the promoters in each repeat of IR1 may transcribe large nuclear RNAs extending through a variable number of copies of IR1 into US2.

Two separate approaches led simultaneously to the discovery that EBV specifies small RNAs in latently infected cells. One group, working with serum from patients with lupus erythematosus to identify ribonucleoprotein complexes in human cells, found that ribonucleoprotein antigens from Raji cells contained a new small RNA, slightly larger than other small human cellular RNAs (Lerner *et al.*, 1981). The RNA was mapped to the 5–7 $\times$ 10^6 d region of EBV DNA and shown to be two distinct species. A second group was confronted with the seemingly anomalous results that iodinated cytoplasmic RNA mapped to DNA sequences at 5–7 $\times$ 10^6 d (Rymo, 1979), while polyadenylated and polyribosomal RNAs mapped largely elsewhere (Powell *et al.*, 1979; King *et al.*, 1980, 1981). In resolving this discrepancy, it was demonstrated that the most abundant viral RNA in Raji and IB-4 cells is a 160-bp, nonpolyadenylated RNA, encoded by the 5–7 $\times$ 10^6 d region (van Santen *et al.*, 1981; King *et al.*, 1981). By both size and oligonucleotide fingerprint, the small RNA has been shown to be two distinct molecular species (Lerner *et al.*, 1981). The function of the RNA is unknown. Some of the small RNA remains associated with polyadenylated RNA through two cycles of chromatography on oligo(dT) columns and subsequently separates from the polyadenylated RNA only after heat denaturation, indicating that the small RNA hydrogen bonds to some polyadenylated RNAs (van Santen and Kieff, in preparation). The small RNA is encoded by the same DNA strand that encodes the 2.3-kb polyadenylated RNA from this region. No homology of this region to other regions in EBV DNA is detectable by Southern-blot hybridization. Therefore, the small RNA cannot bind to an EBV polyadenylated RNA by means of long regions of homology. The nature and significance of the binding of the small RNA to polyadenylated RNA remain unknown.

African Burkitt-tumor biopsies from four patients were analyzed for EBV RNA (Dambaugh *et al.*, 1979). In contrast to the results obtained with the RNA from tumor-cell lines grown in culture, both polyadenylated and nonpolyadenylated RNA from tumor tissue hybridized to only 3–7% of EBV DNA. These results suggest that either there is less extensive transcription of EBV DNA in tumor tissue or there is more rapid degradation of RNAs. By Southern-blot hybridization analysis and by hybridization of RNA in solution to labeled fragments of viral DNA, the polyadenylated RNA in Burkitt-tumor tissue was loosely mapped primarily to two regions: the region around and including IR1 and the region 90–115 $\times$ 10^6 d. Some hybridization to other EBV DNA fragments was found as well.

C. Proteins and Antigenic Reactivities in Latently Infected, Growth-Transformed Cells

The concentration of EBNA in latently infected cells is too low to be detected by direct or indirect immunofluorescence with most immune human sera. Detection of EBNA by fluorescence microscopy usually requires amplification of the antigen–antibody reactions using complement and fluorescent-labeled anticomplement antibody (Reedman and Klein, 1973). EBNA is associated with cellular chromosomes, but rapidly diffuses away from the chromosomes if the cells are not promptly fixed. Much of the soluble antigen that can be extracted from latently infected cells and detected with immune human sera seems to be associated with EBNA (Lenoir *et al.*, 1976; Ohno *et al.*, 1977; Luka *et al.*, 1977, 1978; Baron and Strominger, 1978). Several lines of evidence indicate that EBNA is encoded by EBV: First, EBNA has been detected only in lymphocytes infected with EBV. Second, antibody to EBNA has been detected only in individuals with antibody to EBV structural antigens. Third, cells nonpermissively infected with herpesvirus pan (HVPan) and herpesvirus papio (HVPapio) also have chromosomal binding antigens (Gerber *et al.*, 1976b; Ohno *et al.*, 1978, 1979; Rabin *et al.*, 1980). The HVPan antigen can be detected in the nuclei of nonpermissively infected cells by complement-enhanced immunofluorescence using chimpanzee immune sera. The HVPapio antigen cannot be detected in HVPapio-infected cells by this method, but can be bound to acid-fixed chicken or salamander chromosomes, where it can be detected by immunofluorescence using baboon or human immune sera. Fourth, papio lymphocytes latently infected with HVPapio can be superinfected with EBV. EBNA can then be detected in the papio cells using human, but not papio, antisera. Fifth, human cells can be infected with HVPan and chimpanzee lymphocytes can be infected with EBV. Both infections are nonpermissive. The nuclear antigen induced by EBV is recognized only by human sera, while that induced by HVPan is recognized preferentially by chimpanzee sera.

The amount of EBNA in cells correlates loosely with the number of copies of the EBV genome. Unlike the synthesis of histones, that of EBNA does not seem to be cell-cycle-dependent. The amount of EBNA in the cell increases through G_1 phase. After infection of normal B lymphocytes with EBV, EBNA can be detected within 12 hr. Cellular DNA synthesis begins approximately 20 hr after detection of EBNA (Einhorn and Ernberg, 1978).

The soluble complement-fixation and the acid-fixed chromosome assays for EBNA have been used for following antigenic activity during purification (Lenoir *et al.*, 1976; Ohno *et al.*, 1977; Luka *et al.*, 1977, 1978; Baron and Strominger, 1978). The antigenic activity is stable at 80°C for 30 min and can be further purified by DNA cellulose and hydroxyapatite chromatography and by immunoprecipitation. The molecular weight of the antigenic activity in crude cellular extracts is estimated to be 180,000 d by ultrafiltration and ultracentrifugation. After denaturation, the purified material consists of two polypeptides of 48×10^3 and 53×10^3 d (Luka *et al.*, 1980). The V8 protease, cyanogen bromide products, and glycine content of the 53×10^3 d polypeptide may be similar to those of the cellular 53×10^3 d polypeptide that binds to SV40 T antigen.

Three lines of evidence indicate that there are additional polypeptides specified by EBV in the nonpermissively infected cell: First, as noted in Section III.B, at least five polyadenylated cytoplasmic and presumably mRNAs have been identified in latently infected cells (van Santen *et al.*, 1981). Second, sera from latently infected rheumatoid-arthritis patients and other sera from latently infected humans recognize another nuclear antigen in latently infected cells. This second nuclear antigen has been termed rheumatoid-arthritis-associated nuclear antigen (RANA) (Alspaugh *et al.*, 1978). The acronym perpetuates a misnomer, since neither the antibody nor the antigen is strongly associated with rheumatoid disease (Catalano *et al.*, 1980). RANA activity is maximal early in G_1 phase and is not confined to the nucleus (Slovin *et al.*, 1980). The antigen can be detected without complement enhancement and is not altered by deoxyribonuclease treatment. Sera with high anti-RANA titers also react with latently infected cell extracts in immunodiffusion tests. There is a high correlation between the development and titer of anti-RANA and anti-EBNA antibodies. In an attempt to segregate EBNA and RANA, latently infected lymphocytes were fused with mouse or hampster fibroblasts (Slovin *et al.*, 1981). Nine clones of hybrid cells were EBNA-positive, RANA-negative, and 16 were RANA-positive, EBNA-negative. RANA-positive, EBNA-negative clones contained DNA from at least 80% of the EBV genome. The segregation of antigens in the hybrid cells suggests that RANA and EBNA are different antigens. The third line of evidence that there are additional EBV polypeptides in the latently infected, growth-transformed cell is a recent study using radioimmuno- and fluoroimmunoelectrophoresis procedures with sodium dodecyl sulfate extracts and immune human sera (Strnad *et al.*, 1981). Polypeptides of 73 and 81 $\times$ 10^3 d are detected in nonpermissively infected cells. Subcloning of a mixed population of EBNA-positive and EBNA-negative cells revealed that the 73 $\times$ 10^3 d polypeptide segregated with an EBNA-positive clone and was absent from an EBNA-negative clone. Furthermore, partially purified EBNA preparations from Raji cells are enriched for the 65 and 81 $\times$ 10^3 d polypeptides. The 63–67 $\times$ 10^3 d region of denaturing polyacrylamide gels of Raji extracts removes anti-EBNA activity from human sera, suggesting that the 65 $\times$ 10^3 d polypeptide is an important antigenic component of EBNA.

The data are adequate to conclude that EBV specifies at least one chromatin-binding polypeptide antigen. RANA could be a separate viral gene product, a cellular protein specifically induced by EBV, or a modification of EBNA. Similarly, the 48, 53, 65, and 81 $\times$ 10^3 d polypeptides could be separate viral gene products, related posttranslational modifications of a common polypeptide, or cellular polypeptides induced by EBV infection. The simplest hypothesis that would account for most of the existing data is that EBNA and RANA are separate EBV-induced polypeptides, that the 48 $\times$ 10^3 d viral and 53 $\times$ 10^3 d cellular polypeptides are the principal components of EBNA, and that the 65 and 81 $\times$ 10^3 d polypeptides are other viral polypeptides that are components of RANA. This hypothesis does not encompass all the existing data, such as the specific absorption of EBNA antibody with 63–65 $\times$ 10^3 d polypeptides from extracts of Raji cells. The latter suggests that the 48 $\times$ 10^3 d polypeptide may be a partial protease digestion product of the 65 $\times$ 10^3 polypeptide.

IV. ABORTIVE AND PRODUCTIVE INFECTION

Some latently infected B lymphocytes continuously give rise to progeny cells that are permissive of viral replication. As described in Section I, cell type is an important determinant of the frequency of permissive infection. One report put forward the hypothesis that virus type is also a determinant of the frequency of permissive infection and that virus from the Burkitt endemic region of Africa might yield more permissive infection than virus from the United States (Gerber *et al.*, 1976a). This hypothesis has not been rigorously evaluated.

The frequency of permissive infection in cultures of partially permissive infected cell lines can be increased by the addition of chemical inducers. A tumor-promoting phorbol ester, 12-*O*-tetradecanoyl phorbol-13 acetate (TPA), and an inducer of differentiation, sodium butyrate, are the most consistently effective inducers of permissive infection (zur Hausen *et al.*, 1978; Kallin *et al.*, 1979). Many other chemicals, including inhibitors of cellular DNA, RNA, and protein synthesis, hormones, cyclic nucleotides, and antibody against IgM on the cell surface, have some inducing activity (Gerber, 1972; Hampar *et al.*, 1972, 1974, 1976). The variety of these chemicals makes it difficult to suggest a common mechanism for their effects. If anything, the data suggest that regulation of viral replication in partially permissive cells is not a stringent process and is influenced by changes in cellular metabolism. A hypothesis that has been suggested is that the extent of methylation of viral DNA could be a final common path for regulating the frequency of productive viral infection. Available data dealing with CG methylation as described in Section II.D.1 indicate that the level of CG methylation of DNA is not determinative of expression. No studies have been done of methylation at specific CG sites near promoters in Epstein–Barr virus (EBV) DNA or of the level of methylation of other nucleotides.

Following exposure to butyrate or TPA, as many as 20–40% of the cells in a partially permissive culture become productively infected (zur Hausen *et al.*, 1979; Kallin *et al.*, 1979). The diffuse (D) and restricted (R) components of early antigen (EA) can be detected with some immune human sera. In addition, cellular macromolecular synthesis is inhibited, viral DNA is synthesized within the nucleus, the viral capsid antigen (VCA) complex can be detected in the cytoplasm, viral nucleocapsids assemble in the nucleus, and the membrane antigen (MA) complex can be detected on the cell surface and nuclear membrane (Edson and Thorley-Lawson, 1981). Mature virus is produced by budding through the nuclear membrane. Treatment of induced cultures with phosphonoacetic acid, phosphonoformic acid, acycloguanosine, or other inhibitors of viral DNA synthesis results in diminished synthesis of viral DNA, VCA, and MA. The number of cells containing EA is unchanged or increased (Gergely *et al.*, 1971; Mueller-Lantzsch *et al.*, 1979, 1980a,b).

Some latently infected cell lines that are totally nonpermissive of viral replication can be induced to a state of abortive infection by treatment of the cultures with chemical inducers. TPA is the most effective chemical inducer. The abortive infection induced in latently infected cells by TPA is characterized by the expression of EA-D and EA-R in 10–40% of the cells in a culture of previously latently infected Raji cells and by the inhibition of cellular DNA, RNA, and protein synthesis. TPA does not induce the synthesis of viral DNA, VCA, or MA in latently infected cultures such as the Raji cell line (Mueller-Lantzsch *et al.*, 1979 Zur Hausen *et al.*, 1978, 1979).

Superinfection of some latently infected cultures, including Raji, with the virus produced by the P3HR-1 cell line results in a more permissive state of viral infection than can be induced by TPA. In addition to EA expression, viral DNA synthesis is increased at least 100-fold (Yajima and Nonoyama, 1976). Low levels of viral VCA and MA can be detected in a small fraction of the cells, and some virus is produced. The released virus can be distinguished from the superinfecting P3HR-1 virus, since the virus produced following superinfection cam induce growth transformation of B lymphocytes (Yajima *et al.*, 1978). The Raji cell line has been the prototype for most studies on the abortive and partially permissive states of EBV infection. The extent to which the viral DNA, RNA, proteins, and virus produced in superinfected Raji cells are products of the endogenous Raji EBV DNA, of the superinfecting P3HR-1 DNA, or of putative recombinants between Raji and P3HR-1 EBV DNAs is unknown.

A. Viral Deoxyribonucleic Acid in Abortive and Productive Infection

Productive infection is usually studied by induction of viral replication in partially permissive, but largely latently infected, cell cultures. Since viral DNA in such cells is in a closed circular form before induction, replication must proceed from this as an initial substrate. The DNA could replicate as a circular molecule or go through a linear intermediate. Since viral DNA that can circularize usually replicates through a circular intermediate, it is likely that EBV would follow this rule. Some attempt has been made to examine EBV replication in superinfected Raji cells (Siegel *et al.*, 1981). The system is favorable in that there is an enormous burst of viral DNA replication within several hours of infection of Raji cells with the P3HR-1 virus. Potential problems include inaccuracy in determining the multiplicity of superinfection and the possibility that defective P3HR-1 DNA could replicate differently than nondefective DNA. In earlier studies, restriction-endonuclease digests of DNA from superinfected cells could not be differentiated from P3HR-1 viral DNA, possibly for technical reasons (Shaw *et al.*, 1977). Using pulse labels and gradient separation, first on isopyknic CsCl gradients and then on velocity glycerol gradients, 65 S and possibly 80 S forms of viral DNA were detected between 4 and 24 hr after superinfection of Raji cells (Siegel *et al.*, 1981). After a chase period, the 80 S form was diminished and 55–65 S DNA increased, suggesting that the 80 S DNA is a replication intermediate. Longer molecules were not detected with either short or long labeling periods. Evidence that longer molecules would have survived experimental manipulation was not presented. The organization of viral sequences in the 80 S "intermediate" and in the 65 S DNA has not been determined. The 65 S DNA is presumed to be open circles. The accumulation of open-circular progeny molecules and the absence of longer molecules suggests that replication does not proceed through the generation of linear concatamers.

B. Viral Ribonucleic Acids in Abortive and Productive Infection

Some of the changes in viral gene expression following chemical induction have been studied by comparing viral RNAs from cultures in which abortive infection was induced with RNAs from latently infected cultures (Tanaka *et al.*, 1977; King *et al.*, 1981). In these studies, which are summarized in Fig. 6, Raji cells were induced with 5-iodo-2-deoxyuridine (IUdR). EA was detected in 10% of the cells. The complexity and abundance of nuclear RNA increased slightly. Nuclear RNA from abortively infected cultures hybridized to 40% of labeled EBV DNA, while nuclear RNA from latently infected cultures hybridized to 35% of labeled EBV DNA. These figures correspond to 80 and 70% of the genome, respectively, assuming that the RNAs are asymmetric transcripts. In contrast to the slight increase in the complexity of nuclear RNAs with the transition from latent to abortive infection, there was a marked increase in the complexity and abundance of stable polyadenylated and polyribosomal RNAs. The complexity of polyadenylated RNA increased from 18 to 33% and polyribosomal RNA from 10 to 30% of the DNA. Thus, while in latent infection there was considerable disparity in the complexity of nuclear, polyadenylated, and polyribosomal RNAs, there was little disparity in abortive infection. Hybridization of ^{32}P-labeled cDNA synthesized from nuclear or polyribosomal RNAs to Southern blots of fragments of viral DNA confirms that nuclear RNA from abortively infected cells is encoded by most of the same fragments that encode nuclear RNA in latently infected cells. While a restricted set of viral RNAs accumulates on the polyribosomes of latently infected cells, polyribosomal RNA from abortively infected cells is encoded by the same DNA fragments that encode nuclear RNAs. Thus, regions of the viral genome that encode only nuclear RNA before induction encode both nuclear and polyribosomal RNAs after induction. These data indicate that there is a major change in the posttranscriptional processing of viral RNAs follow-

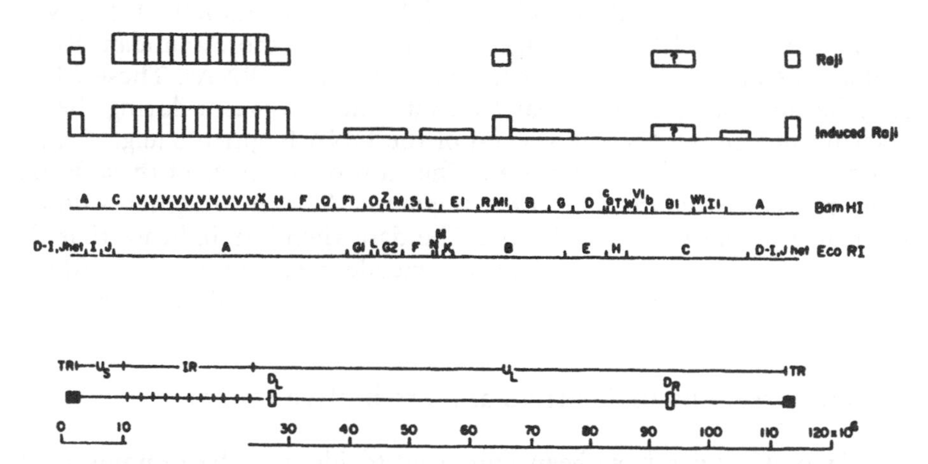

FIGURE 6. Physical map of *Bam*HI and *Eco*RI restriction-endonuclease fragments of the standard form of Raji EBV DNA with the location of DNA sequences that encode stable polyribosomal RNA in Raji and abortively infected Raji cells. The height of each bar is proportional to the extent of hybridization, to blots of viral DNA, of ^{32}P-labeled DNA complementary to Raji RNA. It is uncertain whether the duplications D_L and D_R both encode RNA (King *et al.*, 1981).

ing induction. The changes could be in cellular processing of viral transcripts or in the size of the primary transcript. The comparative maps of viral polyribosomal RNAs from abortively and latently infected Raji cells are shown in Fig. 6. Studies of the effect of actinomycin D on the induction of abortive infection indicate that the drug blocks induction (Hampar *et al.*, 1974b). This suggests that new transcription is required for expression of early antigens, but does not indicate whether the new transcripts differ from previously existing nuclear RNAs or encode a protein that alters RNA processing. Analysis of the complexity of the viral RNAs in Raji cells superinfected with P3HR-1 virus has yielded results similar to those obtained with RNA from abortively infected Raji cells induced with IUdR (Orellana and Kieff, 1977; Tanaka *et al.*, 1977). These studies have not been extended to the point where new RNAs associated with the enhanced expression of the viral genome in superinfected Raji cells have been detected.

At 3 days after TPA induction of productive infection in partially permissive B95-8 cultures, MA is detected on the surface of 20% of the cells and VCA is detected in the cytoplasm of 10% of the cells. Cytoplasmic polyadenlyated RNAs in these cells are encoded by almost every fragment of UL and part of US1 (Hummel and Kieff, 1982). The sizes of the RNAs encoded by each region of the EBV genome were determined by hybridization of labeled EBV DNA fragments to Northern blots of cytoplasmic polyadenylated RNAs. The data are summarized in Fig. 7. Several points should be made about these data: First, some RNAs encoded by a fragment are present in high abundance (e.g., the 2.8-, 3.6-, 4.0- and 4.5-kb RNAs identified by labeled *Bam*HI F), while other RNAs encoded by the same fragment are present in lower abundance (e.g., 9.9-, 8.0-, 6.9-, and 5.8-kb RNAs identified by labeled *Bam*HI F). Many of the less abundant RNAs are larger in size than the more abundant RNAs and are also identified in Northern blots of nuclear RNAs. These RNAs may be nuclear precursors that leak into the cytoplasm during the extraction procedure. Second, several of the DNA fragments (e.g., *Bam*HI F) encode several abundant RNAs. The sum of the sizes of these RNAs is more than twice the size of the fragment. Adjacent fragments do not hybridize to most of these RNAs. The data therefore indicate that the same or complementary nucleotides encode a portion of these RNAs. Work is in progress to identify the polypeptides encoded by these RNAs.

C. Viral Proteins in Abortive and Productive Infection

Two strategies have been employed to identify the proteins made early in permissive EBV infection. One is to induce abortive infection in Raji cells with TPA. The second is to block late protein synthesis in permissively infected cells using an inhibitor of viral DNA synthesis. As might be anticipated from data cited previously—that the viral infection induced by chemicals aborts at an earlier stage than the semipermissive

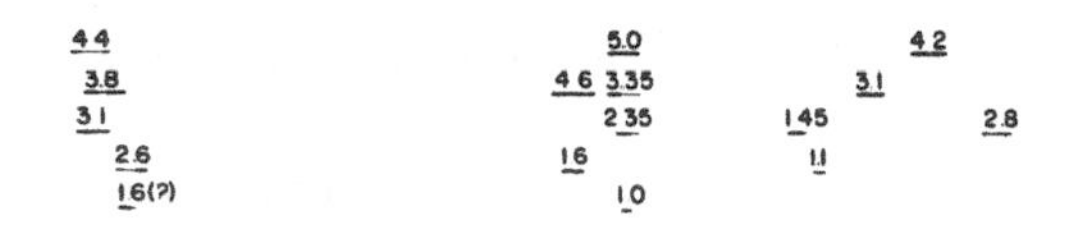

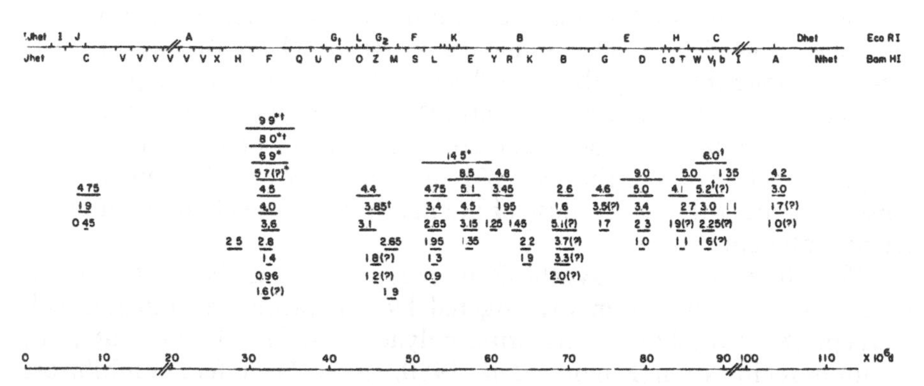

FIGURE 7. Transcriptional map indicating the size (in kb) and map location of cytoplasmic polyadenylated RNAs expressed during productive infection of TPA-induced partially permissive B95-8 cells (Hummel and Kieff, 1982). RNA species indicated above the *Eco*RI and *Bam*HI linkage maps of B95-8 EBV DNA were detected by hybridizations of labeled B95-8 *Eco*RI DNA fragments with Northern blots of polyadenylated cytoplasmic RNAs from TPA-treated B95-8 cells; RNAs indicated below the linkage map were detected by hybridizations with labeled *Bam*HI fragments. * Denotes an RNA that is a probable nuclear precursors; † denotes an RNA that is larger in size than the homologous DNA fragment, but is not detected by hybridizations with adjacent DNA fragments; (?) denotes an RNA that is barely detected by hybridization of Northern gels. The discontinuity in the DNA linkage map before 20 × 10^6 d reflects the variability in number of copies of IR1 (*Bam*HI V fragment). The discontinuity at 90–100 × 10^6 d indicates the deletion from B95-8 DNA of sequences present in all other EBV isolates examined.

infection induced by superinfection—the two approaches have yielded different results. Two new soluble polypeptides are induced in Raji cells by TPA and immunoprecipitate with human anti-EA sera. These two polypeptides are 85–90 and 30–40 × 10^3 d (Mueller-Lantzsch *et al.*, 1979).

P3HR-1-superinfected Raji cells and chemically induced P3HR-1 cultures have been prototype cultures for studies of polypeptide synthesis in permissive infection (Mueller-Lantzsch *et al.*, 1980b; Feighny *et al.*, 1980; Kallin *et al.*, 1979; Bodemer *et al.*, 1980). In permissively infected cultures, the 85–90 and 30–40 × 10^3 d polypeptides are predominantly cytoplasmic in location and are synthesized in cells blocked for late polypeptide synthesis by inhibitors of viral DNA replication. Additional immunoprecipitable polypeptides are present in P3HR-1-superinfected Raji cells and in permissively infected P3HR-1 cells blocked for viral DNA replication. Human anti-EA sera immunoprecipitate 140 and 125 × 10^3 d polypeptides from superinfected Raji cells and 140, 125, 100, and

50×10^3 d polypeptides from induced P3HR-1 cultures. These polypeptides are all unaffected by inhibitors of viral DNA synthesis and are therefore early polypeptides that may also be components of the EA complexes.

Late viral antigen components have been identified in extracts of superinfected Raji cells and in permissively infected P3HR-1 cells (Mueller-Lantzsch *et al.*, 1980b; Kallin *et al.*, 1980; Feighny *et al.*, 1980, 1981). Anti-VCA and anti-MA human sera immunoprecipitate polypeptides of 150–165, 145–158, 75, 46, 34, and 18×10^3 d. The synthesis of these latter components and of the MAs described in Section II.C is prevented by treatment with inhibitors of viral DNA synthesis. All these late antigen polypeptides have been detected in extracts of both nuclei and cytoplasm. The sizes of the 150–165 and $145–158 \times 10^3$ d late components are compatible with those of the major viral capsid and membrane matrix polypeptides.

The time course of synthesis of polypeptides in P3HR-1-superinfected Raji cells has been investigated by separating the infected cell polypeptides directly on denaturing polyacrylamide gels without prior immunoprecipitation (Feighny *et al.*, 1980, 1981). The effect of inhibitors of viral DNA synthesis on the synthesis of each of the polypeptides, the partitioning of the polypeptides between nucleus and cytoplasm, and the phosphorylation of the polypeptides have been evaluated. The results are summarized in Table I. Some polypeptides similar in size to preexisting cellular polypeptides continue to be made after infection, while the synthesis of most cellular polypeptides is inhibited. These polypeptides are designated as "persistent" polypeptides and are believed to be cellular. However, the possibility that some of these polypeptides are viral and are similar in size to cellular polypeptides has not been rigorously excluded. The synthesis of these polypeptides throughout infection at the same rate as before infection favors the hypothesis that these are cellular polypeptides. During the 48-hr period following superinfection, 31 new polypeptides appear. The synthesis of only 7 of these polypeptides is inhibited by inhibitors of viral DNA synthesis. Therefore, within the context of the preceding discussion, 24 polypeptides are defined as being early polypeptides. Several polypeptides are synthesized only at specific time intervals after superinfection, suggesting that the synthesis of these polypeptides is highly regulated. Various inhibitors of cellular or viral DNA synthesis, or of both, were tested for their differential effect on viral DNA and viral protein synthesis. Several of the drugs that inhibited viral DNA synthesis resulted in overproduction of 6 early polypeptides, further suggesting that in productive infection, later polypeptides repress the synthesis of earlier polypeptides. Of the early viral-induced polypeptides, 4 are phosphorylated and accumulate to varying extents in the nucleus. Other early nuclear polypeptides are not detectably phosphorylated. The time course of the synthesis and that of the phosphorylation of specific polypeptides do not correlate well. Some proteins made early

TABLE I. Polypeptides Synthesized in Raji Cells after Superinfection by P3HR-1 Virus[a]

Polypeptide	Molecular weight ($\times 10^3$)	Time of maximum synthesis (hr postinfection)	Effect of ACV and PAA on synthesis[b]	Intracellular location[c]
1	155	16	—	N
2	145	16	Inhibition	N
2A	142	24	—	—
3	140	10	—	N + C
3A	135	24	—	—
4	125	10	—	C
5	123	24	—	—
6	122	24	Inhibition	C
7A	117	6	—	—
8	116	24	—	—
10	110	24	Inhibition	N + C
10A	105	10	Stimulation[d]	—
11E	100	10	—	C
11A	98	10	—	—
12	95	24	Inhibition	C
13[e]	92	24	—	N
14	86	24	—	—
16[e]	75	10	—	N + C
18[e]	68	24	—	N + C
19[e]	62	6	—	N
21	56	24	Inhibition	N
24[e]	48	6	—	N + C
26	30	10	—	C
27[e]	26	10	—	C
28	25	10	Inhibition	—
28A	24	24	—	—
28D	22	24	—	—
29	21	24	—	—

[a] Adapted from Feighny *et al.* (1980, 1981).
[b] (ACV) Acyclovir; (PAA) phosphonoacetic acid. [c] (N) Nucleus; (C) cytoplasm; (—) not determined.
[d] Synthesis observed only in presence of ACV and PAA. [e] Phosphoproteins.

after superinfection are intensely phosphorylated very early and continue to be synthesized after their phosphorylation ceases.

V. SUMMARY

It is evident from this discussion that a great deal is now known about the biochemistry of Epstein–Barr virus (EBV). Because of the unusual biological properties of EBV, the direction and scope of biochemical and genetic analyses have not followed classic approaches. Nevertheless, the conclusions drawn from these studies have provided a general framework for more refined studies.

Recent technical advances have made invaluable contributions to our understanding of the molecular biology of EBV. The availability of libraries of cloned DNA fragments from different EBV isolates has led to new insights into the organization of the viral genome. EBV DNA is a linear double-stranded molecule of approximately 170×10^3 bp that is organized into short and long stretches of DNA of unique complexity separated by two nonhomologous sets of internal repeats. The DNA is bounded at both termini by direct tandem repeats. The conservation of this same general structure [with the possible exclusion of the short internal repeat (IR2)] in two lymphotropic primate herpesviruses, HVPapio and HVPan, that have biological properties similar to those of EBV, suggests that the structure has functional significance.

There is a limited degree of sequence variability among EBV isolates. Furthermore, there is no apparent correlation between the clinical origin of EBV isolates and the variations observed. P3HR-1, a subclone of the Jijoye Burkitt-tumor-derived cell line, is unique among EBV strains in having lost the ability to growth-transform cells *in vitro*. This is the only case in which a change in the DNA is associated with an altered phenotype. In infected, transformed cells, the direct terminal repeats at both ends of EBV DNA are covalently linked to form a circular episome. In addition to circular EBV DNA molecules with the standard EBV DNA organization, there are minor populations of defective molecules in many infected cell lines.

EBV gene expression in infected cells has been traditionally viewed as having three phases: latent, abortive, and permissive. Latent infection is associated with differential posttranscriptional processing of viral RNA. Thus, latently infected cells contain nuclear RNA transcribed from a significant fraction of the viral genome and mRNAs of lower complexity encoded by three separate regions of the genome. Seven cytoplasmic mRNAs have been tentatively identified and mapped to these three regions. In addition, there are two abundant small "virus associated or VA-like" RNAs in latently infected cells. A small amount of the "VA-like" RNA is noncovalently bound to polyadenylated RNAs. EBV specifies at least one chromatin-binding polypeptide antigen. Two distinct nuclear antigenic activities, EBV intranuclear antigen (EBNA) and rheumatoid-arthritis-associated nuclear antigen (RANA), have been described in latently infected cells. Other lines of evidence suggest the presence of additional viral polypeptides.

The abortive stage of infection can be induced in certain latently infected cell lines by treatment with chemical inducers. The appearance of viral early antigens (EA) is associated with an enhanced level of viral gene expression and inhibition of cellular macromolecular synthesis. There is a slight increase in the sequence complexity of nuclear RNA following chemical induction and a dramatic change in posttranscriptional processing that leads to an increase in the complexity and abundance of stable polyadenylated and polyribosomal RNAs. New polyribosomal RNAs are encoded by DNA in the long unique region (UL). The complexity of mRNA in abortively infected cells is sufficient to encode several additional viral polypeptides. Two major component polypeptides of the EA complex in 12-*O*-tetradecandyl phorbol-13 acetate (TPA)-treated cells have been identified.

The analysis of viral gene expression during permissive infection is dependent on superinfection or TPA treatment of cells to induce viral replication. The pattern of RNA synthesis reflects a significant change in gene expression. Nearly all the DNA sequences in the UL encode RNA, while the abundant mRNAs encoded by the internal reiteration in latent and abortive infection are not detected. Some of the viral polypeptides have been identified as "late" by their sensitivity to inhibition of viral DNA synthesis. The size of some of these proteins is similar to the size of major component polypeptides of the virion.

The restrictive and asynchronous nature of EBV gene expression has necessitated the use of *in vitro* manipulations such as chemical induction, inhibition, or superinfection to study the biochemical events during advanced stages of infection. The paucity of laboratory-derived mutants has hampered efforts to identify specific viral products and their functions. The advent of recent technology, such as the molecular cloning of EBV DNA in bacteria, the development of methods for rapid DNA sequencing, *in vitro* translation systems, and monoclonal antibodies, enables a new course of investigation whereby specific regions of the viral genome can be studied. Viral polypeptides can be mapped by *in vitro* translation of viral-DNA-selected RNA, by transfection of cells with specific cloned DNA fragments, and by analysis of recombinant viruses. Sequencing of viral DNA can provide some insight into the location of specific genes as well as the regulatory sites that control their expression. The inducibility of latently infected cells to more advanced stages of viral gene expression is indicative of stringent control of EBV gene expression. The factors involved may be diverse, but new techniques that will permit the analysis of regulatory sites within the EBV genome and of the steps in processing of EBV RNA should enable major advances in understanding the chemical basis for the unusual biological properties of EBV.

ACKNOWLEDGMENTS. Julia Dowling and Victor Raczkowski assisted in cell culture. Patricia Morrison and Rose Moore assisted in manuscript preparation. This research was supported by Grants CA 19264 and CA 17281 from the National Cancer Institute and by Grant MV 32 G from the American Cancer Society. Elliott Kieff is the recipient of American Cancer Society Faculty Research Award FRA-206.

REFERENCES

Adams, A., Lindahl, T., and Klein, G., 1973, Linear association between cellular DNA and Epstein–Barr virus DNA in a human lymphoblastoid cell line, *Proc. Natl. Acad. Sci. U.S.A.* **70**:2888–2892.

Adams, A., Bjursell, G., Kaschka-Dierich, C., and Lindahl, T., 1977, Circular Epstein–Barr virus genomes of reduced size in a human lymphoid cell line of infectious mononucleosis origin, *J. Virol.* **22**:373–380.

Alspaugh, M., Jensen, F., Rabin, H., and Tan, E., 1978, Lymphocytes transformed by Epstein–Barr virus: Induction of nuclear antigen reactive with antibody in rheumatoid arthritis, *J. Exp. Med.* **147**:1018–1025.

Andersson-Anvret, M., and Lindahl, T., 1978, Integrated viral DNA sequences in Epstein–Barr-virus-converted human lymphoma lines, *J. Virol.* **25**:710–718.

Baron, D., and Strominger, J., 1978, Partial purification and properties of the EBV associated nuclear antigen, *J. Biol. Chem.* **253**:2875–2881.

Bodemer, W., Summers, W., and Niederman, J., 1980, Detection of virus specific antigens in EB(P3HR-1) virus superinfected Raji cells by immunoprecipitation, *Virology* **103**:340–349.

Bornkamm, G.W., Delius, H., Zimber, U., Hudewentz, J., and Epstein, M.A., 1980, Comparison of Epstein–Barr virus strains of different origin by analysis of the viral DNAs, *J. Virol.* **35**:603–618.

Burkitt, D.P. 1962, A tumor safari in east and central Africa, *Br. J. Cancer* **16**:379–386.

Catalano, M., Carson, D., Niederman, J., Feorino, P., and Vaughan, J., 1980, Antibody to the rheumatoid arthritis nuclear antigen: Its relationship to *in vivo* EBV infection, *J. Clin. Invest.* **65**:1238–1245.

Cheung, A., and Kieff, E., 1981, Epstein–Barr virus DNA X: A direct repeat within the internal direct repeat of EBV DNA, *J. Virol.* **40**:501–507.

Cheung, A., and Kieff, E., 1982, Epstein–Barr Virus DNA XI: The nucleotide sequence of the large internal repeat in EBV DNA, *J. Virol.*, in press.

Colby, B.M., Shaw, J.E., Elion, G.B., and Pagano, J.S., 1980, Effect of Acyclovir [9-(2-hydroxyethoxymethyl)guanine] on Epstein–Barr virus DNA replication, *J. Virol.* **34**:560–568.

Dambaugh, T., 1980, Molecular cloning of Epstein–Barr virus DNA and analysis of RNA in human Burkitt's tumor tissue, Ph.D. thesis, University of Chicago.

Dambaugh, T., and Kieff, E., 1982, Two related tandem direct repeat sequences in EBV DNA, *J. Virol.*, in press.

Dambaugh, T., Beisel, C., Hummel, M., King, W., Fennewald, S., Cheung, A., Heller, M., Raab-Traub, N., and Kieff, E., 1980a, Epstein–Barr virus DNA. VII. Molecular cloning and detailed mapping of EBV (B95-8) DNA, *Proc. Natl. Acad. Sci. U.S.A.* **77**:2999–3003.

Dambaugh, T., Raab-Traub, N., Heller, M., Beisel, C., Hummel, M., Cheung, A., Fennewald, S., King, W., and Kieff, E., 1980b, Variations among isolates of Epstein–Barr virus, *Ann. N. Y. Acad. Sci.* **602**:711–719.

Datta, A.K., Feighny, R.J., and Pagano, J.S., 1980, Induction of Epstein–Barr virus-associated DNA polymerase by 12-*O*-tetradecanoylphorbol-13-acetate: Purification and characterization, *J. Biol. Chem.* **255**(11):5120–5125.

Deinhardt, F., Falk, L., Wolfe, L., Paciza, J., and Johnson, D., 1975, Response of marmosets to experimental infection with Epstein–Barr virus, in: *Oncogensis and Herpesviruses II* (G. de Thé, M. Epstein, and H. zur Hausen, eds.), pp. 161–168, IARC, Lyon.

Delius, H., and Bornkamm, G.W., 1978, Heterogeneity of Epstein–Barr virus. III. Comparison of a transforming and non-transforming virus by partial denaturation mapping of their DNAs, *J. Virol.* **27**:81–89.

Desgranges, C., Wolf, H., de Thé, G., Shanmugaratnam, K., Ellouz, R., Cammoun, N., Klein, G., and zur Hausen, H., 1975, Nasopharyngeal carcinoma. X. Presence of Epstein–Barr genomes in epithelial cells of tumors from high and medium risk areas, *Int. J. Cancer* **16**:7–15.

Desrosiers, R., Mulder, C., and Fleckenstein, B., 1979, Methylation of herpesvirus saimiri DNA in lymphoid tumor cell lines, *Proc. Natl. Acad. Sci. U.S.A.* **76**:3839–3843.

de-Thé, G., 1979, Demographic studies implicating the virus in the causation of Burkitt's lymphoma: Prospects for nasopharyngeal carcinoma, in: *The Epstein–Barr Virus* (M. Epstein and B. Achong, eds.), pp. 418–435, Springler-Verlag, Berlin.

Dolyniuk, M., Pritchett, R., and Kieff, E., 1976a, Proteins of Epstein–Barr virus, *J. Virol.* **17**:935–949.

Dolyniuk, M., Wolff, E., and Kieff, E., 1976b, Proteins of Epstein–Barr virus. II. Electrophoretic analysis of the polypeptides of the nucleocapsid and the glucosamine- and polysaccharide-containing components of the enveloped virus, *J. Virol.* **18**:289–297.

Edson, C., and Thorley-Lawson, D., 1981, Epstein–Barr virus membrane antigens: Characterization, distribution and strain differences, *J. Virol.* **39**:172–184.

Einhorn, L., and Ernberg, I., 1978, Induction of EBNA precedes the first cellular S phase after EBV infection of human lymphocytes, *Int. J. Cancer* **21**:157–160.

Epstein, M.A., and Achong, B.G., 1968, Specific immunofluorescence test for the herpes type EB virus of Burkitt lymphoblasts, authenticated by electron microscopy, *J. Natl. Cancer Inst.* **40**:593–607.

Epstein, M.A., and Achong, B.G., 1970, The EB virus, in: *Burkitt's Lymphoma* (D.P. Burkitt and D.H. Wright, eds.), pp. 231–248, Livingstone, Edinburgh.

Epstein, M., and Achong, B.G., 1979, Morphology of the virus and of virus induced cytopathologic changes, in: *The Epstein–Barr Virus* (M. Epstein and B. Achong, eds.), pp. 24–33, Springer-Verlag, Berlin.

Epstein, M.A., Achong, B.G., and Barr, Y., 1964, Virus particules in cultured lymphoblasts from Burkitt's lymphoma, *Lancet* **1**:702–703.

Epstein, M.A., Henle, G., Achong, B.G., and Barr, Y.M., 1965, Morphological and biological studies on a virus in cultured lymphoblasts from Burkitt's lymphoma, *J. Exp. Med.* **121**:761–770.

Epstein, M.A., Achong, B.G., and Pope, J., 1967, Virus in cultured lymphoblasts from a New Guinea Burkitt lymphoma, *Br. Med. J.* **21**:290–291.

Epstein, M.A., Hunt, R.D., and Rabin, H., 1973, Pilot experiments with EB virus in owl monkeys (*Aotus trivirgatus*). I. Reticuloproliferative disease in an inoculated animal, *Int. J. Cancer* **12**:309–318.

Falk, L., Deinhardt, F., Nonoyama, M., Wolfe, L., and Bergholz, C., 1976, Properties of a baboon lymphotropic herpesvirus related to Epstein–Barr virus, *Int. J. Cancer* **18**:798–807.

Feighny, R., Farrell, M., and Pagano, J., 1980, Polypeptide synthesis and phosphorylation in EBV infected cells, *J. Virol.* **34**:455–463.

Feighny, R., Henry, B., and Pagano, J., 1981, EBV polypeptides: Effect of inhibition of viral DNA replication in their synthesis, *J. Virol.* **37**:61–71.

Fischer, D., Miller, G., Gradoville, L., Heston, L., Westrate, M., Maris, W., Wright, J., Brandsma, J., and Summers, W., 1981, Genome of a mononucleosis EBV contains fragments previously regarded to be unique to Burkitt's lymphoma isolates, *Cell* **24**:543–554.

Fresen, K.O., and zur Hausen, H., 1976, Establishment of EBNA-expressing cell lines by infection of Epstein–Barr virus (EBV)-genome-negative human lymphoma cells with different EBV strains, *Int. J. Cancer* **17**:161–166.

Fresen, K.O., and zur Hausen, H., 1977, Transient induction of a nuclear antigen unrelated to Epstein–Barr nuclear antigen in cells of two human B-lymphoma lines converted by Epstein–Barr virus, *Proc. Natl. Acad. Sci. U.S.A.* **74**:363–366.

Fresen, K.O., Cho, M.-S., Gissmann, L., and zur Hausen, H., 1979, NC 37-R1 EB-virus: A possible recombinant between intracellular NC 37 viral DNA and superinfecting P3HR-1 EBV, *Intervirology* **12**:303–310.

Furlong, D., Swift, H., and Roizman, B., 1972, Arrangement of herpes virus deoxyribonucleic acid in the core, *J. Virol.* **10**:1071–1074.

Gerber, P., 1972, Activation of Epstein–Barr virus by 5'-bromo-deoxyuridine in virus free human cells, *Proc. Natl. Acad. Sci. U.S.A.* **69**:83–85.

Gerber, P., and Birch, S.M., 1967, Complement-fixing antibodies in sera of human and non-human primates to viral antigens derived from Burkitt's lymphoma cells, *Proc. Natl. Acad. Sci. U.S.A.* **58**:478–484.

Gerber, P., and Hoyer, B., 1971, Induction of cellular DNA synthesis in human leukocytes by Epstein–Barr virus, *Nature (London)* **231**:46–47.

Gerber, P., Whang-Peng, J., and Monroe, J.H., 1969, Transformation and chromsome changes induced by Epstein–Barr virus in normal human leukocyte cultures, *Proc. Natl. Acad. Sci. U.S.A.* **63**:740–747.

Gerber, P., Nkrumah, F., Pritchett, R., and Kieff, E., 1976a, Comparative studies of Epstein–Barr virus strains from Ghana and the United States, *Int. J. Cancer* **17**:71–81.

Gerber, P., Pritchett, R., and Kieff, E., 1976b, Antigens and DNA of a chimpanzee agent related to Epstein–Barr virus, *J. Virol.* **19**:1090–1100.

Gerber, P., Kalter, S.S., Schidlovsky, G., Peterson, W.D., Jr., and Daniel, M.D., 1977, Biologic and antigenic characteristics of Epstein–Barr virus-related herpesviruses of chimpanzees and baboons, *Int. J. Cancer* **20**:448–459.

Gergely, L., Klein, G., and Ernberg, I., 1971, The action of DNA antagonists on Epstein–Barr virus (EBV)-associated early antigen (EA) in Burkitt lymphoma lines, *Int. J. Cancer* **7**:293–302.

Given, D., and Kieff, E., 1978, DNA of Epstein–Barr virus. IV. Linkage map for restriction enzyme fragments of the B95-8 and W91 strains of EBV, *J. Virol.* **28**:524–542.

Given, D., and Kieff, E., 1979, DNA of Epstein–Barr virus. VI. Mapping of the internal tandem reiteration, *J. Virol.* **31**:315–324.

Given, D., Yee, D., Griem, K., and Kieff, E., 1979, DNA of Epstein–Barr virus. V. Direct repeats at the ends of EBV DNA, *J. Virol.* **30**:852–862.

Glaser, R., and O'Neill, F.J., 1972, Hybridization of Burkitt lymphoblastoid cells, *Science* **176**:1245–1247.

Goldman, N., Landon, H.C., and Reisher, J.I., 1968, Fluorescent antibody and gel diffusion reactions of human and chimpanzee sera with cells cultured from Burkitt tumors and normal chimpanzee blood, *Cancer Res.* **28**:2489–2495.

Graessman, A., Wolf, H., and Bornkamm, G., 1980, Expression of Epstein–Barr virus genes in different cells types after microinjection of viral DNA, *Proc. Natl. Acad. Sci. U.S.A.* **77**:433–436.

Hampar, B., Derge, J.G., Martos, L.M., and Walker, J.L., 1972, Synthesis of Epstein–Barr virus after activation of the viral genome in a "virus-negative" human lymphoblastoid cell (Raji) made resistant to 5-bromodeoxyuridine, *Proc. Natl. Acad. Sci. U.S.A.* **69**:78–82.

Hampar, B., Derge, J.G., and Showalter, D.S., 1974a, Enhanced activation of the repressed Epstein–Barr virus genome by inhibitors of DNA synthesis, *Virology* **58**:298–301.

Hampar, B., Derge, J.G., Nonoyama, M., Chang, S.-Y., Tagamets, M., and Showalter, S.D., 1974b, Programming of events in Epstein–Barr virus-activated cells induced by 5-iododeoxyuridine, *Virology* **62**:71–89.

Hampar, B., Tanaka, A., Nonoyama, M., and Derge, J., 1974c, Replication of the resident repressed EBV genome during early S phase (S-1 period) of non-producer Raji cells, *Proc. Natl. Acad. Sci. U.S.A.* **71**:631–633.

Hampar, B., Lenoir, G., Nonoyama, M., Derge, J.G., and Chang, S.-Y., 1976, Cell cycle dependence for activation of Epstein–Barr virus by inhibitors of protein synthesis or medium deficient in arginine, *Virology* **69**:660–668.

Hayward, S.D., and Kieff, E., 1976, Epstein–Barr virus-specific RNA. I. Analysis of viral RNA in cellular extracts and in the polyribosomal fraction of permissive and non-permissive lymphoblastoid cell lines, *J. Virol.* **18**:518–525.

Hayward, S.D., and Kieff, E., 1977, The DNA of Epstein–Barr virus. II. Comparison of the molecular weights of restriction endonuclease fragments of the DNA of strains of EBV and identification of end fragments of the B95-8 strain, *J. Virol.* **23**:421–429.

Hayward, S.D., Pritchett, R., Orellana, T., King, W., and Kieff, E., 1976, The DNA of Epstein–Barr virus fragments produced by restriction enzymes: Homologous DNA and RNA in lymphoblastoid cells, in: *Animal Virology*, Vol. 4 (D. Baltimore, A. Huang, and C.F. Fox, eds.), pp. 619–639, Academic Press, New York.

Hayward, S., Nogee, L., and Hayward, G., 1980, Organization of repeated regions within the Epstein–Barr virus DNA molecule, *J. Virol.* **33**:507–521.

Heller, M., and Kieff, E., 1981, Co-linearity of Epstein–Barr virus DNA of herpes virus Papio, *J. Virol.* **37**:821–826.

Heller, M., Gerber, P., and Kieff, E., 1982a, The DNA of Herpes Virus Pan, a third member of the Epstein–Barr Virus—Herpes Virus Papco group., *J. Virol.* **41**:931–939.

Heller, M., Dambaugh, T., and Kieff, E., 1981a, Epstein–Barr virus DNA. IX. Variation among viral DNAs, *J. Virol.* **38**:632–648.

Heller, M., Gerber, P., and Kieff, E., 1981b, Herpes virus papio DNA is similar in organization to Epstein–Barr virus DNA, *J. Virol.* **37**:698–709.

Heller, M., van Santen, V., and Kieff, E., 1982b, A simple repeat sequence in Epstein–Barr Virus DNA is transcribed in latent and production infection. *J. Virol.*, in press.

Heller, M., Henderson, A., and Kieff, E., 1982c, A repeat array in Epstein–Barr Virus DNA is related to cell DNA sequences interspersed on human chromosomes. *Proc. Natl. Acad. Sci. U.S.A.*, in press.

Henderson, E., Miller, G., Robinson, J., and Heston, L., 1977a, Efficiency of transformation of lymphocytes by Epstein–Barr virus, *Virology* **76**:152–163.

Henderson, E., Robinson, J., Frank, A., and Miller, G., 1977b, Epstein–Barr virus: Transformation of lymphocytes separated by size or exposed to bromodeoxyuridine and light, *Virology* **82**:196–205.

Henderson, E., Heston, L., Grogon, E., and Miller, G., 1978, Radiobiologic inactivation of Epstein–Barr virus, *J. Virol.* **25**:51–59.

Henle, G., and Henle, W., 1966, Immunofluorescence in cells derived from Burkitt's lymphoma, *J. Bacteriol.* **91**:1248–1256.

Henle, W., Diehl, V., Kohn, G., zur Hausen, H., and Henle, G., 1967, Herpes-type virus and chromosome marker in normal leukocytes after growth with irradiated Burkitt cells, *Science* **157**:1064–1065.

Henle, W., Henle, G., Zajac, B., Pearson, G., Waubke, R., and Scriba, M., 1970, Differential reactivity of human serums with early antigens induced by Epstein–Barr virus, *Science* **169**:188–190.

Hinuma, Y., Konn, M., Yamaguchi, J., Wudarski, D., Blakeslee, J., and Grace, J., 1967, Immunofluorescence and herpes type virus particles in the P3HR-1 Burkitt lymphoma clone, *J. Virol.* **1**:1045–1051.

Hoffman, G., Lazarowitz, S., and Hayward, D., 1980, Monoclonal antibody against a 250,000 dalton glycoprotein of EBV identifies a membrane antigen and a neutralizing antigen, *Proc. Natl. Acad. Sci. U.S.A.* **77**:2979–2983.

Honess, R., and Roizman, B., 1974, Regulation of herpesvirus macromolecular synthesis. I. Cascade regulation of the synthesis of three groups of viral proteins, *J. Virol.* **14**:8–18.

Hummel, M., and Kieff, E., 1982, Epstein–Barr Virus RNA VIII. Viral RNA in permissively infected B95-8 cells. *J. Virol.*, in press.

Hummeler, K., Henle, G., and Henle, W., 1966, Fine structure of a virus in cultured lymphoblasts from Burkitt lymphoma, *J. Bacteriol.* **91**:1366–1368.

Jehn, U., Lindahl, T., and Klein, G., 1972, Fate of virus DNA in the abortive infection of human lymphoid cell lines by Epstein–Barr virus, *J. Gen. Virol.* **16**:409.

Jondal, M., and Klein, G., 1973, Surface markers on human B and T lymphocytes. II. Presence of Epstein–Barr virus receptors on B lymphocytes, *J. Exp. Med.* **138**:1365–1378.

Kallin, B., Luka, J., and Klein, G., 1979, Immunochemical characterization of Epstein–Barr virus-associated early and late antigens in butyrate treated P3HR-1 cells, *J. Virol.* **32**:710–716.

Kaschka-Dierich, C., Adams, A., Lindahl, T., Bornkamm, G., Bjursell, G., and Klein, G., 1976, Intracellular forms of Epstein–Barr virus DNA in human tumor cells *in vivo*, *Nature (London)* **260**:302–306.

Kaschka-Dierich, C., Falk, L., Bjursell, G., Adams, A., and Lindahl, T., 1977, Human lymphoblastoid cell lines derived from individuals without lymphoproliferative disease contain the same latent forms of Epstein–Barr virus DNA as those found in tumor cells, *Int. J. Cancer* **20**:173–180.

Kawai, Y., Nonoyama, M., and Pagano, J., 1973, Reassociation kinetics for Epstein–Barr virus DNA: Non-homology to mammalian DNA and homology of viral DNA in various diseases, *J. Virol.* **12**:1006–1012.

Kieff, E., and Levine, J., 1974, Homology between Burkitt herpes viral DNA and DNA in continuous lymphoblastoid cells from patients with infectious mononucleosis, *Proc. Natl. Acad. Sci. U.S.A.* **71**:355–358.

Kieff, E., Raab-Traub, N., Given, D., King, W., Powell, A.T., Pritchett, R., and Dambaugh, T., 1978, Mapping of putative transforming sequences of EBV DNA, in: *Oncogenesis and Herpesviruses III* (G. de Thé, W. Henle, and F. Rapp, eds.), pp. 527–552, IARC, Lyon.

Kieff, E., Given, D. Powell, A.L.T., King, W., Dambaugh, T., and Raab-Traub, N., Epstein–Barr virus: Structure of the viral DNA and analysis of viral RNA in infected cells, *Biochim. Biophys. Acta Rev.* **560**:355–373.

King, W., Powell, A., Raab-Traub, N., Hawke, M., and Kieff, E., 1980, Epstein–Barr virus RNA. Viral RNA in a restringently infected, growth transformed cell line, *J. Virol.* **36**:506–518.

King, W., van Santen, V., and Kieff, E., 1981, Epstein–Barr virus RNA. VI. Viral RNA in restringently and abortively infected Raji cells, *J. Virol.* **38**:649–660.

King, W., Dambaugh, T., Heller, M., Dowling, J., and Kieff, E., 1982, Epstein–Barr Virus DNA XII. A variable region of the EBV genome is included in the P3 HR-1 deletion. *J. Virol.* **43:** in press.

Kintner, C., and Sugden, B., 1979, The structure of the termini of the DNA of Epstein–Barr virus, *Cell* **17**:661–671.

Kintner, C., and Sugden, B., 1981, Conservation and progressive methylation of EBV DNA sequences in transformed cells, *J. Virol.* **38**:305–316.

Klein, G., Giovanella, B.C., Lindahl, T., Fialkow, P.J., Singh, S., and Stehlin, J., 1974, Direct evidence for the presence of Epstein–Barr virus DNA and nuclear antigen in malignant epithelial cells from patients with anaplastic carcinoma of the nasopharynx, *Proc. Natl. Acad. Sci. U.S.A.* **71**:4737–4741.

Klein, G., Svedmyr, E., Jondal, M., and Persson, P., 1976, EBV determined nuclear antigen (EBNA) positive cells in the peripheral blood of infectious mononucleosis patients, *Int. J. Cancer* **23**:746–751.

Landon, J.C., Ellis, L.B., Zene, V.H., and Frabrizio, D.P.A., 1968, Herpes-type virus in cultured leukocytes from chimpanzees, *J. Natl. Cancer Inst.* **40**:181–192.

Lee, Y., Tanaka, A., Law, R., Nonoyama, M., and Rabin, H., 1981, Linkage map of the fragments of herpes virus papio DNA, *J. Virol.* **37**:710–720.

Lenoir, G., Berthelon, M.-C., Favre, M.-C., and de Thé, G., 1976, Characterization of Epstein–Barr virus antigens. I. Biochemical analysis of the complement-fixing soluble antigen and relationship with Epstein–Barr virus-associated nuclear antigens, *J. Virol.* **17**:672–674.

Lerner, M., Andrews, N., Miller, G., and Steitz, J., 1981, Two small RNAs encoded by EBV and complexed with protein are precipitated by antibodies from patients with systemic lupus erythematosus, *Proc. Natl. Acad. Sci. U.S.A.* **78**:805–809.

Levy, J.A.S., Levy, S.B., Hirshaut, Y., Kafuko, G., and Prince, A., 1971, Presence of EBV antibodies in sera from wild chimpanzees, *Nature (London)* **233**:559–560.

Lindahl, T., Adams, A., Bjursell, G., Bornkamm, G.W., Kascha-Dierich, C., and Jehn, U., 1976, Covalently closed circular duplex DNA of Epstein–Barr virus in a human lymphoid cell line, *J. Mol. Biol.* **102**:511–530.

Luka, J., Siegert, W., and Klein, G., 1977, Solubilization of the Epstein–Barr virus-determined nuclear antigen and its characterization as a DNA-binding protein, *J. Virol.* **22**:1–8.

Luka, J., Lindahl, T., and Klein, G., 1978, Purification of the Epstein–Barr virus-determined nuclear antigen from EBV transformed human lymphoid cells, *J. Virol.* **27**:604–611.

Luka, J., Jornvall, H., and Klein, G., 1980, Purification and biochemical characterization of the EBV determined nuclear antigen and an associated protein with a 53,000 d subunit, *J. Virol.* **35**:592–602.

Manolov, G., and Manolova, Y., 1972, Marker band in one chromosome 14 from Burkitt lymphomas, *Nature (London)* **237**:33–34.

Mayyasi, S., Schidlovsky, G., Bulferre, L., and Buschek, F., 1967, Coating reaction of herpes type virus isolated from malignant tissues with an antibody present in sera, *Cancer Res.* **27**:2020–2023.

Menezes, J., Leibold, W., and Klein, G., 1975, Biological differences between different Epstein–Barr virus (EBV) strains with regard to lymphocyte transforming ability, *Exp. Cell Res.* **92**:478–484.

Miller, G., and Lipman, M., 1973a, Release of infectious Epstein–Barr virus by transformed marmoset leukocytes, *Proc. Natl. Acad. Sci. U.S.A.* **70**:190–194.

Miller, G., and Lipman, M., 1973b, Comparison of the yield of infectious virus from clones of human and simian lymphoblastoid lines transformed by EBV, *J. Exp. Med.* **138**:1398–1412.

Miller, G., Shope, T., Lisco, H., Still, D., and Lipman, M., 1972, Epstein–Barr virus: Transformation cytopathic changes, and viral antigens in squirrel monkey and marmoset leukocytes, *Proc. Natl. Acad. Sci. U.S.A.* **69**:383–387.

Miller, G., Robinson, J., Heston, L., and Lipman, M., 1974, Differences between laboratory strains of Epstein–Barr virus based on immortalization, abortive infection and interference, *Proc. Natl. Acad. Sci. U.S.A.* **71**:4006–4010.

Miller, G., Coope, D., Niederman, J., and Pagano, J., 1976, Biological properties and viral surface antigens of Burkitt lymphoma and mononucleosis derived strains of EBV released from transformed marmoset cells, *J. Virol.* **18**:1071–1080.

Moss, D., and Pope, J., 1972, Assay of the infectivity of Epstein–Barr virus by transformation of human leukocytes *in vitro*, *J. Gen. Virol.* **17**:233–236.

Moss, D., and Pope, J., 1975, EB virus associated nuclear antigen production and cell proliferation in adult peripheral blood leukocytes inoculated with the QiMR-WIL strain of EB virus, *Int. J. Cancer* **15**:503–511.

Mueller-Lantzsch, N., Yamamoto, N., and zur Hausen, H., 1979, Analysis of early and late Epstein–Barr virus associated polypeptides by immunoprecipitation, *Virology* **97**:378–387.

Mueller-Lantzsch, N., Georg, B., Yamamoto, N., and zur Hausen, H., 1980a, Epstein–Barr virus induced proteins. III. Analysis of polypeptides from P3HR-1 EBV-superinfected NC 37 cells by immunoprecipitation, *Virology* **102**:231–233.

Mueller-Lantzsch, N., Georg, B., Yamamoto, N., and zur Hausen, H., 1980b, Epstein–Barr virus-induced proteins. II. Analysis of surface polypeptides from EBV-producing and superinfected cells by immunoprecipitation, *Virology* **102**:401–411.

Neubauer, R., Rabin, B., Strnad, M., Nonoyama, M., and Nelson, R., 1979, Establishment of a lymphoblastoid cell line and isolation of an Epstein–Barr related virus of gorilla origin, *J. Virol.* **31**:845–853.

Nilsson, K., 1971, High frequency establishment of human immunoglobulin-producing lymphoblastoid lines from normal and malignant lymphoid tissue and peripheral blood, *Int. J. Cancer* **8**:432–442.

Nilsson, K., Klein, G., Henle, W., and Henle, G., 1971, The establishment of lymphoblastoid lines from adult and fetal human lymphoid tissue and its dependence on EBV, *Int. J. Cancer* **8**:443–450.

Nilsson, K., Klein, G., Henle, G., and Henle, W., 1972, The role of EBV in the establishment of lymphoblastoid cell lines from adult and fetal lymphoid tissue, in: *Oncogenesis and Herpesviruses* (P.M. Biggs, G. de Thé, and L.N. Payne, eds.), pp. 285–290, IARC, Lyon.

Nonoyama, M., and Pagano, J.S., 1971, Detection of Epstein–Barr viral genome in nonproductive cells, *Nature (London) New Biol.* **233**:103–106.

Nonoyama, M., and Pagano, J.S., 1972, Separation of Epstein–Barr virus DNA from large chromosomal DNA in non-virus producing cells, *Nature (London) New Biol.* **238**:169–171.

Nonoyama, M., and Pagano, J.S., 1973, Homology between Epstein–Barr virus DNA and viral DNA from Burkitt's lymphoma and nasopharyngeal carcinoma determined by DNA–DNA reassociation kinetics, *Nature (London)* **242**:44–47.

Nonoyama, M., Huang, C.H., Pagano, J.S., Klein, G., and Singh, S., 1973, DNA of Epstein–Barr virus detected in tissue of Burkitt's lymphoma and nasopharyngeal carcinoma, *Proc. Natl. Acad. Sci. U.S.A.* **70**:3265–3268.

O'Conor, G.T., and Rabson, A.S., 1965, Herpes-like particles in an American lymphoma: Preliminary note, *J. Natl. Cancer Inst.* **35**:899–903.

Ohno, S., Luka, J., Lindahl, T., and Klein, G., 1977, Identification of a purified complement-fixing antigen as the EBV determined nuclear antigen (EBNA) by its binding to metaphase chromosomes, *Proc. Natl. Acad. Sci. U.S.A.* **74**:1605–1609.

Ohno, S., Luka, J., Falk, L., and Klein, G., 1978, Serologic reactivities of human and baboon sera against EBNA and herpes virus papio determined nuclear antigen, *Eur. J. Cancer* **14**:955–960.

Ohno, S., Luka, J., Falk, L., and Klein, G., 1979, Detection of a nuclear, EBNA-type antigen in apparently EBNA negative herpesvirus papio (HVP)-transformed lymphoid lines by the acid fixed nuclear binding technique, *Int. J. Cancer* **20**:941–946.

Orellana, T., and Kieff, E., 1977, Epstein–Barr virus specific RNA. II. Analysis of polyadenylated viral RNA in restringent, abortive and productive infection, *J. Virol.* **22**:321–330.

Pagano, J.S., and Huang, C.-H., 1976, Epstein–Barr virus genome in infectious mononucleosis, *Nature (London)* **263**:787–789.

Pagano, J.S., Huang, C.-H., Klein, G., de Thé, G., Shanmugaratnam, K., and Yang, C.-S., 1975, Homology of Epstein–Barr virus DNA in nasopharyngeal carcinomas from Kenya, Taiwan, Singapore and Tunisia, in: *Oncogenesis and Herpesviruses II* (G. de Thé, M.A. Epstein, and H. zur Hausen, eds.), pp. 179–190, IARC Scientific Publication No. 11, Lyon.

Pearson, G., and Qualtière, L., 1978, Papain solubilization of the Epstein–Barr virus induced membrane antigen, *J. Virol.* **28**:344–351.

Pizzo, P.A., Magrath, I.T., Chattopadhyay, S.K., Biggar, R.J., and Gerber, P., 1978, A new tumor-derived transforming strain of Epstein–Barr virus, *Nature (London)* **272**:629–631.

Pope, J., Achong, B., and Epstein, M., 1968a, Cultivation and pure structure of virus bearing lymphoblasts from a second New Guinea Burkitt lymphoma, *Int. J. Cancer* **3**:171–182.

Pope, J., Horne, M., and Scott, W., 1968b, Transformation of fetal human leukocytes *in vitro* in filtrates of a human leukaemic cell line containing herpes-like virus, *Int. J. Cancer* **3**:857–866.

Pope, J.H., Scott, W., Reedman, B.M., and Water, M.K., 1971, EB virus as a biologically active agent, in: *Recent Advances in Human Tumor Virology and Immunology* (W. Nakahara, K. Nishioka, T. Hirayama, and Y. Ito, eds., pp. 177–188, University of Tokyo Press, Tokyo.

Powell, A.L.T., King, W., and Kieff, E., 1979, Epstein–Barr virus specific RNA. III. Mapping of the DNA encoding viral specific RNA in restringently infected cells, *J. Virol.* **29**:261–274.

Pritchett, R.F., Hayward, S.D., and Kieff, E., 1975a, DNA of Epstein–Barr virus. I. Comparison of DNA of virus purified from HR-1 and B95-8 cells, *J. Virol.* **15**:556–569.

Pritchett, R.F., Hayward, S.D., and Kieff, E., 1975b, Analysis of the DNA of Epstein–Barr viruses and transcriptional products in transformed cells, in: *Oncogenesis and Herpesviruses* (H. zur Hausen, G. de Thé, and M. Epstein, eds.), pp. 177–191, IARC Monograph, Lyon.

Pritchett, R., Pedersen, M., and Kieff, E., 1976, Complexity of EBV homologous DNA in continuous lymphoblastoid cell lines, *Virology* **74**:227–231.

Qualtière, L., and Pearson, G., 1979, Epstein–Barr virus induced membrane antigens: Immuno-chemical characterization of Triton X-100 solubilized viral membrane antigens from EBV-superinfected Raji cells, *Int. J. Cancer* **23**:808–817.

Qualtière, L., and Pearson, G., 1980, Radioimmune precipitation study comparing the Epstein–Barr virus membrane antigens expressed on P3HR-1 virus superinfected Raji cells to those expressed on cells in a B95-8 virus transformed producer culture activated with tumor promoting agent (TPA), *Virology* **102**:360–369.

Raab-Traub, N., Pritchett, R., and Kieff, E., 1978, DNA of Epstein–Barr virus. III. Identification of restriction enzyme fragments which contain DNA sequences which differ among strains of EBV, *J. Virol.* **27**:388–398.

Raab-Traub, N., Dambaugh, T., and Kieff, E., 1980, DNA of Epstein–Barr virus. VIII. B95-8, the previous prototype, is an unusual deletion derivative, *Cell* **22**:257–267.

Rabin, H., Neubauer, R., Hopkins, F., and Nonoyama, M., 1978, Further characterization of a herpes virus-positive orangutan cell line and comparative aspects of *in vitro* transformation with lymphotropic Old World primate herpesviruses, *Int. J. Cancer* **21**:762–767.

Rabin, H., Strnad, B.C., Neubauer, R.H., Brown, A.M., Hopkins, R.F., and Mazur, R.A., 1980, Comparisons of nuclear antigens of Epstein–Barr virus (EBV) and EBV-like simian viruses, *J. Gen. Virol.* **48**:265–272.

Ragona, G., Ernberg, I., and Klein, G., 1980, Induction and biological characterization of the Epstein–Barr virus (EBV) carried by the Jijoye lymphoma line, *Virology* **101**:553–557.

Reedman, B.M. and Klein, G., 1973, Cellular localization of an Epstein–Barr virus (EBV)-associated complement-fixing antigen in producer and non-producer lymphoblastoid cell lines, *Int. J. Cancer* **11**:499–520.

Reedman, B.M., Klein, G., Pope, J.H., Walters, M.K., Hilgers, J., Singh, S., and Johansson, B., 1974, Epstein–Barr virus-associated complement-fixing and nuclear antigens in Burkitt lymphoma biopsies, *Int. J. Cancer* **13**:755–763.

Robinson, J. Smith, D., and Niederman, J., 1981, Plasmacytic differentiation of circulating EBV infected B lymphocytes during acute infectious mononucleosis, *J. Exp. Med.* **153**:235–244.

Rymo, L., 1979, Identification of transcribed regions of Epstein–Barr virus DNA in Burkitt lymphoma-derived cells, *J. Virol.* **32**:8–18.

Rymo, L., and Forsblum, S., 1978, Cleavage of Epstein–Barr virus DNA by restriction endonucleases EcoRI, Hind III and Bam I, *Nucleic Acid Res.* **5**:1387–1402.

Shaw, J., Seebeck, T., Li, J.-L., and Pagano, J., 1977, Epstein–Barr virus DNA synthesized in superfected Raji cells, *Virology* **77**:762–771.

Shope, T., Dechairo, D., and Miller, G., 1973, Malignant lymphoma in cotton-top marmosets following inoculation of Epstein–Barr virus, *Proc. Natl. Acad. Sci. U.S.A.* **10**:2487–2491.

Siegel, P. Clough, W., and Strominger, J., 1981, Sedimentation characteristics of newly synthesized Epstein–Barr virus viral DNA in superinfected cells, *J. Virol.* **38**:880–885.

Skare, J., and Strominger, J., 1980, Cloning and mapping of Bam HI endonuclease fragments from the transforming B95-8 strain of Epstein–Barr virus, *Proc. Natl. Acad. Sci. U.S.A.* **77**:3860–3864.

Slovin, S., Vaughan, J., and Carson, D., 1980, Expression of EBNA and RANA during different phases of the cell growth cycle, *Int. J. Cancer* **26**:9–15.

Slovin, S., Glassy, M., Dambaugh, T., Catalano, M., Curry, R., Ferrone, S., Kieff, E., Vaughan, J., and Carson, D., 1981, Discordant expression of two Epstein–Barr virus associated antigens EBNA and RANA in man-rodent somatic cell hybrids, *J. Immunol.* **127**:585–590.

Spear, P., 1980, Composition and organization of herpes virus virions and properties of some of the structural proteins, in: *Oncogenic and Herpesviruses*, Vol. 1 (F. Rapp, ed.), CRC Press, Baca Raton, Florida.

Stevens, D.A., Pry, T.W., Blackman, E.A., and Manaker, R.A., 1970, Comparison of antigens from human and chimpanzee herpes-type virus-infected hemic cell lines, *Proc. Soc. Exp. Biol. Med.* **133**:678–683.

Stewart, S., Lovelace, E., Whang, J., and Ngu, V., 1965, Burkitt tumor, tissue culture, cytogenetic and viral studies, *J. Natl. Cancer Inst.* **34**:319–328.

Strnad, B., Neubauer, R., Rabin, H., and Mazur, R., 1979, Correlation between EBV membrane antigen and three large cell surface glycoproteins, *J. Virol.* **32**:885–894.

Strnad, B., Schuster, T., Hopkins, R., Neubauer, R., and Rabin, H., 1981, Identification of an EBV nuclear antigen by fluoroimmuno-electrophoresis and radioimmunoelectrophoresis, *J. Virol.* **38**:996–1004.

Sugden, B., and Mark, W., 1977, Clonal transformation of adult human leukocytes by Epstein–Barr virus, *J. Virol.* **23**:503–508.

Sugden, B., Summers, W.C., and Klein, G., 1976, Nucleic acid renaturation and restriction endonuclease cleavage analyses show that the DNAs of a transforming and non-transforming strain of Epstein–Barr virus share approximately 90% of their nucleotide sequences, *J. Virol.* **18**:765–775.

Sugden, B., Phelps, M., and Domoradzki, J., 1979, EBV DNA is amplified in transformed lymphocytes, *J. Virol.* **31**:590–595.

Summers, W., and Klein, G., 1976, Inhibition of Epstein–Barr virus DNA synthesis and late gene expression by phosphonoacetic acid, *J. Virol.* **18**:151–155.

Svedmyr, E., and Jondal, M., 1975, Cytotoxic effector cells specific for B cell lines transformed by Epstein–Barr virus are present in patients with infectious mononucleosis, *Proc. Natl. Acad. Sci. U.S.A.* **72**:1622–1666.

Tanaka, A., Nonoyama, M., and Glaser, R., 1977, Transcription of latent Epstein–Barr virus genomes in human epithelial/Burkitt hybrid cells, *Virology* **82**:63–68.

Thorley-Lawson, D., 1979, Characterization of cross-reacting antigens on the EBV envelope and plasma membranes of producer cells, *Cell* **16**:33–42.

Thorley-Lawson, D., and Edson, C., 1979, Polypeptides of the EBV membrane antigen complex, *J. Virol.* **32**:458–467.

Thorley-Lawson, D., and Geilinger, K., 1980, Monoclonal antibodies against the major glycoprotein (gp 350/220) of EBV neutralize infectivity, *Proc. Natl. Acad. Sci. U.S.A.* **77**:5307–5311.

Thorley-Lawson, D., and Strominger, J.L., 1976, Transformation of human lymphocytes by Epstein–Barr virus is inhibited by phosphonoacetic acid, *Nature (London)* **263**:332–334.

Toplin, I., and Schidlovsky, G., 1966, Partial purification and electron microscopy of the virus in the EB-3 cell line derived from a Burkitt lymphoma, *Science* **152**:1084–1085.

Van Santen, V., Cheung, A., and Kieff, E., 1981, Epstein–Barr virus (EBV) RNA. VII. Viral cytoplasmic RNA in a restringently infected cell line transformed *in vitro* by EBV, *Proc. Natl. Acad. Sci. U.S.A.* **78**:1930–1934.

Volsky, D.J., Shapiro, I.M., and Klein, G., 1980, Transfer of Epstein–Barr virus receptors to receptor-negative cells permits virus penetration and antigen expression, *Proc. Natl. Acad. Sci. U.S.A.* **77**(9):5453–5457.

Wagner, E.K., Roizman, B., Savage, T., Spear, P., Mizell, M., Darr, F., and Sypowicz, D., 1970, Characterization of the DNA of the herpesvirus associated with Lucke adenocarcinoma of the frog and Burkitt lymphoma of man, *Virology* **42**:257–261.

Wolf, H., zur Hausen, H., Klein, G., Becker, Y., Henle, G., and Henle, W., 1975, Attempts to detect virus-specific DNA sequences in human tumors. III. EBV DNA in nonlymphoid nasopharyngeal carcinoma cells, *Med. Microbiol. Immunol.* **161**:15–21.

Yajima, Y., and Nonoyama, M., 1976, Mechanisms of infection with Epstein–Barr virus. I. Viral DNA replication and formation of non-infectious virus particles in superinfected Raji cells, *J. Virol.* **19**:187–194.

Yajima, Y., Marczynska, B., and Nonoyama, M., 1978, Transforming activity of Epstein–Barr virus obtained by superinfection of Raji cells, *Proc. Natl. Acad. Sci. U.S.A.* **75**:2008–2010.

Yamamoto, N., and Hinuma, Y., 1976, Clonal transformation of human leukocytes by Epstein–Barr virus in soft agar, *Int. J. Cancer* **17**:191–196.

Zur Hausen, H., and Schulte-Holthausen, H., 1970, Presence of EB virus nucleic acid homology in a "virus free" line of Burkitt's tumor cells, *Nature (London)* **227**:245–248.

Zur Hausen, H., Schulte-Holthause, H., Klein, G. Henle, W., Henle, G., Clifford, P., and Santesson, L., 1970, EB virus DNA in biopsies of Burkitt tumors and anaplastic carcinomas of the nasopharynx, *Nature (London)* **228**:1056–1057.

Zur Hausen, H., Diehl, V., Wolf, H., Schulte-Holthausen, H., and Schneider, U., 1972, Occurrence of Epstein–Barr virus genomes in human lymphoblastoid cell lines, *Nature (London) Biol.* **237**:189–190.

Zur Hausen, H., O'Neill, F.J., and Freese, U., 1978, Persisting oncogenic herpesvirus induced by the tumor promotor TPA, *Nature (London)* **272**:373–375.

Zur Hausen, H., Bornkamm, G., Schmidt, R., and Hecker, E., 1979, Tumor initiators and promotors in the induction of EBV, *Proc. Natl. Acad. Sci. U.S.A.* **76**:782–785.

CHAPTER 4

Biology of Lymphoid Cells Transformed by Epstein–Barr Virus

JAMES E. ROBINSON, M.D., AND GEORGE MILLER, M.D.

I. ESTABLISHMENT OF LYMPHOBLASTOID-CELL LINES FROM PATIENTS

A. Historical Background

In 1964, M.A. Epstein and Barr (1964) and Pulvertaft (1964a) reported the first successful attempts to establish continuous lymphoblastoid-cell lines (LCLs) from explants of Burkitt lymphoma (BL). Subsequent electron-microscopic examination of thin sections of the EB1 cell line revealed virus particles that were morphologically similar to the herpesvirus group (M.A. Epstein *et al.*, 1964a). In lines derived shortly thereafter from other BL tumors (M.A. Epstein *et al.*, 1964b; Stewart *et al.*, 1965; Rabson *et al.*, 1966; Minowada *et al.*, 1967), viral particles resembling those found in EB1 cells were seen. Such particles were not confined to lines originating from BL tissue, but were also seen in a fraction of cells in lymphoid lines established from patients with various malignancies (Armstrong, 1966; Moore *et al.*, 1966; Zeve *et al.*, 1966), from patients with infectious mononucleosis (IM) (Pope, 1967; Diehl *et al.*, 1968), and from apparently normal individuals (Moore *et al.*, 1967).

The newly observed virus failed to cause cytopathic effects in tissue-culture systems susceptible to known human herpesviruses, nor did it

JAMES E. ROBINSON, M.D., and GEORGE MILLER, M.D., • Departments of Pediatrics and Epidemiology and Public Health, Yale University School of Medicine, 333 Cedar Street, New Haven, Connecticut 06510.

produce lesions in embryonated hens' eggs or affect infant mice inoculated intracerebrally. Indeed, the virus associated with BL failed to show biological activity in any conventional assay system (M.A. Epstein *et al.*, 1965).

Cells that produced viral particles were found to contain antigens detectable by indirect immunofluorescence using sera from patients with BL (G. Henle and W. Henle, 1966). Virus particles in lines from diverse sources appeared to be immunologically identical (M.A. Epstein and Achong, 1968) and serologically distinct from most known animal herpesviruses (G. Henle and W. Henle, 1966, 1967). Because it appeared to be truly unique, the virus was named Epstein–Barr virus (EBV) after its discoverers. It was proposed that the virus might be the agent responsible for maintaining the continuous growth of lymphoid-cell lines. This hypothesis proved to be correct.

Cocultivation of X-irradiated cells from a BL-cell line, Jijoye, or from a line, LS, from a leukemic that produced EBV particles, with normal peripheral-blood leukocytes resulted in the establishment of new LCLs (W. Henle *et al.*, 1967; Miller *et al.*, 1969). Filtered supernatant fluids from a cell line (QIMR-WIL) induced transformation of fetal leukocytes (Pope *et al.*, 1968). Likewise, cell lines were established after infection of leukocytes with partially purified virus from the Jijoye producer BL-cell line (Gerber *et al.*, 1969). That the transforming principle was in fact the new virus was demonstrated by the neutralization of transformation by sera containing antibody to the virus by the indirect immunofluorescence test (Pope *et al.*, 1969; Gerber *et al.*, 1969; Miller *et al.*, 1971).

The spontaneous outgrowth of cell lines from the blood or lymphoid tissues of patients with malignancy or of apparently normal individuals also depends on EBV. In this case, the virus is already present in a portion of the cells placed in culture as a result of a lifelong carrier state established during primary EBV infection. Such cells are not present in patients who lack antibodies to EBV (Gerber *et al.*, 1969; Chang *et al.*, 1971; Nilsson *et al.*, 1971).

In addition to its association with BL, EBV was shown to be the etiological agent of IM (G. Henle *et al.*, 1968; A.S. Evans *et al.*, 1968). Subsequently, the virus was found to be associated with another human malignancy, nasopharyngeal carcinoma (NPC). Since malignant epithelial cells, and not lymphoid cells, appear to harbor the EBV genome in this disease, a discussion of the infected cells in NPC is beyond the scope of this chapter; the subject has been reviewed recently (G. Klein, 1979).

B. Burkitt-Lymphoma Lines

Cell lines have been established readily from BL biopsies. Usually, cells are dispersed and grown in stationary suspension cultures with or without fibroblastic "feeder cells" (M.A. Epstein and Barr, 1964; Pulvertaft, 1964a,b; M.A. Epstein *et al.*, 1964a; Stewart *et al.*, 1965; Epstein and Barr, 1965; Rabson *et al.*, 1966; Minowada *et al.*, 1967; Pope *et al.*, 1967; Nadkarni *et al.*, 1969; van Furth *et al.*, 1972; A.L. Epstein *et al.*, 1976).

BL lines have also been grown from organ culture using either the Trowell grid system (Jensen *et al.*, 1964) or the Spongostan grid culture technique (Nilsson, 1971). The latter system has been reported to be highly efficient in the establishment of cell lines (Nilsson and Ponten, 1975). Outgrowth of cell lines from BL biopsies is usually evident in 2–4 weeks, although in some instances up to 100 days have been required for outgrowth (Nilsson and Ponten, 1975). The establishment of a cell line is evidenced by the formation of large clumps of cells, increased acid production, and a rapid increase in cell number. Burkitt-cell lines appear to grow directly from tumor cells. The morphology, pattern of immunoglobulin synthesis, and chromosome abnormalities of BL lines resemble the tumors from which they are derived.

The success rate of establishing lines from cultured biopsies is far less than 100% regardless of the method used (van Furth *et al.*, 1972; Nilsson and Ponten, 1975). This observation is somewhat puzzling, since tumors from African patients generally contain the EBV genome and can be considered to be already transformed *in vivo* (Nilsson and Ponten, 1975). Two reasons may be put forward to explain a lower than expected efficiency of BL-cell-line outgrowth. The number of tumor cells that are viable by the time they are put into culture may be small, perhaps due to the necessity of transporting the tissue on ice by air from Africa (van Furth *et al.*, 1972; Gunven *et al.*, 1980). Alternatively, many of the cells may enter a lytic viral cycle when placed in culture. This would not be unexpected, since it is known that EBV-infected cells in the circulation of patients with IM do not express lytic infection *in vivo*, but do undergo viral replication when placed in culture (Rickinson *et al.*, 1974; Robinson *et al.*, 1980a). The extent to which BL cells enter lytic viral replication after culture initiation is not known.

C. Lymphoblastoid-Cell Lines from Patients with Lymphoproliferative Disease and from Normals

Continuously growing LCLs have been established from the peripheral blood or lymphoid tissues from patients with a variety of diseases, in particular IM, or from normal individuals.

Many cell lines were initially named for their source. Thus, lines were called leukemia- or lymphoma-cell lines. However, it soon became apparent that cells in all the lines, with the exception of BL lines, had similar characteristics regardless oi origin and usually lacked the characteristics of the original tumor cells. Therefore, it was recognized early on that LCLs from patients with malignant disease do not necessarily reflect an origin from malignant tissue (Clarkson, 1967; Moore *et al.*,

1968; Nilsson *et al.*, 1970; Belpomme *et al.*, 1972; Nilsson and Ponten, 1975). For example, cell lines derived from NPC were clearly lymphoblastoid (de Thé *et al.*, 1970), as were EBV-transformed lines from myeloma (Nilsson *et al.*, 1970). Clearly, EBV-infected lymphocytes within the tumor tissues were responsible for the initiation of these cell lines.

To establish lines from peripheral blood of patients with leukemia (Moore *et al.*, 1966) or from normal individuals (Gerber and Monroe, 1968), relatively large volumes of blood (500–1000 ml) were required. Cultures were initiated at high cell densities ranging from 4 to 15 $\times 10^6$ cells/ml in stationary suspension cultures (Iwakata and Grace, 1964; Moore *et al.*, 1966). Lines were established from patients with various malignancies (Iwakata and Grace, 1964; Armstrong, 1966; Zeve *et al.*, 1966; Clarkson, 1967). The outgrowth of cell lines did not become evident until 30 days, but usually required 2–3 months.

Organ cultures using the lens paper grid or Spongostan grid systems also proved to be efficient for the establishment of cell lines from normal and malignant lymph nodes (Nilsson *et al.*, 1971) as well as from small volumes of normal peripheral blood (Nilsson, 1971). Mononuclear leukocytes isolated from normal peripheral blood on Ficoll–Hypaque gradients show a higher rate of spontaneous outgrowth into cell lines than do leukocytes separated by gravity sedimentation (Brodsky and Hurd, 1974). Submitogenic doses of phytohemagglutinin markedly enhance spontaneous outgrowth, allowing the use of relatively small volumes of blood and cell cultures with relatively low cell densities.

D. Lymphoblastoid-Cell Lines from Patients with Infectious Mononucleosis

Spontaneously arising cell lines are most readily grown from peripheral blood of patients with IM (Pope, 1967; Diehl *et al.*, 1968; Moses *et al.*, 1968). The ease with which cell lines can be established from IM blood is due to the presence of relatively large numbers of EBV-infected cells in the circulation during the acute phase of the disease. Rocchi *et al.* (1977), using a quantitative culture assay in microtiter plates, estimated that the minimum number of blood lymphocytes needed to give rise to cell lines during acute disease was 2×10^3 cells. In convalescence, it required the cultivation of a minimum of 10^7 cells to give rise to cell lines.

On the basis of the detection of EBV nuclear antigen (EBNA) in B-cell-enriched fractions of peripheral-blood lymphocytes during IM, it is now known that the number of circulating EBV-infected cells in acute disease is actually much higher than the estimates of Rocchi and colleagues, suggest (G. Klein *et al.*, 1976; Katsuki *et al.*, 1979; Robinson *et al.*, 1981). During the 1st week of illness, up to 19% of T-cell-depleted lymphocytes are EBNA-positive, while in the 2nd and 3rd weeks, the proportion of EBNA-positive cells declines to very low levels.

Several mechanisms that may account for the outgrowth of EBV-genome-positive cell lines from cultured lymphocytes from patients with IM are illustrated in Fig. 1. Virus may be carried passively on the surface of cells without actually infecting them (Model 1). Such adsorbed virus could be transferred to normal B cells after cultures are initiated. This explanation is unlikely, since the expression of EBNA in a portion of blood lymphocytes indicates that these cells are infected. Model 2 is based on the experiments of Rickinson *et al.* (1974), who showed convincingly that some of the blood lymphocytes infected *in vivo* during mononucleosis become lytically productive of virus when placed in culture. The virus released from these cells then infects normal B cells in the culture, which in turn give rise to spontaneously growing cell lines. To what extent this two-step mechanism accounts for the origin of such lines is not known. However, it is clear that cells infected by EBV *in vivo* are capable of direct proliferation on being placed in culture (Model 3). Peripheral-blood lymphocytes from IM patients will form colonies of EBNA-positive cells in semisolid medium (Hinuma and Katsuki, 1978; Katsuki *et al.*, 1979). Furthermore, a small number of mitotic EBNA-positive cells can be detected in peripheral blood during acute IM. The number of antigen-positive cells in mitosis expands dramatically after 18 hr in cultures (Robinson *et al.*, 1980b). These findings suggest that EBV-infected lymphocytes exhibit altered growth properties *in vivo*. Further-

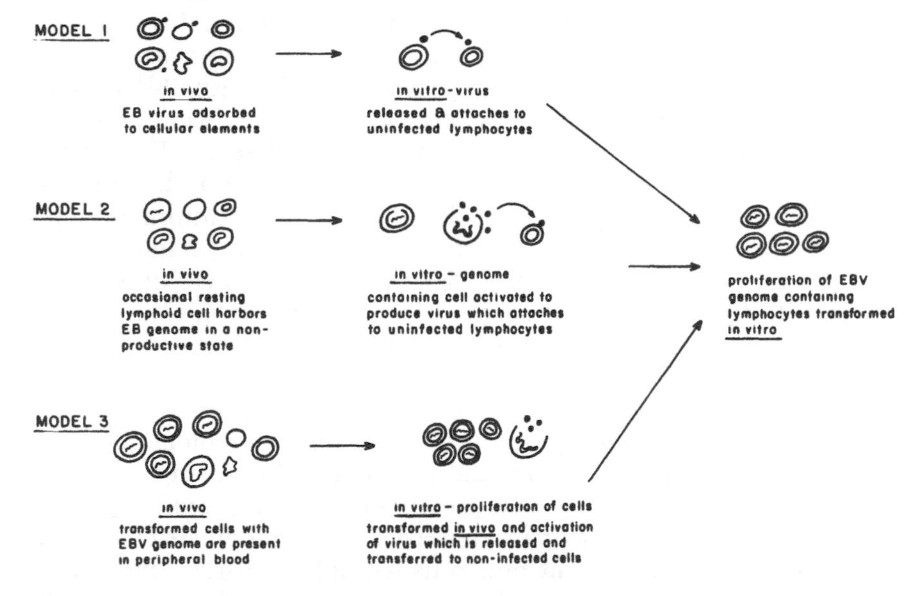

FIGURE 1. Possible mechanisms of formation of LCLs from peripheral-blood leukocytes of mononucleosis patients.

more, such virus-altered cells can give rise to fatal polyclonal tumors in individuals with certain immunodeficiencies (see Section V.C.3). It should be pointed out that Models 2 and 3 as shown in Fig. 1 are not mutually exclusive. The nature of the cell–virus interaction that accounts for the lifelong carrier state in which small numbers of EBV-infected cells are found in blood and lymph node in seropositive individuals (Nilsson *et al.*, 1971) is unknown.

II. ESTABLISHMENT OF LYMPHOBLASTOID CELL LINES BY *IN VITRO* TRANSFORMATION

A. Target Cells of Epstein–Barr Virus

1. B-Cell Characteristics

Epstein–Barr virus (EBV) has a unique tropism for bone-marrow-derived (B) lymphocytes. Regardless of their source, all lymphoblastoid-cell lines (LCLs) that carry the EBV genome express B-cell characteristics, namely, immunoglobulin synthesis (Tanigaki *et al.*, 1966; Fahey *et al.*, 1966; Finegold *et al.*, 1967; Wakefield *et al.*, 1967; Glade and Chessin, 1968; Nilsson *et al.*, 1971) or receptors for fixed complement (Shevach *et al.*, 1972; Jondal and Klein, 1973; Huber *et al.*, 1976). The B-cell characteristics of EBV-transformed cell lines led to an investigation of which cells had receptors for EBV. Only B cells were found to form rosettes with EBV producer cells that expressed viral envelope material on their cytoplasmic membranes (Jondal and Klein, 1973). After removal of B cells, no rosettes were formed. In another type of assay, cells with EBV receptors were identified by immunofluorescence. Lymphocytes were first incubated with supernatant fluids of EBV-producer-cell lines. These fluids presumably contain EB virions or virus envelope material. Cells that bound virus or viral antigens were identified by direct immunofluorescence using sera containing EBV antibodies. In this assay, B lymphocytes or B-cell lines, but not T cells, had EBV receptors (Greaves *et al.*, 1975; Menezes *et al.*, 1976). Surprisingly, a T-cell line (Molt 4) that does not contain the EBV genome was positive for EBV receptors in this assay (Menezes *et al.*, 1977).

In a third type of assay, the presence of EBV receptors was detected by measuring the ability of a group of cells to adsorb out biological activity from a virus preparation (Sairenji and Hinuma, 1973). Using this assay, Einhorn *et al.* (1978) studied the expression of virus receptors in various subpopulations of human lymphocytes. Their results indicate that B cells, and perhaps to a small extent null cells (complement-receptor-positive but surface-immunoglobulin- and E-rosette-negative cells), possess EBV receptors, but that T lymphocytes do not.

2. EBV Receptors

The particular proteins on host-cell membranes that are responsible for the highly specific binding of the virus to B cells have not yet been biochemically defined. However, current evidence indicates that EBV receptors are closely associated with C3 receptors. The presence of EBV receptors on transformed lymphoid-cell lines correlates with the presence of C3 receptors; lines lacking C3 receptors also appear to lack virus receptors (Jondal *et al.*, 1976; Yefenof *et al.*, 1977). Preincubation of transformed cells expressing C3 receptors with concentrated EBV blocks rosette formation with erythrocytes coated with antibody and complement (EAC); conversely, adsorption of EBV detectable by immunofluorescence is blocked by preexposure of the cells to purified C3 (Jondal *et al.*, 1976). EBV and C3 receptors have been shown by two-color immunofluorescence to be localized to the same area of the cell membrane and to be redistributed together in cocapping experiments. Neither marker cocaps with constant-fragment (Fc) receptors, surface immunoglobulins, or immune-associated (Ia) antigens (Yefenof *et al.*, 1976).

Finally, removal of EBV or C3 receptors by membrane stripping from transformed cells reduces the capacity of the cells to adsorb EBV, while stripping cells of Fc receptors, surface immunoglobulin, or β_2-microglobulins has no effect on virus adsorptions (Yefenof and Klein, 1977).

3. Population of Cells Transformed by EBV

The tropism of EBV for B lymphocytes has been shown not only by studies on virus receptors but also by functional studies identifying the subpopulation of lymphocytes that are induced to proliferate after virus infection. During the course of *in vitro* transformation of a mixed population of lymphocytes, there is a progressive increase in the number of cells showing B-cell surface characteristics (Pattengale *et al.*, 1974), and the cells in which EBV nuclear antigen (EBNA) first appears bear C3 receptors (Yata *et al.*, 1975). Transformation efficiency is highest in partially purified subpopulations of lymphocytes enriched in B lymphocytes (Schneider and zur Hausen, 1975; Yata *et al.*, 1975; Henderson *et al.*, 1977a; Katsuki *et al.*, 1977; Robinson *et al.*, 1977). EBV does not induce EBNA in, nor does it cause transformation of, T lymphocytes (Menezes *et al.*, 1976; Henderson *et al.*, 1977a; Robinson *et al.*, 1977; Einhorn *et al.*, 1978) or null cells (Robinson *et al.*, 1977; Einhorn *et al.*, 1978). Cells in which DNA synthesis is stimulated appear to be part of the same population that is ultimately transformed (Robinson *et al.*, 1979).

The number of transformed cells in a preparation can be measured using a quantitative culture technique in microtiter plates (Henderson *et al.*, 1977a). Cells are exposed to virus and freed of unadsorbed inoculum by washing and incubating with neutralizing antibody. Serial dilutions containing known numbers of cells are plated in association with feeder cultures. The fraction of transformed cells is calculated from the ratio between the microwells that ultimately show transformation and the total number of cells plated at the limiting dilution. The observed transformation efficiency must be corrected for the plating efficiency of transformed cells, which is usually about 10%. In subpopulations enriched for B lymphocytes, the transformation frequency has been as high as 3%; corrected for plating efficiency, this would be equivalent to a transformation rate of 30% (Henderson *et al.*, 1977b).

The number of cells transformed by EBV can also be estimated by plating inoculated cells in semisolid medium (Mizuno *et al.*, 1976; Sugden and Mark, 1977; Yamamoto and Hinuma, 1976). Transformation frequency in this assay is somewhat lower than in the microtiter system; the assay may selectively detect a population of transformed cells that is characterized by a high cloning efficiency. However, the method has proved useful for isolating clones of transformed cells.

A third means of measuring the number of cells transformed by EBV makes use of the stimulation of DNA synthesis by the virus. The amount of [^{3}H]deoxythymidine ([^{3}H]-dT) incorporated into DNA by virus-exposed cells is directly proportional to the size of the inoculum. This assay is quantitative because it reflects the number of cells stimulated by the virus. Thus, it has been possible to extrapolate the total [^{3}H]-dT incorporated by virus-exposed leukocytes to a standard curve constructed from the [^{3}H]-dT incorporated by known numbers of tranformed cells in log-phase growth. According to this technique, approximately 3% of mixed primary leukocytes from umbilical-cord blood are initially stimulated into DNA synthesis and are presumably transformed (Robinson and Miller, 1975). Similar results were obtained if autoradiography was used to count the number of stimulated cells (Robinson *et al.*, 1979).

4. Physiological State of Target Cells

EBV can stimulate DNA synthesis in and ultimately transform B cells that are in a resting state. Evidence for this is based on several kinds of data. There is no correlation between the level of spontaneous DNA synthesis and susceptibility to transformation (Henderson *et al.*, 1977b). When human umbilical-cord lymphocytes are first depleted of T lymphocytes, thereby enriching for B lymphocytes, and then separated into subpopulations based on size by velocity sedimentation through linear gradients of Ficoll–Hypaque mixtures, cells that are small and show little or no spontaneous DNA synthesis are transformed by EBV with extremely high efficiency. By contrast, the large cells that show a very high rate of spontaneous DNA synthesis are transformed at a very low efficiency (Henderson *et al.*, 1977b). The data show that small, resting B lymphocytes are the usual targets for transformation by EBV. Furthermore, virus-induced DNA synthesis is maximal in those populations that contain only well-differentiated lymphocytes (Thorley-Lawson and Strominger, 1978; Robinson *et al.*, 1979); transformation does not require the presence of any other cell type, such as macrophages as suggested by Pope *et al.* (1974).

A third line of evidence that EBV transforms resting lymphocytes is that pretreatment of umbilical-cord lymphocytes with 5-bromodeoxyuridine followed by exposure to light results in the elimination of cells in DNA synthesis. Such treatment of the cells performed before inoculation with virus does not have any effect on the transformation of lymphocytes measured by either stimulating DNA synthesis or the transformed-centers assay (Henderson *et al.*, 1977b).

While it is clear that resting cells are sufficient targets for EBV transformation, little information is available regarding the susceptibility of B lymphocytes at other stages of the cell cycle or of differentiation. Treatment of lymphocytes with phytohemagglutinin (PHA) causes only a slight increase in transformation frequency (Henderson *et al.*, 1977b). A much greater increase (about 3-fold) in transformation efficiency has been observed when cells have been treated with lipopolysaccharide (LPS) from *Escherichia coli* (Henderson *et al.*, 1978). LPS is not mitogenic for umbilical cord leukocytes in that it has no effect on DNA synthesis in uninfected cells. PHA is mainly a T-cell mitogen, and the slight enhancing effects it showed on transformation efficiency may have been due to the elaboration of growth factors, which allows a greater cloning efficiency of EBV-infected cells. A population of B cells that has been stimulated by a direct mitogen, such as staphylococcal protein A, does not show increased susceptibility to transformation (Anvret and Miller, 1981).

B. Early Events of Transformation

1. Adsorption and Penetration

On the basis of kinetics experiments to determine the time after inoculation that neutralizing antibodies no longer inhibit transformation, it appears that adsorption to and penetration of B cells by EBV is completed within 1–2 hr after inoculation (Thorley-Lawson and Strominger, 1978; Wilson and Miller, 1979). The mechanism by which EBV enters cells is not yet established. In electron-microscopic examination of Raji cells superinfected with P3HR-1 virus, enveloped viral particles can be seen to fuse with cell membranes (Seigneurin *et al.*, 1977). This observation has not been made as yet in primary lymphocytes. However, fluorescence polarization studies show that the high fluorescence polarization of dye-labeled virions shows a rapid decrease after virus adsorption. These results have been interpreted as indicating fusion of the more rigid viral envelope with the more fluid host-cell membrane. Since the half-time for dye transfer for both primary cells and Raji cells is the same, the mechanism for penetration for each type of cell is thought to be similar (Rosenthal *et al.*, 1978).

Further evidence that EBV receptors are intimately related to C3 receptors is the observation showing that preincubation of cells with excess C3 blocks adsorption as measured by the transfer of dye and the reduction in the polarization of fluorescence (Rosenthal *et al.*, 1978).

2. EBNA Synthesis

The earliest EBV-associated antigen that can be detected after infection is EBNA, which appears 6–12 hr after inoculation (Menezes *et al.*, 1976; Einhorn and Ernberg, 1978; Takada and Osato, 1979; Robinson and Smith, 1981). The number of EBNA-positive cells in a culture is low initially but increases rapidly, reaching a plateau 24–36 hr after exposure to virus (Robinson and Smith, 1981). When low multiplicities of virus are used, EBNA induction follows one-hit kinetics; however, with increasingly higher multiplicities of virus, a point is reached at which no further increase in the number of EBNA-positive cells occurs, indicating that all the susceptible cells have been infected (Zerbini and Ernberg, 1982). In human umbilical-cord blood, virtually all (up to 97%) of the B cells can be induced to synthesize EBNA (Zerbini and Ernberg, 1982). At least 35–40% of the T-cell-depleted population of lymphocytes in huma adult blood can be induced to express EBNA (Robinson and Smith, 1981), but whether all B lymphocytes in this population can be infected is not known.

EBNA synthesis in primary lymphocytes occurs before any cellular DNA synthesis can be detected (Einhorn and Ernberg, 1978; Takada and Osato, 1979; Robinson and Smith, 1981). It also appears to be independent of viral DNA synthesis, since inhibitors of viral DNA synthesis such as phosphonoacetic acid (PAA), 9-β-D-arabinofuranosyladenine (ARA-A), or 1-β-D-arabinofuranosylcytosine (ARA-C) do not affect EBNA induction (Thorley-Lawson and Strominger, 1976; Takada and Osato, 1979; Robinson and Smith, 1981). Protein and RNA synthesis are required for EBNA synthesis (Takada and Osato, 1979; Ernberg, 1982). Thus, EBNA is a classic early protein. The function of EBNA in the process of transformation remains speculative. It is the only serologically defined EBV-associated antigen that is invariably expressed by all EBV-transformed lymphoid cells *in vitro* as well as *in vivo* (Lindahl *et al.*, 1974). In addition, EBNA is a DNA-binding protein (Luka *et al.*, 1977) that may function to initiate DNA synthesis in a manner analogous to the role postulated for the simian virus 40 T antigen (G. Klein *et al.*, 1980).

3. DNA Synthesis and Cellular Proliferation

During the process of transformation, EBV stimulates DNA synthesis in infected cells (Gerber and Hoyer, 1971; Gerber and Lucas, 1972; Robinson and Miller, 1975). Increased levels of [^{3}H]thymidine incorporation are found 24–36 hr after exposure to virus or about 20 hr after EBNA synthesis begins (Einhorn and Ernberg, 1978; Takada and Osato, 1978; Robinson and Smith, 1981). The amount of DNA synthesized by infected cells is linearly related to the size of the virus inoculum, a relationship that has been used to assay the biological activity of the virus (Robinson and Miller, 1975). Virus-induced stimulation of DNA synthesis requires the complete viral genome. The target size for UV or X-irradiation is the same as for transformation (Henderson *et al.*, 1979).

The DNA synthesized during transformation is predominantly cellular (Gerber and Hoyer, 1971; Henderson *et al.*, 1977b). The possibility that small amounts of viral DNA are also synthesized cannot be excluded. Cell lines transformed at low multiplicities of infection to assure infection by single infectious particles carry multiple copies of EBV DNA in every cell (Sugden *et al.*, 1979). Thus, it is possible that amplification of the viral genome is a prerequisite for transformation. Recent evidence suggests, however, that at least some of the increase in genome number per cell occurs after cell lines are established (Sugden *et al.*, 1979).

Several inhibitors of DNA synthesis appear to inhibit viral DNA synthesis at doses that do not significantly affect cellular DNA synthesis. At appropriate doses, both ARA-A and PAA exert such a discriminatory effect on EBV DNA synthesis in producer cell lines, reducing expression of viral capsid antigen without inhibiting cellular proliferation (Nyormi *et al.*, 1976; Benz *et al.*, 1978; Henderson *et al.*, 1979). These drugs have different effects on the process of transformation. Low doses of ARA-A reduce EBV-induced [^{3}H]thymidine incorporation significantly more than they do PHA-stimulated lymphocytes. However, established cell lines remain sensitive to the inhibitory effects of ARA-A (Benz *et al.*, 1978; Henderson *et al.*, 1979). By contrast, PAA, when added to cultures within 3 days after inoculation, allows [^{3}H]thymidine incorporation to proceed to a plateau level, but no further (Thorley-Lawson and Strominger, 1978). Outgrowth of transformed cells is prevented as long as the drug is present. This effect is reversible, since removal of the drug results in the resumption of DNA synthesis and the onset of proliferation. If the drug is added more than 3 days after inoculation, no inhibitory effects are seen. These observations have been interpreted as suggesting that PAA inhibits a crucial event in transformation that occurs about 72 hr after inoculation. Strominger and Thorley-Lawson (1979) have suggested that this event may involve the integration of the EBV genome into the host genome. These experiments must be interpreted cautiously, since the dose of PAA (200 μg/ml) needed to produce the results noted above caused B95-8 cells to become abnormally large and inhibited mitosis. Further, the same dose reduced PHA stimulation of normal lymphocytes by 30% (Nyormi *et al.*, 1976).

The physical association of EBNA with the chromosomes of infected cells (Reedman and Klein, 1973) has been used to time the onset of cellular proliferation in virus-exposed B lymphocytes from normal adults (Robinson and Smith, 1981). A small number of mitotic EBNA-positive cells were first seen 36–48 hr after inoculation. The proportion of EBNA-positive mitoses increased to 10% of the total cell population by 96 hr after inoculation; however, during this period, an increase in the actual number of antigen-positive cells was not evident. The high multiplicities of virus used in these experiments may have been inhibitory to cell growth; alternatively, a portion of the infected cells may not be able to proliferate (Robinson and Smith, 1981).

III. SURFACE OF EPSTEIN–BARR-VIRUS-TRANSFORMED CELLS

A. Lymphocyte Surface Markers

Lymphoid-cell lines containing the EBV genome express surface markers that are characteristic of B lymphocytes. Thus, surface immunoglobulin, complement receptors, and Epstein–Barr virus (EBV) receptors are found on most EBV-transformed cells (Shevach *et al.*, 1972; Jondal and Klein, 1973; Greaves *et al.*, 1975; A.L. Epstein *et al.*, 1976; Huber *et al.*, 1976; Jondal *et al.*, 1976; Yefenof *et al.*, 1976; Robinson *et al.*, 1977; Siegert *et al.*, 1977; Yefenof and Klein, 1977; Yefenof *et al.*, 1977). Low-affinity immunoglobulin G (IgG) Fc receptors are frequently detectable in most lines as well (Huber *et al.*, 1976). A group of EBV-positive cell lines derived from patients with a variety of hematopoietic malignancies appear to lack surface immunoglobulins, but express complement receptors (Fu and Hurley, 1979).

A number of Burkitt-lymphoma (BL) lines lack complement receptors (Budzko *et al.*, 1976; A.L. Epstein *et al.*, 1976; Yefenof *et al.*, 1977). In addition, cottontop marmoset lymphocytes transformed *in vitro* by EBV do not express complement receptors, although the original B cells had these receptors prior to transformation (Robinson *et al.*, 1977). It has been suggested that complement receptors are absent in lines that release virus (G. Klein *et al.*, 1978). This may occur if virus production selects against cells that possess receptors for complement (G. Klein *et al.*, 1979). Evidence supporting this hypothesis comes from studies of the BL line P3HR-1, which lacks detectable complement receptors, is devoid of EBV receptors, and produces relatively large quantities of virus (Yefenof *et al.*, 1977). The parent line, Jijoye, and several sublines of P3HR-1 produce little or no virus and express both complement and EBV receptors (G. Klein *et al.*, 1978). However, the correlation is not consistent, since a number of nonproducer lines derived from BL appear to lack detectable complement receptors (Budzko *et al.*, 1976; Yefenof *et al.*, 1977).

A correlation has also been found between the presence of glycoprotein (gp) 27/35 (see Section III.C) and the expression of the receptors shared by EBV and complement (G. Klein *et al.*, 1979). P3HR-1 cells synthesize very low amounts of gp 27/35 and lack these receptors, while Jijoye cells and nonproducer sublines of P3HP-1 synthesize 20–50 times more gp 27/35 (Trowbridge *et al.*, 1977; G. Klein *et al.*, 1979). In addition, antibody reacting with gp 27/35 on nonproducer cells blocks both complement and EBV receptors (Trowbridge *et al.*, 1977). However, one BL line, Rael, expresses high levels of gp 27/35, but lacks complement and EBV receptors (Trowbridge *et al.*, 1977). The possibility that gp 27/35 expression might exert some control over virus production should be studied.

When EBV-transformed cells are incubated at 37°C with fresh human, mouse, or rat serum, complement is activated via the alternate complement pathway. As a result, C3b, detectable by the formation of immune-adherence rosettes or by immunofluorescence, is deposited on the cell surface (Okada and Baba, 1974; Budzko *et al.*, 1976; Theofilopoulos and Perrin, 1977). Two groups of investigators showed that deposition of C3b on the cells after complement activation does not depend on the presence of complement receptors (Okada and Baba, 1974; Budzko *et al.*, 1976). Two other reports showed that this C3b deposition occurs only on cells that bear complement receptors (Theofilopoulos and Perrin, 1977; Yefenof *et al.*, 1977). It should be pointed out that complement activation initiated by zymosan granules results in C3b deposition on the granules, which do not have C3b receptors (Huber *et al.*, 1976).

B. Expression of HLA Antigens and β_2-Microglobulin

EBV-transformed cell lines express antigens of the human leukocyte antigen (*HLA*) complex including its invariable polypeptide subunit β_2-microglobulin (β_2m). The HLA phenotype of a lymphoblastoid-cell line (LCL) generally agrees with that of the cells of origin (Bernoco *et al.*, 1969; Papermaster *et al.*, 1969; Rogentine and Gerber, 1969). Typing of LCLs has provided several technical difficulties. Sera used for complement-mediated lysis frequently have lymphotoxins that react with LCLs and thus must be screened carefully (Crichton *et al.*, 1973). When lines have been examined over long periods of culture, the HLA type changes; sometimes new specificities have been added and other times specificities have been lost (Crichton *et al.*, 1973). The change in HLA antigens could not be explained by cross-contamination of cell cultures (Crichton *et al.*, 1973). Furthermore, newly established lines frequently had specificities not present on the parent cells (Dick *et al.*, 1973). The change in HLA expression observed in some lines has not been fully explained.

Certain anomalous reactivities with LCLs have been found with a number of typing sera (Bernoco *et al.*, 1969; Ferrone *et al.*, 1971; Dick *et al.*, 1973, 1975; Lindblom and Nilsson, 1973; Pious *et al.*, 1974; Bodmer *et al.*, 1975). Some of the antibodies in these typing sera appear to be reacting with immune-associated (Ia) antigens that are expressed on LCL cells and normal B lymphocytes (Bodmer *et al.*, 1975; Dick *et al.*, 1975). However, some of the reactivity in these sera with LCLs is due to the expression of cross-reactive HLA antigens. These new antigen specificities, which appear on cells soon after the LCLs are established and which are not present on the cells of origin (Dick *et al.*, 1973), nearly always belong to closely related groups of HLA antigens (Lindblom and Nilsson, 1973; Bodmer *et al.*, 1975). Since the amount of HLA antigens on LCLs is about 36 times greater than on the original cells (McCune *et al.*, 1975), the appearance of cross-reacting HLA antigens may simply be due to an increase in the concentration of minor determinants already on the cells. Thus, some of the "new" HLA antigens on LCLs may reflect quantitative and not qualitative changes (McCune *et al.*, 1975).

In general, BL lines synthesize quantities of HLA antigens similar to those produced by LCLs (Welsh *et al.*, 1977). However, one BL-cell line, Daudi, expresses neither HLA nor its smaller subunit, β_2-m (Nilsson *et al.*, 1974; Bodmer *et al.*, 1975). Cells from another BL line, P3HR-1, synthesize very little HLA (Ostberg *et al.*, 1975). β_2-Microglobulin is expressed on the cell surface by many BL lines in almost the same quantities as by LCLs. However, LCLs secrete relatively larger quantities of β_2m than do BL lines (Nilsson *et al.*, 1974; Ostberg *et al.*, 1975; Welsh *et al.*, 1977).

C. Ia-Like Antigens

Cells transformed by EBV express on their surface a polypeptide complex that is immunochemically distinct from HLA and β_2m and appears to behave like Ia antigens of mice. The Ia-like antigen complex has been purified from LCLs and consists of two polypeptides with molecular weights of 23,000 and 30,000 in one study (Humphreys *et al.*, 1976) and of 27,000 and 35,000 (gp 27/35) in another (Trowbridge *et al.*, 1977). This antigen complex appears to be pleomorphic (Humphreys *et al.*, 1976) and may correspond to the *HLA-D* locus, which is responsible for stimulation in the mixed-lymphocyte reaction (Geier and Cresswell, 1977). As discussed in Section III.B, the expression of gp 27/35 on transformed cells correlates fairly well with the expression of EBV and complement receptors.

D. Surface Antigens on Lymphoblastoid-Cell Lines Recognized by T Lymphocytes

There are a number of antigens on EBV-transformed cells that are recognized by T lymphocytes in assays of cell-mediated immunity. These antigens are defined functionally according to the reactions they mediate. Their characteristics are summarized in Table I.

TABLE I. Functional Characterization of Surface Antigens on Epstein–Barr-Virus-Transformed Cells Recognized in Mixed-Lymphocyte Stimulation or in Cytoxicity Assays

Assay	Stimulating cells	Target cells	Responding cells	Assay method[a]	EBV immune status of donor[a]	HLA restriction?	Nature of antigen[a]
Mixed-lymphocyte culture	Autologous or allogeneic LCL	Autologous or allogeneic LCL	T lymphocytes	$[^3H]$-TdR, incorporation	Positive or negative	No	Ia ?Neoantigens
Primary T-cell cytotoxicity	EBV-infected cells *in vivo*	Autologous or allogeneic LCL	T-lymphocytes[b] in peripheral blood	^{51}Cr release, growth inhibition	Primary EBV infection (IM)	No	LYDMA
Secondary T-cell cytotoxicity	Autologous LCL	Autologous LCL	T-lymphocytes[b] stimulated by autologous LCL *in vitro*	^{51}Cr release, growth inhibition	EBV-seropositive (5–23 weeks post-IM)	Yes	LYDMA, HLA

[a] ($[^3H]$-TdR) Tritiated thymidine; (IM) infectious mononucleosis; (LYDMA) lymphocyte-detected membrane antigen.
[b] Natural killer cells are first removed.

1. Antigens Recognized in Mixed-Lymphocyte Cultures

EBV-transformed cells are potent stimulators of allogeneic and autologous lymphocytes in mixed-lymphocyte cultures (Hardy *et al.*, 1969; Flier *et al.*, 1970; Green and Sell, 1970; Steel and Hardy, 1970; Han *et al.*, 1971; Junge *et al.*, 1971; Knight *et al.*, 1971; Golub *et al.*, 1972; Steel *et al.*, 1973). Activation of responder cells occurs equally well with lymphocytes from seropositive and seronegative individuals (Junge *et al.*, 1971; Weksler, 1976; Chang and Chang, 1980; Tanaka *et al.*, 1980). The antigens on the surface of LCLs responsible for stimulation are not known. Ia-like antigens on normal B cells will stimulate autologous T cells, and it has been postulated that these Ia antigens on LCLs are also responsible for stimulation (Wernet *et al.*, 1978; Geier and Cresswell, 1977). However, Ia antigens on LCLs may not be the only antigens involved, since autologous LCLs are more potent stimulators in mixed-lymphocyte cultures than are untransformed B cells (van de Stouwe *et al.*, 1977; Chang and Chang, 1980). Perhaps, like expression of HLA antigens, expression of Ia antigens is increased following transformation. Alternatively, other newly acquired antigens, such as EBV-specific or lymphoblast antigens, may be involved. In this regard, LCLs have been shown to express an antigen or antigen complex that cross-reacts with human placental trophoblasts, transformed human epithelial-cell lines (HeLa, Hep_2, L132), and cottontop marmoset erythrocytes (Yeh *et al.*, 1982). This antigen is not found on normal human lymphocytes.

2. Antigens on LCLs Recognized by Cytotoxic T Lymphocytes

T-cell-mediated cytotoxicity reactions against LCLs appear to be of two types, primary and secondary. In primary cytotoxicity reactions, T cells are activated *in vivo* during infectious mononucleosis (IM). The stimulating cells are presumed to be EBV-infected cells. The cytotoxic cells have been demonstrated by ^{51}Cr-release or growth-inhibition assays only in peripheral blood of patients for 2–4 weeks following the onset of IM (Svedmyr and Jondal, 1975; Royston *et al.*, 1975; Hutt *et al.*, 1975; Rickinson *et al.*, 1977; Svedmyr *et al.*, 1978). This type of cytotoxicity is not HLA-restricted; i.e., it does not require the sharing of HLA antigens between target and effector cells (Seeley *et al.*, 1981). Susceptibility of target cells to cytotoxicity in this system depends on the presence of the EBV genome, since only EBV-genome-positive lines are killed and EBV-genome-negative lines are not. The antigen recognized on LCLs by primary cytotoxic T cells has been termed the EBV-associated lymphocyte-detected membrane antigen (LYDMA). Whether this putative antigen is virus-encoded is not known.

In secondary cytotoxicity reactions, an *in vitro* sensitization step is necessary to generate cytotoxic T cells. Cultivation of peripheral-blood lymphocytes from EBV-seropositive individuals with autologous EBV-infected cells, either established LCLs or freshly inoculated B cells, results in the generation of effector T cells that lyse LCLs or inhibit their proliferation (Thorley-Lawson *et al.*, 1977; Moss *et al.*, 1978, 1979; Rickinson *et al.*, 1979; Misko *et al.*, 1980; Sugamura and Hinuma, 1980; Tanaka *et al.*, 1980). Although LCLs stimulate DNA synthesis in autologous mixed-lymphocyte cultures from seronegative individuals, cytotoxic T cells are not generated (van de Stouwe *et al.*, 1977; Moss *et al.*, 1979; Chang and Chang, 1980; Tanaka *et al.*, 1980). These observations seem to indicate that previous EBV infection results in the development of EBV-specific memory T cells that can be reactivated when challenged with antigen *in vitro*. The major antigen recognized by secondary cytotoxicity is also thought to be LYDMA (Rickinson *et al.*, 1981). However, by contrast to the absence of HLA restriction observed with primary cytotoxic T cells during IM, recognition of EBV-transformed cells by cytotoxic T cells generated *in vitro* is, in fact, HLA-restricted. Thus, secondary cytotoxicity measured by cell lysis or growth inhibition is greatest when effector and target cells share one or more HLA-A or -B determinants (Misko *et al.*, 1980; Rickinson *et al.*, 1980b, 1981). In addition, this type of cytotoxicity can be blocked by monoclonal antibodies against an antigenic determinant common to HLA-A, B, and C molecules or against β_2m (Rickinson *et al.*, 1981). Rickinson *et al.* (1980a) have shown that memory T cells that give rise to secondary cytotoxic cells *in vitro* are not present in peripheral blood during IM and do not develop for 5–23 weeks after acute disease.

E. Surface Glycoproteins of Transformed Cells

Surface glycoproteins on cells from EBV-genome-positive and -negative cell lines have been studied using the galactose-oxidase-catalyzed tritiated sodium borohydride labeling technique (Nilsson *et al.*, 1977a). LCLs have a surface glycoprotein pattern that closely resembles that found on mitogen-stimulated B lymphoblasts. By contrast, BL-cell lines have a less differentiated pattern very similar to that of resting B lymphocytes. In addition, BL-cell lines have two pairs of glycoproteins not found on LCLs. These are of approximate molecular weights 87,000–85,000 and 71,000–69,000. These glycoproteins are not virus-specific, since they are also found on cells of EBV-genome-negative lymphoma-cell lines (Nilsson *et al.*, 1977a).

An alteration in sialic-acid-containing fucosyl glycopeptides has been described in many types of tumor cells or in their derived cell lines and appears to be a reliable marker for neoplastic change. This alteration consists of an increase in the molecular size and in sialic acid content of these fucosyl glycopeptides. Recently, van Beek *et al.* (1979) found that EBV-positive LCLs and BL lines and EBV-genome-negative lines all expressed these abnormal glycopeptides. Thus, LCLs appear to have some degree of a neoplastic phenotype. The role that these surface alterations play in the pathogenesis of lymphoid neoplasia such as BL is not known.

IV. IMMUNOGLOBULIN SYNTHESIS BY EPSTEIN–BARR-VIRUS-TRANSFORMED CELLS

Most Epstein–Barr virus (EBV)-transformed cells maintain the principal functional characteristic of the B lymphocytes from which they originated, namely, immunoglobulin synthesis. Burkitt-lymphoma (BL) lines and lymphoblastoid-cell lines (LCLs) differ in their patterns of immunoglobulin synthesis as well as other biological properties. Thus, immunoglobulin production by each category of transformant will be considered separately.

A. Immunoglobulin Synthesis in Burkitt-Lymphoma and Derived Cell Lines

A majority of BL-cell lines show some evidence of immunoglobulin synthesis, but many lines that do not produce immunoglobulins have been described (Fahey *et al.*, 1966; Tanigaki *et al.*, 1966; Wakefield *et al.*, 1967; Hinuma and Grace, 1967; Osunkoya *et al.*, 1968; Takahashi *et al.*, 1969a; van Furth *et al.*, 1972; Bechet *et al.*, 1974). BL-tumor cells also usually show immunoglobulin synthesis, but a variable proportion of tumor biopsies have been immunoglobulin-negative (Osunkoya *et al.*, 1968; van Furth *et al.*, 1972; Fialkow *et al.*, 1973; Gunven *et al.*, 1980).

1. Heavy-Chain Synthesis by BL cells

IgM is the most common class of immunoglobulin expressed in BL tumors and in BL-cell lines; however, IgG appears to be synthesized more commonly by both tumors and established lines than has been appreciated in previous reviews (Nilsson and Ponten, 1975; Nilsson, 1979). Most early studies were limited to small numbers of cell lines. Some investigators encountered IgM expression most frequently (Wakefield *et al.*, 1967; Takahashi *et al.*, 1969a), while others found that a greater proportion of the lines they examined produced IgG (Fahey *et al.*, 1966; Finegold *et al.*, 1967). The accumulated published data on BL lines were summarized by Sherr *et al.* (1971). Of 27 BL lines that had been examined, 12 produced IgM, 5 produced IgG, 3 produced both IgM and IgG, and 6 did not synthesize immunoglobulins. Only one BL line synthesized IgA (Osunkoya *et al.*, 1968). Later reports have shown that high proportions of newly established BL lines produced IgG alone or in combination with IgM (van Furth *et al.*, 1972; Bechet *et al.*, 1974).

The patterns of immunoglobulin production by cells in BL biopsies have been studied using two different methods, and conflicting results have been obtained. One method involves the measurement of immunoglobulin synthesis *de novo* by the incorporation of radiolabeled amino acids that, when secreted, can be detected in supernatant culture fluids by immunoelectrophoresis and autoradiography. The other method is the demonstration of membrane immunoglobulins on cell surfaces by immunofluorescence. These techniques measure different parameters of immunoglobulin synthesis that appear to be under separate cellular control (McCune *et al.*, 1980). Since the class of immunoglobulins expressed on the cell surface may differ from that secreted into the culture medium, neither of these assays when used alone reveals the full repetoire of heavy chains that are produced by a population of cells (Litwin *et al.*, 1974b; van Boxel and Buell, 1974; Premkumar *et al.*, 1975). Surface immunofluorescence, when applied to the examination of tumor biopsies, has the further disadvantage of not being able to distinguish whether immunoglobulins are actually synthesized by the cells on which they are seen or whether such immunoglobulins are reacting with membrane antigens as specific antibody or are passively bound by receptors for immune complexes (E. Klein *et al.*, 1968; Gunven *et al.*, 1980).

By biosynthetic techniques, BL cells were found to secrete relatively little immunoglobulin; only 50% of biopsies produced immunoglobulin (van Furth *et al.*, 1972). IgG was the predominant heavy chain secreted, though IgM was also secreted by some tumors (Osunkoya *et al.*, 1968; van Furth *et al.*, 1972). Tumor cells and the cell lines derived from them had similar patterns of immunoglobulin synthesis, suggesting that the lines were representative of the tumor cells. Thus, it was unlikely that secreted IgG was produced by nonmalignant lymphocytes within the biopsies.

IgM was the most common immunoglobulin class detected on BL cells by surface immunofluorescence (E. Klein *et al.*, 1968; Gunven *et al.*, 1980). The surface IgG seen on some cells generally gave weaker staining and was found to disappear with time in culture. Some of this IgG appeared to be coating the cells (E. Klein *et al.*, 1968; Gunven *et al.*, 1980). In only one instance has information concerning surface and secretory immunoglobulin expression by BL biopsies been cited in the literature, although not as a formal study. Some of the BL biopsies and cell lines that secreted IgG described by van Furth *et al.* (1972) were found to express surface IgM by immunofluorescence. These results emphasize the importance of using more than one type of assay to study immunoglobulin synthesis. For this reason, it is unfortunate that a recent large study of BL biopsies (Gunven *et al.*, 1980) was limited to the examination of surface immunoglobulin.

2. Light-Chain Synthesis by BL Cells; Evidence for Monoclonality

Most BL lines in which immunoglobulin production can be demonstrated synthesize only one type of light chain, lambda or kappa. Since individual cells can produce only one type of light chain, these observations have been used to support the hypothesis that BL arises from a single clone of malignant cells. Further support for the hypothesis that BL is a monoclonal disease comes from the observation that BL biopsies and cell lines generally express only one of two possible glucose-6-phosphate dehydrogenase (G-6-PD) isoenzyme patterns (Fialkow *et al.*, 1973; Bechet *et al.*, 1974). These data, however, are highly biased, since this marker can be evaluated only in females who are heterozygotes at the X-linked *G-6-PD* locus, a relatively small proportion of BL patients. Some BL lines, albeit a minority, synthesize two types of light chains or express both G-6-PD isoenzyme patterns, indicating that they are of multicellular origin (Sherr *et al.*, 1971; van Furth *et al.*, 1972; Fialkow *et al.*, 1973; Bechet *et al.*, 1974).

The critical experiments needed to test the extent to which BL is a monoclonal or a polyclonal disease have not been performed. The interpretation of assays for surface immunoglobulin expression on BL-biopsy cells has been hampered by the problem of antibody coating of tumor cells (Gunven *et al.*, 1980). Furthermore, the evaluation of immunoglobulin synthesis and secretion in cell lines derived from tumors in stationary suspension or organ culture is clouded by the possibility that nontumor lymphoid cells present in BL biopsies may grow out in culture (Fialkow *et al.*, 1973; Nilsson, 1979). The question might best be settled by the direct cloning of tumor cells in soft agar. Those clonal lines that meet the current criteria for BL lines rather than LCLs could then be evaluated with respect to surface and secretory heavy- and light-chain production and with respect to G-6-PD isoenzyme patterns or with other assays of genetic polymorphism.

B. Immunoglobulin Production by Lymphoblastoid-Cell Lines

Essentially all LCLs synthesize immunoglobulins and tend to secrete relatively more immunoglobulins than do the BL lines (Sherr *et al.*, 1971). An exceptional group of EBV-genome-positive cell lines that lack the capacity to synthesize immunoglobulins were reported by Fu and Hurley (1979). These lines, derived from patients with various hematopoietic diseases, resemble BL cells in their morphology and might have been derived by transformation of abnormal rather than normal B lymphocytes.

1. Heavy- and Light-Chain Synthesis by LCLs

The patterns of immunoglobulin production by LCLs are highly variable. Lines arising from peripheral-blood lymphocytes during infectious mononucleosis (IM) or from lymph-node or tonsil cells grown in suspension or organ culture contain several populations of cells that secrete mixtures of immunoglobulins (Glade and Chessin, 1968; Nilsson, 1971; Bechet *et al.*, 1974). These initially polyclonal lines show a gradual decline in immunoglobulin synthetic capacity with time (Wakefield *et al.*, 1967; Finegold *et al.*, 1967; Glade and Chessin, 1968) and usually evolve toward the secretion of a single type and class of both light and heavy chains. This selection process is random and does not favor a particular immunoglobulin class (Nilsson, 1971; Bechet *et al.*, 1974). Some lines maintain the stable production of more than one class of heavy chains. The existence of individual transformed cells that synthesize more than one heavy-chain class (Bloom *et al.*, 1971; Takahashi *et al.*, 1969b; Fahey and Finegold, 1967; van Boxel and Buell, 1974) indicates that immunoglobulin genes in these cells are not mutually exclusive.

As with BL cells, the class of immunoglobulins present on the surface of LCL cells may differ from that found in the cytoplasm or secreted into the medium (Litwin *et al.*, 1974b; van Boxel and Buell, 1974; Premkumar *et al.*, 1975). Many LCLs express surface IgD that, when present, is always found in association with surface IgM (van Boxel and Buell, 1974; Gordon *et al.*, 1977). However, multiple classes of immunoglobulin molecules may be expressed on cells with or without IgD. Within a line, a portion of cells lack detectable surface immunoglobulin; however, studies involving the isolation of clones have shown that each cell is capable of immunoglobulin synthesis (Hinuma and Grace, 1967; Takahashi *et al.*, 1969b; Bloom *et al.*, 1971; Litwin and Lin, 1976).

The immunoglobulins produced by most LCLs are complete molecules identical to those produced by normal B cells (Finegold *et al.*, 1967; Baumal *et al.*, 1971). Excess synthesis of light chains is frequently observed (Wakefield *et al.*, 1967; Fahey *et al.*, 1971; Litwin *et al.*, 1974a), but this has been noted in normal B cells as well (Wakefield *et al.*, 1967). Some cells may produce only light or only heavy chains (Takahashi *et al.*, 1969a).

During logarithmic growth, LCLs secrete immunoglobulin at a rate of 2–3 $\mu g/10^6$ cells per day (Fahey *et al.*, 1971). The proportion of the protein synthesized by LCL cells that is immunoglobulin has been estimated to be 0.8–4.5%, compared to 53% in myeloma cells (Gordon *et al.*, 1977). However, in some lines, as much as 30% of the protein synthesized may be immunoglobulin (Baumal *et al.*, 1971).

2. Polyclonal Activation of B Cells by EBV *in Vitro*

When normal B cells are exposed to EBV *in vitro*, a number of cells become infected and begin to proliferate. It is not surprising, then, that immunoglobulin secreted soon after inoculation is of a polyclonal pattern (Rosen *et al.*, 1977; Kirchner *et al.*, 1979; Bird and Britton, 1979). For this reason, EBV has been termed a polyclonal activator of B cells. The kinetics of immunoglobulin synthesis induced by EBV have been studied using an indirect plaque assay (Bird and Britton, 1979). Plaques become detectable 3–4 days after exposure of cells to virus. The number of plaque-forming cells reaches maximal levels by 7 days and then decreases rapidly to unstimulated levels by 12 days. These results are similar to those obtained when pokeweed mitogen was used to stimulate cells. The decline in the number of antibody-producing cells seen after 7 days in culture may have been due to exhaustion of nutrients in the medium.

The appearance of immunoglobulin-secreting cells detected in plaque assays is independent of helper T cells and depends on an intact EBV genome (Kirchner *et al.*, 1979; Bird and Britton, 1979). Thus, EBV appears to induce some degree of differentiation of infected B cells. However, differentiation is limited, since LCL cells do not become fully mature plasma cells; usually, no more than 20–30% of the cells synthesize cytoplasmic immunoglobulins (Glade and Chessin, 1968; J. Robinson, unpublished). By contrast, a majority (70–80%) of EBV-infected cells in peripheral blood during acute IM have been shown to be differentiated toward plasma cells (Robinson *et al.*, 1981). It is possible that immune regulatory mechanisms operative during IM cause greater numbers of infected cells to differentiate *in vivo* than occurs *in vitro*.

3. Specific Antibody Production by LCLs

Immunoglobulins produced by LCLs have rarely been assigned specific antibody function. Several LCLs producing IgM that agglutinated glutaraldehyde-treated erythrocytes have been reported (Steel *et al.*, 1974; Joss *et al.*, 1976). Recently, Steinitz *et al.* (1977) established cell lines that secrete specific antibody. Antigen-binding B cells were first isolated by rosette formation with antigen-coated autologous erythrocytes. These cells were then transformed by EBV. Cell lines that produce antibodies to the hapten 4-hydroxy-3,5-dinitrophenacetic acid (NNP), the hapten trinitrophenyl (TNP), and Group A streptococcal carbohydrate have been established using the same technique (Kozbor *et al.*, 1979; Steinitz *et al.*, 1979a,b). All the lines produce antibody of the IgM class. Lines that produce mononclonal antibodies were established by the reselection and cloning of antigen-binding cells after the LCLs were established (Steinitz *et al.*, 1979b).

V. MORPHOLOGY AND GROWTH CHARACTERISTICS

A. Comparison of Lymphoblastoid-Cell Lines and Burkitt-Lymphoma Lines

The morphological characteristics of lymphoblastoid-cell lines (LCLs) and Burkitt-lymphoma (BL) lines have been reviewed recently by Nilsson (1979). When examined by light or electron microscopy, cells from both types of lines share many features with mitogen- or antigen-stimulated lymphoblasts (Pulvertaft, 1964b; Moore *et al.*, 1968). Most stained cells contain an immature nucleus, several large nucleoli, and abundant basophilic and pyroninophilic cytoplasm (M.A. Epstein *et al.*, 1966; Moore *et al.*, 1968; Fahey *et al.*, 1971). The size and shapes of cells in LCLs vary greatly; the cell diameter ranges from 8 to 22 μm (Nilsson and Ponten, 1975). A majority of LCL cells resemble immature lymphoblasts, but medium-sized and smaller, more differentiated cells, some resembling plasmacytes, and occasional multinucleated cells may be present (Moore *et al.*, 1968; Gerber and Monroe, 1968). By contrast, BL lines consist mainly of immature lymphoblasts (M.A. Epstein *et al.*, 1966; Pope *et al.*, 1967; Nilsson and Ponten, 1975) that tend to be more uniform in size and somewhat smaller in diameter (range 9–13 μm, mean 10–11 μm) than in LCLs (Nilsson and Ponten, 1975).

In transmission electron microscopy, the nuclei of both LCL and BL cells contain finely granular chromatin and the cytoplasm contains abundant ribosomes and mitochondria. The Golgi apparatus is well developed in LCLs; endoplasmic reticulum is usually sparse, although some cells show more differentiated characteristics (Moore *et al.*, 1968; Heyden and Heyden, 1974; Nilsson and Ponten, 1975). BL cells lack a well-developed Golgi apparatus, and the ribosomes are not attached to endoplasmic reticulum (M.A. Epstein and Achong, 1965; M.A. Epstein *et al.*, 1966; Pope *et al.*, 1967; Moore *et al.*, 1968). Fat droplets, which are numerous in BL cells, are less frequently seen in LCLs (Nilsson, 1979). In cells from both types of lines, cytoplasmic annulate lamellae and projections or doublings of the nuclear membrane can be demonstrated (M.A. Epstein and Achong, 1965; Chandra *et al.*, 1968; Gerber and Monroe, 1968; Recher *et al.*, 1969).

Morphological differences between cells of LCLs and BL lines are most strikingly distinct when cells are viewed directly in culture or by scanning electron microscopy (SEM) (Clarkson, 1967; Nilsson and Ponten, 1975). LCL cells are highly pleomorphic and motile. Cells constantly change shape; uropod formation is common, giving cells a "hand-mirror" configuration. The cells exhibit what has been termed "Ping-Pong" locomotion (Nilsson, 1979), in which a uropod forms at one cell pole and then at another so that each cell continuously changes its direction. By SEM or by immunofluorescence staining for the contractile protein actin, long villous projections are either restricted to one pole of the cell or distributed evenly over the surface (Fagraeus *et al.*, 1975; Nilsson and Ponten, 1975). These cells adhere rapidly to fibroblast feeder cultures, assuming a wide variety of shapes; they do not adhere to glass or plastic surfaces. The morphology of LCLs in suspension cultures is striking. Generally, large dense clumps form that sometimes resemble flakes or doughnuts. Several investigators have reported differences in the density, size, and shape of the cell clumps in lines transformed by different strains of Epstein–Barr virus (EBV) (Katsuki and Hinuma, 1975; Levitt *et al.*, 1980).

B. Chromosomal Abnormalities

In early studies of the cytogenetics of EBV-transformed cells, precise mapping of structural rearrangements on chromosomes was not possible, since banding techniques had not yet become available. Thus, these studies were limited to describing major aberrations of chromosome structure and number. Most BL-cell lines, BL tumor cells, and LCLs from many sources were found to be diploid or near diploid (Stewart *et al.*, 1965; Miles and O'Neill, 1967; W. Henle *et al.*, 1967; Tomkins, 1968), although major changes in chromosome mode were seen in cell lines after more than 1 or 2 years in culture (Huang and Moore, 1969; Steel *et al.*, 1971). Cloning of one BL line appeared to induce chromosomal changes (Ikeuchi *et al.*, 1971). Numerous inconsistent structural abnormalities were found in many cell lines. One of the most frequently observed abnormalities was a marker chromosome that was a prominent subterminal constriction on the long arms of a group C chromosome. This marker was found in variable proportions of the cells of both LCLs and BL lines (W. Henle *et al.*, 1967; Kohn *et al.*, 1967, 1968; Miles and O'Neill, 1967; Pope *et al.*, 1968; Miles *et al.*, 1968; Nadkarni *et al.*, 1969; Huang *et al.*, 1970; Whang-Peng *et al.*, 1970).

The application of chromosome banding techniques to the study of EBV-transformed cells led to the identification of a chromosome marker, a 14q+ translocation, in cells of BL tumor biopsies and cell lines (Manolov and Manolova, 1972). This marker has been found in most but not all BL tumors from Africa, the United States, and Europe, but has not been seen in EBV-positive LCLs (Jarvis *et al.*, 1974; Zech *et al.*, 1976; Kaiser-McCaw *et al.*, 1977; Philip *et al.*, 1977). The translocation has been shown to be reciprocal between the end segments of the long arms of chromosomes 8 and 14 (Zech *et al.*, 1976; Manolova *et al.*, 1979). Identical translocations have been described in several EBV-genome-negative non-Burkitt lymphoid malignancies (Jarvis *et al.*, 1974; Zech *et al.*, 1976; Fukuhara and Rowley, 1978; Mark *et al.*, 1978; Mitelman *et al.*, 1979a).

Not all 14q+ translocations found in lymphoid malignancies are identical; a variety of donor chromosomes and variable break points on chromosome 14 have been described (Fukuhara *et al.*, 1979). Variant translocations involving 2;8 or 8;22 have been described in several cases of BL (van den Berg *et al.*, 1979; Berger *et al.*, 1979; Miyoshi *et al.*, 1979).

Recently, the 8;14 translocation was seen in cells from an inguinal lymph node of a patient with nasopharyngeal carcinoma (Mitelman *et al.*, 1979b). The histology did not show evidence of malignancy, and the node was negative for EBV genomes. It was suggested that the 8;14 translocation in this case might represent a premalignant state. The development of the 14q+ marker may be a critical change in a cell that renders it malignant, or it may be one of many steps in the development of the malignant phenotype.

C. Tumorigenicity of Epstein–Barr-Virus-Transformed Cells

1. Cloning in Semisolid Medium and Heterotransplantation

BL-cell lines are derived by outgrowth of tumor cells *in vitro* and thus represent truly neoplastic cells. LCLs, on the other hand, arise from normal B lymphocytes transformed *in vitro* or *in vivo*, and the malignant potential of these cells is uncertain. LCLs are polyclonal and lack the karyotypic abnormalities found in BL cells. Attempts have been made to test the oncogenic potential of LCLs of diverse origins, but reliable markers for malignancy are lacking. Two parameters that have been tested are the ability of cells to grow as colonies in semisolid media and the capacity of cells to cause tumors after heterotransplantation. These parameters have been used as indicators of neoplastic transformation in several nonlymphoid cell systems (reviewed by Nilsson *et al.*, 1977b).

In general, LCLs have a lower cloning efficiency in agar than do BL lines (Hinuma and Grace, 1967; Imamura and Moore, 1968; Huang *et al.*, 1969; Nilsson and Ponten, 1975; Nilsson *et al.*, 1977b). However, this difference is only relative, since colonies of EBV nuclear antigen (EBNA)-positive cells can be readily grown in agarose from peripheral blood of patients with infectious mononucleosis (IM) (Hinuma and Katsuki, 1978; Katsuki *et al.*, 1979; Robinson *et al.*, 1980a) or from normal blood lymphocytes following *in vitro* infection with EBV (Katsuki *et al.*, 1977; Sugden and Mark, 1977).

In several studies, lethal human lymphoid tumors were induced by heterotransplantation of both BL and LCL cells. The animals used in these experiments were newborn rats (Southam *et al.*, 1969a), rats made tolerant to transformed lymphocytes (Southam *et al.*, 1969b), newborn Syrian hamsters, or newborn Syrian hamsters or mice treated with antilymphocyte serum (ALS) (Adams *et al.*, 1971; Deal *et al.*, 1971). ALS-treated newborn hamsters developed serially transplantable tumors after inoculation with cells from an IM-derived cell line. These tumors produced human IgG and IgM in quantities that could be detected in serum by immunodiffusion (Adams *et al.*, 1971). Similar serially transplantable immunoglobulin-secreting tumors developed in ALS-treated newborn hamsters after inoculation with peripheral-blood leukocytes from three patients with IM (Adams *et al.*, 1973). Unfortunately, whether these tumors contained the EBV genome was not established.

BL-cell lines have been shown to induce tumors when injected subcutaneously into nude mice; LCLs lack this capacity (Nilsson *et al.*, 1977b). The failure of LCLs to grow in subcutaneously inoculated nude mice may be due to locally efficient natural killer (NK)-cell activity (Herberman, 1978). However, in a more recent study (Giovanella *et al.*, 1979), BL lines as well as LCLs induced tumors after intracerebral inoculation in adult nude mice or after subcutaneous inoculation in newborn nude mice (Nilsson *et al.*, 1979). The results of the studies on the heterotransplantability of LCL cells suggest that the immune mechanisms in the test animals are important determinants in the development of tumors. Thus, both LCLs and BL lines cause tumors when the host lacks a fully developed immune system or has been further compromised by treatment with ALS or when cells are inoculated in immunologically privileged sites. However, in the subcutaneous space of nude mice, BL cells appear to have a significantly greater oncogenic potential than do LCLs. Thus, LCL cells seem to be conditionally neoplastic by heterotransplantation assays.

2. Tumorgenicity of EBV in New World Primates

Inoculation of animals of several species of New World primates with EBV or autologous cells transformed by EBV *in vitro* results in a spectrum of responses. In cottontop marmosets (CTMs), cell-free virus induces fatal malignant lymphoma in about one third of the animals inoculated. Other CTMs inoculated with EBV develop either clinically inapparent infection or a transient nonmalignant lymphoproliferative disease (Shope *et al.*, 1973; Miller *et al.*, 1977). Lymphoma can also be induced in CTMs by inoculation with autologous transformed cells. In this case, it is not clear whether the tumor is caused by the cells transformed *in vitro* or by virus that is released from them (see Section VI.D). CTMs appear to be uniquely susceptible to oncogenesis by EBV, since *Callithrix* marmosets develop only mild lymphoproliferative disease after virus exposure and frank malignant lymphoma does not occur (Falk *et al.*, 1976). Disseminated lymphoproliferative disease has been described in one EBV-inoculated owl monkey (M.A. Epstein *et al.*, 1973). Neither squirrel monkeys nor woolly monkeys develop tumors when injected with cell-free virus or autologous transformed cells (Shope and Miller, 1973; Andiman and Miller, 1978). However, EBV-transformed squirrel monkey cells appear to be conditionally neoplastic, since immunosuppressed animals with malaria developed disseminated lymphoproliferative disease shortly after inoculation with autologous cells (Leibold *et al.*, 1976).

The lymphomas induced in CTMs by EBV morphologically resembled immunoblastic sarcoma in humans (Miller *et al.*, 1978). Each tumor cell contains EBNA and multiple copies of the EBV genome (Werner *et al.*, 1975; Miller *et al.*, 1977). It is not known whether these tumors are monoclonal or polyclonal. In one instance, a marker chromosome was identified in the cells of a tumor, suggesting a monoclonal or oligoclonal origin (Rabin *et al.*, 1977).

3. Tumorigenicity of EBV-Transformed Lymphoid Cells in Man

BL cells are unquestionably malignant in man; however, an etiological role for EBV in BL has not yet been established. If BL is caused by EBV, tumors develop months or years after primary infection (de Thé *et al.*, 1978). It has been suggested that cells transformed by EBV during primary infection undergo an evolution during a period of chronic proliferation. As a result of multiple cell divisions, a critical cytogenetic change, the 8;14 reciprocal translocation, may render one or more clones of EBV-genome-positive cells malignant (G. Klein, 1980). The factors that could be responsible for the proposed chronic proliferation may be environmental, but remain hypothetical.

During IM, there is a limited polyclonal proliferation of EBV-infected cells. These cells express EBNA, show plasmacytic differentiation during acute disease, can undergo mitosis *in vivo*, and can grow directly into LCLs in semisolid medium (G. Klein *et al.*, 1976; Katsuki *et al.*, 1979; Robinson *et al.*, 1980b, 1981). Usually, the proliferation of EBV-infected cells is stringently controlled by the host environment, but in certain individuals, EBV causes unchecked proliferation resulting in polyclonal B-cell lymphoma. Tumors consisting of EBV-genome-positive, EBNA-positive cells have developed in patients with subtle immunodeficiencies (Robinson *et al.*, 1980a; Thestrup-Pedersen *et al.*, 1980), in renal-transplant recipients (Marker *et al.*, 1979; Crawford *et al.*, 1980; Hanto *et al.*, 1981), and in male members of kindreds with an X-linked disorder (Purtilo, 1980).

The 8;14 translocation associated with BL is not found in these tumors, although several other chromosomal abnormalities have been observed (Robinson *et al.*, 1980a; Thestrup-Pedersen *et al.*, 1980; Hanto *et al.*, 1981). The development of tumors in these groups of patients appears to depend on the transformation of B lymphocytes by EBV *in vivo* and a defect in host immune mechanisms that normally control the proliferation of EBV-infected cells. Patients with the X-linked lymphoproliferative syndrome have decreased NK-cell activity (Sullivan *et al.*, 1980); however, the nature of the defect or defects in immunity responsible for the growth of tumors in these groups of patients has not been clearly defined. Polyclonal B-cell lymphomas induced by EBV may be regarded as opportunistic neoplasms.

VI. EPSTEIN–BARR VIRAL GENOME EXPRESSION IN TRANSFORMED CELLS

In this section, we shall review the Epstein–Barr virus (EBV)–cell relationship that occurs in cells that have been infected *in vivo* and in cell lines that have been established directly from blood or lymphoid cells or by means of *in vitro* transformation.

A. Types of Cell–Virus Interactions

The expression of EBV in the cell that harbors it is influenced by two general factors: various properties of the *host* cell and the constitution of the EBV genome contained therein. At this time, however, it is not possible to go beyond this generalization to a more specific molecular definition of those host or viral functions that modulate expression. It is possible only to define in operational terms several different types of cell–virus interactions.

1. Permissive Cell

The cell type that permits *full virus expression* and replication of infectious progeny has not been cultivated *in vitro*, and only hints are available about its identity *in vivo*. EBV is regularly found in the saliva of infected individuals (Morgan *et al.*, 1979). It is present in the oropharyngeal secretions of 20% of healthy individuals who possess antibodies to EBV and in approximately 50% of seropositive persons who have received immunosuppressive drugs to maintain renal allografts or as treatment for collagen vascular disease (Strauch *et al.*, 1974). It is likely that some cell in the salivary gland is the source of this virus and therefore permits a full cycle of viral growth. Samples taken by cannulation of the parotid duct are positive for virus. Cells in the salivary gland are positive for EBV DNA by cytohybridization (Wolf *et al.*, personal communication). However, the exact identity of this cell is still unclear; it may be a ductal or glandular epithelial cell or a lymphoid cell.

As noted above, there is no cell fully permissive of viral replication that has been cultivated *in vitro*. Most tissue-culture cells appear to lack virus receptors. This block to infection has recently been overcome by several different techniques. DNA that is microinjected will lead to the expression of early antigen (EA) (Graessmann *et al.*, 1980). Virus receptors can be transplanted onto tissue-culture cells by a process of cell fusion (Volsky *et al.*, 1980). Such cells will now accept infection by mature virus and permit the viral replicative cycle to occur fully or partially. It has recently been possible to transfect cultures of human placental cells with EBV DNA (Miller *et al.*, 1981). Some of these placental cells release infectious EBV; thus, they permit a full cycle of viral replication *in vitro*.

2. Producer Cell Lines

Certain types of EBV-transformed cells permit a full cycle of viral replication to occur in a proportion of the cells. These virus *producer cell lines* are derived from several sources. A few human lines established from patients are available that release virus in amounts adequate for study of their biological and structural properties. These lines are QIMR-WIL from a patient with leukemia, 883L from a patient with mononucleosis, and AG876 from a Ghanaian child with Burkitt lymphoma (BL) (Pope *et al.*, 1969; Blacklow *et al.*, 1971; Pizzo *et al.*, 1978). Another human line, Jijoye, originally from a patient with BL, is a virus producer (Ragona *et al.*, 1980; Sairenji and Hinuma, 1980). A subclone of this line,

HR-1, has been widely studied because the virus it releases has lost the transforming phenotype and is biologically and structurally complex (Hinuma *et al.*, 1967; Miller *et al.*, 1974; Fresen *et al.*, 1977; Delius and Bornkamm, 1978).

Another source of virus producer lines is cottontop marmoset cells that are transformed *in vitro* (Miller and Lipman, 1973a,b). A variety of marmoset lines are now available containing virus that originated from patients with infectious mononucleosis (IM), BL, and nasopharyngeal carcinoma (NPC). Virus released from marmoset cells has been the main source of structural comparison of EBV isolates from different diseases and ecological niches.

A third type of virus producer is epithelial lines derived from NPC. These lines cannot be cultivated *in vitro*, but are propagated by passage in athymic nude mice (G. Klein *et al.*, 1974a). Virus is produced by such lines and has been propagated by transformation of human and marmoset cells (Trumper *et al.*, 1977; Crawford *et al.*, 1979).

3. Nonproducer Cell Lines

Most cells that harbor the EBV genome, both *in vivo* and in cell culture, do not permit mature virus production and are thus referred to as "nonproducers." This type of cell–virus relationship probably represents a wide spectrum of controls on the expression of the EBV genome. All the cells studied *in vivo* contain the EBV nuclear antigen (EBNA), which is an invariant marker for the presence of the EBV genome (Reedman and Klein, 1973; Lindahl *et al.*, 1974). These include biopsies of NPC, BL, diffuse polyclonal lymphoma, and circulating infected B cells present in the blood of patients with IM. Furthermore, most cell lines derived from the blood or tumors of man are nonproducers. Likewise, cell lines established by *in vitro* immortalization of human lymphocytes are, with certain exceptions, nonproducers. It is likely that some nonproducer lines, of which the prototype is Raji, never produce any mature virus. Others may intermittently shed minute amounts that may be demonstrated by electron microscopy or by X-irradiation of the cell line and cocultivation with primary lymphoid cells (Wilson and Miller, 1979).

Among the nonproducer cells, there seems to be a spectrum of virus expression. Some contain only the EBNA, others also produce EA, and still others synthesize viral capsid antigens (VCAs). There is also variation in the proportion of cells in different lines that synthesize viral DNA (Moar and Klein 1979). These results suggest the existence of controls of several different viral genes.

4. Converted Cell Lines

Two additional types of cell–virus interactions have been produced in cell culture. They may also occur in man, but this is not well documented. One process, termed *conversion*, describes the relationship between EBV and cell lines that were derived from BL but that initially lacked the EBV genome. These cell lines have the characteristics of B lymphocytes, but do not express EBNA (G. Klein *et al.*, 1974b). A few cells with EBNA appear early after the genome-free lines are exposed to EBV. The majority of cells in the line remain EBNA-negative. Gradually, the proportion of cells that are EBNA-positive increases with time, for it appears that EBNA-positive cells have selective growth advantages *in vitro* (Steinitz and Klein, 1975). Lines converted by the HR-1 virus spontaneously synthesize low levels of EA, and a rare cell shows VCA. These antigens have not been seen in cells converted by B95-8. Since none of the converted lines has been found to release virus, they resemble nonproducer lines. The important difference is that the EBV genome did not provide the initial immortalizing signal to these cells. One important implication of these lines is that some EBV-genome-negative B-cell lymphomas may become superinfected with EBV *in vivo*.

5. Superinfected Cell Lines

Still another type of cell–virus relationship is represented by a group of lymphoid lines that were originally nonproducers and that were subsequently *superinfected* with a different strain of EBV. When the NC37 or Raji line is superinfected with high multiplicities of the HR-1 strain of EBV, it is converted into a *producer* cell line (Yajima and Nonoyama, 1976; Traul *et al.*, 1977; Fresen *et al.*, 1978). The virus produced by these superinfected cells exhibits altered biological properties. For example, some transforming viruses have been recovered from Raji and NC37 cells superinfected with the nontransforming HR-1 strain of EBV. Structural analyses of the viruses released from the superinfected cells suggest that they are recombinants between the superinfecting virus and the EBV genomes originally resident in the nonproducer cell lines (Fresen *et al.*, 1979).

The implication of the phenomenon of superinfection is that a virus released by a cell line may not necessarily represent the virus that was originally responsible for transformation, but rather another viral variant that superinfected the established line. It is clear from the superinfection experiments that a continuous lymphoblastoid-cell line (LCL) may harbor several EBVs that differ in their structure and biological properties.

6. Hybrid Cell Lines

A final type of cell–virus relationship is that represented by hybrid cell lines (Glaser and Rapp, 1972). A variety of different kinds of somatic-cell hybrids have been formed between different LCLs harboring EBV genomes or between EBV-carrying lymphoblasts and other monolayer cell cultures. These hybrids have been used to explore the controls on the expression of the EBV genome (Glaser and Nonoyama, 1974). For example, in a hybrid between the HR-1 producer line and the D98 epithelial line, no spontaneous expression of VCA or virions occurs, although one parent, the HR-1 line, does produce mature virus. Conversely, a hybrid between D98 and Raji cells is more likely to undergo VCA and virion synthesis following "induction" with 5-iodo-2-deoxyuridine (IUdR) than is the parental Raji strain. These types of results with somatic-cell hybrids suggest a complex network of controls on the latent state of the EBV genome.

B. Genome Content and Complexity

1. Genome Copy Number

Biopsies of BL, NRC, and diffuse polyclonal lymphoma contain multiple copies of the EBV genome (zur Hausen *et al.*, 1970). Multiple copies of EBV DNA are also found in cell lines established following *in vitro* exposure to EBV (Sugden *et al.*, 1979). In the *in vitro* system, it is possible to transform lymphocytes clonally and at very low multiplicities of infection, conditions that make it unlikely that a cell line resulted from infection with more than one virus particle. Even under these conditions, the resultant transformed lines still contain many DNA copies per cell. Cells transformed under different physiological conditions, i.e., with and without mitogens, large cells and small ones, all have multiple copies of EBV DNA. Most of this DNA is episomal (Anvret and Miller, 1981). In some instances, it has been found that clones of cells vary in their copy number over a period of months after they were initially established, usually with an upward drift of copy number (Sugden *et al.*, 1979). However, other clones have a constant level of EBV genomes. It has not yet been established whether genome copy number increases during the transformation process, i.e., whether there is synthesis of viral DNA before transformation.

Genome copy number is usually measured in total cellular DNA extracted from a mass culture. It is clear, however, that what is being measured is an "average" among cells in a given line. This is most clearly the case in virus producer lines, which usually have an average of several hundred copies per cell (Miller *et al.*, 1976). This average results from some cells that are virus factories and presumably contain many thousands of EBV DNA copies and some cells that contain only a few copies, probably only in the latent form. The clearest proof of this idea comes from experiments with inhibitors of viral DNA synthesis such as phosphonoacetic acid (PAA). Treatment with the drug seems to abolish viral DNA synthesis in the virus factories without affecting the synthesis of the latent genomes. Producer cell lines that have been grown continuously in the presence of PAA exhibit a dramatic drop in copy number (Summers and Klein, 1976). Since PAA seems to affect viral DNA polymerases more drastically than cellular ones, the implication is that in the virus factories, viral DNA is synthesized by means of viral DNA polymerases, but that latent viral genomes are replicated under control of cellular enzymes.

In certain cell lines, such as Raji, that have been examined for their EBV DNA copy number over periods of several years, the number of EBV genomes is remarkably stable at about 50–100 copies per cell (zur Hausen *et al.*, 1972; Adams, 1979). This suggests the existence of some regulatory event that permits viral DNA to replicate in synchrony with host-cell DNA. This is not to say, however, that every cell in the Raji line contains the same number of EBV DNA copies, for single cell clones of Raji cells vary in their content of EBV DNA (Bister *et al.*, 1979).

An exception to the rule that cell lines that contain EBV DNA contain multiple copies is some of the converted lines. Most of these lines seem to have relatively small numbers of EBV DNA copies (Andersson and Lindahl, 1976).

The function served by the many copies of EBV DNA present in transformed cells is still obscure. The most obvious explanation is the requirement for a "gene-dosage" effect to transform the cells, but this is not proven.

2. Genome Complexity

Unlike cells transformed by herpes simplex virus or adenovirus, which contain only a fragment of the entire viral genome, LCLs derived by exposure to EBV seem to contain a nearly complete copy of viral DNA. Several types of evidence, both biological and biochemical in nature, support this conclusion. Total cellular DNA of several clones of cells transformed *in vitro* by EBV was analyzed by a combination of restriction-endonuclease digestion followed by Southern transfer and nucleic acid hybridization (Sugden, 1977). Nearly all the DNA fragments found in EBV DNA taken from virions were detected in the total DNA of transformed cells. Episomal EBV DNA has been isolated from BL lines as well as from cells transformed *in vitro* (Lindahl and Adams, 1976). While some variation is observed in the size of the intracellular episomes, as measured by electron microscopy, the circular intracellular DNA is very close in size to the DNA present in virions (Adams, 1979).

A biological approach has also been used to demonstrate that information encoding mature virions is present in some nonproducer cells. Sublines of EBV-transformed neonatal B cells, which were apparent nonproducers since they did not express VCA or release biologically active virus into the culture medium, were found to release infectious virus when they were X-irradiated and cocultivated with primary leukocytes (Wilson and Miller, 1979). Recovery of virus by the technique of X-irradiation and cocultivation was blocked if antiserum was present during the first few hours of cocultivation, but not thereafter. These findings indicate that EBV in biologically active form is present on or near the surface of a small number of cells in "nonproducer" human umbilical-cord lymphocytes transformed *in vitro*. Approximately 1 in 10^4 cells in such lines contains complete virus.

Radiobiological experiments also provide indirect evidence for the participation of a large part of the EBV genome in the initiation of transformation (Henderson *et al.*, 1978). The lethal dose of X-irradiation required to inactivate the immortalizing properties of EBV (LD_{37} = 60,000 rads) is approximately the same as the dose needed to destroy plaque formation by herpes simplex virus, which has a genome of approximately the same size. No dose of X-irradiation or UV irradiation increases the transforming potential of EBV, a finding that is in marked contrast to observations in similar experiments with papovaviruses, adenoviruses, and herpes simplex viruses. It is clear in the case of the cytolytic DNA viruses that the enhancing effect of irradiation on the transformation process is in part related to the inactivation of cytopathic functions of the genome. The initial interaction of transforming strains of EBV with their target cells is not cytolytic; therefore, cell transformation and persistence of the entire viral genome are compatible.

One can only speculate on the reasons for the requirement for large portions of EBV DNA to initiate transformation (Fig. 2). Three general hypotheses can be offered: (1) One is a requirement for the presence of the circular viral plasmid to ensure viral DNA replication. Since circularity is presumably maintained by means of the direct repeats at the termini (Kintner and Sugden, 1979), damage anywhere along the genome impairs the ability to maintain circularity. (2) A second explanation is that amplification of the plasmid, possibly via viral-encoded DNA-synthesizing enzymes, is related to cell transformation through a gene-dosage

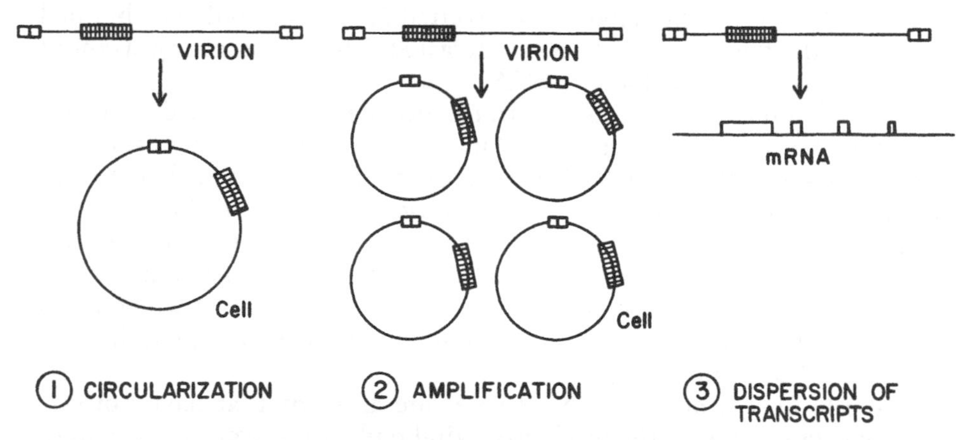

FIGURE 2. Three hypotheses to account for the apparent participation of the entire EBV genome in transformation: (1) The formation of circular intracellular genomes may require intact DNA. (2) The replication and amplification of the episome may require intact DNA. (3) Transcripts representing transforming functions are widely dispersed on the genome.

effect. Transformation selects for cells that contain many EBV DNA copies as the result of an intact set of genes involved in viral DNA synthesis. (3) A third explanation, and one for which there is the most direct experimental evidence, is that transcripts representing functions that may immortalize the cell are coded from genes that are widely dispersed on the genome (Dambaugh *et al.*, 1979; Rymo, 1979).

C. Epstein–Barr Viral Antigen Expression in Transformed Cells

1. Types of Antigens

A series of antigen systems have been defined in EBV-transformed cells using immunofluorescence techniques. A nuclear antigen (EBNA) is found in all cells that harbor EBV DNA, both *in vivo* and *in vitro* (G. Klein, 1975). It appears to be associated with the chromatin and is found on metaphase chromosomes. It has DNA-binding properties, but otherwise its function is not known (G. Klein, 1980). Viruses with immortalizing and with nonimmortalizing biotypes (see Section VI.C.4) contain the capacity to induce EBNA, for both types of viruses can cause the conversion of genome-negative BJAB cells to lines that express EBNA (Clements *et al.*, 1975; Fresen and zur Hausen, 1976). The detection of EBNA requires amplification of the antigen–antibody reaction by use of anticomplement immunofluorescence (Reedman and Klein, 1973). Whether this is due to low amounts of the antigen in the cells or to peculiar structural properties of the antigen is not now understood.

Biological experiments indicate that there is more than one species of EBNA. When BJAB cells were "converted" by addition of the HR-1 virus, two types of EBNA were observed, "faint granular" and "brilliant" (Fresen *et al.*, 1977). Subclones of converted BJAB cells could be obtained in which the two forms of EBNA segregated. The transforming strain of EBV induced only "brilliant" EBNA.

Another nuclear antigen system recently defined is termed "RANA" for "rheumatoid-arthritis-associated nuclear antigen" (Alspaugh and Tan, 1976; Catalano *et al.*, 1979). RANA appears to be distinguishable from EBNA. Its detection does not require anticomplement immunofluorescence, but it can be demonstrated by classic indirect immunofluorescence. The RANA antigen is relatively stable to heating, but, unlike EBNA, is destroyed by fixation in acetone or methanol. Antibodies to RANA are not limited to patients with rheumatoid arthritis, but develop during IM (Catalano *et al.*, 1980).

While EBNA, and possibly RANA, are antigens that have nothing whatsoever to do with the productive viral replicative cycle, all the other EBV antigen systems seem to be part of the life cycle of the virus leading ultimately to viral maturation. An antigen system called early antigen

(EA) appears spontaneously and can also be induced (see Section VI.C.3) in may LCLs. It is termed "early antigen" because its expression does not require viral or cellular DNA synthesis (Gergely *et al.*, 1971). There are at least two different components of EA that are defined on the basis of their distribution within the cell (diffuse and restricted) and on the basis of their sensitivity to fixation in methanol. When EA is spontaneously expressed in a cell line, only a fraction of the cells display the antigen, unlike EBNA, which is present in every cell. The biological function of EA is not known.

Two other antigen systems correlate with viral particle formation and are found only in cell lines that are synthesizing virions. These are "membrane antigen" (MA) and viral capsid antigen (VCA).

MA is present on the surface of that subpopulation of cells in a line that is synthesizing virions. It is spontaneously expressed in producer cell lines such as B95-8 and HR-1; it also appears on the surface of Raji cells following superinfection with the HR-1 virus strain (Qualtière and Pearson, 1979). MAs are also likely to be shared with antigens on the virus envelope that are responsible for eliciting neutralizing antibodies to the virus (Thorley-Lawson, 1979). Cells that are synthesizing MA can absorb neutralizing antibodies from antibody-containing sera; nonproducer cells, which are not synthesizing MA, will not absorb neutralizing antibody (Coope *et al.*, 1979).

VCA is present in both the nucleus and cytoplasm of cells producing virus. It is not found in nonproducer cell lines. VCA also appears in Raji cells when they are superinfected with a high multiplicity of HR-1 virus. Cells synthesizing VCA have probably undergone cytopathic changes and are no longer able to replicate. It has been found that cells that contain VCA also contain virions detectable by electron microscopy.

2. Spontaneous Expression of Viral Antigen

Most EBV-transformed cells studied *in vivo*, whether blood lymphocytes from patients with IM, BL, or NPC cells, express only the EBNA antigen and do not produce the antigens associated with the viral replicative cycle. However, once cells infected *in vivo* are placed in culture, they begin to produce additional viral antigens such as EA, VCA, and MA. Apparently, those cells that produce these antigens are destroyed through cytopathic effects presumably mediated by viral-encoded proteins. In a portion of cells taken from blood or biopsy viral, replicative functions are not manifest. Such cells are able to form continuous cell lines.

In the continuous lines, all the cells display EBNA, and a portion of the cells spontaneously produce the antigen series represented by EA, VCA, and MA. Exactly which antigens are produced, and the relative proportions of the cells that are activated to synthesize these antigens, vary from one LCL to the next. Age and the species of donor whose cells are transformed influence the extent of antigen expression (Fig. 3). Neonatal human cells rarely produce significant amounts of VCA, while adult cells generally have a low level of VCA. Human cells are less productive than squirrel monkey or marmoset cells when transformed by the same virus strain. It is clear from this type of evidence that the host cell plays an important role in regulating viral antigen expression, but the level at which this control operates is not understood. Although only a portion of cells in certain lines express VCA or MA, it is evident that all the cells in a line possess the capacity to do so. Single cell clones isolated from lines resemble their parent—each clone contains a portion of cells with viral antigen (Fig. 3) (Miller and Lipman, 1973b).

There is a general impression that with time in culture, there is often a decrease in the number of cells in a line that are expressing antigens associated with the productive cycle. This is particularly true of the HR-1 line from which nonproducer sublines have readily been obtained. This observation suggests, first, that spontaneous changes occur, either in host or virus, that favor nonexpression of virus. Second, it seems that cells that do not express the antigens of the productive cycle have a selective growth advantage *in vitro* over those that do.

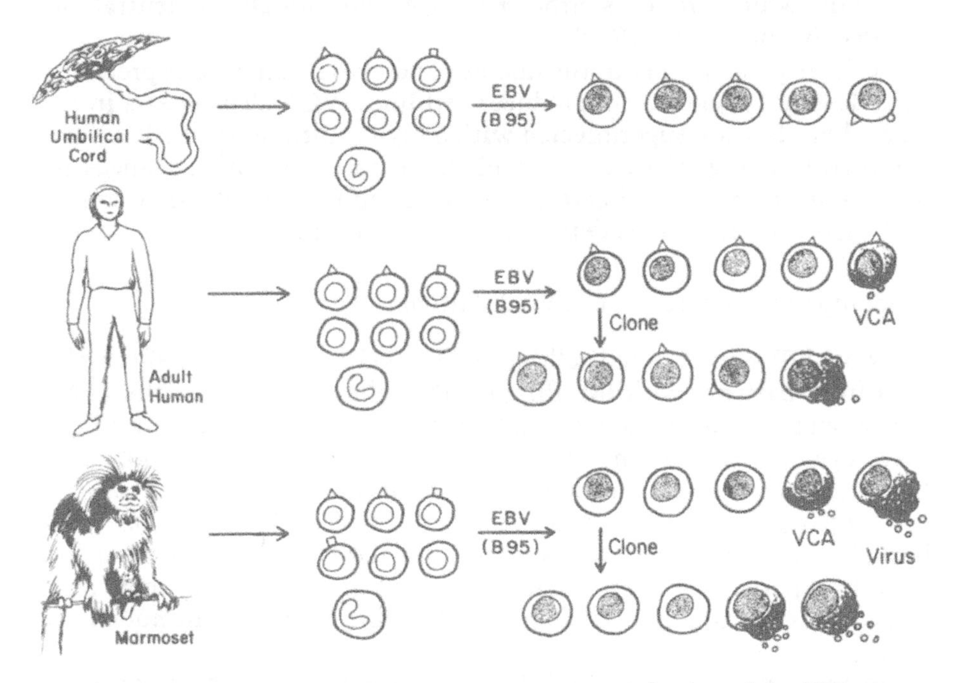

FIGURE 3. Host-determined differences in spontaneous induction of the EBV genome. Transformed cells from human umbilical-cord lymphocytes contain minute amounts of cell-associated virus, but they do not make VCA or release virus. Adult human cells express VCA and release small amounts of virus. Marmoset lines have higher levels of VCA and release the most extracellular virus.

3. Induction of Viral Antigen Expression in Established Lines

A variety of treatments can be shown to affect the extent of viral antigen expression in LCLs. For the most part, the inducing agents do not activate a *new* set of antigens, but increase the level of expression of antigens that are already being spontaneously produced. A prototype for such studies has been the Raji cell line, which spontaneously expresses a very low level of EA.

Halogenated pyrimidines are potent inducing agents in certain non-producing cell lines (Hampar *et al.*, 1972; Gerber and Lucas, 1972). IUdR and BUdR induce EA in Raji cells, but not VCA or virions. When induction of EA synthesis has been examined in synchronized cultures, it has been shown that the critical time for the presence of the inducing agent is the first 60 min of S phase. EA appears 5–7 hr later (Hampar *et al.*, 1974).

Halogenated pyrimidines do not seem to affect antigen synthesis in lines that are engaged in virion production. However, tumor promoters derived form croton oil [such as 12-*O*-tetradecanoyl phorbol-13 acetate (TPA)] are potent inducers of latent EBV (zur Hausen *et al.*, 1978). In nonproducer lines, such as Raji, TPA increases the level of expression of EA (Bister *et al.*, 1979). In cell lines that are already virus producers, TPA induces additional VCA expression and virion release. TPA has proven very useful in increasing the yield of virions for biochemical analysis.

Other inducing agents belong to a variety of different classes of chemical and physical agents. Some of them appear specific for certain EBV-carrying cell lines. For example, the EB_3 line increases its production of VCA in a medium with low arginine content (W. Henle and G. Henle, 1968). Antibody to surface IgM causes induction of EA expression in Raji and other lines with surface IgM (Tovey *et al.*, 1978). Anti-IgM is synergistic with IUdR. An inhibitor of DNA synthesis, *n*-butyrate, is also an inducer of EA expression (Luka *et al.*, 1979; Saemundsen *et al.*, 1980). Some lines are induced to produce VCA by heat, UV irradiation, and even platinum compounds (Lai *et al.*, 1973; Vonka and Kutinova, 1973; Vonka *et al.*, 1972). In no instance is there clear understanding of the mechanisms behind induction.

In certain studies, attempts have been made to probe the influence of the host cell on the induction process by placing the viral genome in a slightly different host-cell environment. For example, sublines of Raji with polyploid genomes have been selected by colcemid treatment, and the ease of induction of EA has been studied in the diploid parent line and in the tetraploid sublines (Shapiro *et al.*, 1978). The ease of induction of EA by IUdR was reduced in all the tetraploid lines even though they contained slightly increased numbers of EBV genomes. These experiments suggested that the restrictive effects of RAji cells on the EBV genomes resident within them could be increased by amplifying the number of host-cell genes.

Another approach to the problem of host-cell control of antigen expression has involved the use of somatic-cell hybrids. For example, IUdR treatment of a Raji–D98 cell hybrid leads to VCA and virion synthesis, whereas similar treatment of the Raji parent leads only to EA synthesis (Glaser and Nonoyama, 1974). In this instance, the results suggest that restrictions imposed by one component of the hybrid (Raji) are overcome by the other (D98).

4. Antigen Expression following Infection of Lymphoid Cells with EBV

EBV antigens have been sought in three different types of *in vitro* infection systems, namely, the infection of primary lymphocytes, the superinfection of EBV-genome-carrier lymphoid lines, and the infection of continuous B-lymphocyte lines that lack the EBV genome (see Table II).

In primary lymphocytes, the transforming strains induce EBNA, but no other antigens associated with the productive replicative cycle, even though such antigens may ultimately be expressed once a cell line has been established. Although the nontransforming HR-1 strain is cytolytic to primary lymphocytes, thus far no viral-induced antigens have been detected in fresh lymphocytes inoculated with this strain.

The cytolytic HR-1 virus induces a full range of antigens associated with the viral replicative cycle in Raji cells. If low multiplicities of virus are added (< 1–10 particles per cell), only EA expression is detected. If larger amounts of virus are added, multiplicities in the range of 1000 particles per cell, new viral DNA, VCA, and virions are made as well. The evidence that the virus produced by infection of Raji cells at high multiplicity is newly made, rather than input virus, comes from a detailed electron-microscopic study in which evidence of herpesvirus maturation was found in superinfected cells examined serially (Seigneurin *et al.*, 1977). New capsids were made as early as 7 hr after infection and were enveloped at various sites such as the nuclear membrane, Golgi membranes, and membranes of cytoplasmic vacuoles.

TABLE II. Cytopathic Effects and Antigen Expression in Lymphoid Cells by Nontransforming and Transforming Strains of Epstein–Barr Virus

	HR-1 virus			B95-8 virus		
Effects and antigens	Primary human lymphocytes	Raji cells	BJAB cells	Primary human lymphocytes	Raji cells	BJAB cells
Cytopathic effects	+	+	(+)	–	–	–
EBNA	–	NA	+	+	NA	+
EA	–	+	+	–	–	–
VCA	–	+	–	–	–	–
MA	–	+	–	–	–	–

Even though Raji cells contain receptors for transforming virus, these cells do not seem to express any new antigen following addition of the B95-8 or other transforming strains of virus.

Conversion of EBV-genome-negative continuous B-cell lines, such as BJAB and Ramos, to permanent EBV carriage and expression of EBNA can be achieved by both biotypes of EBV, transforming and nontransforming. However, lines that are converted to EBV carriage by HR-1 virus spontaneously synthesize low levels of EA, and a rare cell shows VCA, whereas EA synthesis is not associated with conversion by B95-8 (Klein *et al.*, 1974c).

D. Release of Mature Virus

1. Factors That Affect Yield of Extracellular EBV

Only a few human cell lines transformed by EBV release appreciable amounts of extracellular virus. Most lines are nonproducers or at best shed minute amounts of virus into the supernatant phase of the culture. The available evidence suggests that host-cell functions, rather than a specific composition of the viral genome, are important in viral maturation and release. The best evidence on this question comes from experiments in which the same virus strain (B95-8) was used to transform cells of different species of New World primates. Woolly monkey cells transformed by the virus were nonproducers, or very low-level producers, as were human cells; squirrel monkey lymphocytes shed intermediate amounts of virus, while cottontop marmoset (CTM) cells released the most virus (Frank *et al.*, 1976). Since EBV-transformed CTM cells release virus, this system has been widely used to propagate different EBV strains from various geographic regions and from patients with different EBV-associated diseases to serve as a basis for the structural comparison of different virus strains.

Enhanced virus release by EBV-transformed CTM cells is not solely the result of increased activation of the viral genome, but is more likely due to some process of maturation, envelopment, or release. For example, in a comparison of transformed human, squirrel monkey, and CTM clones, there was only a 2- to 10-fold difference in the number of activated cells, as detected by VCA expression, but there was a 10^3- to 10^4-fold increase in virus yield by CTM as contrasted with human cells.

Once a producer cell line has been established, various conditions of culture affect virus release. For example, the source of serum influences the amount of virus shed by some lines. Apparently, an inhibitor of virus release is present in some batches of calf serum (Sairenji and Hinuma, 1975). Temperature and the duration of culture since the last feeding of medium are also important. For most producer lines, virus release is optimal 10–14 days after the cells have been subcultured. Virus release by the Jijoye line and its HR-1 subline is apparently facilitated if the cells are grown at 33°C (Sairenji and Hinuma, 1980).

Apparently, a number of cell lines produce mature virus that remains at or near the cell surface and is not released efficiently. Transforming virus can be recovered from certain of these producer cell lines when they are lethally X-irradiated and cocultivated with fresh lymphocytes or after they are repeatedly frozen and thawed and then cocultivated (Wilson and Miller, 1979). The earliest demonstrations of the immortalizing properties of EBV were made using, as a source of virus, lethally X-irradiated or frozen extracts of producer cell lines (W. Henle *et al.*, 1967; Miller *et al.*, 1969). The mechanisms behind this cell-associated behavior of mature virus have not been studied, but it appears to be a characteristic of most viruses in the herpes group. It might be due to some defect in the final stages of virus budding. Alternatively, some virus-producer lines may still retain virus receptors, and any virus that is made reattaches to the cell membrane.

2. Biological Properties of Released Virus

a. Transforming Virus

The virus released by most producer lines has the transforming phenotype; i.e., it is capable of converting normal human B lymphocytes into continuous LCLs. The biological activity of this virus can be quantitated in one of several assays (Miller, 1980). The level of transforming activity as judged by the morphological appearance of the cells can be titrated by end-point dilution, usually in a microtiter system (Moss and Pope, 1972; Henderson *et al.*, 1977a). Each microwell contains the same number of cells (about $2–4 \times 10^5$), and serial dilutions of virus are added. Transformation is scored at 6–8 weeks. Marmoset lines such as B95-8 that release considerable amounts of transforming virus usually have about 10^5 transforming units/ml culture fluid. There are about 5×10^6 physical particles and about 5×10^6 EBV DNA molecules (about 1 ng) per milliliter of such cultures (Sugden and Mark, 1977). Therefore, approximately 1 in 50 particles or DNA molecules is a transforming unit. There is no evidence that these 50 particles or genomes need to collaborate to initiate one transformation event, for transformation kinetics indicate a "one-hit" event.

Another quantitative assay for EBV of the transforming biotype measures stimulation of DNA synthesis (Robinson and Miller, 1975). This is a much more rapid assay for virus, since the results are known after several days. The extent of incorporation of [^{3}H]deoxythymidine into acid-insoluble material is proportional to the amount of virus added to the leukocytes. By autoradiography, the number and types of cells stimulated into DNA synthesis by the virus can be measured (Robinson *et al.*, 1979).

b. Nontransforming Virus

The cytolytic EBV variant P3J-HR-1 has been extensively studied, and there are a number of different assays available to measure its biological activity. The cell-killing properties of the virus can be measured by following its ability to inhibit colony formation by Raji cells, an assay that has even formed the basis for an EBV microneutralization (Rocchi and Hewetson, 1973).

HR-1 virus induces EA in both Raji and BJAB cells, another basis for measuring its activity (G. Henle *et al.*, 1970) Radiobiological inactivation curves and dilution curves provide evidence that the proportion of the HR-1 genome that is required to induce EA differs in genome-positive and genome-negative target cells. In the former, there is apparently some sort of interaction between the superinfecting and resident genome. In the genome-negative cells, antigen induction is solely the result of input virus. The new proteins synthesized in superinfected Raji cells have been characterized (see Chapter 3), but the biological function of these proteins is not understood.

If high multiplicities of the HR-1 virus variant are used to superinfect certain cell lines that ordinarily do not release virus, new virions are made. These new virions have shown a spectrum of biological properties. Some are transforming, others are cytolytic, still others do not seem to have either property (Yajima and Nonoyama, 1976). It is not yet clear what is the origin of these new viruses. A number of possibilities may be mentioned: They may be recombinants between the input HR-1 virus and the endogenous EBV in the cells. They may be minor components of the input HR-1 virus; they may be endogenous viruses that have been activated. Much further correlative work is needed on the structure and biological properties of the viruses produced as the result of superinfection.

The HR-1 virus is cytolytic to fresh B lymphocytes and interferes with transformation by EBV strains with this property (Miller *et al.*, 1974). Cell killing is specific for B cells, and there is no effect on T cells (Steinitz *et al.*, 1978). Surprisingly, no new identifiable antigens are detectable in the fresh lymphocytes infected with HR-1 virus.

The HR-1 virus seems to have lost many of the properties of the transforming biotypes. It does not stimulate DNA synthesis, and it is not tumorigenic in marmosets (Falk *et al.*, 1974). It is of interest, however, that HR-1 genomes can still code for EBNA, for various EBV-genome-negative cell lines express EBNA after superinfection with HR-1 virus. This finding indicates that a failure to code for EBNA is *per se* not the defect that is characteristic of the HR-1 virus. Several hypotheses might be proposed to account for the nontransforming behavior of the HR-1 variant: The genome is deleted in a region adjacent to the internal repeats (Bornkamm *et al.*, 1980). However, there is no evidence yet from biological experiments that this region codes for a transforming function. The HR-1 genome may have sequences that are "antitransforming," particularly those genes that are responsible for the cytolytic function of the virus. Alternatively, the HR-1 genome may contain transforming genes that are misordered, thereby allowing viral replicative rather than transforming functions to predominate.

The resolution of this important difference between the two strains, through a combination of biological, genetic, and structural studies of the genome, should ultimately prove useful in defining the mechanisms of lymphoid-cell immortalization by Epstein–Barr virus.

REFERENCES

Adams, A., 1979, The state of the virus genome in transformed cells and its relationship to host cell DNA, in: *The Epstein–Barr Virus* (M.A. Epstein and B.G. Achong, eds.), pp. 155–183, Springer-Verlag, Berlin, Heidelberg, New York.

Adams, R.A., Hellerstine, E.E., Pothier, L., Foley, G.E., Lazarus, H., and Stuart, A.B., 1971, Malignant potential of a cell line isolated from the peripheral blood in infectious mononucleosis, *Cancer* **27**:651–657.

Adams, R.A., Pothier, L., Hellerstein, E.E., and Boileau, G., 1973, Malignant immunoblastoma, immunoglobulin synthesis and the progression to leukemia in heterotransplanted acute lymphoblastic leukemia, chronic lymphatic leukemia, lymphoma, and infectious mononucleosis, *Cancer* **31**:1397–1407.

Alspaugh, M.A., and Tan, E.M., 1976, Serum antibody in rheumatoid arthritis reactive with a cell associated antigen, *Arthritis Rheum.* **19**:711–719.

Andersson, M., and Lindahl, T., 1976, EBV DNA in human lymphoid cell lines: *In vitro* conversion, *Virology* **73**:96–105.

Andiman, W., and Miller, G., 1978, Properties of Epstein–Barr virus transformed woolly monkey lymphocytes, *Proc. Soc. Exp. Biol. Med.* **157**:489–493.

Anvret, M.A., and Miller, G., 1981, Copy number and location of Epstein–Barr viral genomes in neonatal human lymphocytes transformed after separation by size and treatment with mitogens, *Virology* **111**:47–55.

Armstrong, D., 1966, Serial cultivation of human leukemic cells, *Proc. Soc. Exp. Biol. Med.* **122**:475–481.

Baumal, R., Bloom, B. and Scharff, M.D., 1971, Induction of long term lymphocyte lines from delayed hypersensitive human donors using specific antigen plus Epstein–Barr virus, *Nature (London) New Biol.* **230**:20–21.

Bechet, J.M., Fialkow, P., Nilsson, K., and Klein, G., 1974, Immunoglobulin synthesis and glucose-6-phosphate dehydrogenase as cell markers in human lymphoblastoid cell lines, *Exp. Cell Res.* **89**:275–282.

Belpomme, D., Minowada, J., and Moore, G.E., 1972, Are some human lymphoblastoid cell lines established from leukemic tissues actually derived from normal leukocytes?, *Cancer* **30**:282–287.

Benz, W., Siegel, P., and Baer, J., 1978, Effects of adenine arabinoside on lymphocytes infected with Epstein–Barr virus, *J. Virol.* **27**:475–482.

Berger, R., Bernheim, A., Weh, H.-J., Flandrin, G., Daniel, M.T., Broust, J.C., and Colbert, N., 1979, A new translocation in Burkitt's tumor cells, *Hum. Genet.* **53**:11–12.

Bernoco, D., Glade, P.R., Broder, S., Miggiano, V.C., Hirschhorn, K., and Ceppellini, R., 1969, Stability of HL-A and appearance of other antigens (LIVA) at the surface of lymphoblasts grown *in vitro*, *Folia Haematol.* **54**:795–895.

Bird, A.G., and Britton, S., 1979, A new approach to the study of human B-lymphocyte function using an indirect plaque assay and a direct B cell activator, *Immunol. Rev.* **45**:41–67.

Bister, K., Yamamoto, N., and zur Hausen, H., 1979, Differential inducibility of Epstein–Barr virus in cloned, non-producer Raji cells, *Int. J. Cancer* **23**:818–825.

Blacklow, N.R., Watson, B.K., Miller, G., and Jacobson, B.M., 1971, Mononucleosis with heterophile antibodies and EB virus infection: Acquisition by an elderly patient in hospital, *Am. J. Med.* **51**:549–552.

Bloom, A.D., Choi, K.W., and Lamb, B.J., 1971, Immunoglobulin production by human lymphocytoid lines and clones: Absence of genic exclusion, *Science* **172**:382–384.

Bodmer, W.F., Jones, E.A., Young, D., Goodfellow, P.N., Bodmer, J.G., Dick, H.M., and Steel, G.M., 1975, Serology of human Ia antigens detected in lymphoid lines: An analysis of the 6th workshop sera, in: *Histocompatibility Testing* (F. Kiessmeyer-Nielsen, ed.), pp. 677–684, Munksgaard, Cophehagen.

Bornkamm, G.W., Delius, H., Zimber, U., Hudewentz, J., and Epstein, M.A., 1980, Comparison of Epstein–Barr virus strains of different origin by analysis of the viral DNAs, *J. Virol.* **35**:603–618.

Brodsky, A.L., and Hurd, E.R., 1974, Enhanced establishment of lymphoblastoid cell lines with the Ficoll Hypaque density gradient, *Proc. Soc. Exp. Biol. Med.* **147**:612–615.

Budzko, D.B., Lachmann, P.J., and McConnell, I., 1976, Activation of the alternative complement pathway by lymphoblastoid cell lines from patients with Burkitt's lymphoma and infectious mononucleosis, *Cell. Immunol.* **22**:98–109.

Catalano, M.A., Carson, D.A., Slovin, S.F., Richman, D.D., and Vaughan, J.H., 1979, Antibodies to Epstein–Barr virus determined antigens in normal subjects and in patients with seropositive rheumatoid arthritis, *Proc. Natl. Acad. Sci. U.S.A.* **76**:5825–5828.

Catalano, M.A., Carson, D.A., Niederman, J.D., Feorino, P., and Vaughan, J.H. 1980, Antibody to the rheumatoid arthritis nuclear antigen: Its relationship to *in vivo* Epstein–Barr virus infection, *J. Clin. Invest.* **65**:1238–1242.

Chandra, S., Moore, G.E., and Brandt, P.M., 1968, Similarity between leukocyte cultures from cancerous and noncancerous human subjects: An electron microscopic study, *Cancer Res.* **28**:1982–1989.

Chang, R., and Chang, Y., 1980, Activation of lymphocytes from Epstein–Barr virus-seronegative donors by autologous Epstein–Barr virus-transformed cells, *J. Infect. Dis.* **142**:156–162.

Chang, R.S., Hsieh, M.-W., and Blankenship, W., 1971, Initiation and establishment of lymphoid cell lines from blood of healthy persons, *J. Natl. Cancer Inst.* **47**:469–477.

Clarkson, B., 1967, Formal discussion: On the cellular origins and distinctive features of cultured cell lines derived from patients with leukemias and lymphomas, *Cancer Res.* **26**:2483–2488.

Clements, G.B., Klein, G., and Povey, S., 1975, Production by EBV infection of an EBNA-positive subline from an EBNA-negative human lymphoma cell line without detectable EBV DNA, *Int. J. Cancer* **16**:125–133.

Coope, D., Heston, L., Brandsma, J., and Miller, 1979, Cross neutralization of infectious mononucleosis and Burkitt lymphoma strains of Epstein–Barr virus with hyperimmune rabbit antisera, *J. Immunol.* **123**:232–238.

Crawford, D.H., Epstein, M.A., Bornkamm, G.W., Achong, B.G., Finerty, S., and Thompson, J.L., 1979, Biological and biochemical observations on isolates of EB virus from the malignant epithelial cells of two nasopharyngeal carcinomas, *Int. J. Cancer* **24**:294–302.

Crawford, D.H., Thomas, J.A., Janossy, G., Sweny, P., Fernando, O.N., Moorhead, J.F., and Thompson, J.H., 1980, Epstein–Barr virus nuclear antigen positive lymphoma after cyclosporin A treatment in a patient with renal allograft, *Lancet* **1**:1355–1356.

Crichton, W.B., Dick, H.M., and Steel, C.M., 1973, Detection of HL-A antigens on cultured lymphoblastoid cells, *Symp. Ser. Immunobiol. Standardization* **18**:11–115.

Dambaugh, T., Nkrumah, F.K., Biggar, R.J., and Kieff, E., 1979, Epstein–Barr virus RNA in Burkitt tumor tissue, *Cell* **16**:313–322.

Deal, D.R., Gerber, P., and Chisar, F.V., 1971, Heterotransplantation of two human lymphoid cell lines transformed *in vitro* by Epstein–Barr virus, *J. Natl. Cancer Inst.* **47**:771–780.

Delius, H., and Bornkamm, G.W., 1978, Heterogeneity of Epstein–Barr virus. III. Comparison of a transforming and a non-transforming virus by partial denaturation mapping of their DNAs, *J. Virol.* 27:81–89.

De Thé, G., Ho, H.C., Kwan, H.C., Desgranges, C., and Favre, M.C., 1970, Nasopharyngeal carcinoma (NPC). 1. Types of cultures derived from tumor biopsies and nontumor tissues of Chinese patients with special reference to lymphoblastoid transformation, *Int. J. Cancer* **6**:189–206.

De Thé, G., Geser, A., Day, N.E., Tukei, P.M., Williams, E.H., Beri, D.P., Smith, P.G., Dean, A.G., Bornkamm, G.W., Feorino, P., and Henle, W., 1978, Epidemiological evidence for causal relationship between Epstein–Barr virus and Burkitt's lymphoma from Ugandan prospective study, *Nature (London)* **274**:756–761.

Dick, H.M., Steel, C.M., Crichton, W.B., and Hutton, M.M., 1973, Anomalous reactivity of some HL-A typing sera, *Symp. Ser. Immunobiol. Standardization* **18**:116–123.

Dick, H.M., Bodmer, W.F., Bodmer, J.G., Steel, C.M., Crichton, W.B., and Evans, J., 1975, HL-A typing of lymphoblastoid cell lines, *in: Histocompatibility Testing* (F. Kiessmeyer-Nielsen, ed., pp. 671–676, Munksgaard, Copenhagen.

Diehl, V., Henle, W., and Kohn, G., 1968, Demonstration of a herpes group virus in cultures of peripheral leucocytes from patients with infectious mononucleosis, *J. Virol.* **2**:663–669.

Einhorn, L., and Ernberg, I., 1978, Induction of EBNA precedes the first cellular S-phase after EBV infection of human lymphocytes, *Int. J. Cancer* **21**:157–160.

Einhorn, I., Steinitz, M., Yefenof, E., Ernberg, I., Bakacs, T., and Klein, G., 1978, Epstein–Barr Virus (EBV) receptors, complement receptors, and EBV infectivity of different lymphocyte fractions of human peripheral blood. II. Epstein–Barr virus studies, *Cell Immunol.* **35**:43–58.

Epstein, A.L., Henle, W., Henle, G., Hewetson, J.F., and Kaplan, H.S., 1976, Surface marker characteristics and Epstein–Barr virus studies of two established North American Burkitt's lymphoma cell lines, *Proc. Natl. Acad. Sci. U.S.A.* **73**:228–232.

Epstein, M.A., and Achong, B.G., 1965, Fine structural organization of human lymphoblasts of a tissue culture strain (EB1) from Burkitt's lymphoma, *J. Natl. Cancer Inst.* **34**:241–253.

Epstein, M.A., and Achong, B.C., 1968, Specific immunofluorescence test for the herpes-type EB virus of Burkitt lymphoblasts authenticated by electron microscopy, *J. Natl. Cancer Inst.* **40**:593–607.

Epstein, M.A., and Barr, Y.M., 1964, Cultivation *in vitro* of human lymphoblasts from Burkitt's malignant lymphoma, *Lancet* **1**:252–253.

Epstein, M.A., and Barr, Y.M., 1965, Characteristics and mode of growth of a tissue culture strain (EB1) of human lymphoblasts from Burkitt's lymphoma, *J. Natl. Cancer Inst.* **34**:231–240.

Epstein, M.A., Achong, B.G., and Barr, Y.M., 1964a, Virus particles in cultured lymphoblasts from Burkitt's lymphoma, *Lancet* **1**:702–703.

Epstein, M.A., Barr, Y.M., and Achong, B.G., 1964b, A second virus-carrying tissue culture strain (EB2) of lymphoblasts from Burkitt's lymphoma, *Pathol. Biol. (Paris)* **12**:1233–1234.

Epstein, M.A., Henle, G., Achong, B.G., and Barr, Y.M., 1965, Morphological and biological studies on a virus in cultured lymphoblasts from Burkitt's lymphoma, *J. Exp. Med.* **121**:761–770.

Epstein, M.A., Achong, B.G., Barr, Y.M., Zajac, B., Henle, G., and Henle, W., 1966, Morphological and virological investigations on cultured Burkitt tumor lymphoblasts (strain Raji), *J. Natl. Cancer Inst.* **37**:547–555.

Epstein, M.A., Hunt, R., and Rabin, H., 1973, Pilot experiments with EB virus in owl monkeys (*Aotus trivirgatus*). I. Reticuloproliferative disease in an inoculated animal, *Int. J. Cancer* **12**:309–318.

Ernberg, I., 1982, Requirements for macromolecular synthesis during primary Epstein–Barr virus infection of lymphocytes (submitted).

Evans, A.S., Niederman, J.C., and McCollum, R.W., 1968, Seroepidemiologic studies of infectious mononucleosis with EB virus, *N. Engl. J. Med.* **279**:1121–1127.

Evans, J., Steel, M., and Arthur, E., 1974, A hemagglutination inhibition technique for detection of immunoglobulins in supernatants of human lymphoblastoid cell lines, *Cell* **3**:153–158.

Fagraeus, A., Nilsson, K., Lidman, K., and Norberg, R., 1975, Reactivity of smooth-muscle antibodies, surface ultrastructure, and mobility in cells of human hematopoietic cell lines, *J. Natl. Cancer Inst.* **55**:783–789.

Fahey, J.L., and Finegold, I., 1967, Synthesis of immunoglobulins in human cell lines, *Cold Spring Harbor Symp. Quant. Biol.* **32**:283–289.

Fahey, J.L., Finegold, I., Rabson, A.S., and Manaker, R.A., 1966, Immunoglobulin synthesis *in vitro* by established human cell lines, *Science* **152**:1259–1261.

Fahey, J.L., Buell, D.N., and Sox, H.C., 1971, Proliferation and differentiation of lymphoid cells: Studies with human cell lines and immunoglobulin synthesis, *Ann. N. Y. Acad. Sci.* **190**:221–234.

Falk, L., Wolfe, L., Deinhardt, F., Paciga, J., Dombos, L., Klein, G., Henle, W., and Henle, G., 1974, Epstein–Barr virus: Transformation of non-human primate lymphocytes *in vitro*, *Int. J. Cancer* **13**:363–376.

Falk, L., Deinhardt, F., Wolfe, L., Johnson, D., Hilgers, J., and de Thé, G., 1976, Epstein–Barr virus: Experimental infection of *Callothrix jacchus* marmosets, *Int. J. Cancer* **17**:785–788.

Ferrone, S., Pellegrino, M.A., and Reisfeld, R.A., 1971, A rapid method for direct HL-A typing of cultured lymphoid cells, *J. Immunol.* **107**:613–615.

Fialkow, P.J., Klein, E., Klein, G., Clifford, P., and Singh, S., 1973, Immunoglobulin and glucose-6-phosphate dehydrogenase as markers of cellular origin in Burkitt lymphoma, *J. Exp. Med.* **138**:89–102.

Finegold, I., Fahey, J.L., and Granger, H., 1967, Synthesis of immunoglobulins by human cell lines in tissue culture, *J. Immunol.* **99**:839–848.

Flier, S.J., Glade, P.R., Broder, S.W., and Hirschhorn, K., 1970, Lymphocyte stimulation by allogeneic and autochthonous cultured lymphoid cells, *Cell. Immunol.* **1**:596–602.

Frank, A., Andiman, W., and Miller, G., 1976, Epstein–Barr virus and non-human primates: Natural and experimental infection, *Adv. Cancer Res.* **23**:171–210.

Fresen, K.O., and zur Hausen, H., 1976, Establishment of EBNA-expressing cell lines by infection of Epstein–Barr virus (EBV)-genome-negative human lymphoma cells with different EBV strains, *Int. J. Cancer* **17**:161–166.

Fresen, K.-O., Merkt, B., Bornkamm, G.W., and zur Hausen, H., 1977, Heterogeneity of Epstein–Barr virus originating from P_3HR-1 cells. I. Studies on EBNA induction, *Int. J. Cancer* **19**:317–323.

Fresen, K.-O., Cho, M.-S., and zur Hausen, H., 1978, Recovery of transforming EBV from non-producer cells after superinfection with non transforming P_3HR-1 EBV, *Int. J. Cancer* **22**:378–383.

Fresen, K.-O., Cho, M.-S., Gissmann, L., and zur Hausen, H., 1979, NC37-R1 Epstein–Barr virus (EBV), a possible recombinant between intracellular NC37 viral DNA and superinfecting P3HR-1 EBV, *Intervirology* **12**:303–310.

Fu, S.M., and Hurley, J.N., 1979, Human cell lines containing Epstein–Barr virus but distinct from the common B cell lymphoblastoid lines, *Proc. Natl. Acad. Sci. U.S.A.* **76**:6637.

Fukuhara, S., and Rowley, J.D., 1978, Chromosome 14 translocations in non-Burkitt lymphomas, *Int. J. Cancer* **22**:14–21.

Fukuhara, S., Ueshima, Y., Shirakawa, S., Uchino, H., and Morikawa, S., 1979, 14q translocations, having a break point at 14q 13 in lymphoid malignancy, *Int. J. Cancer* **24**:739–743.

Geier, S.G., and Cresswell, P., 1977, Rabbit antisera to human B cell alloantigens: Effects on the mixed lymphocyte response, *Cell. Immunol.* **28**:341–354.

Gerber, P., and Hoyer, B., 1971, Induction of cellular DNA synthesis in human leukocytes by Epstein–Barr virus, *Nature (London)* **231**:46–47.

Gerber, P., and Lucas, S., 1972, Epstein–Barr virus-associated antigens activated in human cells by 5-bromodeoxyuridine, *Proc. Soc. Exp. Biol. Med.* **141**:431–435.

Gerber, P., and Monroe, J.H., 1968, Studies on leukocytes growing in continuous culture derived from normal human donors, *J. Natl. Cancer Inst.* **40**:855–864.

Gerber, P., Whang-Peng, J., and Monroe, J.H., 1969, Transformation and chromosome changes induced by Epstein–Barr virus in normal human leukocyte cultures, *Proc. Natl. Acad. Sci. U.S.A.* **63**:740–747.

Gergely, L., Klein, G., and Ernberg, I., 1971, The action of DNA antagonists on Epstein–Barr virus (EBV)-associated early antigen (EA) in Burkitt lymphoma lines, *Int. J. Cancer* **7**:293–302.

Giovanella, B., Nilsson, K., Zech, L., Yim, O., Klein, G., and Stehlin, J., 1979, Growth of diploid, Epstein–Barr virus-carrying human lymphoblastoid cell lines heterotransplanted into nude mice under immunologically privileged conditions, *Int. J. Cancer* **24**:103–113.

Glade, P.R., and Chessin, L.N., 1968, Infectious mononucleosis: Immunoglobulin synthesis by cell lines, *J. Clin. Invest.* **47**:2391–2401.

Glaser, R., and Nonoyama, M., 1974, Host cell regulation of induction of Epstein–Barr virus, *J. Virol.* **14**:174–176.

Glaser, R., and Rapp, F., 1972, Rescue of Epstein–Barr virus from somatic cell hybrids of Burkitt lymphoblastoid cells, *J. Virol.* **10**:288–296.

Golub, S.H., Svedmyr, E.A.J., Hewetson, J.F., and Klein, G., 1972, Cellular reactions against Burkitt lymphoma cells. III. Effector cell activity of leukocytes stimulated *in vitro* with autochthonous cultured lymphoma cells, *Int. J. Cancer* **10**:157–164.

Gordon, J., Hough, D., Karpas, A., and Smith, J.L., 1977, Immunoglobulin expression and synthesis by human haemic cell lines, *Immunology* **32**:559–565.

Graessman, A., Wolf, H., and Bornkamm, G.W., 1980, Expression of Epstein–Barr virus genes in different cell types after microinjection of viral DNA, *Proc. Natl. Acad. Sci. U.S.A.* **77**:433–436.

Greaves, M.F., Brown, G., and Rickinson, A.B., 1975, Epstein–Barr virus binding sites on lymphocyte subpopulations and the origin of lymphoblasts in cultured lymphoid cell lines and in the blood of patients with infectious mononucleosis, *Clin. Immunol. Immunopathol.* **3**:514–524.

Green, S.S., and Sell, K.E., 1970, Mixed leukocyte stimulation of normal peripheral leukocytes by autologous lymphoblastoid cells, *Science* **170**:989–999.

Gunven, P., Klein, G., Klein, E., Norin, T., and Singh, S., 1980, Surface immunoglobulins on Burkitt's lymphoma biopsy cells from 91 patients, *Int. J. Cancer* **25**:711–719.

Hampar, B., Derge, J.G., Martos, L.M., and Walker, J.L., 1972, Synthesis of Epstein–Barr virus after activation of the viral genome in a virus negative human lymphoblastoid cell (Raji) made resistant to five bromodeoxyuridine, *Proc. Natl. Acad. Sci. U.S.A.* **69**:78–82.

Hampar, B., Derge, J.G., Nonoyama, M., Chang, S.Y., Tagameto, M.A., and Showalter, S.D., 1974, Programming of events in Epstein–Barr virus-activated cells induced by 5-iododeoxyuridine, *Virology* **62**:71–89.

Han, T., Moore, G.E., and Sokal, J., 1971, *In vitro* lymphocyte response to autologous cultured lymphoid cells, *Proc. Soc. Exp. Biol. Med.* **136**:976–979.

Hanto, D., Frizzera, G., Purtilo, D., Sakamoto, K., Sullivan, J., Saemundsen, A., Klein, G., Simmons, R., and Najarian, J., 1981, Lymphoproliferative disorders in renal transplant recipients. II. A clinical spectrum and evidence for the role of Epstein–Barr virus, *Cancer Res.* **41**:4253–4261.

Hardy, E.A., Ling, N.R., and Knight, S., 1969, Exceptional lymphocyte stimulating capacity of cells from lymphoid cell lines, *Nature (London)* **223**:511–512.

Henderson, E., Miller, G., Robinson, J., and Heston, L., 1977a, Efficiency of transformation of lymphocytes by Epstein–Barr virus, *Virology* **76**:152–163.

Henderson, E., Robinson, J., Frank, A., and Miller, G., 1977b, Epstein-Barr virus transformation of lymphocytes separated by size or exposed to bromodeoxyuridine and light, *Virology* **82**:196–205.

Henderson, E., Heston, L., Grogan, E., and Miller, G., 1978, Radio-biological inactivation of Epstein–Barr virus, *J. Virol.* **25**:51–59.

Henderson, E., Long, W., and Ribecky, R., 1979, Effects of nucleoside analogs on Epstein–Barr virus induced transformation of human umbilical cord leukocytes and Epstein–Barr virus expression in transformed cells, *Antimicrob. Agents Chemother.* **15**:101–110.

Henle, G., and Henle, W., 1966, Immunofluorescence in cells derived from Burkitt's lymphoma, *J. Bacteriol.* **91**:1248–1256.

Henle, G., and Henle, W., 1967, Immunofluorescence, interference and complement fixation technics in the detection of the herpes-type virus in Burkitt tumor cell lines, *Cancer Res.* **27**:2442–2446.

Henle, G., Henle, W., and Diehl, V., 1968, Relation of Burkitt's tumor-associated herpes-type virus to infectious mononucleosis, *Proc. Natl. Acad. Sci. U.S.A.* **59**:94–101.

Henle, G., Henle, W., Zajac, B., Pearson, G., Waubke, R., and Scriba, M., 1970, Differential reactivity of human serums with early antigens induced by Epstein–Barr virus, *Science* **169**:188–190.

Henle, W., and Henle, G., 1968, Effect of arginine-deficient media on the herpes-type virus associated with cultured Burkitt tumor cells, *J. Virol.* **2**:182–191.

Henle, W., Diehl, V., Kohn, G., zur Hausen, H., and Henle, G., 1967, Herpes-type virus and chromosome marker in normal leukocytes after growth with irradiated Burkitt cells, *Science* **157**:1065–1065.

Herberman, R.B., 1978, Natural cell-mediated cytotoxicity in nude mice, in: *The Nude Mouse in Experimental and Clinical Research* (J. Fogh and B.C. Giovanella, eds.), pp. 136–166, Academic Press, New York.

Heyden, H.W.V., and Heyden, D.v., 1974, Criteria for the differentiation of lymphoid cell lines, *Blut* **24**:323–334.

Hinuma, Y., and Grace, J.T., 1967, Cloning of immunoglobulin-producing human leukemic and lymphoma cells in long term cultures, *Proc. Soc. Exp. Biol. Med.* **124**:107–111.

Hinuma, Y., and Katsuki, T., 1978, Colonies of EBNA-positive cells in soft agar from peripheral leukocytes of infectious mononucleosis patients, *Int. J. Cancer* **21**:426.

Hinuma, Y., Konn, M., Yamaguchi, J., and Grace, J.T., Jr., 1967, Replication of herpes-type virus in a Burkitt lymphoma cell line, *J. Virol.* **1**:1203–1206.

Huang, C.C., and Moore, G.E., 1969, Chromosomes of 14 hematopoietic cell lines derived from peripheral blood of persons with and without chromosome anomalies, *J. Natl. Cancer Inst.* **43**:1119–1128.

Huang, C.C., Imamura, T., and Moore, G.E., 1969, Chromosomes and cloning efficiencies of hematopoietic cell lines derived from patients with leukemia, melanoma, myeloma and Burkitt lymphoma, *J. Natl. Cancer Inst.* **43**:1129–1146.

Huang, C.C., Minowada, J., Smith, R.T., and Osunkoya, B.O., 1970, Reevaluation of relationship between C chromosome marker and Epstein–Barr virus: Chromosome and immunofluorescence analysis of 16 human hematopoietic cell lines, *J. Natl. Cancer Inst.* **45**:815–829.

Huber, C., Sundstrom, C.C., Nilsson, K., and Wigzell, H., 1976, Surface receptors on human haematopoietic cell lines, *Clin, Exp. Immunol.* **25**:367–376.

Humphreys, R.E., McCune, J.M., Chess, L., Herrman, H.C., Malenka, D.J., Mann, D.L., Parham, P., Schlossman, S.F., and Strominger, J.L., 1976, Isolation and immunologic characterization of a human B-lymphocyte specific, cell surface antigen, *J. Exp. Med.* **144**:98–112.

Hutt, L.M., Huang, Y.T., Dascomb, H.E., and Pagano, J.S., 1975, Enhanced destruction of lymphoid cell lines by peripheral blood leukocytes taken from patients with acute infectious mononucleosis, *J. Immunol.* **115**:243–248.

Ikeuchi, T., Minowada, J., and Strandberg, A.A., 1971, Chromosomal variability in ten cloned sublines of a newly established Burkitt's lymphoma cell line, *Cancer* **28**:499–512.

Imamura, R., and Moore, G.W., 1968, Ability of human hematopoietic cell lines to form colonies in soft agar, *Proc. Soc. Exp. Biol. Med.* **128**:1179–1183.

Iwakata, S., and Grace, J.T., Jr., 1964, Cultivation *in vitro* of myeloblasts from human leukemia *N. Y. J. Med.* **64**:2279–2282.

Jarvis, J.E., Ball, G., Rickinson, A.B., and Epstein, M.A., 1974, Cytogenetic studies on human lymphoblastoid cell lines from Burkitt's lymphomas and other sources, *Int. J. Cancer* **14**:716–721.

Jensen, F.C., Gwatkin, R.B.L., and Biggers, J.D., 1964, A simple culture which allows simultaneous isolation of specific types of cells, *Exp. Cell Res.* **34**:440–447.

Jondal, M., and Klein, G., 1973, Surface markers on human B and T lymphocytes. II. Presence of Epstein–Barr virus receptors on B lymphocytes, *J. Exp. Med.* **138**:1365–1378.

Jondal, M., Klein, G., Oldstone, M.B.A., Bokish, V., and Yefenof, E., 1976, An association between complement and Epstein–Barr virus receptors on human lymphoid cells, *Scand. J. Immunol.* **5**:401–410.

Joss, A.W.L., Evans, J., Veitch, D.T., Arthyr, E., and Steel, C.M., 1976, Haemagglutinins produced *in vitro* by human lymphoblastoid cells, *J. Immunogenet.* **3**:323–328.

Junge, U., Hoekstra, J., and Deinhardt, F., 1971, Stimulation of peripheral lymphocytes by allogeneic and autochthonous mononucleosis lymphocyte cell line, *J. Immunol.* **106**:1306–1315.

Kaiser-McCaw, B., Epstein, A.L., Kaplan, H.S., and Hecht, F., 1977, Chromosome 14 translocation in African and North American Burkitt's lymphoma, *Int. J. Cancer* **19**:482–486.

Katsuki, T., and Hinuma, Y., 1975, Characteristics of cell lines derived from human leukocytes transformed by different strains of Epstein–Barr virus, *Int. J. Cancer* **15**:203–210.

Katsuki, T., Hinuma, Y., Yamamoto, N., Aho, T., and Kumagai, K., 1977, Identification of the target cells in human B lymphocytes for transformation by Epstein–Barr virus, *Virology* **83**:287–294.

Katsuki, T., Hinuma, Y., Saito, T., Yamamoto, J., Hirashima, Y., Sudoh, H., Deguchi, M., and Motokawa, M., 1979, Simultaneous presence of EBNA-positive and colony-forming cells in peripheral blood of patients with infectious mononucleosis, *Int. J. Cancer* **23**:746–750.

Kintner, C.R., and Sugden, B., 1979, The structure of the termini of the DNA of Epstein–Barr virus, *Cell* **17**:661–672.

Kirchner, H., Tosato, G., Blaese, M., Bruder, S., and Magrath, I., 1979, Polyclonal immunoglobulin secretion by human B lymphocytes exposed to Epstein–Barr virus *in vitro*, *J. Immunol.* **122**:1310–1313.

Klein, E., Klein, G., Nadkarni, J.S., Wigzell, H., and Clifford, K., 1968, Surface IgM-kappa specificity on a Burkitt lymphoma cell *in vivo* and in derived culture lines, *Cancer Res.* **28**:1300–1310.

Klein, G., 1975, Studies on the Epstein–Barr virus genome and the EBV-determined nuclear antigen in human malignant disease, *Cold Spring Harbor Symp. Quant. Biol.* **39**:783–790.

Klein, G., 1979, The relationship of the virus to nasopharyngeal carcinoma, in: *The Epstein–Barr Virus* (M.A. Epstein and B.G. Achong, eds.), pp. 339–350, Springer-Verlag, Berlin, Heidelberg, New York.

Klein, G., 1980, Immune and non-immune control of neoplastic development: Contrasting effects of host and tumor evolution, *Cancer* **45**:2486–2499.

Klein, G., Giovanella, B.C., Lindahl, T., Fialkow, P.J., Singh, S., and Stehlin, J.S., 1974a, Direct evidence for the presence of Epstein–Barr virus DNA and nuclear antigen in malignant epithelial cells from patients with poorly differentiated carcinoma of the nasopharynx, *Proc. Natl. Acad. Sci. U.S.A.* **71**:4737–4741.

Klein, G., Lindahl, T., Jondal, M., Leibold, W., Menezes, J., Nilsson, K., and Sundstrom, C., 1974b, Continuous lymphoid cell lines with characteristics of B cells (bone marrow-derived) lacking the EBV genome and derived from 3 human lymphomas, *Proc. Natl. Acad. Sci. U.S.A.* **71**:3283–3286.

Klein, G., Sugden, B., Leibold, W., and Menezes, J., 1974c, Infection of EBV-genome-negative and -positive human lymphoblastoid cell lines with biologically different preparations of EBV, *Intervirology* **3**:232–244.

Klein, G., Svedmyr, E., Jondal, M., and Person, P.O., 1976, EBV-determined nuclear antigen (EBNA)-positive cells in peripheral blood of infectious mononucleosis patients, *Int. J. Cancer* **17**:21–26.

Klein, G., Yefenof, E., Falk, K., and Westman, A., 1978, Relationship between Epstein–Barr virus (EBV)-production and the loss of the EBV receptor/complement receptor complex in a series of sublines derived from the same original Burkitt's lymphoma, *Int. J. Cancer* **21**:552–560.

Klein, G., Hyman, R., and Trowbridge, I., 1979, Correlated expression of a B-lymphocyte-specific glycoprotein (gp 27/35) and the EBV receptor/C3 receptor complex in sublines from the same Burkitt lymphoma, *Int. J. Cancer* **23**:37–41.

Klein, G., Luka, J., and Zeuthan, J., 1980, Transformation induced by Epstein–Barr virus and the role of the nuclear antigen, *Cold Spring Harbor Symp. Quant. Biol.* **44**:253–261.

Knight, S.C., Moore, G.E., and Clarkson, B.D., 1971, Stimulation of autochthonous lymphocytes by cells from normal and leukemic lines, *Nature (London)* **229**:185–187.

Kohn, G., Mellman, W.J., Moorhead, P.S., Loftus, J., and Henle, G., 1967, Involvement of C group chromosomes in five Burkitt lymphoma cell lines, *J. Natl. Cancer Inst.* **38**:209–222.

Kohn, G., Diehl, V., Mellman, W.J., Henle, W., and Henle, G., 1968, C-group chromosome marker in long-term leucocyte cultures, *J. Natl. Cancer Inst.* **41**:795–804.

Kozbor, D., Steinitz, M., Klein, G., Koskimies, S., and Mäkalä, O., 1979, Establishment of anti-TNP antibody-producing human lymphoid lines by preselection for hapten binding followed by EBV transformation, *Scand. J. Immunol.* **10**:187–194.

Lai, P.K., Mackay-Scollay, E.M., and Alpers, M.P., 1973, Synthesis of virus-capsid antigen (VCA) enhanced by ultraviolet irradiation of a lymphoblastoid cell line carrying Epstein–Barr virus, *J. Gen. Virol.* **21**:135–143.

Leibold, W., Huldt, G., Flanagan, T., Andersson, M., Dalens, M., Wright, D., Voller, A., and Klein, G., 1976, Tumorigenicity of Epstein–Barr virus (EBV)-transformed lymphoid line cells in autologous squirrel monkeys, *Int. J. Cancer* **17**:533–541.

Levitt, M., Shungu, D., and Henderson, E., 1980, Colony morphology, cell growth and viral radiosensitivity as parameters for detecting strain differences in Epstein–Barr virus (EBV), *Int. J. Cancer* **26**:509–515.

Lindahl, T., and Adams, A., 1976, Covalently closed circular duplex DNA of Epstein–Barr virus in a human lymphoid cell line, *J. Mol. Biol.* **102**:511–530.

Lindahl, T., Klein, G., Reedman, B.M., Johansson, B., Singh, and S., 1974, Relationship between Epstein–Barr virus (EBV) DNA and the EBV-determined nuclear antigen (EBNA) in Burkitt lymphoma biopsies and other lymphoproliferative malignancies, *Int. J. Cancer* **13**:764–772.

Lindblom, B., and Nilsson, K., 1973, HL-A antigens in established human cell lines: International Symposium on HL-A Reagents, *Symp. Ser. Immunobiol. Standardization* **18**:124–128.

Litwin, S.D., and Lin, P.K., 1976, Phenotypic heterogeneity for cells bearing surface immunoglobulin in human lymphoid lines, *Cell. Immunol.* **24**:270–276.

Litwin, S.D., Hütteroth, T.H., Lin, P.K., Kennard, J., and Cleve, H., 1974a, Immunoglobulin expression of cells from human lymphoblastoid lines. I. Heavy and light chain antigens of the cell surface, *J. Immunol.* **113**:661–667.

Litwin, S.D., Hütteroth, T.H., Lin, P.K., Kennard, J., and Cleve, H., 1974b, Immunoglobulin expression of cells from human lymphoblastoid lines. II. Interrelationship among surface, cellular, and secreted immunoglobulins, *J. Immunol.* **113**:668–672.

Luka, J., Kallin, B., and Klein, G., 1979, Induction of the Epstein–Barr Virus (EBV) cycle in latently infected cells by *n*-butyrate, *Virology* **94**:228–231.

Luka, J., Siegert, W., and Klein, G., 1977, Solubilization of the Epstein–Barr virus-determined nuclear antigen and its characterization as a DNA-binding protein, *J. Virol.* **22**:1–8.

Manolov, G., and Manolova, Y., 1972, Marker band in one chromosome 14 from Burkitt lymphomas, *Nature (London)* **237**:33–34.

Manolova, Y., Manolov, G., Kieler, J., Levan, A., and Klein, G., 1979, Genesis of the 14q+ marker in Burkitt's lymphoma, *Hereditas* **90**:5–10.

Mark, J., Ekedahl, C., and Dahlenfors, R., 1978, Characteristics of the banding patterns in non-Hodgkin and non-Burkitt lymphomas, *Hereditas* **88**:229–242.

Marker, S.C., Ascher, N.L., Kalis, J.M., Simmons, R.L., Najarian, J.S., and Balfour, H.H., Jr., 1979, Epstein–Barr virus antibody responses and clinical illness in renal transplant recipients, *Surgery* **85**:433–440.

McCune, J.M., Humphreys, R.E., Yocum, R.R., and Strominger, J.L., 1975, Enhanced representation of HL-A antigens on human lymphocytes after mitogenesis induced by phytohemagglutinin or Epstein–Barr virus, *Proc. Natl. Acad. Sci. U.S.A.* **72**:3206–3209.

McCune, J.M., Lingappa, V.R., Fu, S.M., Bloble, G., and Kunkel, H.G., 1980, Biogenesis of membrane-bound and secreted immunoglobulins. I. Two distinct translation products of human mu-chain, with identical N-termini and different C-termini, *J. Exp. Med.* **152**:463–468.

Menezes, J., Jondal, M., Leibold, W., and Dorval, G., 1976, Epstein–Barr virus interactions with human lymphocyte subpopulations, virus adsorptions, kinetics of expression of Epstein–Barr virus-associated nuclear antigen, and lymphocyte transformation, *Infect. Immun.* **13**:303–310.

Menezes, J., Seigneurin, J.M., Patel, P., Bourkas, A., and Lenoir, G., 1977, Presence of Epstein–Barr virus receptors, but absence of virus penetration in cells of an Epstein–Barr virus genome-negative human lymphoblastoid T line (Molt 4), *J. Virol.* **22**:816–821.

Miles, C.P., and O'Neill, F., 1967, Chromosome studies of 8 *in vitro* lines of Burkitt's lymphoma, *Cancer Res.* **27**:392–402.

Miles, C.P., O'Neill, F., Armstrong D., Clarkson, B., and Keane, J., 1968, Chromosome patterns of human leukocyte established cell lines, *Cancer Res.* **28**:481–490.

Miller, G., 1980, Biology of Epstein–Barr virus, in: *Viral Oncology* (G. Klein, ed.), pp. 713–737, Raven Press, New York.

Miller, G., and Lipman, M., 1973a, Release of infectious Epstein–Barr virus by transformed marmoset leukocytes, *Proc. Natl. Acad. Sci. U.S.A.* **70**:190–194.

Miller, G., and Lipman, M., 1973b, Comparison of the yield of infectious virus from clones of human and simian lymphoblastoid lines transformed by Epstein–Barr virus, *J. Exp. Med.* **138**:1398–1412.

Miller, G., Enders, J.F., and Lisco, H.I., 1969, Establishment of lines from normal human blood leukocytes by co-cultivation with a leukocyte line derived from a leukemic child, *Proc. Soc. Exp. Biol. Med.* **132**:247–250.

Miller, G., Lisco, H., Kohn, H.L., and Stitt, D., 1971, Establishment of cell lines from normal adult human blood leukocytes by exposure to Epstein–Barr virus and neutralization by human sera with Epstein–Barr virus antibody, *Proc. Soc. Exp. Biol. Med.* **137**:1459–1465.

Miller, G., Robinson, J., Heston, L., and Lipman, M., 1974, Differences between laboratory strains of Epstein–Barr virus based on immortalization, abortive infection, and interference, *Proc. Natl. Acad. Sci. U.S.A.* **71**:4006–4010.

Miller, G., Coope, D., Niederman, J., and Pagano, J., 1976, Biologic properties and viral surface antigens of Burkitt lymphoma and mononucleosis derived strains of Epstein–Barr virus released from transformed marmoset cells, *J. Virol.* **18**:1071–1080.

Miller, G., Shope, T., Coope, D., Waters, L., Pagano, J., Bornkamm, G., and Henle, W., 1977, Lymphoma in cotton-top marmosets after inoculation with Epstein–Barr virus: Tumor incidence, histologic spectrum, antibody responses, demonstration of viral DNA, and characterization of viruses, *J. Exp. Med.* **145**:948–967.

Miller, G., Grogan, E., Heston, L., Robinson, J., and Smith, D., 1981, Epstein–Barr viral DNA: Infectivity for human placental cells, *Science* **211**:452–455.

Minowada, J., Klein, G., Clifford, P., Klein, E., and Moore, G.E., 1967, Studies of Burkitt lymphoma cells. I. Establishment of a cell line (B35M) and its characteristics, *Cancer* **20**:1439–1437.

Misko, I., Moss, D., and Pope, J., 1980, HLA antigen restriction of T lymphocyte cytotoxicity to Epstein–Barr virus, *Proc. Natl. Acad. Sci. U.S.A.* **77**:4247–4250.
Mitelman, F., Andersson-Anvret, M., Brandt, L., Catovsky, D., Klein, G., Manolov, G., Manolova, Y., Mark-Vendel, E., and Nilsson, P.G., 1979a, Reciprocal 8;14 translocation in EBV-negative B-cell acute lymphocyte leukemia with Burkitt-type cells, *Int. J. Cancer* **24**:27–33.
Mitelman, F., Klein, G., Andersson-Anvret, M., Forsby, N., and Johansson, B., 1979b, 14q + marker chromosome in an EBV-genome-negative lymph node without signs of malignancy in a patient with EBV-genome-positive nasopharyngeal carcinoma, *Int. J. Cancer* **23**:32–36.
Miyoshi, I., Hiraki, S., Kimura, I., Myamoto, K., and Sato, J., 1979, 2/8 translocation in a Japanese Burkitt's lymphoma, *Experientia* **35**:742.
Mizuno, F., Aya, T., and Osato, T., 1976, Growth in semisolid agar medium of human cord leukocytes freshly transformed by Epstein–Barr virus, *J. Natl. Cancer Inst.* **56**:171–173.
Moar, M.H., and Klein, G., 1979, Abortive expression of the Epstein–Barr virus (EBV) cycle in a variety of EBV DNA-containing cell lines as reflected by nucleic acid hybridization *in situ*, *Int. J. Cancer* **24**:679–687.
Moore, G.E., Grace, J.T., Jr., Citron, P., Gerner, R.E., and Burns, A., 1966, Leukocyte cultures of patients with leukemia and lymphomas, *N. Y. State J. Med.* **66**:2757–2764.
Moore, G.E., Gerner, R.E., and Franklin, H.A., 1967, Culture of normal leukocytes, *J. Am. Med. Assoc.* **199**:519–524.
Moore, G.E., Kitamura, H., and Toshima, S., 1968, Morphology of cultured hematopoietic cells, *Cancer* **22**:245–267.
Morgan, D.G., Niederman, J.C., Miller, G., Smith, H.W., and Dowaliby, J.M., 1979, Site of Epstein–Barr virus replication in the oropharynx, *Lancet* **1**:1154–1157.
Moses, H.L., Glade, P.R., Kasel, J.A., Rosenthal, A.S., Hirshaut, Y., and Chessin, L.N., 1968, Infectious mononucleosis: Detection of herpes-like virus and reticular aggregates of small cytoplasmic particles in continuous lymphoid cell lines derived from peripheral blood, *Proc. Natl. Acad. Sci. U.S.A.* **60**:489–496.
Moss, D.J., and Pope, J.H., 1972, Assay of the infectivity of Epstein–Barr virus by transformation of human leucocytes *in vitro*, *J. Gen. Virol.* **17**:233–236.
Moss, D., Rickinson, A., and Pope, J., 1978, Long-term T-cell-mediated immunity to Epstein–Barr virus in man. I. Complete regression of virus-induced transformation in cultures of seropositive donor leukocytes, *Int. J. Cancer* **22**:662–668.
Moss, D.J., Rickinson, A., and Pope, J., 1979, Long-term T-cell-mediated immunity to Epstein–Barr virus in man. III. Activation of cytotoxic T cells in virus-infected leukocyte cultures, *Int. J. Cancer* **23**:618–625.
Nadkarni, J.S., Nadkarni, J.J., Clifford, P., Manolov, G., Fenyö, E.M., and Klein, E., 1969, Characteristics of new cell lines derived from Burkitt lymphomas, *Cancer* **23**:64–79.
Nilsson, K., 1971, High frequency-establishment of human immunoglobulin producing lymphoblastoid lines from normal and malignant lymphoid tissue and peripheral blood, *Int. J. Cancer* **8**:432–442.
Nilsson, K., 1979, The nature of lymphoid cell lines and their relationship to the virus, in: *The Epstein–Barr Virus* (M.A. Epstein and B.G. Achong, eds.), pp. 225–281, Springer-Verlag, Berlin, Heidelberg, New York.
Nilsson, K., and Ponten, J., 1975, Classification and biological nature of established human hematopoietic cell lines, *Int. J. Cancer* **15**:321–341.
Nilsson, K., Bennich, H., Johansson, S.G.O., and Ponten, J., 1970, Established immunoglobulin producing myeloma (IgE) and lymphoblastoid (IgG) cell lines from an IgE myeloma patient, *Clin. Exp. Immunol.* **7**:477–459.
Nilsson, K., Klein, G., Henle, G., and Henle, W., 1971, The establishment of lymphoblastoid cell lines and its dependence on EBV, *Int. J. Cancer* **8**:443–450.
Nilsson, K., Evrin, P.R., and Welsh, K.I., 1974, Production of β_2-microglobulin by normal and malignant human cell lines and peripheral lymphocytes, *Transplantation* **21**:53–84.
Nilsson, K., Anderson, L.C., Gahmberg, C.G., and Wigzell, H., 1977a, Surface glycoprotein

patterns of normal and malignant human lymphoid cells. II. B cells, B blasts and Epstein–Barr virus (EBV)-positive and -negative B lymphoid cell lines, *Int. J. Cancer* **20**:708–718.

Nilsson, K., Giovanella, B.C., Stehlin, J.S., and Klein, G., 1977b, Tumorigenicity of human hematopoietic cell lines in athymic nude mice, *Int. J. Cancer* **19**:337–342.

Nyormi, O., Thorley-Lawson, D.A., Elkington, J., and Strominger, J.L., 1976, Differential effect of phosphonoacetic acid on the expression of Epstein–Barr viral antigens and virus production, *Proc. Natl. Acad. Sci. U.S.A.* **73**:1745–1748.

Okada, H., and Baba, T., 1974, Rosette formation of human erythrocytes on cultured cells of tumor origin and activation of complement by cell membrane, *Nature (London)* **248**:521–522.

Ostberg, L., Rask, L., Nilsson, K., and Peterson, P.A., 1975, Independent expression of the two HL-A antigen polypeptide chains, *Eur. J. Immunol.* **5**:462–468.

Osunkoya, B.O., McFarlane, H., Luzatto, L., Udeozo, I.O.K., Mottram, F.C., Williams, A.I.O., and Ngu, V.A., 1968, Immunoglobulin synthesis by fresh cells and established cell lines from Burkitt's lymphoma, *Immunology* **14**:851–860.

Papermaster, V.M., Papermaster, B.W., and Moore, G.E., 1969, Histocompatibility antigens of human lymphocytes in long term culture, *Fed. Proc. Fed. Am. Soc. Exp. Biol.* **28**:379.

Pattengale, P.K., Smith, R.W., and Gerber, P., 1974, B-cell characteristics of human peripheral and cord blood lymphocytes transformed by Epstein–Barr virus, *J. Natl. Cancer Inst.* **52**:1081–1086.

Philip, P., Krogh Jensen, M., and Pallesen, G., 1977, Marker chromosome 14q+ in nonendemic Burkitt's lymphoma, *Cancer* **39**:1495–1499.

Pious, D., Bodmer, J., and Bodmer, W., 1974, Antigenic expression and cross reactions in HL-A variants of lymphoid cell lines, *Tissue Antigens* **4**:247–256.

Pizzo, P.A., Magrath, I.T., Chattopadhyay, S.K., Biggar, R.J., and Gerber, P., 1978, A new tumour-derived transforming strain of Epstein–Barr virus, *Nature (London)* **272**:629–631.

Pope, J.H., 1967, Establishment of cell lines from peripheral leukocytes in infectious mononucleosis, *Nature (London)* **216**:810–811.

Pope, J.H., Achong, B.G., and Epstein, M.A., 1967, Burkitt lymphomas in New Guinea: Establishment of a line of lymphoblasts *in vitro* and description of their fine structure, *J. Natl. Cancer Inst.* **39**:933–945.

Pope, J.H., Horne, M.K., and Scott, W., 1968, Transformation of foetal human leucocytes *in vitro* by filtrates of a human leukemic cell line containing herpes-like virus, *Int. J. Cancer* **3**:857–866.

Pope, J.H., Horne, M.K., and Scott, W., 1969, Identification of the filtrable leukocyte-transforming factor of QIMR,WIL cells as herpes-like virus, *Int. J. Cancer* **4**:255–260.

Pope, J.H., Scott, W., and Moss, D.J., 1974, Cell relationships in transformation of human leukocytes by Epstein–Barr virus, *Int. J. Cancer* **14**:122–129.

Premkumar, E., Singer, P.A., and Williamson, A.R., 1975, A human lymphoid cell line secreting immunoglobulin G and retaining immunoglobulin M in the plasma membrane, *Cell* **5**:87–92.

Pulvertaft, R.J.V., 1964a, Cytology of Burkitt's tumour (African lymphoma), *Lancet* **1**:238–240.

Pulvertaft, R.J.V., 1964b, Phytohemagglutinin in relation to Burkitt's tumor (African lymphoma), *Lancet* **2**:552–553.

Purtilo, D.T., 1980, Epstein-Barr virus induced oncogenesis in immunodeficient individuals, *Lancet* **1**:300–303.

Qualtière, L.F., and Pearson, G.R., 1979, Epstein–Barr virus-induced membrane antigens: Immunochemical characterization of Triton X-100 solubilized viral membrane antigens from EBV-superinfected Raji cells, *Int. J. Cancer* **23**:808–817.

Rabin, H., Neubauer, R., Hopkins, R., and Levy, B., 1977, Characterization of lymphoid cell lines established from multiple Epstein–Barr virus (EBV)-induced lymphomas in a cotton-topped marmoset, *Int. J. Cancer* **20**:44–50.

Rabson, A.S., O'Conor, G.T., Baron, S., Whang, J.J., and Legallais, F.Y., 1966, Morphologic, cytogenetic and virologic studies *in vitro* of a malignant lymphoma from an African child, *Int. J. Cancer* **1**:89–106.

Ragona, G., Ernberg, I., and Klein, G., 1980, Induction and biological characterization of the Epstein–Barr virus (EBV) carried by the Jijoye lymphoma line, *Virology* **101**:553–557.

Recher, L., Sinkovics, J.G., Sykes, J.A., and Whitescarver, J., 1969, Electron microscopic studies of suspension cultures derived from human leukemic and non-leukemic sources, *Cancer Res.* **29**:271–285.

Reedman, B.M., and Klein, G., 1973, Cellular localization of an Epstein–Barr virus (EBV)-associated complement-fixing antigen in producer and non-producer lymphoblastoid cell lines, *Int. J. Cancer* **11**:499–520.

Rickinson, A.B., Jarvis, J.E., Crawford, D.H., and Epstein, M.A., 1974, Observations on the type of infection by Epstein–Barr virus in peripheral lymphoid cells of patients with infectious mononucleosis, *Int. J. Cancer* **14**:704–715.

Rickinson, A.B., Crawford, D., and Epstein, M.A., 1977, Inhibition of the *in vitro* outgrowth of Epstein–Barr virus-transformed lymphocytes by thymus-dependent lymphocytes from infectious mononucleosis patients, *Clin. Exp. Immunol.* **28**:72–79.

Rickinson, A., Moss, D., and Pope, J., 1979, Long-term T-cell-mediated immunity to Epstein–Barr virus in man. II. Components necessary for regression in virus-infected leukocyte cultures, *Int. J. Cancer* **23**:610–717.

Rickinson, A., Moss, D., Pope, J., and Ahlberg, N., 1980a, Long-term T-cell-mediated immunity to Epstein–Barr virus in man. IV. Development of T-cell memory in convalescent infectious mononucleosis patients, *Int. J. Cancer* **25**:59–65.

Rickinson, A., Wallace, L., and Epstein, M., 1980b, HLA-restricted T-cell recognition of Epstein–Barr virus-infected B cells, *Nature (London)* **283**:865–867.

Rickinson, A., Moss, D., Wallace, L., Misko, I., Epstein, M., and Pope, J., 1981, Long-term T-cell-mediated immunity to Epstein–Barr virus, *Cancer Res.* **41**:4216–4221.

Robinson, J., and Miller, G., 1975, Assay for Epstein–Barr virus based on stimulation of DNA synthesis in mixed leukocytes from human umbilical cord blood, *J. Virol.* **15**:1065–1072.

Robinson, J., and Smith, D., 1981, Infection of human B-lymphocytes with high multiplicities of Epstein–Barr virus: Kinetics of EBNA expression, cellular DNA synthesis and mitosis, *Virology* **109**:336–343.

Robinson, J.E., Andiman, W.A., Henderson, E., and Miller, G., 1977, Host determined differences in expression of surface marker characteristics on human and simian lymphoblastoid cell lines transformed by Epstein–Barr virus, *Proc. Natl. Acad. Sci. U.S.A.* **74**:749–753.

Robinson, J., Frank, A., Henderson, E., Schweitzer, J., and Miller, G., 1979, Surface markers and size of lymphocytes in human umbilical cord blood stimulated into deoxyribonucleic acid synthesis by Epstein–Barr virus, *Infect. Immun.* **26**:225–231.

Robinson, J., Brown, N., Andiman, W., Halliday, K., Francke, U., Robert, M., Andersson-Anvret, M., Horstmann, D., and Miller, G., 1980a, Diffuse polyclonal B-cell lymphoma during primary infection with Epstein–Barr virus, *N. Engl. J. Med.* **302**:1293–1297.

Robinson, J., Smith, D., and Niederman, J., 1980b, Mitotic EBNA-positive lymphocytes in peripheral blood during infectious mononucleosis, *Nature (London)* **287**:334–335.

Robinson, J., Smith, D., and Niederman, J., 1981, Plasmacytic differentiation of circulating Epstein–Barr virus-infected B lymphocytes during acute infectious mononucleosis, *J. Exp. Med.* **153**:235–244.

Rocchi, G., and Hewetson, J.F., 1973, A practical and quantitative microtest for detection of neutralizing antibodies against Epstein–Barr virus, *J. Gen. Virol.* **13**:385–391.

Rocchi, G., de Felici, A., Ragona, G., and Heinz, A., 1977, Quantitative evaluation of Epstein–Barr-virus-infected mononuclear peripheral blood leukocytes in infectious mononucleosis, *N. Engl. J. Med.* **296**:132–134.

Rogentine, G.N., and Gerber, P., 1969, HL-A antigens of human lymphoid cells in long term tissue culture, *Transplantation* **8**:28–37.

Rosen, A., Gergely, P., Jondal, M., Klein, G., and Britton, S., 1977, Polyclonal Ig production after Epstein–Barr virus infection of human lymphocytes *in vitro*, *Nature (London)* **267**:52–56.

Rosenthal, K., Yamovich, S., Inbar, M., and Strominger, J.L., 1978, Translocation of a hydrocarbon fluorescent probe between Epstein–Barr virus and lymphoid cells: An assay for early events in viral infection, *Proc. Natl. Acad. Sci. U.S.A.* **75**:5076–5080.

Royston, I., Sullivan, J., Periman, P., and Perlin, E., 1975, Cell-mediated immunity to Epstein–Barr-virus transformed lymphoblastoid cells in acute infectious mononucleosis, *N. Engl. J. Med.* **293**:1159–1163.

Rymo, L., 1979, Identification of transcribed regions of Epstein–Barr virus DNA in Burkitt lymphoma-derived cells, *J. Virol.* **32**:8–18.

Saemundsen, A.K., Kallin, B., and Klein, G., 1980, Effect of *n*-butyrate on cellular and viral DNA synthesis in cells latently infected with Epstein–Barr virus, *Virology* **107**:557–561.

Sairenji, T., and Hinuma, Y., 1973, Modes of Epstein–Barr virus in human floating cell lines, *Gann* **64**:583–590.

Sairenji, T., and Hinuma, Y., 1975, Inhibitory effect of adult bovine serum on release of infectious Epstein–Barr virus from a virus-carrier cell line, *J. Natl. Cancer Inst.* **55**:339–343.

Sairenji, T., and Hinuma, Y., 1980, Re-evaluation of a transforming strain of Epstein–Barr virus from the Burkitt lymphoma cell line, Jijoye, *Int. J. Cancer* **26**:337–342.

Schneider, U., and zur Hausen, H. 1975, Epstein–Barr virus-induced transformation of human leukocytes after cell fractionation, *Int. J. Cancer* **15**:59–66.

Seeley, J., Svedmyr, E., Wieland, O., Klein, G., Möller, E., Eriksson, E., Andersson, K., and van der Waal, L., 1981, Epstein–Barr selective T cells in infectious mononucleosis are not restricted to HLA-A and B antigens, *J. Immunol.* **127**:293–300.

Seigneurin, J.M., Vuillaume, M., Lenoir, G., and de Thé G., 1977, Replication of Epstein–Barr virus: Ultrastructural and immunofluorescent studies of P_3HR-1 superinfected Raji cells, *J. Virol.* **24**:836–845.

Shapiro, I.M., Andersson-Anvret, M., and Klein, G., 1978, Polyploidization of Epstein–Barr virus (EBV) carrying lymphoma lines decreases the inducibility of EBV-determined early antigen following P_3HR-1 virus superinfection or iododeoxyuridine treatment, *Intervirology* **10**:94–101.

Sherr, C.J., Schenkein, I., and Uhr, J.W., 1971, Synthesis and intracellular transport of immunoglobulin in secretory and nonsecretory cells, *Ann. N. Y. Acad. Sci.* **190**:250–267.

Shevach, E.M., Herberman, R., Frank, M.M., and Green, I., 1972, Receptors for complement and immunoglobulin on human leukemic cells and human lymphoblastoid cell lines, *J. Clin. Invest.* **51**:1933–1938.

Shope, T., and Miller, G., 1973, Epstein–Barr virus: Heterophile responses in squirrel monkeys inoculated with virus transformed autologous leukocytes, *J. Exp. Med.* **137**:140–147.

Shope, T., Dechairo, D., and Miller, G., 1973, Malignant lymphoma in cotton-top marmosets after inoculation with Epstein–Barr virus, *Proc. Natl. Acad. Sci. U.S.A.* **70**:2487–2491.

Siegert, W., Moar, M.H., Bell, C., and Klein, G., 1977, Demonstration of complement receptors on lymphoblastoid cells by radiolabeled antibodies and *in situ* autoradiography, *Cell. Immunol.* **31**:234–241.

Southam, C.M., Burchenal, J.H., Clarkson, B., Tanzi, A., Mackey, R., and McComb, V., 1969a, Heterotransplantation of human cell lines from Burkitt's tumors and acute leukemia into newborn rats, *Cancer* **23**:281–299.

Southam, C.M., Burchenal, J.H., Clarkson, B., Tanzi, A., Mackey, R., and McComb, V., 1969b, Heterotransplantability of human cell lines derived from leukemia and lymphomas into immunologically tolerant rats, *Cancer* **24**:211–222.

Steel, C.M., and Hardy, D.A., 1970, Evidence of altered antigenicity in cultured lymphoid cells from patients with infectious mononucleosis, *Lancet* **1**:1322–1323.

Steel, C.M., McBeath, S., and O'Riordan, M.L., 1971, Human lymphoblastoid cell lines. II. Cytogenetic studies, *J. Natl. Cancer Inst.* **47**:1203–1214.

Steel, C.M., Hardy, D.A., Ling, N.R., Dick, H.M., Macintosh, P., and Crichton, W.B., 1973, The interaction of normal lymphocytes and cells from lymphoid cell lines. III. Studies on activation in an autochthonous system, *Immunology* **24**:177–189.

Steel, C.M., Evans, J., Joss, A.W.L., and Arthus, M.E., 1974, Antibody activity associated with immunoglobulins secreted by human lymphoblastoid cell lines, *Nature* (*London*) **252**:604–605.

Steinitz, M., Klein, G., 1975, Comparison between growth characteristics of an Epstein–Barr virus (EBV)-genome-negative lymphoma line and its EBV-converted subline *in vitro*, *Proc. Natl. Acad. Sci. U.S.A.* **72**:3518–3520.

Steinitz, M., Klein, G., Koskimies, S., and Mäkelä, O., 1977, EB virus-induced B lymphocyte cell lines producing specific antibody, *Nature* (*London*) **269**:429–422.

Steinitz, M., Bakacs, T., and Klein, G., 1978, Interaction of the B95–8 and P_3HR-1 substrains of Epstein–Barr virus (EBV) with peripheral human lymphocytes, *Int. J. Cancer* **22**:251–257.

Steinitz, M., Koskimies, S., Klein, G., and Mäkelä, O., 1979a, Establishment of specific antibody producing human lines by antigen pre-selection and Epstein–Barr (EBV)-transformation, *J. Clin. Lab. Immunol.* **2**:1–7.

Steinitz, M., Seppälä, I., Eichmann, K., and Klein, G., 1979b, Establishment of a human lymphoblastoid cell line with specific antibody production against Group A streptococcal carbohydrate, *Immunobiology* **156**:41–47.

Stewart, S.E., Lovelace, E., Whang, J.J., and Ngu, V.A., 1965, Burkitt tumor, tissue culture, cytogenetic and virus studies, *J. Natl. Cancer Inst.* **34**:319–327.

Strauch, B., Siegel, N., Andrews, L., and Miller, G., 1974, Oropharyngeal excretion of Epstein–Barr virus by renal transplant recipients and other patients treated with immunosuppressive drugs, *Lancet* **1**:234–237.

Strominger, J., and Thorley-Lawson, D., 1979, Early events in transformation of human lymphocytes by the virus, in: *The Epstein–Barr Virus* (M.A. Epstein, and B.G. Achong, eds.), pp. 185–204, Springer-Verlag, Berlin, Heidelberg, New York.

Sugamura, K., and Hinuma, Y., 1980, *In vitro* induction of cytotoxic T lymphocytes specific for Epstein–Barr virus-transformed cells: Kinetics of autologous restimulation, *J. Immunol.* **124**:1045–1049.

Sugden, B., 1977, Comparison of Epstein–Barr viral DNA's in Burkitt lymphoma biopsy cells and in cells clonally transformed *in vitro*, *Proc. Natl. Acad. Sci. U.S.A.* **74**:4651–4655.

Sugden, B., and Mark, W., 1977, Clonal transformation of adult human leukocytes by Epstein–Barr virus, *J. Virol.* **23**:503–508.

Sugden, B., Phelps, M., and Domoradzki, J., 1979, Epstein–Barr virus DNA is amplified in transformed lymphocytes, *J. Virol.* **31**:590–595.

Sullivan, J.L., Byron, K.S., Brewster, F.E., and Purtilo, D.T., 1980, Deficient natural killer cell activity in the X-linked lymphoproliferative syndrome, *Science* **210**:543–545.

Summers, W.C., and Klein, G., 1976, Inhibition of EBV DNA synthesis and late gene expression by phosphonoacetic acid, *J. Virol.* **18**:151–155.

Svedmyr, E., and Jondal, M., 1975, Cytotoxic effector cells specific for B cell lines transformed by Epstein–Barr virus are present in patients with infectious mononucleosis, *Proc. Natl. Acad. Sci. U.S.A.* **72**:1622–1626.

Svedmyr, E., Jondal, M., Henle, W., Weiland, O., Rombo, L., and Klein, G., 1978, EBV specific killer T cells and serologic responses after onset of infectious mononucleosis, *J. Clin. Lab. Immunol.* **1**:225–232.

Takada, K., and Osato, T., 1979, Analysis of the transformation of human lymphocytes by Epstein–Barr virus. I. Sequential occurrence from the virus-determined nuclear antigen synthesis, to blastogenesis, to DNA synthesis, *Intervirology* **11**:30–39.

Takahashi, M., Yagi, Y., Moore, G.F., and Pressman, D., 1969a, Pattern of immunoglobulin production in individual cells of human hematopoietic origin in established culture, *J. Immunol.* **102**:1274–1283.

Takahashi, M., Tanigaki, N., Yagi, Y., Moore, G.E., and Pressman, D., 1969b, Immunoglobulin production in cloned sublines of a human lymphocytoid cell line, *J. Immunol.* **102**:1388–1393.

Tanaka, Y., Sugamura, K., Hinuma, Y., Sato, H., and Okochi, K., 1980, Memory of Epstein–Barr virus-specific cytotoxic T cells in normal seropositive adults as revealed by an *in vitro* restimulation method, *J. Immunol.* **125**:1426–1431.

Tanigaki, N., Yagi, Y., Moore, G.E., and Pressman, D., 1966, Immunoglobulin production in human leukemia cell lines, *J. Immunol.* **97**:634–646.

Theofilopoulos, A.N., and Perrin, L.H., 1977, Lysis of human cultured lymphoblastoid cells by cell-induced activation of the properdin pathway, *Science* **195**:878–880.

Thestrup-Pedersen, K., Esmann, V., Bisballe, S., Jensen, J.R., Pallesen, G., Hastrup, J., Madsen, M., Thorling, K., Grazia-Masucci, M., Saemundsen, A.K., and Ernberg, I., 1980, Epstein–Barr-virus-induced lymphoproliferative disorder converting to fatal Burkitt-like lymphoma in a boy with interferon-inducible chromosomal defect, *Lancet* **2**:997–1001.

Thorley-Lawson, D.A., 1979, Characterization of cross-reacting antigens on the Epstein–Barr virus envelope and plasma membranes of producer cells, *Cell* **16**:33–42.

Thorley-Lawson, D., and Strominger, J.L., 1976, Transformation of human lymphocytes by Epstein–Barr virus is inhibited by phosphonoacetic acid, *Nature (London)* **263**:332–334.

Thorley-Lawson, D., and Strominger, J., 1978, Reversible inhibition by phosphonoacetic acid of human B lymphocyte transformation by Epstein-Barr virus, *Virology* **86**:423–431.

Thorley-Lawson, D., Chess, L., and Strominger, J., 1977, Suppression of *in vitro* Epstein–Barr virus infection: A new role for adult human T lymphocytes, *J. Exp. Med.* **146**:495–508.

Tomkins, G.A., 1968, Chromosome studies on cultured lymphoblastoid cells from cases of New Guinea Burkitt lymphoma, myeloblastic and lymphoblastic leukemia and infectious mononucleosis, *Int. J. Cancer* **3**:644–653.

Tovey, M.G., Lenoir, G., Bergen-Lours, J., 1978, Activation of latent Epstein-Barr virus by antibody to human IgM, *Nature (London)* **276**:270–272.

Traul, K.A., Stephens, R., Gerber, P., and Peterson, W.D., 1977, Productive Epstein–Barr viral infection of the human lymphoblastoid cell line, 6410, with release of early antigen inducing and transforming virus, *Int. J. Cancer* **20**:247–255.

Trowbridge, I.S., Hyman, R., and Klein, G., 1977, Human B cell line deficient in the expression of B cell-specific glycoproteins (GP 27, 35), *Eur. J. Immunol.* **7**:640–645.

Trumper, P.A., Epstein, M.A., Giovanella, B.C., and Finerty, S., 1977, Isolation of infectious EB virus from the epithelial tumor cells of nasopharyngeal carcinoma, *Int. J. Cancer* **20**:655–662.

Van Beek, W., Nilsson, K., Klein, G., and Emmelot, P., 1979, Cell surface glycoprotein changes in Epstein–Barr virus-positive and -negative human hematopoietic cell lines, *Int. J. Cancer* **23**:464–473.

Van Boxel, J.A., and Buell, D.N., 1974, IgD on cell membranes of human lymphoid cell lines with multiple immunoglobulin classes, *Nature (London)* **251**:443–444.

Van den Berg, H., Parloir, H., Gossy, S., Englebienne, V., Cornu, G., and Sokal, G., 1979, Variant translocation in Burkitt lymphoma, *Cancer Genet. Cytogenet.* **1**:9–14.

Van de Stouwe, R., Kunkel, H., Halper, J., and Weksler, M., 1977, Autologous mixed lymphocyte culture reactions and generation of cytotoxic T cells, *J. Exp. Med.* **146**:1809–1814.

Van Furth, R., Gorte, H., Nadkarni, J.S., Klein, E., and Clifford, P., 1972, Synthesis of immunoglobulins by biopsied tissues and cell lines from Burkitt's lymphoma, *Immunology* **22**:847–857.

Volsky, D.J., Shapiro, I.M., and Klein, G., 1980, Transfer of Epstein–Barr virus receptors to receptor-negative cells permits virus penetration and antigen expression, *Proc. Natl. Acad. Sci. U.S.A.* **77**:5453–5457.

Vonka, V., and Kutinova, L., 1973, Increase of EB virus positive cells in cultured Burkitt lymphoma cells after short-time exposure to heat, *Neoplasma* **20**:349–352.

Vonka, V., Kutinova, L., Drobnik, J., and Bräuerova, J., 1972, Increase of Epstein–Barr virus positive cells in EB 3 cultures after treatment with cisdichlorodiammine-platinum (11), *J. Natl. Cancer Inst.* **48**:1277–1281.

Wakefield, J.D., Thorbecke, G.J., Old, L.J., and Boyse, E.A., 1967, Production of immunoglobulins and their subunits by human tissue culture cell lines, *J. Immunol.* **99**:308–319.
Weksler, M., 1976, Lymphocyte transformation induced by autologous cells. III. Lymphoblast-induced lymphocyte stimulation does not correlate with EB viral antigen expression or immunity, *J. Immunol.* **116**:310–314.
Welsh, K.I., Dorval, G., Nilsson, K., Clements, G., and Wigzell, H., 1977, Quantitation of β_2-microglobulin and HLA on the surface of human cells. II. *In vitro* cell lines and their hybrids, *Scand. J. Immunol.* **6**:265–271.
Werner, J., Wolf, H., Apodaca, J., and zur Hausen, H., 1975, Lymphoproliferative disease in a cotton-top marmoset after inoculation with infectious mononucleosis-derived Epstein–Barr virus, *Int. J. Cancer* **15**:1000–1008.
Wernet, P., 1976, Human Ia-type alloantigens, methods of detection, aspects of chemistry and biology, markers of disease states, *Transplant. Rev.* **30**:271–298.
Whang-Peng, J., Gerber, P., and Knutsen, T., 1970, So-called C marker chromosome and Epstein–Barr virus, *J. Natl. Cancer Inst.* **45**:831–839.
Wilson, G., and Miller, G., 1979, Recovery of Epstein–Barr virus from nonproducer human lymphoid cell lines, *Virology* **95**:351–358.
Yajima, Y., and Nonoyama, M., 1976, Mechanisms of infection with Epstein–Barr virus. I. Viral DNA replication and formation of non-infectious virus particles in superinfected Raji cells, *J. Virol.* **19**:187–194.
Yamamoto, N., and Hinuma, Y., 1976, Clonal transformation of human leukocytes by Epstein–Barr virus in soft agar, *Int. J. Cancer* **17**:191–196.
Yata, J., Desgranges, C., Nakagawa, T., Favre, M.D., and de Thé, G., 1975, Lymphoblastoid transformation and kinetics of appearance of viral nuclear antigen (EBNA) in cordblood lymphocytes infected by Epstein–Barr virus (EBV), *Int. J. Cancer* **15**:377–384.
Yefenof, E., and Klein, G., 1977, Membrane receptor stripping confirms the association between EBV receptors and complement receptors on the surface of human B lymphoma lines, *Int. J. Cancer* **20**:347–352.
Yefenof, E., Klein, G., Jondal, M., and Oldstone, M.B.A., 1976, Surface markers on human B- and T-lymphocytes. IX. Two-color immunofluorescence studies on the association between EBV receptors and complement receptors on the surface of lymphoid cell lines, *Int. J. Cancer* **17**:693–700.
Yefenof, E., Klein, G., and Kvarnung, K., 1977, Relationships between complement activation, complement binding and EBV absorption by human hematopoietic cell lines, *Cell. Immunol.* **31**:225–233.
Yeh, C.G., Robinson, J., and Faulk, W.P., 1982, Antigens of Epstein–Barr virus, infected marmoset lymphocytes, identification on marmoset erythrocytes, human extra-embryonic membranes and transformed cells (submitted).
Zech, L., Haglund, U., Nilsson, K., and Klein, G., 1976, Characteristic chromosomal abnormalities in biopsies and lymphoid-cell lines from patients with Burkitt and non-Burkitt lymphomas, *Int. J. Cancer* **17**:47–56.
Zerbini, M., and Ernberg, I., 1982, The Epstein–Barr virus (EBV): Infection and growth stimulating effect in human B lymphocytes (submitted).
Zeve, V.H., Lucas, S.L., and Manaker, R.A., 1966, Continuous cell culture from a patient with chronic myelogenous leukemia. II. Detection of a herpes-like virus by electron-microscopy, *J. Natl. Cancer Inst.* **37**:761–773.
Zur Hausen, H., Schulte-Holthausen, H., Klein, G., Henle, W., Henle, G., Clifford, P., and Sanesson, L., 1970, EBV DNA in biopsies of Burkitt tumours and anaplastic carcinomas of the nasopharynx, *Nature (London)* **228**:1056–1058.
Zur Hausen, H., Diehl, V., Wolf, H., Schulte-Holthausen, H., and Schneider, U., 1972, Occurrence of Epstein–Barr virus genomes in human lymphoblastoid cell lines, *Nature (London) New Biol.* **237**:189–190.
Zur Hausen, H., O'Neill, F.J., Freese, U.K., and Hecher, E., 1978, Persisting oncogenic herpesvirus induced by tumor promoter TPA, *Nature (London)* **272**:373–375.

CHAPTER 5

Immunology of Epstein–Barr Virus

WERNER HENLE, M.D., AND GERTRUDE HENLE, M.D.

I. INTRODUCTION

The Epstein–Barr virus (EBV) was initially detected by electron microscopy in a small percentage of cells cultured from Burkitt lymphomas (BL), the most frequent malignancy of African children (M.A. Epstein *et al.*, 1964). This tumor had become a special target for virologists because epidemiological observations made in endemic regions pointed to participation of an infectious agent either directly or as a cofactor in its development (cf. Burkitt and Wright, 1970). Despite this background, the virus found in the tumor cells was considered by most virologists to be a harmless passenger because it belonged morphologically to the herpes group of viruses and none of its members had as yet been shown to be potentially oncogenic. The fact that it could not be transmitted to any routine cell cultures, chick embryos, or experimental animals suggested that it was not one of the then known three human herpesviruses (herpes simplex, cytomegalovirus, and varicella virus) and thus provided the first clue that it was a previously unknown agent (M.A. Epstein *et al.*, 1965).

Confronted with a possibly new virus, we found it essential to develop immunological procedures to determine (1) whether the virus was antigenically related to any other herpesviruses; (2) whether the patient from whom the virus was derived and other BL patients had antibodies to it and, if they did, (a) whether patients with other diseases or healthy controls had no such antibodies or, if they had, (b) whether

WERNER HENLE, M.D., and GERTRUDE HENLE, M.D. • The Joseph Stokes, Jr., Research Institute of The Children's Hospital of Philadelphia, University of Pennsylvania School of Medicine, 34th Street and Civic Center Boulevard, Philadelphia, Pennsylvania 19104.

the BL patients had higher titers of antibodies than the control groups. By indirect-immunofluorescence tests with fixed cell smears from the BL cultures, the virus was shown to be antigenically distinct from all other herpesviruses (G. Henle and W. Henle, 1966a,b), and it was thenceforth called the Epstein–Barr virus after the EB-1 culture in which it was first seen. All sera from BL patients elicited brilliant fluorescence in the virus-producing cells of the test smears, but so did many of the African control sera and, indeed, sera from other parts of the world (G. Henle and W. Henle, 1966a; Levy and Henle, 1966). The BL sera had, however, a 10-fold higher geometric mean antibody titer (GMT) than the controls (G. Henle *et al.*, 1969). Other studies had shown that a large proportion of BL cells derived from biopsies or cultures expressed an antigen in their membrane that was absent in other cells from the patients (Klein *et al.*, 1966, 1967). The cell-membrane antigen(s) turned out to be EBV-determined (Klein *et al.*, 1968a). Thus, EBV remained a candidate for the etiology of BL.

The virus also became a candidate for the cause of undifferentiated nasopharyngeal carcinomas (NPCs) through a chance observation. Old *et al.* (1966, 1968) noted that sera from African NPC patients, represented among the controls, reacted as well as BL sera with extracts of cultured BL cells in double-immunodiffusion tests, yielding up to five lines of precipitation. Following this lead, we titrated sera from African, Chinese, and European NPC patients in indirect-immunofluorescence tests, and they yielded 8- to 10-fold higher GMTs than sera from patients with carcinomas at other sites of head and neck, from patients with other tumors of the nasopharynx, or from healthy controls (W. Henle and G. Henle, 1968; de Schryver *et al.*, 1969; W. Henle *et al.*, 1970a).

The apparent association of EBV with BL and NPC was further established by the demonstration of EBV DNA in nearly all biopsies from such tumors (zur Hausen *et al.*, 1970; Nonoyama *et al.*, 1973; Lindahl *et al.*, 1974; Pagano *et al.*, 1975; Andersson-Anvret *et al.*, 1977) as well as the detection of the EBV-associated nuclear antigen in the tumor cells (Reedman *et al.*, 1974; Wolf *et al.*, 1973; Klein *et al.*, 1974; Huang *et al.*, 1974).

Both BL and NPC have a high incidence in certain, but different, geographic regions of the world (10–40/100,000 per year), yet overall these tumors are rare considering that practically nobody escapes infection by EBV anywhere in the world (cf. W. Henle and G. Henle, 1979). We thought, therefore, that the virus might be primarily the cause of a common benign disease that by another chance event was found to be infectious mononucleosis (IM) (G. Henle *et al.*, 1968). A seronegative technician in our laboratory provided the initial clue when she developed the disease and seroconverted in its course. Subsequent prospective studies amply confirmed the initial observations, so that there remains no doubt that EBV is the cause of IM (cf. G. Henle and W. Henle, 1979).

On a worldwide basis, by far the greatest number of primary EBV infections occur in early childhood when they are mostly inapparent but, like IM, establish a permanent, latent viral carrier state that can become unbalanced by immunosuppressive diseases or therapy (cf. W. Henle and G. Henle, 1980). Before discussion of the immune responses encountered under these various conditions, it is necessary to review briefly the virus-specific antigen–antibody systems that have been most useful for seroepidemiological surveys and the serodiagnosis of the EBV-associated diseases.

II. EPSTEIN–BARR-VIRUS-SPECIFIC ANTIGENS

Epstein–Barr virus (EBV) has thus far been shown to replicate only in cultured lymphoblasts, whether derived from Burkitt-lymphoma (BL) biopsies or the peripheral blood of infectious mononucleosis (IM) patients or viral carriers. Even these cells are of only low permissiveness because no more than about 10%, but usually many less, are producing virus at any given time (M.A. Epstein *et al.*, 1965; G. Henle and W. Henle, 1966a; Klein *et al.*, 1968a; Diehl *et al.*, 1968). Furthermore, only lymphocytes of bone-marrow derivation are infectible (Jondal and Klein, 1973; Pattengale *et al.*, 1974b). While oropharyngeal epithelial cells have been claimed to be primary targets of EBV (Lemon *et al.*, 1977) and salivary glands to be one of the persistent habitats of the virus (Niederman *et al.*, 1976; Morgan *et al.*, 1979), neither of these suggestions has been irrefutably proven.

The lymphoblastoid-cell lines (LCLs) are divided into "producer" and "nonproducer" cultures depending on whether or not they shed infectious virus into the media. Because of the restricted replication of the virus, the immunological approach to its detection and identification, to the determination of its geographic dissemination by antibody surveys, and to the evaluation of its role in human disease by specific antibody responses had of necessity to be based mainly on immunofluorescence techniques. Despite these handicaps, several distinct groups of EBV-specific antigens have been differentiated to date, and tests for detection and titration of the corresponding antibodies have been developed that are useful in the serodiagnosis and clinical assessment of EBV-associated diseases.

A. Viral Capsid Antigen

Viral capsid antigen (VCA) was the first EBV-specific antigen described (G. Henle and W. Henle, 1966a). It was detected by indirect immunofluorescence in smears of cultured BL cells and subsequently also in smears of LCLs derived from peripheral leukocytes of various types of donors. The proportion of immunofluorescent cells corresponded closely to the proportion of cells shown by electron microscopy to harbor virus particles (G. Henle and W. Henle, 1966a; Hinuma *et al.*, 1967; Klein *et al.*, 1968a). By pulse–labeling of lymphoblasts with [^{3}H]thymidine and combined immunofluorescence and autoradiography, the isotope was found spread to the cytoplasm from the nucleus of immunofluorescent cells, whereas in the nonfluorescent cells it was limited to the nucleus, if present at all (zur Hausen *et al.*, 1967). Individually picked fluorescent cells, after being embedded and thin-sectioned, were found on electron-microscopic examination to contain numerous virus particles, whereas nonfluorescent cells, similarly prepared, were free of virus (zur Hausen *et al.*, 1967; M.A. Epstein and Achong, 1968). As determined by negative-contrast electron microscopy, nonenveloped, naked nucleocapsids extracted and concentrated from cells of producer LCLs acquired a coat of antibodies and became in part agglutinated when exposed to sera that were positive in the immunofluorescence test, but not when negative sera were used (W. Henle *et al.*, 1966; Mayyasi *et al.*, 1967). Antibodies to other herpesviruses failed to coat the nucleocapsids. The fluorescence was clearly due to an antigen produced by the indigenous virus, and because the antibodies reacted evidently with nucleocapsids, it was called viral capsid antigen when it became necessary to distinguish it from other, subsequently identified, EBV-specific antigens.

Anyone who has experienced a primary EBV infection at any time in the past has antibodies to VCA at usually moderate titers (cf. W. Henle and G. Henle, 1978). Since practically nobody escapes infection by EBV, commercially available human γ-globulin can be used after conjugation with fluorescein–isothiocyanate for detection of VCA-positive cells by the direct-immunofluorescence technique (G. Henle and W. Henle, 1966a).

B. Epstein–Barr-Virus-Induced Early Antigens

When nonproducer LCLs (Raji or RPMI 64-10) were inoculated with EBV from a producer LCL (P3HR-1), further growth of the culture was partially or totally inhibited, yet indirect-immunofluorescence tests with standard sera from healthy donors, containing antibodies to VCA, stained only very few, if any, of the exposed lymphoblasts (W. Henle *et al.*, 1970b). However, when sera from patients with IM, BL, or nasopharyngeal carcinoma (NPC) were substituted for the standard sera, brilliantly fluorescent cells were observed within 24–48 hr in numbers clearly proportional to the dose of virus used for inoculation of the nonproducer LCL. Obviously, the cells had synthesized antigen(s) to which patients with EBV-associated diseases usually do have antibodies (W. Henle *et al.*, 1971; G. Henle *et al.*, 1971b), whereas healthy donors after long-past primary EBV infections usually do not. The antigen(s) appeared well in advance of VCA in those few cells that synthesized this antigen, suggesting the name "early antigen(s)" (EA). It was shown subsequently (Gergely *et al.*, 1971a,b) that EA synthesis did indeed proceed without restraint when viral DNA replication was prevented by exposure of the cells to 1-β-D-arabinofuranosylcytosine (ARA-C) or other DNA inhibitors. Evidently, only a single, abortive cycle of viral replication was induced by P3HR-1 virus in the cells from nonproducer LCLs, yielding EA but rarely VCA (W. Henle *et al.*, 1970b). Consequently, very few if any infectious virus particles were produced, so that 24 hr postinfection, no increase in EA-synthesizing cells was noted, although the intensity of staining increased up to 48 hr due to an increase in the amount of antigen synthesized within infected cells. The abortive cycle, once initiated, leads invariably to the death of the cell (W. Henle *et al.*, 1970b; Gergely *et al.*, 1971c; Rocchi *et al.*, 1973).

As will be discussed in Section II.C, every cell in producer or nonproducer LCLs carries EBV genomes and expresses the EBV-associated nuclear antigen (EBNA). Exposure of these cells to P3HR-1 virus thus constitutes a superinfection that might, and has indeed been shown to, lead to recombinations between the DNAs of the resident and superinfecting viruses (Cho *et al.*, 1980). The resident viral genomes are usually in a repressed state, but they can be derepressed by exposure of nonproducer cultures to 5-bromodeoxyuridine (BUdR), 5-iodo-2-deoxyuridine (IUdR), *n*-butyrate, phorbol esters, and other substances (Gerber, 1972; Hampar *et al.*, 1972; Luka *et al.*, 1978; zur Hausen *et al.*, 1978). Under these conditions, synthesis of EA is induced in considerable numbers of cells, often progressing to synthesis also of VCA and infectious virus particles.

On examination of numerous sera from patients with EBV-associated diseases for antibodies to EA, two distinct patterns of immunofluorescent staining were discerned, showing the existence of at least two EA components (G. Henle *et al.*, 1971a). Some of the patients' sera caused diffuse fluorescence of nucleus and cytoplasm of abortively EBV-superinfected cells, whereas other sera stained a mass restricted to the cytoplasm adjacent to the nucleus. Antibodies to the diffuse (D) component of the EA complex were shown to arise transiently in IM and to reach high titers in NPC, whereas antibodies to the restricted (R) component were noted to develop to often high titers in BL. Antibodies to EA-D or EA-R were found to be usually absent in sera from healthy individuals or to be present at most at low titers among those who maintain relatively high anti-VCA titers. They may be observed to reach higher levels following activation of persistent EBV infections due to immunosuppressive diseases or therapy (cf W. Henle and G. Henle, 1980). The D component resists fixation by acetone and methanol, whereas the R component is denatured by methanol fixation (G. Henle *et al.*, 1971a). It has not been possible to denature EA-D without also destroying EA-R. The parallel use of acetone and methanol-fixed smears serves to confirm the visual differentiation between antibodies to EA-D or EA-R or both, which at low antibody titers at times is difficult. The nature and functions of these two antigens are not known. Two distinct EA components are induced in nonproducer cells also after their exposure to halogenated pyrimidines (Sugawara and Osato, 1973).

C. Epstein–Barr-Virus-Associated Nuclear Antigen

A complement-fixation soluble (CF/S) antigen has been extracted from cells of producer as well as nonproducer cultures that reacted with many human sera (Armstrong *et al.*, 1966; Pope *et al.*, 1969a; Gerber and Deal, 1970). Its relationship to EBV was not immediately recognized because it was present in nonproducer cells, although the suggestion was made that it could be a nuclear or membrane neoantigen induced by the virus (Armstrong *et al.*, 1966). Only sera with, not sera without, antibodies to VCA reacted with the CF/S antigen, establishing its relationship to EBV (Pope *et al.*, 1969a). In efforts to localize this antigen within cells of producer and nonproducer cultures, Reedman and Klein (1973) used the anti-C′ immunofluorescence technique and showed that the nucleus of nearly every cultured lymphoblast became brilliantly stained by this technique using almost any human serum with antibodies to VCA. Anti-VCA-negative sera failed to stain the nuclei. This EBNA is invariably present in cells from LCLs that are positive for EBV DNA, and, vice versa, BL or NPC biopsies or cell cultures that reveal the presence of viral DNA invariably contain EBNA-positive cells (Reedman *et al.*, 1974; Lindahl *et al.*, 1974; Klein *et al.*, 1974; Huang *et al.*, 1974). Biopsies or cultures free of EBV DNA do not contain EBNA-positive cells.

Antibodies to EBNA are almost uniformly present in anti-VCA-positive sera except in sera from patients in the acute phase of IM (G. Henle *et al.*, 1974) or sera from patients with some T-cell defects or dysfunctions (Berkel *et al.*, 1979). There is good evidence that the major component in CF/S antigen preparations derived from nonproducer LCLs is identical with EBNA because the titers of antibodies to either antigen are closely similar (Klein and Vonka, 1974; Ohno *et al.*, 1977). Antibodies to various nonspecific nuclear antigens yield in part patterns of staining in the anti-C′ immunofluorescence test that clearly differ from the typical EBNA immunofluorescence, but others cause staining that is practically indistinguishable from EBNA-specific immunofluorescence (Yoshida, 1971). The anti-EBNA test thus needs to be controlled by including cell smears from EBNA-negative LCLs as controls. Because anticomplementary sera may yield false-negative reactions or prozones when they are mixed with the complement (two-stage procedure), it is essential to charge the test-cell smears consecutively with serum, C′, and the anti-C′ conjugate (three-stage procedure) (W. Henle *et al.*, 1974b).

D. Epstein–Barr-Virus-Determined Cell-Membrane Antigens

By indirect-immunofluorescence tests with BL sera and live cells derived from BL biopsies or cultured therefrom, Klein *et al.* (1966, 1967) demonstrated antigens in the cell membranes that were not detectable on peripheral leukocytes, lymph-node, or bone-marrow cells of the patients. When it was noted that the percentages of cells with positive membrane immunofluorescence in given BL cultures correlated closely with the percentages of VCA-positive cells, it appeared that the antigen(s) involved were not tumor-specific but were induced by EBV (Klein *et al.*, 1968a). Indeed, membrane antigen (MA) was also detectable on cultured lymphoblasts derived from peripheral lymphocytes of IM patients or viral carriers (Klein *et al.*, 1968b). It appeared, furthermore, in the membranes of cells from nonproducer LCLs after exposure to EBV *in vitro* (Gergely *et al.*, 1971a; Pearson *et al.*, 1971), but to show this reliably, superficially attached viral components (Dunkel and Zeigel, 1970) had to be removed after the adsorption period by trypsin treatment of the cells (Sairinji and Hinuma, 1975; Dölken and Klein, 1976). MA was proven to be distinct from VCA by absorption of sera with cells from a producer line that removed all MA reactivity but hardly affected the antibody titers to VCA, and by the existence of "discordant" sera, showing high anti-VCA but low anti-MA titers and vice versa (Pearson *et al.*, 1969, 1971; Silvestre *et al.*, 1971).

Antibodies to MA are present only in sera that also contain antibodies to VCA (Klein *et al.*, 1968a 1969), and they appear *de novo* in patients during the acute phase of IM (Klein *et al.*, 1968b). The detection of antibodies to MA by indirect immunofluorescence is complicated, however, by the not infrequent presence in sera of antibodies to isoantigens or lymphoblast antigens other than MA. To avoid this problem, a blocking test has been devised in which the target cells are first exposed to test serum dilutions and then to fluorescein–isothiocyanate-conjugated antibodies to MA from a donor shown to have no isoantibodies or other nonspecific antibodies to lymphoblast membranes (Klein *et al.*, 1969).

Two sets of EBV-determined cell MAs have been differentiated. The early MA (EMA) is thought to be synthesized independently of prior viral DNA replication, whereas the late MA (LMA) is not (Ernberg *et al.*, 1974; Silvestre *et al.*, 1974; Sairinji *et al.*, 1977). At least three different EMA specificities have been discerned in blocking tests with human sera having antibodies to one, two, or all three of them, which may represent different determinants of a given macromolecule (Klein *et al.*, 1969; A. Svedmyr *et al.*, 1970). Whether EMA is synthesized without prior viral DNA replication has been questioned by Pearson and Orr (1976), who noted that inhibitors of DNA synthesis significantly reduced production of this antigen. LMA is found only on cells that also synthesize VCA. The envelopes of mature virus particles are derived from cell membranes that contain both EMA and LMA. Thus, antibodies to these antigens are probably in part identical with EBV-neutralizing antibodies (Pearson *et al.*, 1970; de Schryver *et al.*, 1974b). They are also probably identical with the antibodies reactive in antibody-dependent cellular cytotoxicity (ADCC) using EBV-antigen-producing cells as targets (Pearson, 1978; Patarroyo *et al.*, 1980).

Another virus-determined antigen to which no antibodies have as yet been demonstrated is present in the membranes of producer as well as nonproducer cells. This antigen renders nonproducer cells targets for EBV-specific killer T cells that become detectable transiently in the course of IM and specifically attack EBNA-positive lymphoblasts (E. Svedmyr and Jondal, 1975; Jondal *et al.*, 1976; E. Svedmyr *et al.*, 1978). It is being referred to as the lymphocyte-detected membrane antigen (LYDMA).

E. Epstein–Barr-Virus-Neutralizing Antibodies

As mentioned, these are probably at least in part identical with antibodies to MA components. They are measured by a number of different techniques that prevent either superinfection of nonproducer LCLs or transformation of peripheral or cord-blood lymphocytes by EBV. Two different virus populations are required (Miller *et al.*, 1974). The lytic P3HR-1 virus is needed for the first type of test, and its neutralization becomes evident from prevention of EA synthesis in the exposed cells or of death of the abortively superinfected cells so that they retain their capacity to form colonies in the wells of microtiter plates (Pearson *et al.*, 1970; Rocchi *et al.*, 1973). For the second type of test, transforming virus, such as B95-8 virus, is needed, and the criteria for neutralization of the virus are prevention of establishment of LCLs, failure of the exposed cultured lymphocytes to show enhanced incorporation of tritiated thymidine, or nonemergence of EBNA-positive cells (Pope *et al.*, 1969b; Miller *et al.*, 1971, 1972; Robinson and Miller, 1975). These various techniques for measuring neutralizing antibodies yield closely comparable results (de Schryver *et al.*, 1974b). Determination of neutralizing-antibody titers is not extensively practiced because of the complexities of the procedures. It would be the method of choice for determination of the immune status of given populations, but fortunately, anti-VCA-positive sera also have, with very rare exceptions, EBV-neutralizing antibodies, and sera with neutralizing antibodies but lacking anti-VCA have never been observed (Hewetson *et al.*, 1973; Rocchi *et al.*, 1973).

F. Antibody-Dependent Cellular Cytotoxicity

The antibodies involved in this reaction are, like EBV-neutralizing antibodies, directed against EBV-determined EMA or LMA or both (Jondal, 1976; Pearson and Orr, 1976; Pearson *et al.*, 1978, 1979). The effector cells are lymphocytes devoid of B- and T-cell markers that possess Fc receptors (cf. Pearson, 1978). They are equally effective whether they are derived from anti-VCA-positive or -negative donors. The ADCC reaction is measured by chromium-51 release from destroyed labeled target cells, these target cells being lymphoblasts from nonproducer lines *per se* (control) or after superinfection with sufficient EBV to induce EA synthesis in at least 30% of the cells. The ADCC test is more sensitive for measuring antibodies to MA than the immunofluorescence assay. If such reactions also occur *in vivo*, they would eliminate producer cells as soon as EBV determined MA appears on their surface. LMA seems to be the target for the ADCC reaction (Takaki *et al.*, 1980). This could explain why EA- and VCA-positive cells are not detectable in BL or NPC biopsies.

G. Further Antigens and Other Serological Tests

There are additional antigens that are coded for by EBV, but they have not as yet played significant roles in the analysis of immune responses in EBV-associated diseases. Further antigens will probably be found in the future. There are also techniques other than immunofluorescence for measuring antibodies to EBV-determined antigens, i.e., immunodiffusion, complement fixation, immune-adherence hemagglutination, passive hemagglutination, enzyme-linked immunosorbentassay, radioimmunoassay, and other assays (cf. Ernberg and Klein, 1979). Most of these tests require the availability of reasonably pure antigens to be sure that the antibodies measured are directed against the desired antigens and not against other virus-specific antigens or normal host-cell components. Sufficient amounts of any purified antigen are not yet available. It is an advantage of the immunofluorescence techniques that they have built-in controls in that only the appropriate percentages of cells ought to become fluorescent and the pattern of staining must conform to that of the antigen under test. Nonspecific fluorescence due to antibodies unrelated to EBV on the cell surface, in the cytoplasm, or in the nucleus, as found for instance in autoimmune conditions, is generally immediately recognized as such on microscopic reading of the tests.

III. EPSTEIN–BARR-VIRUS-SPECIFIC IMMUNE RESPONSES

The immunofluorescence techniques described in the preceding section have permitted delineation of Epstein–Barr virus (EBV)-specific antibody spectra (Fig. 1) that are characteristic for each of the EBV-associated diseases as well as for silent primary or persistent latent infections. The salient features of the immune responses seen under these various conditions will be discussed in the following sections.

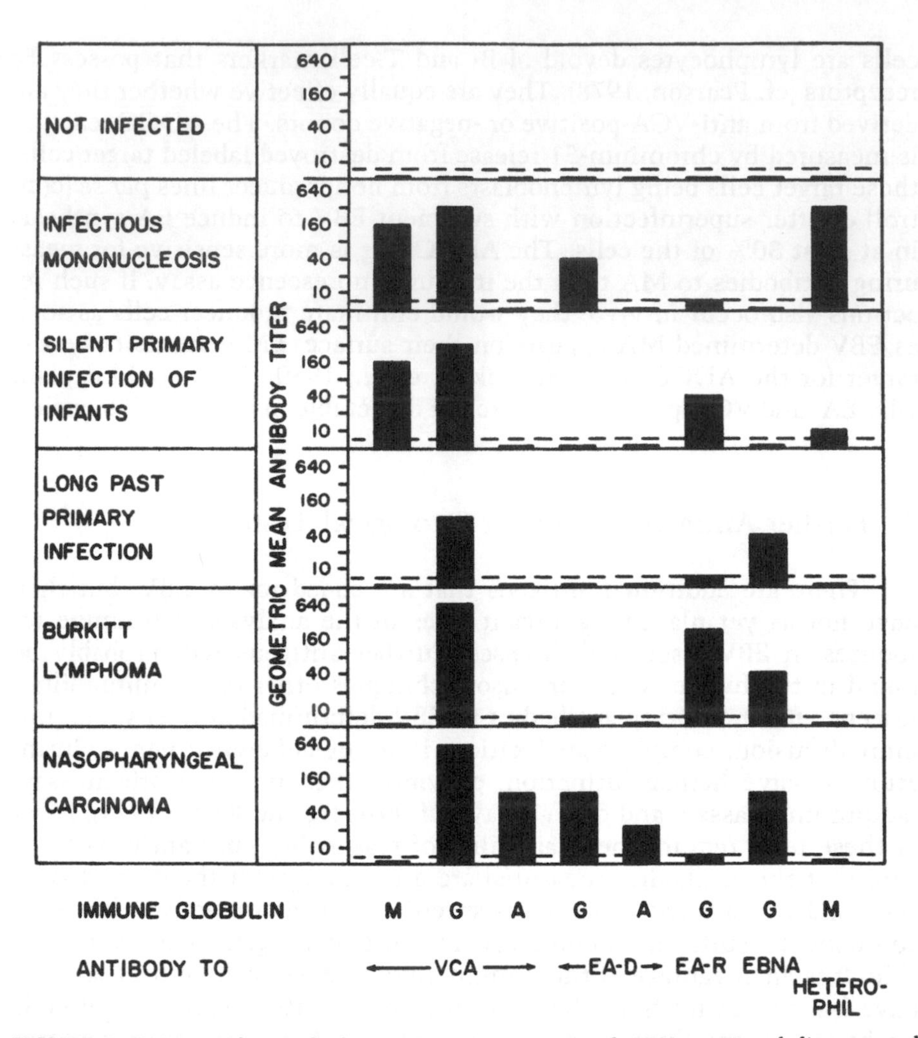

FIGURE 1. EBV-specific antibody patterns in patients with EBV-associated diseases and healthy controls. Reprinted from W. Henle and G. Henle (1982) by courtesy of Marcel Dekker, Inc.

A. Silent Infections

By far the greatest number of primary EBV infections on a worldwide basis remain silent. Seroepidemiological surveys have shown that in most parts of the world, including the remotest regions, antibodies to EBV are acquired generally before the age of 3–5 years (Levy and Henle, 1966; Tischendorf *et al.*, 1970; Black *et al.*, 1970; Lang *et al.*, 1977; W. Henle and G. Henle, 1979). Yet, infectious mononucleosis (IM) is not observed in such areas (Diehl *et al.*, 1969a), suggesting that at an early age, the vast majority of primary EBV infections are unaccompanied by any illness or induce such mild respiratory signs and symptoms that they become submerged among the numerous viral insults children have to experience during their early years. To provide information on this question, a prospective study of infants was carried out in Ghana (Biggar *et al.*, 1978a,b). A total of 31 newborn infants were visited monthly for 15 months and once more at 21 months of age. Depending on the initial height of titers, maternal antibodies became nondetectable (<1 : 10) in 2–7 months after birth. Primary EBV infections occurred at the earliest 2 months after maternal antibodies were no longer measurable. By 1 year of age, 44% of the infants had seroconverted, and by 21 months of age, 84%. No clinical signs or hematological changes suggestive of IM were noted during the period of seroconversion. There were no or at most barely significant heterophil antibody responses. Otherwise, the EBV-specific antibody pattern conformed to that seen in IM with one notable exception (Fig. 1). While the infants, like IM patients, developed high immunoglobulin M (IgM) and IgG antibody titers to viral capsid antigen (VCA) and showed no antibodies to EBV-associated nuclear antigen (EBNA) during the early period after seroconversion, the antibodies to the early antigen (EA) complex in the infants were directed against the restricted component (EA-R), whereas in IM they are, as a rule, directed against the diffuse component (EA-D). Evidence has been obtained that primary EBV infections in American infants under 2 years of age take an essentially similar course (Fleisher *et al.*, 1979a).

These observations confirmed that EBV infections remain silent in the great majority of infants, but this does not preclude that an occasional infant develops full-blown IM as will be discussed in the next section. It is unknown at present over what age range primary EBV infections convert from dominantly silent to dominantly overt, as observed in college students, of whom usually between 50 and 67% develop IM (Niederman *et al.*, 1968, 1970; Evans *et al.*, 1968; Sawyer *et al.*, 1971; University Health Physicians and PHS Laboratories, 1971; Hallee *et al.*, 1974). It is also unknown whether silent infections at any age above 2 years induce only minor, if any, heterophil antibody responses, and whether they evoke antibodies to EA-R, as in the infants, or to EA-D, as in IM.

Certain observations suggest that the emergence of antibodies to EA-D is related to a considerable extent to lymph-node involvement. Lymphadenopathy is a key feature of IM, and high titers of antibodies to EA-D are among the characteristic serological features of nasopharyngeal carcinoma (NPC), in which invasion of draining cervical lymph nodes by the tumor is a common event. In Burkitt lymphoma (BL), lymph nodes are usually not affected initially, but they are often invaded in the final stages, when antibodies to EA-D frequently emerge.

All primary EBV infections end in a permanent, latent viral carrier state in the lymphoreticular system that is responsible for the lifelong maintenance of IgG antibodies to VCA and EBNA as well as of neutralizing antibodies at moderate titers. The viral carriers intermittently excrete EBV into the oropharynx (Gerber *et al.*, 1972; Miller *et al.*, 1973;

Chang *et al.*, 1973; Miller, 1975), from where its transmission to susceptibles by salivary contact is accomplished. Thus, under the primitive conditions of prehistoric times that still exist in some parts of the world today, an almost perfect relationship has evolved between EBV and its host that permits ready transmission of the virus to infants followed by its lifelong persistence without causing significant disease. With improved hygienic standards and living conditions, primary EBV infections are often delayed to an age when they may cause illness of varying severity.

B. Infectious Mononucleosis

For reasons just stated, this disease is largely restricted to economically advanced societies and to their more prosperous social strata. The EBV-specific serology serves to determine who is susceptible to IM and to aid in the serodiagnosis of this disease. It has also contributed to the understanding of the pathogenesis of IM and the control of the ensuing viral carrier state.

1. Determination of Susceptibility

IM occurs only in individuals who previously had no antibodies to VCA (G. Henle *et al.*, 1968; Niederman *et al.*, 1968; Evans *et al.*, 1968; Gerber *et al.*, 1968; Tischendorf *et al.*, 1970; Hirshaut *et al.*, 1971; University Physicians and PHS Laboratories, 1971; Hallee *et al.*, 1974). The indirect-immunofluorescence test for IgG antibodies to VCA is therefore a dependable indicator of immunity to the disease. While antibodies to VCA are not protective, it is rare to find an anti-VCA-positive serum that does not also contain EBV-neutralizing antibodies (Rocchi *et al.*, 1973; Hewetson *et al.* 1973). Should a safe and potent EBV vaccine ever become available, it would be easy to identify by a simple screening test for anti-VCA those students in schools, colleges, military academies, and other institutions who require protection.

2. Serodiagnosis of IM

The serodiagnosis of many viral infections rests on the demonstration of 4-fold or greater increments in antibody titers between the acute and convalescent phases of illness. This requirement cannot be fulfilled in IM because most cases (>80%) show already near-peak antibody titers when first examined, but it also is unnecessary because primary EBV infections evoke the characteristic antibody pattern shown in Fig. 1, which permits a reliable serodiagnosis with a single serum specimen (cf. W. Henle *et al.*, 1974a). In the acute phase of IM, high titers of IgM and IgG antibodies are measured; about 85% of patients transiently develop antibodies to EA-D, which in prolonged cases may in time be replaced occasionally by low titers of antibodies to EA-R (Horwitz *et al.*, 1975), but antibodies to EBNA are as yet absent because they arise only several weeks or months after onset (G. Henle *et al.*, 1974). EBV-neutralizing antibodies, like anti-VCA and anti-EA-D, appear early in the acute phase (Hewetson *et al.*, 1973), whereas antibodies to the complement-fixation soluble antigen emerge, like anti-EBNA, only during convalescence (Vonka *et al.*, 1972). Should confirmation of the diagnosis be needed, subsequent sera from the patient should reveal a decline or disappearance of the IgM antibodies to VCA and of the IgG antibodies to EA-D as well as emergence of antibodies to EBNA.

EBV-specific serodiagnostic tests are often not required in typical cases of IM because the heterophil antibody response of the Paul–Bunnell–Davidsohn type observed in most patients is pathognomic for the disease (Davidsohn and Lee, 1969). They are required for heterophil-antibody-negative cases of IM or IM-like illnesses to separate those caused by EBV from those caused by cytomegalovirus and other viruses or by *Toxoplasma gondii* (Klemola *et al.*, 1967, 1970; Evans *et al.*, 1968, 1975; Wahren *et al.*, 1969; Nikoskelainen *et al.*, 1974; Horwitz *et al.*, 1977). EBV-specific serodiagnostic tests are also essential for heterophil-antibody-positive patients who fail to show some of the characteristic clinical or hematological signs of IM, or for cases with severe complications, except possibly those due to mechanical factors, e.g., obstruction of the airways or rupture of the spleen. Some of the complications are due to autoimmune responses, i.e., acquired hemolytic anemia, thrombocytopenia, agranulocytosis, aplastic anemia, agammaglobulinemia, immune-complex deposition in the kidney and elsewhere, and others. Other complications arise from extensive lymphoproliferation in various organs leading in the CNS to meningoencephalitis, transverse myelitis, Guillain–Barré syndrome, Bell's palsy, or cerebellar ataxia (Grose *et al.*, 1975; Lange *et al.*, 1976; Cleary *et al.*, 1980); in the liver, to hepatic necrosis and now also Reye's syndrome (Fleisher *et al.*, 1980); in the heart, to peri- and myocarditis; in the kidney, to nephritis; and to other conditions. Almost any organ can be affected. These complications may occur simultaneously with, or subsequent to, typical signs of IM, but on rare occasions even in their absence, and they also can be unaccompanied by heterophil-antibody responses. They account for the fatalities observed in IM. A special situation obtains in the autosomal recessive X-linked lymphoproliferative syndrome (Bar *et al.*, 1974; Purtilo *et al.*, 1975, 1978), in which a combined B- and T-cell deficiency or dysfunction prevents or delays the usual self-limitation of IM (see Section III.C).

While IM usually affects adolescents and young adults, typical cases may be observed in older individuals, the oldest recorded being 78 years of age (Horwitz *et al.*, 1976). Typical cases have also been seen in children and infants as young as 10 months (Tischendorf *et al.*, 1970; Joncas and Mitnyan, 1970; Schmitz *et al.*, 1972; Sutton *et al.*, 1974; Ginsburg *et al.*, 1977; Fleisher *et al.*, 1979b; Horwitz *et al.*, 1981). The EBV-specific antibody responses in early childhood cases of IM compare with those of older patients, except that a few infants may respond with anti-EA-R instead of with anti-EA-D. Heterophil-antibody responses are absent more frequently and their titers are lower in infants than in older children. Thus, the geometric mean heterophil-antibody titers (GMTs) increase with the age of the patients (Fleisher *et al.*, 1979b; Horwitz *et al.*, 1981).

The parallel performance of several separate tests for the serodiagnosis of IM seem overly complex, but it has the advantage that the tests not only complement but also control each other. If, for instance, a serum reacts strongly in both the test for IgM antibodies to VCA and that for antibodies to EBNA, the chances are that the serum contains either rheumatoid factor (RF) or nonspecific antinuclear antibodies (ANAs). RF, an IgM antibody against the Fc region of IgG, can mimic a positive test for VCA-specific IgM antibodies (Nikoskelainen *et al.*, 1974; G. Henle *et al.*, 1979), and ANAs can yield EBNA-like staining patterns (Yoshida, 1971). Either possibility can be excluded by appropriate tests for RF or an EBNA-negative cell control in the anticomplement immunofluorescence test for antibodies to EBNA. Very occasionally, a patient might respond unusually fast with EBNA-specific antibodies.

3. Pathogenesis of IM

There are few hard facts covering the 4- to 7-week period from transmission of EBV to onset of illness. A course of events can be surmised, however, on the basis of the limited biological activities of the virus, the restricted permissiveness of B lymphocytes for the virus, the long incubation period of IM, the early peak titers of antibodies to membrane antigen (MA), EA, and VCA contrasted with the late emergence of antibodies to EBNA, and, finally, observations concerning cellular immune responses in the course of the disease. The events are schematically presented in Fig. 2.

It has been suggested that some, possibly fully permissive, epithelial cells in the oropharynx are the primary targets of EBV (Lemon *et al.*, 1977), but this has yet to be irrefutably demonstrated. B lymphocytes would then become infected and transformed as secondary targets, but these cells, in fact, could themselves be the initial targets. If there were fully permissive oropharyngeal epithelial cells, one would wonder why the incubation period of IM should be as long as 4–7 weeks. If B lymphocytes, which are abundantly present in Waldeyer's ring, were the first and only targets, their low degree of permissiveness would impede the spread of the infection, yet in time permit release of sufficient MA, EA, and VCA from occasional productively infected cells to explain the long incubation period and the near-peak titers of the corresponding antibodies at onset of illness. Obviously, insufficient amounts of EBNA are released during the incubation period for early formation of antibodies to this

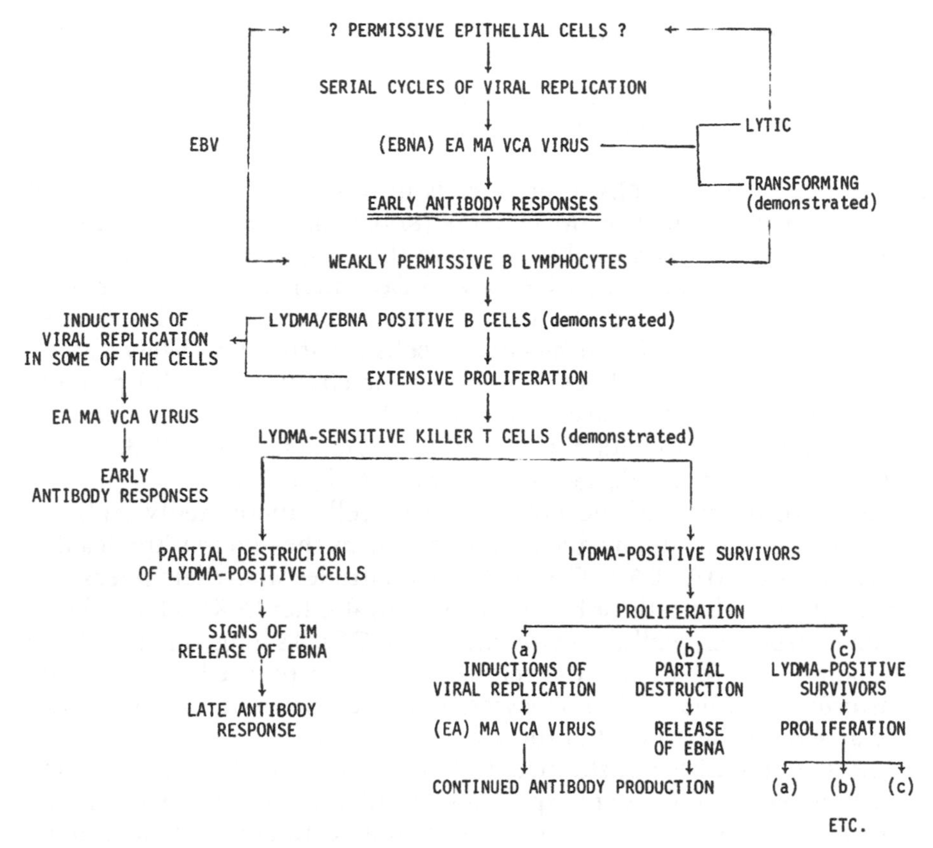

FIGURE 2. Suggested scheme of events that occur during and following primary EBV infections.

antigen as well. The question regarding the initial target cells awaits further study.

B lymphocytes are transformed by EBV *in vitro* into permanently growing lymphoblasts (W. Henle *et al.*, 1967; Pope *et al.*, 1968, 1969b; Gerber *et al.*, 1969; Miller *et al.*, 1971). It is certain that such immortalization of B lymphocytes occurs also *in vivo* and is responsible for part of the lymphoproliferation observed in IM. EBNA-positive lymphoid cells are transiently demonstrable, although with difficulty and generally in only small numbers, in the circulating blood during the early acute phase of the disease (Klein *et al.*, 1976; Katsuki *et al.*, 1979; Robinson *et al.*, 1980b). Their detection is less difficult in lymph nodes obtained by biopsy or at autopsy from unusually severe cases. Under such conditions, as many as 50% of the cells separated from lymph nodes were shown to express EBNA (Virelizier *et al.*, 1978; Britton *et al.*, 1978; Crawford *et al.*, 1979; Robinson *et al.*, 1980a; Look *et al.*, 1982). Such EBNA-positive cells have also been found dispersed throughout various organs of fatal cases. When EBV-genome-carrying lymphocytes are placed in culture, some of them are spontaneously induced to replicate virus that then goes on to transform normal B cells within the cultured population into EBNA-positive lymphoblasts (Rickinson *et al.*, 1974, 1977; Rocchi *et al.*, 1977). However, some EBV-genome-carrying lymphocytes, kept continuously exposed to antibodies to EBV, form colonies in soft agar (Hinuma and Katsuki, 1978), and EBNA-positive cells in mitosis have been observed in the peripheral blood of IM patients (Robinson *et al.*, 1980b). EBNA positive LCL have been established from the peripheral lymphocytes of a susceptible (anti-VCA negative) boy 13 days after he kissed a girl who two days later developed IM; i.e., 25 days before onset of his disease (Svedmyr *et al.*, 1982). EBV-transformed cells constitute probably a small part of the atypical lymphocytes found in the circulation of IM patients that in the early acute phase have mostly B-cell, but later on mostly T-cell, characteristics (Sheldon *et al.*, 1973; Virolainen *et al.*, 1973; Mangi *et al.*, 1974; Enberg *et al.*, 1974; Pattengale *et al.*, 1974a).

A small fraction of the EBNA-positive cells undoubtedly replicate virus spontaneously at any given time *in vivo*, as they do *in vitro*, leading to synthesis of MA, EA, VCA, and virus particles and their release by degeneration of the cells or by action of natural killer (NK) cells or, later, antibody-dependent cellular cytotoxicity (ADCC). As long as neutralizing antibodies are still absent, the released virus can proceed to infect and transform additional B lymphocytes, but once neutralizing antibodies have arisen, further spread of the virus is unlikely.

Most of the EBV-transformed, EBNA-positive B cells do not become virus producers and thus are expected to proliferate in the patient as they do in culture. Since they express the lymphocyte-detected membrane antigen (LYDMA) (E. Svedmyr and Jondal, 1975; Jondal *et al.*, 1976), they become targets for T cells induced to proliferate in response. Indeed, a subpopulation of T cells that specifically attacks cultured EBNA-positive, but not -negative, lymphoblasts is detectable in the circulation of IM patients during the acute phase of illness, but no longer in convalescence (E. Svedmyr and Jondal, 1975; Royston *et al.*, 1975; Hutt *et al.*, 1975). The emergence of suppressor T lymphocytes in the course of IM has been studied in detail (Tosato *et al.*, 1979; Haynes *et al.*, 1979). The EBV-specific killer T cells are best demonstrable after their separation from NK cells (Svedmyr *et al.*, 1978). The activated T lymphocytes probably constitute a large part of the atypical lymphocytes that 2 or more weeks after onset of IM have mostly T-cell characteristics (Sheldon *et al.*, 1973; Virolainen *et al.*, 1973; Pattengale *et al.*, 1974a; Mangi *et al.*, 1974).

It is the violent interaction between the EBV-genome-carrying B lymphocytes and the LYDMA-sensitive T cells that probably accounts for many of the clinical signs of IM: the tonsillitis, pharyngitis, lymphadenopathy, and hepatosplenomegaly. In silent primary EBV infections, these events are presumably kept at a low level so that they do not register clinically. Some of the complications of the disease are thought to be due to seeding of proliferating EBV-genome-carrying B lymphocytes throughout various organs, which then attract attacks from the activated T cells. There is no evidence that EBV multiplies in parenchymal or stromal cells of the affected organs. The destruction of EBV-genome-carrying lymphoblasts by LYDMA-sensitive T cells leads finally to the release of sufficient amounts of EBNA to induce detectable production of antibodies to this antigen. This course of events would account for the late emergence of this antibody.

The EBV-carrying B lymphocytes are never totally eradicated in the course of IM, but persist in small numbers in the lymphoreticular system, evidently for life. EBV-positive lymphoblastoid-cell lines (LCLs) can be established from peripheral leukocytes of anti-VCA positive donors at frequencies directly proportional to the number of cells seeded in culture, and nearly uniformly from lymph nodes (Diehl *et al.*, 1969b; Nilsson *et al.*, 1971; Nilsson, 1979). It has been estimated that among 10^8 or more peripheral lymphocytes, one may find at least one that will produce virus *in vitro* and, in turn, transform cord-blood lymphocytes (Rocchi *et al.*, 1977). While the virus thus persists clearly in lymphoid tissue, it has been suggested that virus excreted into the oropharynx is derived, at least in part, from salivary glands (Niederman *et al.*, 1976; Morgan *et al.*, 1979). It should be remembered that salivary glands are interspersed by numerous small lymph nodes. Thus, while EBV DNA has been found recently in parotid glands and salivary gland tumors (Wolf *et al.*, 1981; Lanier *et al.*, 1981), it remains to be determined whether the viral genomes are located in salivary gland cells or in the abundantly present lymphocytes.

The EBV-specific killer T cells are also probably never completely eliminated after recovery from IM, but they persist in numbers too small to be detectable by present means. It has been shown recently (Thorley-Lawson *et al.*, 1977; Moss *et al.*, 1978, 1979; Rickinson *et al.*, 1979) that when mononuclear cells from peripheral blood of viral carriers are exposed to EBV *in vitro*, within 1–2 weeks there appear proliferating, EBNA-positive cells that may then disappear again depending on the number of cells seeded in culture, thus failing to give rise to LCLs. Exposure to EBV of mononuclear cells from antibody-negative donors or from cord blood led to the establishment of LCLs at a high frequency, as did exposure of the separated B-lymphocyte fraction of viral carriers (Rickinson *et al.*, 1979). The regression phenomenon was shown to be an *in vitro* expression of long-term T-cell-mediated immunity and due to virus-specific memory T cells capable of becoming cytotoxic when exposed to the appropriate antigen(s) *in vitro* (Moss *et al.*, 1979; Tanaka *et al.*, 1980; Rickinson *et al.*, 1981; Schooley *et al.*, 1981).

On the basis of the observations discussed above, one can visualize the following further interactions: Some of the EBV-carrying B lymphocytes are induced at any given time spontaneously to synthesize enough MA, VCA, and virus particles, but usually not enough EA, to maintain production of the corresponding antibodies. As soon as early MA (EMA) or late MA (LMA) or both appear on the cell surface, they may be destroyed by NK cells or by ADCC before antigen synthesis has run its full course. Others of the EBV-carrying cells are continuously destroyed by LYDMA-sensitive killer T cells, thereby releasing enough EBNA for persistent antibody stimulation. However, a fraction of the EBV-carrying cells will always escape and proliferate further to replenish the EBNA-positive subpopulation of B cells, some of which will enter into the lytic cycle of viral replication and others fall prey to specific killer T cells. These interactions appear to be in a constant equilibrium because the antibody titers to EBV-specific antigens are maintained, as a rule, at nearly constant levels for many decades (G. Henle and W. Henle, 1970; W. Henle and G. Henle, 1979).

These virus–lymphocyte interactions probably explain a good part of the pathogenesis of IM and of the persistent viral infection, even if it should turn out that lymphocytes are not initial targets of the virus and that salivary-gland cells share the persistent viral carrier state with lymphocytes. According to the scheme presented, the equilibrium of the persistent infection is maintained essentially by cellular immune responses. Changes in the control mechanisms due to immunosuppressive conditions should alter the humoral antibody pattern as reviewed in the next section. Another late but very rare consequence of the persistent, latent EBV infections is the emergence of BL or NPC (discussed in Section III.D).

C. Primary and Persistent Epstein–Barr Virus Infections in Immunologically Compromised Individuals

As discussed in the preceding section, B lymphocytes are the secondary if not the initial targets of EBV in primary infections and also one habitat, if not the sole habitat, of the virus in the ensuing persistent viral carrier state. In both situations, cellular immune responses are instrumental in controlling the infectious process and the proliferation of viral-genome-carrying B lymphocytes. One may anticipate, therefore, that primary and persistent EBV infections in immunologically compromised individuals evoke more or less unusual EBV-specific antibody patterns, depending on the immunological defect involved.

1. Primary EBV Infections in Patients with Immunodeficiencies

These have not attracted particular attention judging by the scarcity of pertinent reports. Consequently, one may assume either that they are not particularly severe and even silent or that they may present with such highly unusual clinical pictures that a diagnosis of IM is not readily considered. Both alternatives seem to apply under certain circumstances. While cases of IM have been seen among patients with acute lymphocytic leukemia, Hodgkin's disease, or renal transplants (Taylor, 1953; Lampkin *et al.*, 1967; Ragab and Vietti, 1969; Stevens *et al.*, 1971; Feldman and Yarvis, 1974; Lange *et al.*, 1978; Grose *et al.*, 1977), the reports did not mention unusual features or increased severity of the disease. The EBV-specific serological responses, when studied, showed some abnormalities. Antibodies to EBNA appeared, if at all, only after unusually long delays and then reached only barely deectable titers (Grose *et al.*, 1977; Lange *et al.*, 1978). In a juvenile Hodgkin's disease patient, the titers of IgG antibodies to VCA did not decline during convalescence to a lower persistent level and antibodies to EA-D did not disappear. Instead, both antibodies were retained at high titers over an observation period of 16 months (Lange *et al.*, 1978). The X-linked recessive lymphoproliferative syndrome (Purtilo *et al.*, 1975, 1978), a combined B- and T-cell deficiency, serves as an example of unusually severe, often fatal cases of IM. Such patients may show high heterophil-antibody responses, yet no emergence of EBV-specific antibodies (Bar *et al.*, 1974). A systematic search for primary EBV infections in individuals with various types of immunodeficiencies should be rewarding because it would elucidate the effect(s) of given defects on the antibody response and, in turn, provide information on the immunological controls that call a halt to the infectious process and control the ensuing persistent viral infections.

2. Changes in Antibody Patterns of Viral Carriers Due to Immunodeficiencies

Unusual spectra and titers of EBV-specific antibodies may develop as a consequence of a great variety of clinical conditions (cf. W. Henle and G. Henle, 1973, 1980). These include Hodgkin's disease and other lymphomas, chronic lymphocytic leukemia and other leukemias, and carcinomas and sarcomas (Johansson *et al.*, 1970, 1971; Levine *et al.*, 1971a,b; de Schryver *et al.*, 1972; Hesse *et al.*, 1973; Dumont *et al.*, 1976), as well as nonmalignant diseases, such as sarcoidosis, systemic lupus erythematosus, rheumatoid arthritis, and other connective-tissue diseases (Hirshaut *et al.*, 1970; Wahren *et al.*, 1971; Rothfield *et al.*, 1973; Alspaugh *et al.*, 1981). Common to all are immunosuppressive effects of the diseases *per se* or of the therapy they require. In line with this interpretation is the fact that inherited or acquired immunodeficiencies also result in unusual persistent patterns of EBV-specific antibodies, e.g., ataxia–telangiectasia (Joncas *et al.*, 1977; Berkel *et al.*, 1979), Behçet's disease, Nezelof's syndrome, agammaglobulinemias (Berkel *et al.*, 1979), or the immunosuppressive therapy administered to organ-transplant patients (Cheeseman *et al.*, 1980; Lange *et al.*, 1980).

The EBV-specific antibody patterns may be affected in several ways, suggesting different effects of various immunological defects. In some conditions, patients may show elevated titers of IgG antibodies to VCA, with or without emergence of antibodies to EA-R or to EA-D, whereas the titers of antibodies to EBNA remain within the normal range. Such an overrepresentation of elevated anti-VCA titers ($\geqq$1:320) is seen in most of the conditions listed above, so that the GMTs are 2–3 times higher than those of the controls (W. Henle and G. Henle, 1980). While the ratio between the anti-VCA and anti-EBNA titers of individual patients or controls may vary over a wide range (>64-fold), the ratio between the GMTs of these antibodies may provide some significant information. In healthy control groups, the ratio between the GMTs ranges from 1.5 to 3, whereas in immunodeficient patients, it may be increased substantially due to overall increased anti-VCA or to overall decreased anti-EBNA titers or to both.

In renal-transplant patients, the injection of antithymocyte globulin (ATG), in addition to routine immunosuppression, was shown to produce frequently within 1 month a 4-fold or greater rise in VCA-specific IgG titers, often accompanied by emergence of low levels of anti-EA-R but not by increases in anti-EBNA titers (Cheeseman *et al.*, 1980). In patients not receiving ATG, such antibody responses became evident only after many months of routine immunosuppressive therapy. ATG injections also increased the titers of preexisting antibodies to other viruses of the herpes group or nonherpes viruses. Furthermore, it led to an earlier enhancement of oropharyngeal excretion of EBV than observed in the absence of additional immunosuppression by ATG.

In as-yet-untreated juvenile patients with Hodgkin's disease, the anti-VCA titers were not significantly different from those of healthy viral carriers (Lange *et al.*, 1978). However, the anti-EBNA titers of the patients were reduced as compared to the controls, so that the ratio between the GMts of the two antibodies was 9 instead of 2. In the course of therapy and thereafter, the ratio rose to 17, mainly due to increased anti-VCA titers.

In about 1% of the patients with Hodgkin's disease or chronic lymphocytic leukemia followed at the Karolinska Hospital, Stockholm (Masucci *et al.*, 1981), as well as in patients with bone-marrow transplants (Lange *et al.*, 1980), the VCA-specific IgG antibody titers may rise gradually and steadily to excessively high titers, i.e., up to 1:20,480. In such cases, antibodies to EA are usually directed against EA-D and their titers may also reach unusually high levels. Furthermore, IgA antibodies to VCA and EA-D may become detectable. In contrast, anti-EBNA titers again remain within normal limits. The appearance of IgM antibodies to VCA has not been documented in these patients or in any other conditions discussed in this section. Any apparently positive results were subsequently found to be due to intervention of RF (G. Henle *et al.*, 1979).

In ataxia–telangiectasia, Behçet's disease, Nezelof's syndrome, acute lymphocytic leukemia, or marrow recipients, antibodies to EBNA are often of unusually low titer or nondetectable (Joncas *et al.*, 1977; Berkel *et al.*, 1979; Lange *et al.*, 1980). In some of these patients, high VCA-specific IgG antibody titers may be found, but in others, the levels of these antibodies remain within the range normally seen in healthy individuals after long-past primary EBV infections.

The changes in EBV-specific antibody patterns seen in some of the conditions mentioned suggest an activation of the persistent, latent viral carrier state. In line with this suggestion is the increased frequency of oropharyngeal excretion of EBV in immunologically compromised patients (Strauch *et al.*, 1974; Chang *et al.*, 1978; Lange *et al.*, 1978; Cheeseman *et al.*, 1980). Such activation apparently is rarely accompanied by signs of illness that are referable to the virus. However, patients with immunodeficiencies show an increased incidence of B-cell lymphomas (Kersey *et al.*, 1973), some of which could well be associated with EBV. It is of interest, therefore, that EBV DNA has been demonstrated recently in a B-cell lymphoma of a patient with ataxia–telangiectasia (Saemundsen *et al.*, 1981) and in five lymphomas thus far of renal-transplant patients (Saemundsen *et al.*, 1981; Hanto *et al.*, 1981). It could not be determined whether the lymphoproliferation in these patients was mono- or polyclonal or whether the tumor cells showed a translocation between chromosomes 8 and 14.

The immunological defects that enhance anti-VCA and anti-EA or reduce anti-EBNA production, or have both effects, are yet to be determined. One may surmise the following: (1) A deficiency or dysfunction of suppressor T cells prevents the normal shutoff of antibody synthesis by B lymphocytes. (2) Absence or dysfunction of NK cells or of ADCC-reactive cells lets the production of MA, EA, and VCA in spontaneously induced EBV-carrying B cells go to completion, whereas normally such producer cells would be destroyed by NK cells or the ADCC reaction as soon as virus-specific MA components appear in their membranes. (3) A loss or dysfunction of LYDMA-sensitive killer T cells permits an increase in the EBV-genome-carrying B-cell population that, in turn, would have two possible consequences: (a) no or little EBNA is released by T-cell action, so that antibodies to EBNA are not replenished and thus decline to low or nondetectable levels; (b) overall, more EBV-genome-carrying B cells may enter spontaneously into lytic cycles of viral replication, even though the percentage of induced cells may remain unchanged, thus providing increased amounts of MA, EA, and VCA, but not EBNA, for enhanced antibody production.

The suggested defects may be operative singly or in combination in a given condition. In ataxia–telangiectasia, in which a T-cell deficiency is a prominent feature (Berkel *et al.*, 1979), absence or dysfunction of LYDMA-sensitive T cells would lead to a decline of anti-EBNA levels, whereas absence or dysfunction of suppressor T cells could be responsible for elevated anti-VCA and anti-EA levels. In Hodgkin's disease, high anti-VCA titers correlated with the degree of T-cell dysfunction (Johansson *et al.*, 1975). Efforts to correlate the excessively high anti-VCA titers seen in a few cases of Hodgkin's disease or chronic lymphocytic leukemia with deficiencies in various lymphocyte subpopulations are now in progress. Preliminary observations indicate that different immunological impairments as measured by NK and memory T cell activities, lymphocyte migration inhibition, etc, may account for the unusual antibody responses of given patients. (Masucci *et al.*, 1981; Szigeti *et al.*, 1982). The effects of immunosuppressive conditions on primary and persistent EBV infections have been reviewed in a recent symposium published 1981 in Cancer Research (vol. 41, pp. 4209–4304). It is conceivable that the EBV-specific serology may serve as another parameter in the assessment of cellular immune competence.

D. Epstein–Barr-Virus-Associated Malignancies

1. Burkitt Lymphoma

a. Endemic BL in Africa

As pointed out in Section I, the association of EBV with African BL rests not only on serological evidence but also, most convincingly, on the nearly regular demonstration of EBV DNA in biopsies from the tumor and of EBNA in the lymphoma cells (zur Hausen *et al.*, 1970; Nonoyama *et al.*, 1973; Lindahl *et al.*, 1974; Reedman *et al.*, 1974; Olweny *et al.*, 1977). EA-, VCA-, or virus-particle-containing cells are not observed in fresh biopsies as a rule, but they become readily detectable within 1 or 2 days once the tumor cells have been placed in culture (Nadkarni *et al.*, 1970). This does not necessarily denote that no lytic cycles of viral replication occur within the tumor; it is likely, rather, that cells synthesizing EA and VCA escape detection because they are removed by host defenses as soon as virus-determined antigens appear in their membranes.

BL is not an immediate consequence of primary EBV infections, but a rare, late event in viral carriers. In African regions where BL is endemic, children acquire antibodies to the virus generally before they reach the age of 3 years (Kafuko *et al.*, 1972; de Thé *et al.*, 1978), yet the peak incidence of the tumor falls between the ages of 6 and 8 years (Burkitt and Wright, 1970). A prospective study of BL, initiated 8 years ago in the West Nile district of Uganda under the auspices of the International Agency for Research on Cancer, excluded the possibility that BL was an immediate consequence of rare, delayed primary EBV infections. Serum was obtained from over 42,000 children, aged 1–5 years initially, and up to 8 years later (de Thé *et al.*, 1978). Among the pre-bled children, 14 cases of BL emerged in the ensuing 5 years, of which 12 were EBV-associated. All had antibodies to EBV in the preserum obtained from 5 to 56 months prior to diagnosis. The anti-VCA titers in most of the presera from the EBV-associated cases were higher than the titers of their 5 age- and sex-matched controls, and also higher than the anti-VCA GMTs of numerous children of the same age as the patients. While the presera failed, with one exception, to reveal antibodies to EA components, the elevated anti-VCA titers nevertheless suggest that a somewhat heavier than usual persistent EBV infection might predispose to later development of BL.

It has been surmised that a primary EBV infection in early infancy, as commonly observed in African regions of BL prevalence (Biggar *et al.*, 1978a,b), could be another important factor in the origin of BL because as yet immature B lymphocytes might be transformed by the virus under these conditions and prevented from further differentiation, resulting in a subpopulation of immature, EBV-genome-carrying lymphoid cells. Such cells could well show an enhanced susceptibility to later conversion into fully malignant cells by an as yet undetermined event.

The considerations discussed above imply that EBV initiates the process leading to BL by establishing a polyclonal, EBNA-positive subpopulation of B cells that is already endowed with a permanent growth potential, an attribute of malignant cells, but that remains under control of cell-mediated immune defenses of their hosts. These cells retain a normal karyogram. Only one of them needs to be converted at a later time by a second event to give rise to a monoclonal tumor of EBNA-positive BL cells (Fialkow *et al.*, 1970). Since BL cells regularly show a reciprocal translocation between chromosomes 8 and 14 (Manolov and Manolova, 1972; Jarvis *et al.*, 1974; Zech *et al.*, 1976; Manolova *et al.*, 1979), the factor that induces this abnormality might also be responsible for the malignant conversion of the precursor cells. This interpretation is supported by the fact that BL cells induce tumors in nude mice when injected by any route, whereas cultured lymphoblasts transformed by EBV *in vitro* or derived from peripheral lymphocytes of IM patients or viral carriers do so only in the immunologically privileged site of the CNS or in immunologically incompetent, newborn animals (Nilsson *et al.*, 1977). It is conceivable that an active, persistent EBV infection with enhanced lymphoproliferation involving especially immature types of lymphoid cells might cause mutations, as might the lymphoproliferative and immunosuppressive effects of holoendemic malaria present in regions of endemic BL (O'Conor, 1970; Greenwood *et al.*, 1972). However, other factors prevalent in those regions, whether of an immunological, nutritional, chemical, parasitic, or other nature, must be considered as possible contributing factors. To sort them out will be a most difficult task.

With the emergence of BL in pre-bled children of the prospective study, rises in anti-VCA titers became evident in only two who had relatively low initial titers, whereas in the others the high titers remained unchanged (de Thé *et al.*, 1978). However, in all but one patient, antibodies to EA appeared with the tumor, these being directed in nine against EA-R and in two against EA-D. A high anti-EA-R titer was also noted to arise with the development of BL in a patient who earlier had been successfully treated for Hodgkin's disease (Magrath *et al.*, 1975).

The results of these studies conformed to previous reports that had shown (see Fig. 1) that BL patients at admission generally have elevated anti-VCA titers with a GMT 8–10 times higher than that of control children (G. Henle *et al.*, 1969, 1971b). These patients also usually have antibodies to the R component that may exceed even anti-VCA in titer (G. Henle *et al.*, 1971b; W. Henle *et al.*, 1973a; W. Henle and G. Henle, 1977). In a few patients, antibodies to the D component are dominant, and in such cases it cannot be determined whether or not anti-EA-R is also present at a titer equal to or lower than that of anti-EA-D because the D reactions outshine the R staining. Antibodies to EBNA in BL cover a wide range of titers, exceeding those of control children at the high end of the range and falling below those of controls at the low end of the range. Excess antigenic stimulation by large partly necrotic tumors may account for high titers and antigen–antibody complex formation (Mukojima *et al.*, 1973; Heimer and Klein, 1976) or T-cell deficiencies or dysfunctions for the low titers. An inverse relationship between anti-EBNA titers and dermal responses to Raji-cell extracts has been reported (Nkrumah *et al.*, 1976), as has a straight relationship between anti-EBNA titers and dermal reactivity to recall antigens (Olweny *et al.*, 1977).

There is evidence that the height of anti-VCA and anti-EA-R titers is related to the total tumor burden (Nkrumah *et al.*, 1976; Magrath *et al.*, 1980c) and that the emergence of anti-EA-D, as mentioned earlier, depends on invasion of lymph nodes by the tumor. The antibody patterns observed also reflect, to a considerable extent, the prognosis of the patients (W. Henle *et al.*, 1973a; Nkrumah *et al.*, 1976). If patients, after being brought to remission, showed no antibodies to EA-R or a gradual decline over many months in the titer of anti-EA-R to lower or ultimately nondetectable levels, they had at most one or two early relapses that again responded to chemotherapy, but none thereafter during observation periods from 5 to 10 years. However, if the patients maintained or developed high anti-EA-R titers, they had several, ultimately fatal relapses. Recurrent tumors were observed in some patients after remission periods as long as 78 months (Biggar *et al.*, 1981). Also, these patients retained high anti-EA-R titers ($\geqq 1:40$) during the long remission periods. These very late relapses raise the question whether they are the result of new tumor inductions or whether they arise from surviving, dormant tumor cells held in check by host defenses. On the basis of isozyme determinations, new inductions have been demonstrated in a few cases (Fialkow *et al.*, 1973). The persistently high antibody titers during the remission presumably denote a highly active viral carrier state that, as suggested by the prospective study in the West Nile district of Uganda, may predispose to induction of BL.

Antibodies to EA-R, an intracellular antigen, serve merely as indicators of the prognosis, but they do not directly influence it. Antibodies to MA, however, might play a role in the control of BL. Patients in remission have shown a decline in the titers of antibodies to MA prior to recurrence of the tumor (Gunvén *et al.*, 1970, 1973). Bacille Calmette–Guérin therapy of BL patients increases the titer of antibodies to MA, but not of antibodies to EA or VCA, as a rule (Gunvén *et al.*, 1973, 1978). High titers of antibodies to MA components as determined by ADCC correlated well with regression of the tumor, whereas very low titers correlated with progression (Pearson *et al.*, 1979). Also, LYDMA-sensitive T lymphocytes have been observed in BL biopsies (Jondal *et al.*, 1975).

b. Sporadic BL in Caucasian Patients

Less than 3% of African cases with a diagnosis of BL are unassociated with EBV; biopsies from their tumors do not contain EBV DNA, the tumor cells do not express EBNA, and the patients show an antibody spectrum and titers consistent with a primary EBV infection in the distant past (Andersson-Anvret *et al.*, 1977; de Thé *et al.*, 1978). The percentage of cases unassociated with EBV is vastly greater (over 80%) among American patients with a diagnosis of BL, and these either have no antibodies to virus-specific antigens or show antibody patterns comparable to those of healthy individuals after long-past primary EBV infections (Pagano *et al.*, 1973; Andersson-Anvret *et al.*, 1976; Ziegler *et al.*, 1976; A.L. Epstein *et al.*, 1976). As compared to African patients, EBV-associated American or European cases of BL cover an older age range, including young adults. These patients show a spectrum and titers of EBV-specific antibodies characteristic of their African counterparts (Andersson-Anvret *et al.*, 1976; Ziegler *et al.*, 1976; A.L. Epstein *et al.*, 1976; Bornkamm *et al.*, 1976; Gravell *et al.*, 1976). Even a cluster of four young adult cases of EBV-associated BL living within 30 miles of each other has been observed in Pennsylvania within 1 year (Judson *et al.*, 1977).

While earlier studies failed to discern significant differences between the histology of EBV-associated and that of non-EBV-associated BL (Berard *et al.*, 1969) or between cultured cells derived from both types of tumors (Nilsson, 1979), recent observations indicate the existence of some differences with respect to the size of the tumor cells and their growth rates in culture (Magrath *et al.*, 1980a,b). Furthermore, African and American BL differ with respect to the peak age incidence, the organ distribution of tumors, and the effectiveness of therapy (Mann *et al.*, 1976). The possibility remains, therefore, that the two types of tumors are different entities with separate causes, although the evidence for such a conclusion is as yet tenuous.

2. Nasopharyngeal Carcinoma

The interval between primary EBV infection and development of NPC is even longer than that observed in the case of BL. Primary EBV infections in regions where NPC is most frequent (southern China, North and East Africa, Alaska) occur again mostly under the age of 3–5 years, whereas the peak incidence of the tumor falls into the 5th decade of life, although some cases arise already during adolescence and others later (Clifford, 1970; Ho, 1972; de Thé, 1972). No prospective studies on NPC have been reported, but sera from seven Alaskan natives were discovered in collections assembled for other purposes, which had been obtained from 17 months to 10 years before the diagnosis of NPC (Lanier *et al.*, 1980). All but one of the presera showed EBV-specific antibody patterns in no way distinct from those seen normally in healthy individuals after long-past primary EBV infections. There was thus no clue to the future fate of these donors, which should not be surprising because even in stage I of NPC, some patients do not yet show the antibody spectrum characteristic of this tumor. However, two presera from the seventh Alaskan NPC patient, obtained 17 and 21 months before diagnosis, showed antibody patterns clearly compatible with NPC (see Fig. 1). Since this patient was admitted in stage IV and died soon thereafter, it seems clear that he already carried the tumor when the early sera were collected. As will be discussed later, screening for VCA-specific IgA antibodies turned out to be useful for early identification of NPC patients in regions where the tumor is endemic.

It is unknown at present how EBV becomes associated with the NPC cells. Since NPC, like BL, is thought to have a monoclonal origin (Fialkow *et al.*, 1972), viral genomes must have been present already in the very first malignant cell. Whether in regions of high (southern China) or low frequency of NPC (Europe, America), EBV DNA has been detected regularly in undifferentiated or poorly differentiated carcinomas of the postnasal space in amounts equivalent to multiple viral genomes per cell (zur Hausen *et al.*, 1970; Nonoyama *et al.*, 1973; Pagano *et al.*, 1975; Andersson-Anvret *et al.*, 1977, 1979; Lanier *et al.*, 1978; Glaser *et al.*, 1980), and the carcinoma cells express EBNA (Wolf *et al.*, 1973, 1975; Klein *et al.*, 1974; Huang *et al.*, 1974, 1978; Lanier *et al.*, 1978, 1981). Well-differentiated carcinomas of the nasopharynx are not associated with EBV. Also, biopsies from carcinomas at other sites of head and neck or elsewhere, whether undifferentiated or well differentiated, have been uniformly negative for EBV DNA and EBNA-positive carcinoma cells. Thus, detection of EBV DNA in a cervical-lymph-node metastasis of an as yet undetected primary carcinoma—or, more readily testable, the demonstration of EBNA-positive carcinoma cells in touch preparations from such a lymph-node biopsy—points clearly to the postnasal space as the primary site of the tumor (Coates *et al.*, 1978; Lanier and colleagues, unpublished).

To date, no epithelial cells have been shown to be infectible or transformable, or both, *in vitro* by EBV. Although such cells may exist, they have not been established in culture. Also, NPC cells have not been maintained in permanent culture, most likely because they require special, as yet unknown, growth factors. However, microinjection of EBV

DNA into normal epithelial cells has led to synthesis of EA and VCA (Graessmann *et al.*, 1980), and transplantation of EBV receptors from lymphoblasts to epithelial cells has permitted productive infection of these altered cells by EBV (Volski *et al.*, 1980). These observations have shown that while epithelial cells lack EBV receptors, they nevertheless have the capacity to replicate EBV antigens. Neither of these manipulations has led as yet to transformation of epithelial cells and their endowment with a permanent growth potential *in vitro*. Like NPC cells, any transformed cells obtained might require special culture conditions for permanent growth. NPC cells survive in culture long enough, however, to allow derepression of the viral genomes by BUdR and to induce viral replication (Trumper *et al.*, 1976), and infectious virus particles have been obtained from NPC biopsies that, after serial passage in nude mice, were capable of transforming cord-blood lymphocytes into permanently growing, EBNA-positive lymphoblasts (Trumper *et al.*, 1977). These results proved that whole viral genomes are present in the carcinoma cells.

With the emergence of NPC, most patients anywhere in the world develop high titers of IgG antibodies to VCA (W. Henle and G. Henle, 1968; de Schryver *et al.*, 1969; W. Henle *et al.*, 1970a; Kawamura *et al.*, 1970; Lin *et al.*, 1972, 1973; Lynn *et al.*, 1973a,b; de Thé *et al.*, 1975). Furthermore, they develop high titers to other EBV-specific antibodies (see Fig. 1), i.e., IgG antibodies to EA-D (G. Henle *et al.*, 1971a; W. Henle *et al.*, 1973b; de Schryver *et al.*, 1974a; Levine *et al.*, 1977) and, especially noteworthy, IgA antibodies to VCA and to EA-D, which is an almost unique feature of NPC (G. Henle and W. Henle, 1976; Ho *et al.*, 1978a, Neel *et al.*, 1980) because such high titers of VCA- and EA-D-specific IgA antibodies are observed only rarely in other conditions (see Section III.C.2). Antibodies to MA components (de Schryver *et al.*, 1969, 1974a) and to EBNA (de Thé *et al.*, 1975) are also often elevated as compared to controls. Recently, a high incidence of antibodies that neutralize EBV-specific deoxyribonuclease was reported to be another characteristic feature of NPC (Cheng *et al.*, 1980).

The characteristic EBV-specific antibody pattern of NPC may not yet be fully apparent in patients who are in stage I of the disease, when the carcinoma is still limited to the nasopharynx, because the antibody responses depend to a considerable degree on the total tumor burden. Thus, the incidences of VCA-specific IgA and of EA-D-specific IgA and IgG antibodies increase with an advance of the disease from stage I to IV, with consequent 4- to 10-fold increases in the GMTs (W. Henle *et al.*, 1973b, 1977).

The EBV-specific serology has three practical applications in NPC. As already mentioned, it may serve to detect patients bearing the tumor before they are sufficiently disturbed to seek medical help (Ho *et al.*, 1978b; Lanier *et al.*, 1980). This has been demonstrated in two large-scale surveys in areas of the People's Republic of China where NPC is most frequent (Zeng *et al.*, 1980). A total of about 160,000 person over 30 years of age were screened for IgA antibodies to VCA. In the first study during 1978, 117 reactors were found among over 56,500 individuals, and 20 of these were subsequently shown to have NPC. In the second study during 1979, using a more sensitive but evidently less specific technique, 1183 reactors were recorded among 91,500 persons tested, and 28 of these were carrying NPC. The attack rates were thus similar for the two years, 35 and 31/100,000. Most of the patients so discovered were in stage II, III, or IV. These patients probably already had symptoms referable to the tumor and also the characteristic antibody spectrum. The anti-VCA-IgA-positive individuals who were evidently free of NPC are being followed further, but it must be remembered that IgA antibodies to VCA at low titers are found occasionally in patients with other diseases and even, though rarely, in healthy donors (G. Henle and W. Henle, 1976).

The second use for the EBV-specific serology is in the diagnosis of NPC, especially in cases of lymph-node metastases in which the primary site of the carcinoma has remained obscure. An antibody spectrum characteristic of NPC would point to the postnasal space as the primary site, as would the detection of EBV DNA or of EBNA-positive carcinoma cells in the lymph node biopsy, as already discussed.

The third use for the EBV-specific serology is the monitoring of NPC patients after radiation or other therapy. Earlier studies had shown that newly diagnosed NPC patients had generally higher antibody titers than long-term survivors (W. Henle *et al.*, 1973b; Lynn *et al.*, 1973b; de Schryver *et al.*, 1974a; de Thé *et al.*, 1975). Accordingly, a 5-year longitudinal follow-up study was initiated in Hong Kong (W. Henle *et al.*, 1977). On the basis of the results, the more than 100 NPC patients could be divided into four groups. The largest, comprising about half the patients, showed a steady, slow decline in the titers of all antibodies, thereby approaching the pattern normally seen in healthy viral carriers, i.e., VCA-specific IgG reaching lower, persistent levels, with the other antibodies (EA-D-specific IgG and IgA and VCA-specific IgA) tending to disappear in time. These patients remained free of detectable disease during the entire observation period. The second group, comprising about one third of the patients, responded only transiently to the radiation therapy. Sooner or later, they showed increases in the titers of one or more of the antibodies, these increases often becoming evident months before clinical recognition of recurrent tumors or metastases. These patients did not survive the 5-year observation period. The third group, comprising about 15% of the patients, showed declines in antibody titers initially, seemingly heralding a good prognosis, but then at some time later a reversal of this trend. The renewed broadening of the antibody spectrum and the increases in titers became evident again months before clinical detection of recurrent tumor activity. Finally, in the few remaining patients, the tumor invaded the CNS by direct extension from the nasopharynx without significant involvement of cervical lymph nodes. The antibody patterns of these patients were not particularly characteristic of NPC because the total tumor burden remained small until shortly before death, when finally some increases in antibody titers became detectable. It thus became evident that the EBV-specific serology can alert to relapses or metastases long before they are clinically detectable. A broadening of the antibody spectrum and rises in the antibody titers should therefore lead to an intensified search for the location(s) of recurrent tumors and, when the tumors are found, to early reinstitution of appropriate therapy.

The EBV-specific serodiagnostic tests reveal the prognosis of patients only slowly. It takes many months before the prognostically favorable downward trend of antibody titers becomes clearly evident. Furthermore, the trend can be reversed. It is noteworthy, therefore, that the prognosis might be established by another test procedure, i.e., by ADCC (Pearson *et al.*, 1978). NPC patients showing high titers of ADCC-reactive antibodies, which are directed against EBV-determined MAs, are likely to become long-term survivors without evidence of disease, whereas patients with very low titers seem to be destined to succumb to the tumor.

IV. CONCLUDING REMARKS

It is evident from the foregoing discussion that much more is known about humoral than cellular immune responses in primary and persistent EBV infections. The Epstein–Barr virus (EBV)-specific serology was instrumental in identifying the virus as the cause of infectious mononucleosis (IM), a common clinical response to primary EBV infections essentially limited to adolescents in Western societies, and in establishing its intimate association with Burkitt lymphoma (BL) and nasopharyngeal carcinoma (NPC), which arise in EBV carriers due to as yet unknown secondary factors that apparently are concentrated in the geographic regions where the tumors are endemic. Each of the three diseases evokes its own characteristic spectrum of antibodies of different immunoglobulin classes against several EBV-determined antigens, and this spectrum serves to establish or to confirm the clinical diagnosis. The EBV-specific serology also affords prognostic information on the tumor patients and may serve to monitor the effectiveness of therapy. Yet, none of the antibodies commonly measured is likely to cause the self-limitation of primary EBV infections or to affect cells infected and transformed by the virus. These are functions of the cellular immune system or of a combination of humoral and cellular immune responses, as seen in the antibody-dependent cellular cytotoxicity reaction. Only a beginning has been made in the identification of subpopulations of leukocytes that become operative during overt primary EBV infections and terminate the disease process or that keep persisting clones of EBV-transformed B lymphocytes under control. Much of the knowledge has been gained and will be further gained from studies of primary and persistent EBV infections in immunologically compromised patients with identified immunodeficiencies.

On the basis of the available information, it would be difficult to consign EBV to a mere passenger role in BL and NPC, yet ultimate proof for a causal relationship of the virus to the genesis of these two tumors is still lacking. Prevention of the two malignancies by an EBV vaccine would go far in closing this remaining gap, but such an approach is faced with numerous obstacles (cf. M.A. Epstein, 1979). It will be necessary to find a convenient, abundant, and safe source of the virus from cells untainted by malignant traits; to develop reliable tests for exclusion of an enhanced oncogenic potential of the virus; to induce a long-lasting immunity not only against overt infections, which would suffice for the prevention of IM, but also against the establishment of an EBV carrier state, which is considered a requirement for protection against BL and NPC but raises the question whether a killed viral or viral subunit vaccine can accomplish this since live viral vaccines could not be used because they presumably *per se* establish persistent infections; and, finally, to solve the logistics of administering a vaccine under the field conditions prevailing in endemic regions of BL and NPC during the short, uncertain intervals between the loss of maternal antibodies and the acquisition of natural infections (Biggar *et al.*, 1978a). These problems should not deter work toward development of a vaccine, with due caution, if for no other goal than the prevention of IM, which probably can be achieved most readily, because this disease can be very prolonged, extremely severe, and occasionally fatal.

REFERENCES

Alspaugh, M.A., Henle, G., Lennette, E.T., and Henle, W., 1981, The significance of elevated levels of antibodies to Epstein–Barr virus antigens in patients with rheumatoid arthritis, *J. Clin. Invest.* **66**:113–1140.

Andersson-Anvret, M., Ziegler, J.L., Klein, G., and Henle, W., 1976, Association of Epstein–Barr viral genomes with American Burkitt lymphoma, *Nature (London)* **260**:357–359.

Andersson-Anvret, M., Forsby, N., Klein, G., and Henle, W., 1977, Studies on the occurrence of Epstein–Barr virus-DNA in nasopharyngeal carcinomas, in comparison with tumors of other head and neck regions, *Int. J. Cancer* **20**:486–494.

Andersson-Anvret, M., Forsby, N., Klein, G., Henle, W., and Bjorklund, A., 1979, Relationship between the Epstein–Barr virus genome and nasopharyngeal carcinoma in Caucasian patients, *Int. J. Cancer* **23**:762–767.

Armstrong, D., Henle, G., and Henle, W., 1966, Complement fixation tests with cell lines derived from Burkitt's lymphoma and acute leukemias, *J. Bacteriol.* **91**:1257–1262.

Bar, R., DeLor, C.L., Clausen, K.P., Hurtubise, P., Henle, W., and Hewetson, J., 1974, Fatal infectious mononucleosis in a family, *N. Engl. J. Med.* **290**:363–367.

Berard, C., O'Conor, G.T., Thomas, L.B., and Torloni, H., 1969, Histopathological definition of Burkitt's tumour, *Bull. WHO* **40**:601–609.

Berkel, A.I., Henle, W., Henle, G., Klein, G., Ersoy, F., and Sanal, O., 1979, Epstein–Barr virus related antibody patterns in ataxia–telangiectasia, *Clin. Exp. Immunol.* **35**:196–201.

Biggar, R.J., Henle, W. Fleisher, G., Böcker, J., Lennette, E.T., and Henle, G., 1978a, Primary Epstein–Barr virus infections in African infants. I. Decline of maternal antibodies and time of infection, *Int. J. Cancer* **22**:239–243.

Biggar, R.J., Henle, G., Böcker, J., Lennette, E.T., Fleisher, G., and Henle, W., 1978b, Primary Epstein–Barr virus infections in African infants. II. Clinical and serological observations during seroconversion, *Int. J. Cancer* **22**:244–250.

Biggar, R.J., Nkrumah, F.K., Henle, W., and Levine, P.H., 1981, Very late relapses in patients with Burkitt's lymphoma: Clinical and serological studies, *J. Natl. Cancer Inst.* **66**:439–444.

Black, F.L., Woodall, J.P., Evans, A.S., Liebhaber, H., and Henle, G., 1970, Prevalence of antibody against viruses in the Tiriyo, an isolated Amazon tribe, *Am. J. Epidemiol.* **91**:430–438.

Britton, S., Andersson-Anvret, M., Bergely, P., Henle, W., Jondal, M., Klein, G., Sandstedt, B., and Svedmyr, E., 1978, EB virus immunity and tissue distribution in a fatal case of infectious mononucleosis, *N. Engl. J. Med.* **298**:89–92.

Burkitt, D., and Wright, D.H., 1970, *Burkitt's Lymphoma*, Livingstone, Edinburgh.

Chang, R.S., Lewis, J.P., and Abildgaard, C.F., 1973, Prevalence of oropharyngeal excretors of leukocyte-transforming agents among a human population, *N. Engl. J. Med.* **289**:1325–1329.

Chang, R.S., Lewis, J.P., Reynold, R.D., Sullivan, M.J., and Neuman, J., 1978, Oropharyngeal excretion of Epstein–Barr virus by patients with lymphoproliferative disorders and renal homografts, *Ann. Intern. Med.* **88**:34–40.

Cheeseman, S.H., Henle, W., Rubin, R.H., Tolkoff-Rubin, N.E., Cosimi, A.B., Cantell, K., Winkle, S., Herrin, J.T., Black, P.H., Russell, P.S., and Hirsch, M.S., 1980, Epstein–Barr virus infection in renal transplant recipients: Effects of antithymocyte globulin and interferon, *Ann. Intern. Med.* **93**:39–42.

Cheng, Y.-C., Chen, J.-Y., Glaser, R., and Henle, W., 1980, Demonstration of high frequency and high titer of antibody to Epstein–Barr virus-specific DNase in serum from patients with nasopharyngeal carcinoma, *Proc. Natl. Acad. Sci. U.S.A.* **77**:6162–6165.

Cho, M.-S., Fresen, K.O., and zur Hausen, H., 1980, Multiplicity dependent biological and biochemical properties of Epstein–Barr virus rescued from non-producer lines after superinfection with P3HR-1 EBV, *Int. J. Cancer* **26**:357–363.

Cleary, T.G., Henle, W., and Pickering, L.K., 1980, Acute cerebellar ataxia associated with Epstein–Barr virus infection, *J. Am. Med. Assoc.* **243**:148–149.

Clifford, P., 1970, On the epidemiology of nasopharyngeal carcinoma, *Int. J. Cancer* **5**:287–309.

Coates, H.L., Pearson, G., Neel, H.B., III, Weiland, L.H., and Devine, K.D., 1978, An immunologic basis for detection of occult primary malignancies of the head and neck, *Cancer* **41**:912–918.

Crawford, D.H., Epstein, M.A., Achong, B.G., Finerty, S., Newman, J., Liversedge, S., Tedder, R.S., and Stewart, J.W., 1979, Virological and immunological studies on a fatal case of infectious mononucleosis, *J. Infect.* **1**:37.

Crawford, D.H., Thomas, J.A., Janossy, G., Sweeny, P., Fernando, O.N., Moorhead, J.F., and Thompson, J.H., 1980, Epstein–Barr virus nuclear antigen positive lymphoma after cyclosporin A treatment in a patient with renal allograft. *Lancet* **1**:1355–1356.

Davidsohn, I., and Lee, C.L., 1969, The clinical serology of infectious mononucleosis, in: *Infectious Mononucleosis* (R.L. Carter and H.G. Penman, eds.), pp. 177–200, Blackwell, Oxford.

De Schryver, A., Friberg, S., Jr., Klein, G., Henle, W., Henle, G., de Thé, G., Clifford, P., and Ho, H.C., 1969, Epstein–Barr virus (EBV)-associated antibody patterns in carcinoma of the post-nasal space, *Clin. Exp. Immunol.* **5**:443–459.

De Schryver, A., Klein, G., Henle, G., Henle, W., Cameron, H., Santesson, L., and Clifford, P., 1972, EB-virus associated serology in malignant disease: Antibody levels to viral capsid antigens (VCA), membrane antigens (MA) and early antigens (EA) in patients with various neoplastic conditions, *Int. J. Cancer* **9**:353–364.

De Schryver, A., Klein, G., Henle, W., and Henle, G., 1974a, EB virus-associated antibodies in Caucasian patients with carcinoma of the nasopharynx and in long-term survivors after treatment, *Int. J. Cancer* **13**:319–325.

De Schryver, A., Klein, G., Hewetson, J., Rocchi, G., Henle, W., Henle, G., and Pope, J., 1974b, Comparison of EBV neutralization tests based on abortive infection or transformation of lymphoid cells and their relation to membrane reactive antibodies (anti-MA), *Int. J. Cancer* **13**:353–362.

De Thé, G., 1972, The etiology of nasopharyngeal carcinoma, *Pathobiol. Annu.* **2**:235–254.

De Thé, G., Ho, J.H.C., Ablashi, D.V., Day, N.E., Macario, A.J.L., Martin-Berthelon, M.C., Pearson, G., and Sohier, R., 1975, Nasopharyngeal carcinoma. IX. Antibodies to EBNA and correlation with response to other EBV antigens in Chinese patients, *Int. J. Cancer* **16**:713–721.

De Thé, G., Geser, A., Day, N.E., Tukei, P.M., Williams, E.H., Beri, D.P., Smith, P.G., Dean, A.G., Bornkamm, P., Feorino, P., and Henle, W., 1978, Epidemiological evidence for causal relationship between Epstein–Barr virus and Burkitt's lymphoma from Ugandan prospective study, *Nature (London)* **274**:756–761.

Diehl, V., Henle, G., Henle, W., and Kohn, G., 1968, Demonstration of a herpes group virus in cultures of peripheral leukocytes from patients with infectious mononucleosis, *J. Virol.* **2**:663–669.

Diehl, V., Taylor, J., Parlin, J.A., Henle, G., and Henle, W., 1969a, Infectious mononucleosis in East Africa, *East Afr. Med. J.* **46**:407–413.

Diehl, V., Henle, G., Henle, W., and Kohn, G., 1969b, Effect of a herpes group virus (EBV) on growth of peripheral leukocyte cultures, *In Vitro* **4**:92–99.

Dölken, G., and Klein, G., 1976, Expression of Epstein–Barr virus (EBV) associated membrane antigen (MA) in Raji cells superinfected with two different strains of virus, *Virology* **70**:210–213.

Dumont, J., Liabeuf, A., Henle, W., Feingold, N., and Kourilsky, F.M., 1976, Anti-EBV antibody titers in non-Hodgkin's lymphomas, *Int. J. Cancer* **18**:14–23.

Dunkel, V.C., and Zeigel, R.F., 1970, Studies on the relationship of membrane fluorescence to Epstein–Barr virus infection, *J. Natl. Cancer Inst.* **44**:133–138.

Enberg, R.N., Eberle, B.J., and Williams, C.R., Jr., 1974, T- and B-cells in peripheral blood during infectious mononucleosis, *J. Infect. Dis.* **130**:104–114.

Epstein, A.L., Henle, W., Henle, G., Hewetson, J.F., and Kaplan, H.S., 1976, Surface marker characteristics and Epstein–Barr virus studies of two established North American Burkitt's lymphoma cell lines, *Proc. Natl. Acad. Sci. U.S.A.* **73**:228–232.

Epstein, M.A., 1979, Vaccine control of EB virus-associated tumors, in: *The Epstein–Barr Virus* (M.A. Epstein and B.G. Achong, eds.), pp. 439–448, Springer-Verlag, Berlin, Heidelberg, New York.

Epstein, M.A., and Achong, B.G., 1968, Specific immunofluorescence tests for the herpes-type EB virus of Burkitt lymphoblasts, authenticated by electron microscopy, *J. Natl. Cancer Inst.* **40**:593–607.

Epstein, M.A., Achong, B.G., and Barr, Y.M., 1964, Virus particles in cultured lymphoblasts from Burkitt's lymphoma, *Lancet* **1**:702–703.

Epstein, M.A., Henle, G., Achong, B.G., and Barr, Y.M., 1965, Morphological and biological studies on a virus in cultured lymphoblasts from Burkitt's lymphoma, *J. Exp. Med.* **121**:761–770.

Ernberg, I., and Klein, G., 1979, EB virus-induced antigens, in: *The Epstein–Barr Virus* (M.A. Epstein and B.G. Achong, eds.), pp. 39–60, Springer-Verlag, Berlin, Heidelberg, New York.

Ernberg, I., Klein, G., Kourilsky, F.M., and Silvestre, D., 1974, Differentiation between early and late membrane antigen on human lymphoblastoid cell lines infected with Epstein–Barr virus. I. Immunofluorescence, *J. Natl. Cancer Inst.* **53**:61–65.

Evans, A.S., Niederman, J.C., and McCollum, R.W., 1968, Seroepidemiologic studies of infectious mononucleosis with EB virus, *N. Engl. J. Med.* **279:**1121–1127.

Evans, A.S., Niederman, J.C. Canabre, L.C., West, B., and Richards, V.A., 1975, A prospective evaluation of heterophile and Epstein–Barr virus-specific IgM antibody tests in clinical and subclinical infectious mononucleosis: Specificity of and sensitivity of the tests and persistence of antibody, *J. Infect. Dis.* **132:**546–554.

Feldman, F., and Yarvis, J., 1974, Manifestations of hemolytic phenomena and infectious mononucleosis in a case of lymphatic leukemia, *N.Y. State J. Med.* **44:**1693–1694.

Fialkow, P.J., Klein, G., Gartler, S.M., and Clifford, P., 1970, Clonal origin for individual Burkitt tumours, *Lancet* **1:**384–386.

Fialkow, P.J., Martin, G.M., Klein, G., Clifford, P., and Singh, S., 1972, Evidence for a clonal origin of head and neck tumors, *Int. J. Cancer* **9:**133–142.

Fialkow, P.J., Klein, E., Klein, G., Clifford, P., and Singh, S., 1973, Immunoglobulin and glucose-6-phosphate dehydrogenase as markers of cellular origin in Burkitt's lymphoma, *J. Exp. Med.* **138:**89–102.

Fleisher, G., Henle, W., Henle, G., Lennette, E.T., and Biggar, R.J., 1979a, Primary Epstein–Barr virus infection in American infants: Clinical and serological observations, *J. Infect. Dis.* **139**(5):553–558.

Fleisher, G., Lennette, E.T., Henle, G., and Henle, W., 1979b, Incidence of heterophil antibody responses in children with infectious mononucleosis, *J. Pediatr.* **94**(5):723–728.

Fleisher, G.R., Schwartz, J., and Lennette, E.T., 1981, Primary Epstein–Barr virus infection in association with Reye syndrome, *J. Pediatr.* **97**(6):935–937.

Gerber, P., 1972, Activation of Epstein–Barr virus by bromodeoxyuridine in "virus-free" human cells, *Proc. Natl. Acad. Sci. U.S.A.* **69:**83–85.

Gerber, P., and Deal, D.R., 1970, Epstein–Barr virus-induced viral and soluble complement-fixing antigens in Burkitt's lymphoma cell cultures, *Proc. Soc. Exp. Biol. Med.* **134:**748–751.

Gerber, P., Hamre, D., Moy, R.A., and Rosenblum, E.N., 1968, Infectious mononucleosis: Complement-fixing antibodies to herpes-like virus associated with Burkitt's lymphoma, *Science* **161:**173–175.

Gerber, P., Whang-Peng, J., and Monroe, J.H., 1969, Transformation and chromosome changes induced by Epstein–Barr virus in normal human leukocyte cultures, *Proc. Natl. Acad. Sci. U.S.A.* **63:**740–747.

Gerber, P., Nonoyama, M., Lucas, S., Perlin, E., and Goldstein, L.I., 1972, Oral excretion of Epstein–Barr virus by healthy subjects and patients with infectious mononucleosis, *Lancet* **2:**988–989.

Gergely, L., Klein, G., and Ernberg, I., 1971a, Appearance of Epstein–Barr virus associated antigens in infected Raji cells, *Virology* **45:**10–21.

Gergely, L., Klein, G., and Ernberg, I., 1971b, The action of DNA antagonists on Epstein–Barr virus (EBV)-associated early antigens (EA) in Burkitt lymphoma lines, *Int. J. Cancer* **7:**293–302.

Gergely, L., Klein, G., and Ernberg, I., 1971c, Effect of EBV-induced early antigens on host macromolecular synthesis, studied by combined immunofluorescence and radioautography, *Virology* **45:**22–29.

Ginsburg, C.M., Henle, W., Henle, G., and Horwitz, C.A., 1977, Infectious mononucleosis in children: Evaluation of the Epstein–Barr virus-specific serology, *J. Am. Med. Assoc.* **237:**781–785.

Glaser, R., Nonoyama, M., Scymanowski, R.T., and Graham, W., 1980, Human nasopharyngeal carcinomas positive for Epstein–Barr virus DNA in North America, *J. Natl. Cancer Inst.* **64:**1317–1319.

Graessmann, A., Wolf, H., and Bornkamm, G.W., 1980, Expression of Epstein–Barr virus genes in different cell types after microinjection of viral DNA, *Proc. Nat. Acad. Sci. U.S.A.* **77:**433–436.

Gravell, M., Levine, P.H., McIntyre, R.F., Land, V.J., and Pagano, J.S., 1976, Epstein–Barr virus in an American patient with Burkitt's lymphoma: Detection of viral genome in tumor tissue and establishment of a tumor-derived cell line (NAB), *J. Natl. Cancer Inst.* **56**:701–704.

Greenwood, B.M., Bradley-Moore, A.M., Palit, A., and Bryceson, A.D.M., 1972, Immunosuppression in children with malaria, *Lancet* **1**:169–172.

Grose, C., Henle, W., Henle, G., and Feorino, P.M., 1975, Primary Epstein–Barr virus infections in acute neurologic diseases, *N. Engl. J. Med.* **292**:392–395.

Grose, C., Henle, W., and Horwitz, M.S., 1977, Primary Epstein–Barr virus infection in a renal transplant recipient, *South. Med. J.* **70**:1276–1278.

Gunvén, P., Klein, G., Henle, G., Henle, W., and Clifford, P., 1970, Antibodies to EBV-associated membrane and viral capsid antigens in Burkitt lymphoma patients, *Nature (London)* **228**:1053–1056.

Gunvén, P., Klein, G., Henle, W., Henle, G., Rocchi, G., Hewetson, J.F., Guerra, A., Clifford, P., Singh, S., Demissie, A., and Svedmyr, A., 1973, Immunological studies on a case of recurrent Burkitt's lymphoma during immune stimulant treatment, *Int. J. Cancer* **12**:115–123.

Gunvén, P., Klein, G., Ziegler, J.L., Magrath, I.T., Olweny, C.L., Henle, W., Henle, G., Svedmyr, A., and Demissie, A., 1978, Epstein–Barr virus (EBV) associated and other anti-viral antibodies during intense BCG administration to Burkitt's lymphoma patients in remission, *J. Natl. Cancer Inst.* **60**:31–38.

Hallee, T.J., Evans, A.S., and Niederman, J.C., 1974, Infectious mononucleosis at the United States Military Academy: A prospective study of a single class over 4 years, *Yale J. Biol. Med.* **3**:182–192.

Hampar, B., Derge, J.G. Martos, L.M., and Walker, J.L., 1972, Synthesis of Epstein–Barr virus after activation of the viral genome in a "virus-negative" human lymphoblastoid cell (Raji) made resistant to 5-bromodeoxyuridine, *Proc. Natl. Acad. Sci. U.S.A.* **69**:78–82.

Hanto, D., Frizzera, G., Purtilo, D., Sakamoto, K., Sullivan, J.L, Saemundsen, A.K., Klein, G., Simons, R.L, and Najarian, J.S., 1981, A clinical spectrum of lymphoproliferative disorders in renal transplant recipients and evidence for the role of Epstein–Barr virus, *Cancer Res.* **41**:4253–4261.

Haynes, B.F., Schooley, R.T., Payling-Wright, C.R., Grouse, J.E., Dolin, R., and Fauci, A., 1979, Emergence of suppressor cells of immunoglobulin synthesis during acute Epstein–Barr virus-induced infectious mononucleosis, *J. Immunol.* **123**:2095–2101.

Heimer, R., and Klein, G., 1976, Circulating immune complexes in sera of patients with Burkitt's lymphoma and nasopharyngeal carcinoma, *Int. J. Cancer* **18**:310–316.

Henle, G., and Henle, W., 1966a, Immunofluorescence in cells derived from Burkitt's lymphoma, *J. Bacteriol.* **91**:1248–1256.

Henle, G., and Henle, W., 1966b, Studies on cell lines derived from Burkitt's lymphoma, *Trans. N.Y. Acad. Sci.* **29**:71–79.

Henle, G., and Henle, W., 1970, Observations on childhood infections with the Epstein–Barr virus, *J. Infect. Dis.* **121**:303–310.

Henle, G., and Henle, W., 1976, Epstein–Barr virus-specific IgA serum antibodies as an outstanding feature of nasopharyngeal carcinoma, *Int. J. Cancer* **17**:1–7.

Henle, G., and Henle, W., 1979, The virus as the etiologic agent of infectious mononucleosis, in: *The Epstein–Barr Virus* (M.A. Epstein and B.G. Achong, eds.), pp. 297–320, Springer-Verlag, Berlin, Heidelberg, New York.

Henle, G., Henle, W., and Diehl, V., 1968, Relation of Burkitt tumor associated herpes-type virus to infectious mononucleosis, *Proc. Natl. Acad. Sci. U.S.A.* **59**:94–101.

Henle, G., Henle, W., Clifford, P., Diehl, V., Kafuko, G.W., Kirya, B.G., Klein, G., Morrow, R.H., Munube, G.M.R., Pike, M.C., Tukei, P.M., and Ziegler, J.L., 1969, Antibodies to EB virus in Burkitt's lymphoma and control groups, *J. Natl. Cancer Inst.* **43**:1147–1157.

Henle, G., Henle, W., and Klein, G., 1971a, Demonstration of two distinct components in the early antigen complex of Epstein–Barr virus infected cells, *Int. J. Cancer* **8**:272–282.

Henle, G., Henle, W., Klein, G., Gunvén, P., Clifford, P., Morrow, R.H., and Ziegler, J.L.,

1971b, Antibodies to early Epstein–Barr virus-induced antigens in Burkitt's lymphoma, *J. Natl. Cancer Inst.* **46**:861–871.

Henle, G., Henle, W., and Horwitz, C.A., 1974, Antibodies to Epstein–Barr virus-associated nuclear antigen in infectious mononucleosis, *J. Infect. Dis.* **130**:231–239.

Henle, G., Lennette, E.T., Alspaugh, M.A., and Henle, W., 1979, Rheumatoid factor as a cause of positive reactions in tests for Epstein–Barr virus-specific IgM antibodies, *Clin. Exp. Immunol.* **36**:415–422.

Henle, W., and Henle, G., 1968, Present status of the herpes-group virus associated with cultures of the hematopoietic system, *Perspect. Virol.* **6**:105–117.

Henle, W., and Henle, G., 1973, Epstein–Barr virus (EBV)-related serology in Hodgkin's disease, *Natl. Cancer Inst. Monogr.* **36**:79–84.

Henle, W., and Henle, G., 1977, Antibodies to the R component of EBV-induced early antigens in Burkitt's lymphoma may exceed in titer antibodies to EB viral capsid antigen, *J. Natl. Cancer Inst.* **58**:785–786.

Henle, W., and Henle, G., 1978, The immunological approach to study of possibly virus-induced human malignancies using the Epstein–Barr virus as example, in: *Progress in Experimental Tumor Research* (F. Homburger, ed.), pp. 19–48, S. Karger, Basel.

Henle, W., and Henle, G., 1979, Seroepidemiology of the virus, in: *The Epstein–Barr Virus* (M.A. Epstein and B.G. Achong, eds.), pp. 61–78, Springer-Verlag, Berlin, Heidelberg, New York.

Henle, W., and Henle, G., 1980, Consequences of persistent Epstein–Barr virus infections, in: *Cold Spring Harbor Conferences on Cell Proliferation: Viruses in Naturally Occurring Cancers*, Vol. 7 (M. Essex, G. Todaro, and H. zur Hausen, eds.), pp. 3–10, Cold Spring Harbor Laboratory, Cold Spring Harbor, New York.

Henle, W., and Henle, G., 1982, Epstein–Barr virus and infectious mononucleosis, in: *Herpesvirus Infections of Man* (R. Glaser and T. Stematsky, eds.), pp. 151–167, Marcel Dekker, New York.

Henle, W., Hummeler, K., and Henle, G., 1966, Antibody coating and agglutination of virus particles separated from the EB3 line of Burkitt lymphoma cells, *J. Bacteriol.* **92**:269–271.

Henle, W., Diehl, V., Kohn, G., zur Hausen, H., and Henle, G., 1967, Herpes-type virus and chromosome marker in normal leukocytes after growth with irradiated Burkitt cells, *Science* **157**:1064–1065.

Henle, W., Henle, G., Ho, H.C., Burtin, P., Cachin, Y., Clifford, P., de Schryver, A., de Thé, G., Diehl, V., and Klein, G., 1970a, Antibodies to EB virus in nasopharyngeal carcinoma, other head and neck neoplasms and control groups, *J. Natl. Cancer Inst.* **44**:225–231.

Henle, W., Henle, G., Zajac, B., Pearson, G., Waubke, R., and Scriba, M., 1970b, Differential reactivity of human sera with EBV-induced "early antigens," *Science* **169**:188–190.

Henle, W., Henle, G., Niederman, J.C., Klemola, E., and Haltia, K., 1971, Antibodies to early antigens induced by Epstein–Barr virus in infectious mononucleosis, *J. Infect. Dis.* **124**:58–67.

Henle, W., Henle, G., Gunvén, P., Klein, G., Clifford, P., and Singh, S., 1973a, Patterns of antibodies to Epstein–Barr virus-induced early antigens in Burkitt's lymphoma: Comparison of dying patients with long-term survivors, *J. Natl. Cancer Inst.* **50**:1163–1173.

Henle, W., Ho, H.C., Henle, G., and Kwan, H.C., 1973b, Antibodies to Epstein–Barr virus-related antigens in nasopharyngeal carcinoma: Comparison of active cases and long term survivors, *J. Natl. Cancer Inst.* **51**:361–369.

Henle, W., Henle, G., and Horwitz, C.A., 1974a, Epstein–Barr virus-specific diagnostic tests in infectious mononucleosis, *Hum. Pathol.* **5**:551–565.

Henle, W., Guerra, A., and Henle, G., 1974b, False negative and prozone reactions in tests for antibodies to Epstein–Barr virus-associated nuclear antigen, *Int. J. Cancer* **13**:751–754.

Henle, W., Ho, H.C., Henle, G., Chau, J.C.W., and Kwan, H.C., 1977, Nasopharyngeal carcinoma: Significance of changes in Epstein–Barr virus-related antibody patterns following therapy, *Int. J. Cancer* **20**:663–672.

Hesse, J., Anderson, E., and Levine, P.H., 1973, Antibodies to Epstein–Barr virus and cellular immunity in Hodgkin's disease and chronic lymphocytic leukemia, *Int. J. Cancer* **11**:237–243.

Hewetson, J.F., Rocchi, G., Henle, W., and Henle, G., 1973, Neutralizing antibodies against Epstein–Barr virus in healthy populations and patients with infectious mononucleosis, *J. Infect. Dis.* **128**:283–289.

Hinuma, Y., and Katsuki, T., 1978, Colonies of EBNA-positive cells in soft agar from peripheral leukocytes of infectious mononucleosis patients, *Int. J. Cancer* **21**:426–431.

Hinuma, Y., Kohn, M., Yamaguchi, J., Wudarski, D.J., Blakelee, J.R., Jr., and Grace, J.T., Jr., 1967, Immunofluorescence and herpes-type virus particles in the P3HR-1 Burkitt lymphoma cell line, *J. Virol.* **1**:1045–1051.

Hirshaut, Y., Glade, P., Vieira, L.O., Ainbender, E., Dvorak, B., and Siltzbach, L.E., 1970, Sarcoidosis, another disease associated with serologic evidence for herpes-like virus infection, *N. Engl. J. Med.* **283**:502–506.

Hirshaut, Y., Christenson, W.N., and Perlmutter, J.C., 1971, Prospective study of herpes-like virus role in infectious mononucleosis, *Clin. Res.* **19**:459–463.

Ho, J.H.C., 1972, Nasopharyngeal carcinoma (NPC), *Adv. Cancer Res.* **15**:57–92.

Ho, H.C., Ng, M.H., and Kwan, H.C., 1978a, Factors affecting serum IgA antibodies to Epstein–Barr viral capsid antigens in nasopharyngeal carcinoma, *Br. J. Cancer* **37**:356–362.

Ho, H.C., Kwan, H.C., Ng, M.H., and de Thé, G., 1978b, Serum IgA antibodies to Epstein–Barr virus capsid antigen preceding symptoms of nasopharyngeal carcinoma, *Lancet* **1**:436.

Horwitz, C.A., Henle, W., Henle, G., and Schmitz, H., 1975, Clinical evaluation of patients with infectious mononucleosis and development of antibodies to the R component of the Epstein–Barr virus-induced early antigen complex, *Am. J. Med.* **58**:330–338.

Horwitz, C.A., Henle, W., Henle, G., Segal, M., Arnold, T., Lewis, F.B., Zanick, D., and Ward, P.C.J., 1976, Clinical and laboratory evaluation of elderly patients with heterophil-antibody positive infectious mononucleosis—report of seven patients, ages 40–78, *Am. J. Med.* **61**:333–339.

Horwitz, C.A., Henle, W., Henle, G., Polesky, H., Balfour, H.H., Jr., Siem, R.A., Borken, S., and Ward, P.C.J., 1977, Heterophil-negative infectious mononucleosis and mononucleosis-like illnesses: Laboratory confirmation of 43 cases, *Am. J. Med.* **63**:947–956.

Horwitz, C.A., Henle, W., Henle, G., Goldfarb, M., Gehrz, R.C., Kubec, P., Balfour, H.H., Jr., Fleisher, G.R., and Krivit, W., 1981, Clinical and laboratory evaluation of infants with Epstein–Barr virus-induced infectious mononucleosis: Report of 32 patients (aged 10 to 48 months), *Blood* **57**:933–938.

Huang, D.P., Ho, J.H.C., Henle, W., and Henle, G., 1974, Demonstration of EBV-associated nuclear antigen in NPC cells from fresh biopsies, *Int. J. Cancer* **14**:580–588.

Huang, D.P., Ho, H.C., Henle, W., Henle, G., Saw, D., and Lui, M., 1978, Presence of EBNA in nasopharyngeal carcinoma and control patient tissue related to EBV serology, *Int. J. Cancer* **22**:266–274.

Hutt, L.M., Huang, Y.T., Dascomb, H.E., and Pagano, J.S., 1975, Enhanced destruction of lymphoid cell lines by peripheral blood leukocytes taken from patients with acute infectious mononucleosis, *J. Immunol.* **115**:243–248.

Jarvis, J.E., Ball, G., Rickinson, A.B., and Epstein, M.A., 1974, Cytogenetic studies on human lymphoblastoid cell lines from Burkitt's lymphomas and other sources, *Int. J. Cancer* **14**:716–721.

Johansson, B., Klein, G., Henle, W., and Henle, G., 1970, Epstein–Barr virus (EBV)-associated antibody patterns in malignant lymphoma and leukemia. I. Hodgkin's disease, *Int. J. Cancer* **6**:450–562.

Johansson, B., Klein, G., Henle, W., and Henle, G., 1971, Epstein–Barr virus (EBV)-associated antibody patterns in malignant lymphoma and leukemia. II. Chronic lymphocytic leukemia and lymphocytic lymphoma, *Int. J. Cancer* **8**:475–486.

Johansson, B., Holm, G., Mellstedt, H., Henle, W., Henle, G., Soderberg, G., Klein, G., and Killander, D., 1975, Epstein–Barr virus (EBV)-associated antibody patterns in relation to the deficiency of cell-mediated immunity in patients with Hodgkin's disease (HD), in: *Oncogenesis and Herpes-viruses II* (G. de Thé, M.A. Epstein, and H. zur Hausen, eds.), pp. 237–247, IARC Publication No. 11, Part 2, Lyon.

Joncas, J., and Mitnyan, C., 1970, Serologic response of the EBV antibodies in pediatric cases of infectious mononucleosis and their contacts, *Can. Med. Assoc. J.* **102:**1260–1263.

Joncas, J., Lapointe, N., Gervais, F., Leyfritz, M., and Wills, A., 1977, Unusual prevalence of antibodies to Epstein–Barr virus early antigen in ataxia–telangiectesia, *Lancet* **1:**1160–1163.

Jondal, M., 1976, Antibody-dependent cellular cytotoxicity (ADCC) against Epstein–Barr virus-determined antigens. I. Reactivity in sera from normal persons and from patients with acute infectious mononucleosis, *Clin. Exp. Immunol.* **25:**1–10.

Jondal, M., and Klein, G., 1973, Surface markers on human B and T lymphocytes, *J. Exp. Med.* **138:**1365–1378.

Jondal, M., Svedmyr, E., Klein, E., and Singh, S., 1975, Killer T cells in a Burkitt's lymphoma biopsy, *Nature* (*London*) **225:**405–407.

Jondal, M., Svedmyr, E., Klein, E., and Klein, G., 1976, Epstein–Barr virus (EBV)-specific T and K cell cytotoxicity *in vitro*, in: *Comparative Leukemia Research 1975* (J. Clemmesen and D.S. Yohn, eds.), pp. 265–271, *Biblotheca Haematologica*, No. 43, S. Karger, Basel.

Judson, S.C., Henle, W., and Henle, G., 1977, A cluster of Epstein–Barr virus-associated American Burkitt's lymphoma, *N. Engl. J. Med.* **297:**464–468.

Kafuko, G.W., Day, N.E., Henderson, B.E., Henle, G., Henle, W., Kirya, G., Munube, G., Morrow, R.H., Pike, M.C., Smith, P.G., Tukei, P., and Williams, E.H., 1972, Epstein–Barr virus antibody levels in children from the West Nile District of Uganda: Report of a field study, *Lancet* **1:**706–709.

Katsuki, T., Hinuma, Y., Saito, T., Yamamoto, J., Hirashima, Y., Sudoh, H., Deguchi, M., and Motokawa, M., 1979, Simultaneous presence of EBNA-positive and colony-forming cells in peripheral blood of patients with infectious mononucleosis, *Int. J. Cancer* **23:**746–750.

Kawamura, W.J., Takata, M., and Gotoh, H., 1970, Seroepidemiological studies on nasopharyngeal carcinoma by fluorescent antibody techniques with cultured Burkitt lymphoma cells, *Gann* **61:**55–71.

Kersey, J.H., Spector, B.D., and Good, R.A., 1973, Primary immunodeficiency diseases and cancer: The immunodeficiency–cancer registry, *Int. J. Cancer* **12:**333–347.

Klein, G., and Vonka, V., 1974, Relationship between Epstein–Barr virus determined complement-fixing antigen and nuclear antigen detected by anti-complement immunofluorescence, *J. Natl. Cancer Inst.* **53:**1645–1646.

Klein, G., Clifford, P., Klein, E., and Stjernswärd, J., 1966, Search for tumor-specific immune reactions in Burkitt lymphoma patients by the membrane immunofluorescence reaction, *Proc. Natl. Acad. Sci. U.S.A.* **55:**1628–1635.

Klein, G., Clifford, P., Klein, E., South, R.T., Minowada, J., Kourilsky, F., and Burchenal, J.H., 1967, Membrane immunofluorescence reactions of Burkitt lymphoma cells from biopsy specimens and tissue cultures, *J. Natl. Cancer Inst.* **39:**1027–1044.

Klein, G., Pearson, G., Nadkarni, J.S., Nadkarni, J.J., Klein, E., Henle, G., Henle, W., and Clifford, P., 1968a, Relation between Epstein–Barr viral and cell membrane immunofluorescence of Burkitt tumor cells. I. Dependence of cell membrane immunofluorescence on presence of EB virus, *J. Exp. Med.* **128:**1011–102.

Klein, G., Pearson, G., Henle, G., Henle, W., Diehl, V., and Niederman, J.C., 1968b, Relation between Epstein–Barr viral and cell membrane immunofluorescence in Burkitt tumor cells. II. Comparison of cells and sera from patients with Burkitt's lymphoma and infectious mononucleosis, *J. Exp. Med.* **128:**1021–1030.

Klein, G., Pearson, G., Henle, G., Henle, W., Goldstein, G., and Clifford, P., 1969, Relation between Epstein–Barr viral and cell membrane immunofluorescence in Burkitt tumor cells. III. Comparison of blocking and direct membrane immunofluorescence and anti-EBV reactivities of different sera, *J. Exp. Med.* **129:**697–705.

Klein, G., Giovanella, B.C., Lindahl, T., Fialkow, P.J., Singh, S., and Stehlin, J.S., 1974, Direct evidence for the presence of Epstein–Barr virus DNA and nuclear antigen in malignant epithelial cells from patients with poorly differentiated carcinoma of the nasopharynx, *Proc. Natl. Acad. Sci. U.S.A.* **71**:4737–4741.

Klein, G., Svedmyr, E., Jondal, M., and Persson, P.O., 1976, EBV-determined nuclear antigen (EBNA)-positive cells in the peripheral blood of infectious mononucleosis patients, *Int. J. Cancer* **17**:21–26.

Klemola, E., Kääriäinen, L., von Essen, R., Haltia, K., Koivuniemi, A., and von Bonsdorff, C.H., 1967, Further studies on cytomegalovirus mononucleosis in previously healthy individuals, *Acta Med. Scand.* **182**:311–322.

Klemola, E., von Essen, R., Henle, G., and Henle, W., 1970, Infectious mononucleosis-like disease with negative heterophil agglutination test: Clinical features in relation to Epstein–Barr virus and cytomegalovirus antibodies, *J. Infect. Dis.* **121**:608–614.

Lampkin, B.C., Major, L.C., and Mauer, A.M., 1967, Infectious mononucleosis in a child with acute lymphocytic leukemia, *J. Pediatr.* **71**:876–877.

Lang, D.J., Garruto, R.M., and Gajdusek, D.C., 1977, Early acquisition of cytomegalovirus and Epstein–Barr virus antibody in several isolated Melanesian populations, *Am. J. Epidemiol.* **105**:480–487.

Lange, B.J., Berman, P.H., Bender, J., Henle, W., and Hewetson, J.F., 1976, Encephalitis in infectious mononucleosis: Diagnostic considerations, *Pediatrics* **58**:877–880.

Lange, B., Arbeter, A., Hewetson, J., and Henle, W., 1978, Longitudinal study of Epstein–Barr virus antibody titers and excretion in pediatric patients with Hodgkin's disease, *Int. J. Cancer* **22**:521–527.

Lange, B., Henle, W., Meyers, J.D., Yang, L.C., August, C., Koch, P., Arbeter, A., and Henle, G., 1980, Epstein–Barr virus-related serology in marrow transplant recipients, *Int. J. Cancer* **26**:151–158.

Lanier, A., Talbot, M., Clift, S., Tschopp, C., Dohan, P., Bornkamm, G., and Henle, W., 1978, Epstein–Barr virus D.N.A. in tumour tissue from native Alaskan patients with nasopharyngeal carcinoma, *Lancet* **2**:1095.

Lanier, A.P., Henle, W., Bender, T.R., Henle, G., and Talbot, M.L., 1980, Epstein–Barr virus-specific antibody titers in seven Alaskan natives before and after diagnosis of nasopharyngeal carcinoma, *Int. J. Cancer* **26**:133–138.

Lanier, A.P., Bornkamm, G., Henle, W., Henle, G., Bender, T., Talbot, M.L., and Dohan, P.H., 1981, Association of Epstein–Barr virus with nasopharyngeal carcinoma in Alaskan native patients: serum antibodies and tissue EBNA and DNA, *Int. J. Cancer* **28**:301–305.

Lemon, S.M., Hutt, L.M., Shaw, J.E., Li, J.-L., and Pagano, J.S., 1977, Replication of EBV in epithelial cells during infectious mononucleosis, *Nature (London)* **268**:268–270.

Levine, P.H., Ablashi, D.V., Berard, C.V., Carbone, P.O., Waggoner, D.E., and Malan, L., 1971a, Elevated antibody titers to Epstein–Barr virus in Hodgkin's disease, *Cancer* **27**:416–421.

Levine, P.H., Merril, D.A., Bethlenfalvay, N., Dabich, L., Stevens, D.A., and Waggoner, D.E., 1971b, A longitudinal comparison of antibodies to the Epstein–Barr virus and clinical parameters in chronic lymphocytic leukemia and chronic myelocytic leukemia, *Blood* **38**:479–481.

Levine, P.H., Wallen, W.C., Ablashi, D.V., Granlund, D.J., and Connelly, R., 1977, Comparative studies on immunity to EBV-associated antigens in NPC patients in North America, Tunisia, France and Hong Kong, *Int. J. Cancer* **20**:332–338.

Levy, J.A., and Henle, G., 1966, Indirect immunofluorescence tests with sera from African children and cultured Burkitt lymphoma cells, *J. Bacteriol.* **92**:275–276.

Lin, T.M., Yang, C.S., Ho, S.W., Chiou, T.F., Liu, C.H., Tu, S.M., Chen, K.P., Ito, Y., Kawamura, A., and Hirayama, T., 1972, Antibodies to herpes-type virus in nasopharyngeal carcinoma and control groups, *Cancer* **29**:603–609.

Lin, T.M., Yang, C.S., Chiou, J.F., Tu, S.M., Lin, C.C., Liu, C.H., Chen, K.P., Ito, Y., Kawamura, A., and Hirayama, T., 1973, Seroepidemiological studies on carcinoma of the nasopharynx, *Cancer Res.* **33:**2603–2608.

Lindahl, T., Klein, G., Reedman, B.M., Johansson, B., and Singh, S., 1974, Relationship between Epstein–Barr virus (EBV) DNA and the EBV-determined nuclear antigen (EBNA) in Burkitt lymphoma biopsies and other lymphoproliferative malignancies, *Int. J. Cancer* **13:**764–772.

Look, A.T., Naegele, R.F., Callihan, T., Herrod, H.G., and Henle, W., 1981, Fatal Epstein–Barr virus infection in a child with acute lymphoblastic leukemia in remission *Cancer Res.* **41:**4280–4283.

Luka, J., Kallin, B., and Klein, G., 1978, Induction of the Epstein–Barr virus (EBV) cycle in latently infected cells by *n*-butyrate, *J. Gen. Virology* **40:**511–521.

Lynn, T.C., Tu, S.M., Hirayama, T., and Kawamura, A.J., 1973a, Nasopharyngeal carcinoma and Epstein–Barr virus. I. Factors related to the anti-VCA antibody, *Jpn. J. Exp. Med.* **43:**121–123.

Lynn, T.C., Tu, S.M., Hirayama, T., and Kawamura, A.J., 1973b, Nasopharyngeal carcinoma and Epstein–Barr virus. II. Clinical course and the anti-VCA antibody, *Jpn. J. Exp. Med.* **43:**135–144.

Magrath, I., Henle, W., Owor, R., and Olweny, C., 1975, Antibodies to Epstein–Barr virus antigens before and after the development of Burkitt's lymphoma in a patient treated for Hodgkin's disease, *N. Engl. J. Med.* **292:**621–623.

Magrath, I.T., Pizzo, P.A., Whang-Peng, J., Douglass, E.C., Alabaster, O., Gerber, P., Freeman, C.B., and Novikovs, L., 1980a, Characterization of lymphoma-derived cell lines: Comparison of cell lines positive and negative for Epstein–Barr virus nuclear antigen. I. Physical, cytogenetic, and growth characteristics, *J. Natl. Cancer Inst.* **64:**465–476.

Magrath, I.T., Pizzo, P.A., Whang-Peng, J., Douglass, E.C., Alabaster, O., Gerber, P., Freeman, C.B., and Novikovs, L., 1980b, Characterization of lymphoma-derived cell lines: Comparison of cell lines positive and negative for Epstein–Barr virus nuclear antigen. II. Surface markers, *J. Natl. Cancer Inst.* **64:**477–483.

Magrath, I., Lee, Y.L., Anderson, T., Henle, W., Ziegler, J., Simon, R., and Schein, P., 1980c, Prognostic factors in Burkitt's lymphoma: Importance of total tumor burden, *Cancer* **45:**1507–1515.

Mangi, R.J., Niederman, J.C., Kelleher, J.E., Jr., Dwyer, J.M., Evans, A.S., and Kantor, F.S., 1974, Depression of cell mediated immunity during acute infectious mononucleosis, *N. Engl. J. Med.* **291:**1149–1153.

Mann, R.B., Jaffe, E.S., Braylan, R.C., Naruba, K., Frank, M.M., Ziegler, J.L., and Berard, C.W., 1976, Non-endemic Burkitt's lymphoma: A B cell tumor related to germinal centers, *N. Engl. J. Med.* **295:**685–691.

Manolov, G., and Manolova, Y., 1972, Marker band in one chromosome 14 from Burkitt lymphomas, *Nature (London)* **237:**33–34.

Manolova, Y., Manolov, G., Kieler, G., Levan, A., and Klein, G., 1979, Genesis of the 14q+ marker in Burkitt's lymphoma, *Hereditas* **90:**5–10.

Masucci, M.G., Szigeti, R., Ernberg, I., Bjorkholm, M., Mellstedt, H., Henle, G., Henle, W., Pearson, G., Masucci, G., Svedmyr, E., Johansson, B., and Klein, G., 1981, Cell-mediated immune reactions in three patients with malignant lymphoproliferative diseases in remission and abnormally high EBV-antibody titers, *Cancer Res.* **41:**4292–4301.

Mayyasi, S.A., Schidlovsky, G., Bulfone, L.M., and Buscheck, F.T., 1967, The coating reaction of the herpes-type virus isolated from malignant tissues with antibody present in sera, *Cancer Res.* **27:**2020–2024.

Miller, G., 1975, Epstein–Barr herpesvirus and infectious mononucleosis, *Prog. Med. Virol.* **20:**84–112.

Miller, G., Lisco, H., Kohn, H.I., and Stitt, D., 1971, Establishment of cell lines from normal adult human blood leukocytes by exposure to Epstein–Barr virus and neutralization by human sera with Epstein–Barr virus antibody, *Proc. Soc. Exp. Biol. Med.* **137:**1459–1465.

Miller, G., Niederman, J.C., and Stitt, D.L., 1972, Infectious mononucleosis: Appearance of neutralizing antibody to Epstein–Barr virus measured by inhibition of formation of lymphoblastoid cell lines, *J. Infect. Dis.* **125**:403–406.

Miller, G., Niederman, J.C., and Andrews, L., 1973, Prolonged oropharyngeal excretion of EB virus following infectious mononucleosis, *N. Engl. J. Med.* **288**:229–232.

Miller, G., Robinson, J., Heston, L., and Lipman, M., 1974, Differences between laboratory strains of Epstein–Barr virus based on immortalization, abortive infection and intereference, *Proc. Natl. Acad. Sci. U.S.A.* **71**:4006–4010.

Morgan, D.G., Niederman, J.C., Miller, G., Smith, H.W., and Dowaliby, J.M., 1979, Site of Epstein–Barr virus replication in the oropharynx, *Lancet* **2**:1154–1157.

Moss, D.J., Rickinson, A.B., and Pope, J.H., 1978, Long-term T cell-mediated immunity to Epstein–Barr virus in man. I. Complete regression of virus-induced transformation in cultures of sero-positive donor leukocytes, *Int. J. Cancer* **22**:662–668.

Moss, D.J., Rickinson, A.B., and Pope, J.H., 1979, Long-term T cell-mediated immunity to Epstein–Barr virus in man. III. Activation of cytotoxic T cells in virus-infected leukocyte cultures, *Int. J. Cancer* **23**:618–625.

Mukojima, T., Gunvén, P., and Klein, G., 1973, Circulating antigen–antibody complexes associated with Epstein–Barr virus in recurrent Burkitt's lymphoma, *J. Natl. Cancer Inst.* **51**:1319–132.

Nadkarni, J.S., Nadkarni, J.J., Klein, G., Henle, W., Henle, G., and Clifford, P., 1970, EB viral antigens in Burkitt tumor biopsies and early cultures, *Int. J. Cancer* **6**:10–17.

Neel, H. B., III, Pearson, G.R., Weiland, L.H., Taylor, W.F., Goepfert, H.H., Pilch, B.Z., Lanier, A.P., Huang, A.T., Hyams, V.J., Levine, P.H., Henle, G., and Henle, W., 1980, Anti-EBV serologic tests for nasopharyngeal carcinoma *Laryngoscope* **90**:1981–1990.

Niederman, J.C., McCollum, R.W., Henle, G., and Henle, W., 1968, Infectious mononucleosis, *J. Am. Med. Assoc.* **203**:139–143.

Niederman, J.C., Evans, A.S., Subramanyan, M.S., and McCollum, R.W., 1970, Prevalence, incidence and persistence of EB virus antibody in young adults, *N. Engl. J. Med.* **282**:361–365.

Niederman, J.C., Miller, G., Pearson, H.A., Pagano, J.S., and Dowalibi, J.M., 1976, Infectious mononucleosis, Epstein–Barr virus shedding in saliva and the oropharynx, *N. Engl. J. Med.* **294**:1355–1359.

Nikoskelainen, J., Leikola, J., and Klemola, E., 1974, IgM antibodies specific for Epstein–Barr virus in infectious mononucleosis without heterophil antibodies, *Br. Med. J.* **4**:72–75.

Nilsson, K., 1979, The nature of lymphoid cell lines and their relationship to the virus, in: *The Epstein–Barr Virus* (M.A. Epstein and B.G. Achong, eds.), pp. 225–281, Springer-Verlag, Berlin, Heidelberg, New York.

Nilsson, K., Klein, G., Henle, W., and Henle, G., 1971, The establishment of lymphoblastoid lines from adult and foetal humal lymphoid tissue and its dependence on EBV, *Int. J. Cancer* **8**:443–450.

Nilsson, K., Giovanella, B.C., Stehlin, J.S., and Klein, G., 1977, Tumorigenicity of human hematopoietic cell lines in athymic nude mice, *Int. J. Cancer* **19**:337–344.

Nkrumah, F., Henle, W., Henle, G., Herberman, R., Perkins, V., and Depue, R., 1976, Burkitt's lymphoma: Its clinical course in relation to immunologic reactivities to Epstein–Barr virus and tumor related antigens, *J. Natl. Cancer Inst.* **57**:1051–1056.

Nonoyama, M., Huang, C.H., Pagano, J.S., Klein, G., and Singh, S., 1973, DNA of Epstein–Barr virus detected in tissues of Burkitt's lymphoma and nasopharyngeal carcinoma, *Proc. Natl. Acad. Sci. U.S.A.* **70**:3265–3268.

O'Conor, G., 1970, Persistent immunologic stimulation as a factor in oncogenesis with special reference to Burkitt's tumor, *Am. J. Med.* **48**:279–285.

Ohno, S., Luka, J., Lindahl, T., and Klein, G., 1977, Identification of a purified complement-fixing antigen as the Epstein–Barr virus-determined nuclear antigen (EBNA) by its binding to metaphase chromosomes, *Proc. Natl. Acad. Sci. U.S.A.* **74**:1605–1609.

Old, L.J., Boyse, E.A., Oettgen, H.F., de Harven, E., Geering, G., Williamson, B., and Clifford, P., 1966, Precipitating antibody in human serum to an antigen present in cultured Burkitt's lymphoma cells, *Proc. Natl. Acad. Sci. U.S.A.* **56**:1699–1704.

Old, L.J., Boyse, E.A., Geering, G., and Oettgen, H.F., 1968, Serologic approaches to the study of cancer in animals and man, *Cancer Res.* **28**:1288–1299.

Olweny, C., Atine, I., Kaddu-Mukasa, A., Owor, R., Klein, G., Lindahl, T., Andersson, M., Zech, L., and Henle, W., 1977, Epstein–Barr virus genome studies in Burkitt and non-Burkitt lymphomas in Uganda, *J. Natl. Cancer Inst.* **58**:1191–1196.

Pagano, J.S., Huang, C.-H., and Levine, P.H., 1973, Absence of Epstein–Barr viral DNA in American Burkitt's lymphoma, *N. Engl. J. Med.* **289**:1395–1399.

Pagano, J.S., Huang, C.-H., Klein, G., de Thé, G., Shanmuragatnam, K., and Yang, C.-S., 1975, Homology of Epstein–Barr virus DNA in nasopharyngeal cancinomas from Kenya, Taiwan, Singapore and Tunisia, in: *Oncogenesis and Herpesviruses II* (G. de Thé, M.A. Epstein, and H. zur Hausen, eds.), pp. 177–190, IARC Publication No. 11, Part 2, Lyon.

Patarroyo, M., Blazar, B., Pearson, G., Klein, E., and Klein, G., 1980, Induction of the EBV cycle in B-lymphocyte-derived lines is accompanied by increased natural killer (NK) sensitivity and the expression of EBV-related antigen(s) detected by the ADCC reaction, *Int. J. Cancer* **26**:373–380.

Pattengale, P.K., Smith, R.W., and Perline, E., 1974a, Atypical lymphocytes in acute infectious mononucleosis: Identification by multiple T and B lympocyte markers, *N. Engl. J. Med.* **291**:1145–1148.

Pattengale, P.K., Gerber, P., and Smith R.W., 1974b, B-cell characteristics of human peripheral and cord blood lymphocytes transformed by Epstein–Barr virus, *J. Natl. Cancer Inst.* **52**:1081–1086.

Pearson, G.R., 1978, *In vitro* and *in vivo* investigations on antibody-dependent cellular cytotoxicity, *Curr. Top. Microbiol. Immunol.* **80**:65–96.

Pearson, G.R., and Orr, T.W., 1976, Antibody-dependent lymphocyte cytotoxicity against cells expressing Epstein–Barr virus antigens, *J. Natl. Cancer Inst.* **56**:485–488.

Pearson, G., Klein, G., Henle, G., Henle, W., and Clifford, P., 1969, Relation between Epstein–Barr viral and cell membrane immunofluorescence in Burkitt tumor cells. IV. Differentiation between antibodies responsible for membrane and viral immunofluorescence, *J. Exp. Med.* **129**:707–718.

Pearson, G., Dewey, F., Klein, G., Henle, G., and Henle, W., 1970, Relation between neutralization of Epstein–Barr virus and antibodies to cell membrane antigens induced by the virus, *J. Natl. Cancer Inst.* **45**:989–995.

Pearson, G.R., Henle, G., and Henle, W., 1971, Production of Epstein–Barr virus-associated antigens in experimentally infected lymphoblastoid cell lines, *J. Natl. Cancer Inst.* **46**:1243–1250.

Pearson, G.R., Johansson, B., and Klein, G., 1978, Antibody-dependent cellular cytotoxicity against Epstein–Barr virus-associated antigens in African patients with nasopharyngeal carcinoma, *Int. J. Cancer* **22**:120–125.

Pearson, G.R., Qualtière, L.F., Klein, G., Norin, T., and Bal, I.S., 1979, Epstein–Barr virus-specific antibody-dependent cellular cytotoxicity in patients with Burkitt's lymphoma, *Int. J. Cancer* **24**:402–406.

Pope, J.H., Horne, M.K., and Scott, W., 1968, Transformation of foetal human leukocytes *in vitro* by filtrates of a human leukaemic cell line containing herpes-like virus, *Int. J. Cancer* **3**:857–866.

Pope, J.H., Horne, M.K., and Wetters, E.J., 1969a, Significance of a complement-fixing antigen associated with herpes-like virus and detected in the Raji cell line, *Nature (London)* **222**:166–167.

Pope, J.H., Horne, M.K., and Scott, W., 1969b, Identification of the filtrable leukocyte-transforming factor of QIMR-WIL cells as herpes-like virus, *Int. J. Cancer* **4**:255–260.

Purtilo, D.T., Cassel, C.K., Yang, Y.P.S., Harper, R., Stevenson, S.R., Landing, B.H., and Vawter, G.F., 1975, X-linked recessive progressive combined variable immunodeficiency (Duncan's disease), *Lancet* **1**:935–941.

Purtilo, D.T., Bhawan, J., DeNicola, L., Hutt, L.M., Szymanski, I., Yang, J.P.S., Boto, W., Maier, R., and Thorley-Lawson, D., 1978, Epstein–Barr virus infections in the X-linked recessive lymphoproliferative syndrome, *Lancet* **1**:798–801.

Ragab, A.H., and Vietti, T.J., 1969, Infectious mononucleosis, lymphoblastic leukemia, and the E. B. virus, *Cancer* **24**:261–265.
Reedman, B.M., and Klein, G., 1973, Cellular localization of an Epstein–Barr virus (EBV)-associated complement-fixing antigen in producer and non-producer lymphoblastoid cell lines, *Int. J. Cancer* **11**:499–520.
Reedman, B.M., Klein, G., Pope, J.H., Walters, M.K., Hilgers, G., Singh, S., and Johansson, B., 1974, Epstein–Barr virus-associated complement-fixing and nuclear antigens in Burkitt's lymphoma biopsies, *Int. J. Cancer* **13**:755–763.
Rickinson, A.B., Jarvis, J.E., Crawford, D.H., and Epstein, M.A., 1974, Observations on the type of infection by Epstein–Barr virus in peripheral lymphoid cells of patients with infectious mononucleosis, *Int. J. Cancer* **14**:704–715.
Rickinson, A.B., Finerty, S., and Epstein, M.A., 1977, Comparative studies on adult donor lymphocytes infected by EB virus *in vivo* and *in vitro*: Origin of transformed cells arising in co-cultures with foetal lymphocytes, *Int. J. Cancer* **19**:775–782.
Rickinson, A.B., Moss, D.J., and Pope, J.H., 1979, Long-term T cell-mediated immunity to Epstein–Barr virus in man. II. Components necessary for regression in virus-infected leukocyte cultures, *Int. J. Cancer* **23**:610–617.
Rickinson, A.B., Moss, D.J., Pope, J.H., and Ahlberg, N., 1980, Long-term T-cell-mediated immunity to Epstein–Barr virus in man. IV. Development of T-cell memory in convalescent infectious mononucleosis patients, *Int. J. Cancer* **25**:59–65.
Rickinson, A.B., Moss, D.J., Allen, D.J., Wallace, L.E., Rowe, M., and Epstein, M.A., 1981, Reactivation of Epstein–Barr virus-specific cytotoxic T-cells by *in vitro* stimulation with the autologous lymphoblastoid cell lines, *Int. J. Cancer* **27**:593–601.
Robinson, J., and Miller, G., 1975, Assay for Epstein–Barr virus based on stimulation of DNA synthesis in mixed leukocytes from umbilical cord blood, *J. Virol.* **15**:1065–1072.
Robinson, J.E., Brown, N., Andiman, W., Halliday, K., Francke, U., Robert, M.F., Andersson-Anvret, M., Horstmann, D., and Miller, G., 1980a, Diffuse polyclonal B-cell lymphoma during primary infection with Epstein–Barr virus, *N. Engl. J. Med.* **302**:1293–1297.
Robinson, J., Smith, D., and Niederman, J., 1980b, Mitotic EBNA positive lymphocytes in peripheral blood during infectious mononucleosis, *Nature (London)* **287**:334–335.
Rocchi, G., Hewetson, J., and Henle, W., 1973, Specific neutralizing antibodies in Epstein–Barr virus associated diseases, *Int. J. Cancer* **11**:637–647.
Rocchi, G., de Felici, A., Ragona, G., and Heiviz, A., 1977, Quantitative evaluation of Epstein–Barr-virus-infected mononuclear peripheral blood leukocytes in infectious mononucleosis, *N. Engl. J. Med.* **296**:132–134.
Rothfield, N.F., Evans, A.S., and Niederman, J.C., 1973, Clinical and laboratory aspects of raised viral antibody titers in systemic lupus erythematosus, *Ann. Rheum. Dis.* **32**:238–246.
Royston, I., Sullival, J.L., Perlman, P.O., and Perlin, E., 1975, Cell-mediated immunity to Epstein–Barr virus-transformed lymphoblastoid cells in acute infectious mononucleosis, *N. Engl. J. Med.* **293**:1159–1163.
Saemundsen, A.K., Berkel, A.I., Henle, W. Henle, G., Anvret, M., Sanal, O., Ersoy, F., Caglar, M., and Klein, G., 1981, Epstein–Barr virus carrying lymphoma in an ataxia telangiectasia patient, *Br. Med. J.* **282**:425–427.
Sairenji, T., and Hinuma, Y., 1975, Ultraviolet inactivation of Epstein–Barr virus: Effect on synthesis of virus-associated antigens, *Int. J. Cancer* **16**:1–6.
Sairenji, T., Hinuma, Y., Sekizawa, T., and Yoshida, M., 1977, Appearance of early and late components of Epstein–Barr virus-associated membrane antigen in Daudi cells superinfected with P3HR-1 virus, *J. Gen. Virol.* **38**:111–120.
Sawyer, R.N., Evans, A.S., Niederman, J.C., and McCollum, R.W., 1971, Prospective studied of a group of Yale University freshmen. I. Occurrence of infectious mononucleosis, *J. Infect. Dis.* **123**:263–270.
Schmitz, H., Krainick-Riechert, C., and Scherer, M., 1972, Acute Epstein–Barr virus infections in children, *Med. Microbiol. Immunol.* **158**:58–63.
Schooley, R.T., Haynes, B.F., Grouse, J., Payling-Wright, C., Fauci, A.S., and Dolin, R., 1981, Development of suppressor T lymphocytes for Epstein–Barr virus induced B-lympho-

cyte outgrowth during acute infectious mononucleosis: Assessment by two quantitative systems, *Blood* **57**:510–517.

Sheldon, P.J., Papamichael, M., Hemsted, E.H., and Holborow, E.J., 1973, Thymic origin of atypical lymphoid cells in infectious mononucleosis, *Lancet* **1**:1153–1155.

Silvestre, D., Kourilsky, F.M., Klein, G., Yata, Y., Neauport-Sautes, C., and Levy, J.P., 1971, Relationship between the EBV-associated membrane antigen on Burkitt lymphoma cells and the viral envelope demonstrated by immunoferritin labelling, *Int. J. Cancer* **8**:222–233.

Silvestre, D., Ernberg, I., Neauport-Sautes, C., Kourilsky, F.M., and Klein, G., 1974, Differentiation between early and late membrane antigen on human lymphoid cell lines infected with Epstein–Barr virus. II. Immunoelectron microscopy, *J. Natl. Cancer Inst.* **53**:67–74.

Stevens, D.A., Levine, P.H., Sook, K.L., Sonley, M.J., and Waggoner, D.E., 1971, Concurrent infectious mononucleosis and acute leukemia: Case reports: Review of the literature and serologic studies with the herpes-type virus (EB virus), *Am. J. Med.* **50**:208–217.

Strauch, B., Andrews, L., Siegel, N., and Miller, G., 1974, Oropharyngeal excretion of Epstein–Barr virus by renal transplant recipients and other patients treated with immunosuppressant drugs, *Lancet* **1**:234–237.

Sugawara, K., and Osato, T., 1973, Two distinct antigenic components in an Epstein–Barr virus-related early product induced by halogenated pyrimidines in non-producing human lymphoblastoid cells, *Nature (London) New Biol.* **243**:209–210.

Sutton, R.N.P., Marston, S.D., Almond, E.J.P., and Edmond, R.T.D., 1974, Aspects of Epstein–Barr virus infection in childhood, *Arch. Dis. Child.* **49**:102–106.

Svedmyr, A., Demissie, A., Klein, G., and Clifford, P., 1970, Antibody patterns in different human sera against intracellular and membrane antigen complexes associated with Epstein–Barr virus, *J. Natl. Cancer Inst.* **44**:595–610.

Svedmyr, E., and Jondal, M., 1975, Cytotoxic effector cells specific for B cell lines transformed by Epstein–Barr virus are present within patients with infectious mononucleosis, *Proc. Natl. Acad. Sci. U.S.A.* **72**:1622–1626.

Svedmyr, E., Jondal, M., Henle, W., Weiland, O., Rombo, L., and Klein, G., 1978, EBV specific killer T cells and serologic responses after onset of infectious mononucleosis, *J. Clin. Lab. Immunol.* **1**(3):225–232.

Svedmyr, E., Ernberg, I., Seeley, J., Weiland, O., Masucci, G., Szigeti, R., Masucci, M.G., Tsukuda, H., Blomgren, H., Henle, W., and Klein, G., 1982, Virologic, immunologic, and clinical observations on a patient during the incubation, acute and covalescent phases of infectious mononucleosis (submitted).

Szigeti, R., Masucci, M.G., Henle, W., Henle, G., Purtilo, D., and Klein, G., 1982, Effects of different EBV-determined antigens (EBNA, EA and VCA) on the leukocyte migration of healthy donors and patients with infectious mononucleosis and certain immunodeficiencies, *Clin. Immunol. Immunopath.* **22**:118–127.

Takaki, K., Harada, M., Sairinji, T., and Hinuma, Y., 1980. Identification of the target antigen for antibody-dependent cellular cytotoxicity on cells carrying Epstein–Barr virus genomes. *J. Immunol.* **125**:2112–2117.

Tanaka, Y., Sugamura, K., Hinuma, Y., Sato, H., and Okochi, K., 1980, Memory of Epstein–Barr virus-specific cytotoxic T cells in normal seropositive adults as revealed by an *in vitro* restimulation method, *J. Immunol.* **125**:1426–1431.

Taylor, A.W., 1953, Effects of glandular fever in acute leukemia, *Br. Med. J.* **1**:589–593.

Thorley-Lawson, D.A., Chess, L., and Strominger, J.L., 1977, Suppression of *in vitro* Epstein–Barr virus infection: A new role for adult human T lymphocytes, *J. Exp. Med.* **146**:495–508.

Tischendorf, P., Shramek, G.J., Balagatas, R.C., Deinhardt, F., Knospe, W.H., Noble, G.R., and Maynard, J.E., 1970, Development and persistence of immunity to Epstein–Barr virus in man, *J. Infect. Dis.* **122**:401–409.

Tosato, G., Magrath, J., Koski, J., Dooley, N., and Blaese, M., 1979. Activation of suppressor T cells during Epstein–Barr-virus-induced infectious mononucleosis. *N. Engl. J. Med.* **301**:1133–1137.

Trumper, P.A., Epstein, M.A., and Giovanella, B.C., 1976, Activation *in vitro* by BUdR of a productive EB virus infection in the epithelial cells of nasopharyngeal carcinoma, *Int. J. Cancer* **17**:578–587.

Trumper, P.A., Epstein, M.A., Giovanella, B.C., and Finerty, S., 1977, Isolation of infectious EB virus from the epithelial tumour cells of nasopharyngeal carcinoma, *Int. J. Cancer* **20**:655–662.

University of Health Physicians and PHS Laboratories, 1971, A joint investigation: Infectious mononucleosis and its relation to EB virus antibody, *Br. Med. J.* **4**:643–646.

Virelizier, J.-L., Lenoir, G., and Griscelli, C., 1978, Persistent Epstein–Barr virus infection in a child with hypergammaglobulinaemia and immunoblastic proliferation associated with a selective defect in immune interferon secretion, *Lancet* **2**:231–234.

Virolainen, M., Andersson, L.C., Lalia, M., and Van Essen, R., 1973, T-lymphocyte proliferation in mononucleosis, *Clin. Immunol. Immunopathol.* **2**:114–120.

Volski, D.J., Shapiro, I.M., and Klein, G., 1980, Transfer of Epstein–Barr virus (EBV) receptors to receptor negative cells permits virus penetration and antigen expression, *Proc. Natl. Acad. Sci. U.S.A.* **77**:5453–5457.

Vonka, V., Vlckova, I., Zavadova, H., Kouba, K., Lazonska, Y., and Duben, J., 1972, Antibodies to EB virus capsid antigen and soluble antigen of lymphoblastoid cells in infectious mononucleosis, *Int. J. Cancer* **9**:529–535.

Wahren, B., Espmark, A., and Walden, G., 1969, Serologic studies on cytomegalovirus infection in relation to infectious mononucleosis and similar conditions, *Scand. J. Infect. Dis.* **1**:145–151.

Wahren, B., Carlens, E., Espmark, A., Lundbeck, H., Lufgren, S., Madar, E., Henle, G., and Henle, W., 1971, Antibodies to various herpes viruses in sera from patients with sarcoidosis, *J. Natl. Cancer Inst.* **47**:747–756.

Wolf, H., zur Hausen, H., and Becker, V., 1973, EB viral genomes in epithelial nasopharyngeal carcinoma cells, *Nature (London)* **244**:245–247.

Wolf, H., zur Hausen, H., Klein, G., Becker, V., Henle, G., and Henle, W., 1975, Attempts to detect virus-specific DNA sequences in human tumors. III. Epstein–Barr viral DNA in non-lymphoid nasopharyngeal carcinoma cells, *Med. Microbiol. Immunol.* **161**:15–21.

Wolf, H., Bayliss, G.J., and Wilmes, E., 1981, Biological properties of Epstein–Barr virus, in: *Nasopharyngeal Carcinoma* (E. Grundmann, G.R.F. Krueger, and D.V. Ablashi, eds.), Cancer Campaign Vol. 5, pp. 101–109, Fischer Verlag, Stuttgart, New York.

Yoshida, T.O., 1971, High incidence of antinuclear antibodies in the sera of nasopharyngeal cancer patients, in: *Recent Advances in Human Tumor Virology and Immunology* (W. Nakahara, K. Nishioka, T. Hirayama, and Y. Ito, eds.), pp. 443–460, University of Tokyo Press, Tokyo.

Zech, L., Haglund, U., Nilsson, K., and Klein, G., 1976, Characteristic chromosomal abnormalities in biopsies and lymphoid cell lines from patients with Burkitt and non-Burkitt lymphomas, *Int. J. Cancer* **17**:47–56.

Zeng, Y., Liu, Y., Liu, C.H., Chen, S., Wei, J., Zhu, J., and Zai, H., 1980, Application of an immunoenzymatic and an immunoautoradiographic method for mass survey of nasopharyngeal carcinoma, *Intervirology* **13**:162–168.

Ziegler, J.L., Andersson, M., Klein, G., and Henle, W., 1976, Detection of Epstein–Barr virus DNA in American Burkitt's lymphoma, *Int. J. Cancer* **17**:701–706.

Zur Hausen, H., Henle, W., Hummeler, K., Diehl, V., and Henle, G., 1967, Comparative study of cultured Burkitt tumor cells by immunofluorescence, autoradiography and electron microscopy, *J. Virol.* **1**:830–837.

Zur Hausen, H., Schulte-Holthausen, H., Klein, G., Henle, W., Henle, G., Clifford, P., and Santesson, L., 1970, EB-virus DNA in biopsies of Burkitt tumors and anaplastic carcinomas of the nasopharynx, *Nature (London)* **228**:1056–1058.

Zur Hausen, H., O'Neill, F.J., Freese, U.L., and Hecker, E., 1978, Persisting oncogenic herpesvirus induced by the tumour promoter TPA, *Nature (London)* **272**:373–375.

CHAPTER 6

Herpesvirus saimiri and *Herpesvirus ateles*

BERNHARD FLECKENSTEIN, M.D., AND
RONALD C. DESROSIERS, Ph.D.

I. INTRODUCTION

Herpesvirus saimiri (*H. saimiri*) is a ubiquitous agent of squirrel monkeys (*Saimiri sciureus*), a primate species native to the South American rain forests. The virus can easily be isolated from blood and tissue-culture cells of most healthy squirrel monkeys, and there is no evidence so far that *H. saimiri* is pathogenic in its natural host. After the first report on isolation of *H. saimiri* from primary kidney-cell cultures by Melendez *et al.* (1968) from the New England Primate Research Center, the virus was regarded as a harmless indigenous agent of squirrel monkeys. The interest in this virus arose from the observation that *H. saimiri* is extremely oncogenic in a number of other primate species that are presumably not infected under usual wildlife conditions (Melendez *et al.*, 1969b). Marmoset monkeys of the genus *Saguinus* consistently develop malignant tumors of the lymphatic system, usually within 2 months of infection. The ability to induce neoplastic disease rapidly and reproducibly has made this system attractive for investigation of the biological, immunological, and molecular parameters of cell transformation and tumor induction.

Herpesvirus ateles (*H. ateles*), also first isolated by Melendez *et al.* (1972a), is an analogous virus probably indigenous to spider monkeys

BERNHARD FLECKENSTEIN, M.D. • Institut für Klinische Virologie, University of Erlangen-Nürnberg, Loschgestrasse 7, D-8520 Erlangen, West Germany. RONALD C. DESROSIERS, Ph.D. • New England Regional Primate Research Center, 1 Pine Hill Drive, Harvard Medical School, Southborough, Massachusetts 01772.

(*Ateles* spp.). The virus shares a number of properties with *H. saimiri*. Both viruses seem not to be pathogenic in the natural host species and both are similarly potent tumor inducers in marmosets and other New World primates, the spectrum of susceptible animals possibly being broader for *H. ateles* than for *H. saimiri*. Nevertheless, the viruses are distinct in several features, e.g., antigenicity of viral polypeptides, genome structure, growth behavior in cell culture, *in vitro* transforming potential, and histopathology of virus-induced tumors.

Fascinating similarities exist among *H. saimiri*, *H. ateles*, and the other herpes-group viruses that are proven or suspected to cause neoplastic diseases of lymphatic tissues, mainly Marek's disease virus (MDV) of chickens and the human Epstein–Barr virus (EBV). The term "lymphotropic herpesviruses" was created to subsume *H. saimiri*, *H. ateles*, and MDV with EBV and EBV-related viruses from baboons, chimpanzees, orangutans, and gorillas. The viruses replicate and persist in lymphocytes of infected animals, and some have been shown to produce lymphoma or leukemia or both, in either experimental animals or the natural host. These findings have given rise to expectations that the elucidation of certain molecular biological or immunological mechanisms in the pathogenesis of *H. saimiri*- and *H. ateles*-induced tumor diseases may be relevant to the understanding of some aspects of human viral oncology. In this sense, the herpesvirus neoplasias of nonhuman primates have often been investigated in view of a comparison with human EBV-associated tumors. We would prefer to describe *H. saimiri* and *H. ateles* primarily as distinct entities with many unique biological, genetic, and biochemical characteristics and peculiar pathological and immunological features of tumor disease. We feel that the more important implications of *H. saimiri* and *H. ateles* research, like those of other experimental virus systems, will probably not be immediately relevant for the diagnosis of a specific human cancer disease condition, but may contribute to the recognition of basic principles of oncogenic cell transformation and immune defense mechanisms.

II. NATURAL HISTORY OF ONCOGENIC HERPESVIRUSES FROM NEW WORLD PRIMATES

A. *Herpesvirus saimiri*

In all likelihood, squirrel monkeys (*Saimiri sciureus*) are the main natural hosts for *H. saimiri* (Fig. 1). Subsequent to the first report of isolation of *H. saimiri* from squirrel monkey primary kidney cells (Melendez *et al.*, 1968), several studies confirmed that virtually all squirrel monkeys in the wild and in captivity are infected with the virus (Table I). *Herpesvirus saimiri* can be isolated from more than 80% by cocultivation of peripheral lymphocytes with permissive monolayers (Falk *et*

FIGURE 1. Squirrel monkey (*Saimiri sciureus*).

al., 1972a,b, 1973a; Falk, 1980a,b; Rabin *et al.*, 1973), and according to two of these studies, all animals beyond 1.5 or 2 years of age possess antibodies detectable by indirect immunofluorescence (Falk *et al.*, 1972b; 1973b). The high incidence of virus infection suggests that squirrel monkeys represent the reservoir for *H. saimiri* in the wild, yet does not exclude the possibility that other primate species living in the same ecological niche of South American rain forests may sometimes be infected under natural conditions or even transmit the virus.

Herpesvirus saimiri is horizontally transmitted from adult animals to younger members of squirrel monkey colonies, and there is no indication for prenatal infection. This is apparent from the age distribution of viral antibodies among squirrel monkeys (Deinhardt *et al.*, 1973a; Falk *et al.*, 1972b; Klein *et al.*, 1973). Monkeys born in captivity sometimes have low titers of maternal antibodies that disappear within the first 3 months after birth. Primary infection usually has occurred by the 2nd

TABLE I. Wild-Type Strains of *Herpesvirus saimiri* and *Herpesvirus ateles*

Strain	Species of origin	Tissue/cell type used for isolation	Reference
Herpesvirus saimiri			
S295C (SMKI-83), prime strain, prototype strain)	Squirrel monkey (*Saimiri sciureus*)	Kidney	Melendez *et al.* (1968)
SMHI	Squirrel monkey (*Saimiri sciureus*)	Heart	Daniel *et al.* (1974a)
No. 11	Squirrel monkey (*Saimiri sciureus*)	Peripheral leukocytes	Falk *et al.* (1972a)
OMI	Owl monkey (tumor-bearing) (*Aotus trivirgatus*)	Kidney	Hunt *et al.* (1973)
Herpesvirus ateles			
No. 810	Spider monkey (*Ateles geoffroyi*)	Kidney	Melendez *et al.* (1972a)
No. 73	Spider monkey (*Ateles paniscus*)	Peripheral leukocytes	Falk *et al.* (1974a)

year of life. At the age of 2 years, all or nearly all animals have gone through seroconversion, since antibodies against structural viral polypeptides in titers of at least 1:4 have become detectable by indirect immunofluorescence. Since *in utero* transmission apparently does not occur or at most occurs very rarely, squirrel monkeys can be reared free of the virus if they are separated from adult monkeys early in life. Falk *et al.* (1973a) experimentally infected colony-born seronegative squirrel monkeys with *H. saimiri*, the infection resulting in seroconversion after approximately 3 weeks. About 40 days later, seronegative young cagemates became infected as *H. saimiri* antibody titers appeared and virus could be isolated from blood lymphocytes. At the time of these experiments, it was not possible to differentiate one *H. saimiri* strain from another, to confirm that the experimentally inoculated virus was the one transmitted. It now appears that any two isolates not related epidemiologically can be distinguished by restriction-endonuclease cleavage of virion DNA (Desrosiers and Falk, 1982).

The main route of *H. saimiri* transmission in squirrel monkeys is probably through oral contact with virus-contaminated saliva or exposure of respiratory or conjunctival mucous membranes to virus-containing aerosols. Falk *et al.* (1973b) demonstrated that the virus is secreted into the saliva for prolonged periods of time after primary infection. In a randomly selected group of ten seropositive captive squirrel monkeys, *H. saimiri* was isolated from oral secretions of nine monkeys. Filtration experiments showed that at least some of the infectivity is associated with cell-free virus particles. Apparently, squirrel monkeys, the natural hosts of *H. saimiri*, shed large doses of infectious virus over a long period of their life. It was described repeatedly that cagemate owl monkeys (*Aotus trivirgatus*) could be infected (Barahona *et al.*, 1975; Deinhardt *et al.*, 1974; Hunt *et al.*, 1973, 1975b; Rabin *et al.*, 1975a). The prolonged excretion of infectious *H. saimiri* with saliva is reminiscent of the human Epstein–Barr virus (EBV), which may appear in the oropharynx as an infectious and transforming agent for a period of many months after primary infection.

It has been hypothesized that *H. saimiri* could eventually be transmitted by hematophagous insects; experimental attempts to transmit the virus between susceptible animals (tamarin marmoset monkeys) with stableflies, mosquitos, fleas, and cone-nose bugs have failed (Fischer *et al.*, 1974). This of course does not rigorously exclude arthropod transmission in nature, since the number of potential arthropod vectors in the natural habitat of squirrel monkeys might be large.

It is not clear in what cell types, tissues, or organs *H. saimiri* replicates after entrance into the susceptible natural host. Epithelial organs of the oropharynx (e.g., salivary glands) could possibly be the site of primary infection and the origin of virus excretion, but there is no experimental proof for this hypothesis. Following primary infection, *H. saimiri* persists in white blood cells over years or for the lifetime of the squirrel monkey. This is obvious from the fact that virus can easily be isolated from lymphocytes of adult animals by cocultivation with permissive epithelial monolayers. Only the T-lymphocyte fraction seems to carry persisting *H. saimiri*. Wright *et al.* (1976) separated B and T lymphocytes by a rosette-enrichment technique; the virus could be recovered from the cell subpopulation forming T-cell rosettes but not from B cells. The authors estimate that at least one out of 10^6 T lymphocytes of normal squirrel monkeys will yield virus after cocultivation. The selectivity of *H. saimiri* for T lymphocytes could indicate that this cell population alone possesses the appropriate membrane receptor for adsorption of the virus. The T-cell tropism of *H. saimiri* in its natural host contrasts remarkably to the tropism of EBV. This virus is restricted to B lymphocytes, since primate T lymphocytes are devoid of receptors for EBV (Yefenof *et al.*, 1976).

Cell homogenates of squirrel monkey lymphocytes do not contain infectious virus. Attempts to demonstrate viral antigens of herpesvirus particles in freshly isolated lymphocytes have been unsuccessful (Falk *et al.*, 1972c). This is consistent with the general concept that long-term persistence of herpes-viruses occurs in a nonencapsidated form.

It is a matter of debate whether *H. saimiri* can persist in epithelial kidney cells. The high incidence of virus isolation from primary kidney cultures of squirrel monkeys could be due to persistence of virus in epithelial cells or the presence of virus-carrying lymphocytes in the cultures. So far, it is not known if *H. saimiri* remains in a latent status or in the form of infectious virus particles within the epithelia of squirrel monkey kidneys.

Klein *et al.* (1973) studied the humoral immune response against *H.*

saimiri in experimentally infected young squirrel monkeys and their cagemates. Antibody titers were determined by indirect immunofluorescence, using lytically infected acetone-fixed monolayers to determine antibodies against viral structural polypeptides [late antigen (LA)] and cytosine-arabinoside-arrested cell cultures for determination of antibodies against early antigens (EAs). Squirrel monkeys develop both viral capsid antigen and EA antibodies within 2–3 weeks after infection. Anti-LA antibodies quickly reach high titers, with maximal values from 1:40 to 1:320 after 1–3 months. Thereafter, the titers slowly diminish to reach the intermediate, constant, persistent anti-LA titers that are characteristic of seropositive adult squirrel monkeys. In contrast, the anti-EA reaction is transient. The anti-EA titers reach a peak of about 1:20 within 2 months, then decline within a few months to undetectable levels after 10 months following primary infection. EA-antibody response apparently concides with the phase of maximal virus replication in primary infection, and adult squirrel monkeys usually do not have appreciable antibodies against EAs detected by indirect immunofluorescence (Rabin *et al.*, 1973). Thus, the humoral antibody response in *H. saimiri*-infected squirrel monkeys is clearly similar to the seroreaction in EBV-infected humans, in whom primary infection is usually followed by a transient antibody response against EAs but permanent titers against viral structural polypeptides.

Herpesvirus saimiri does not seem to be pathogenic in the natural host; at least no symptoms of a common acute disease have ever been observed in connection with primary infection. Experimentally infected squirrel monkeys never showed any overt sign of disease (Falk *et al.*, 1973a; Melendez *et al.*, 1969a). Even squirrel monkeys with severe immunosuppression by antilymphocyte globulin, azathioprine, and prednisolone remained clinically well after experimental infection with *H. saimiri*. This does not exclude the possibility that the virus may be related to sporadic forms of neoplastic disease in squirrel monkeys. If such a tumor were to occur with the same low frequency as Burkitt lymphoma in American or European children, it is unlikely that it would have been detected so far in the limited number of monkeys under observation. Spontaneous lymphomas in squirrel monkeys are rare (Anzil *et al.*, 1977). One case was observed in an adult seropositive squirrel monkey of the New England Primate Center colony, and *H. saimiri* was isolated from lymphatic tissues of this animal. However, *H. saimiri*-specific DNA was not found in tumor tissues by (CoT)-hybridization at a detection level of about one viral genome per cell (B. Fleckenstein and M.D. Daniel, unpublished). If an *H. saimiri*-related tumor in the natural host were to follow a pathogenesis similar to that of other herpesvirus-caused lymphomas, multiple genome copies would be present per average tumor-tissue cell (Fleckenstein *et al.*, 1977). Thus, we assume that this tumor case was not related to *H. saimiri*.

B. *Herpesvirus ateles*

In comparison to what is known about the biology of *H. saimiri*, much less is known about that of *H. ateles*. Since *H. ateles* was detected a few years later and most obvious biological parameters of the two viruses appeared to be similar or identical, few experiments on the natural history of *H. ateles* have been performed. *Herpesvirus ateles* was first isolated by Melendez and his group from a primary-kidney-cell culture of a black spider monkey (*Ateles geoffroyi*) (Melendez *et al.*, 1972a–d) (Fig. 2). The isolate was designated *H. ateles* strain No. 810. Falk *et al.* (1974a) were able to isolate four *H. ateles* strains (designated Nos. 73, 87, 93, and 94) from blood samples of 18 spider monkeys (*Ateles paniscus*). The method of isolation was identical with the optimal procedure applied in *H. saimiri* isolation trials, using separated lymphocytes in cocultivation with permissive cells. Thus, largely on the basis of virus-isolation

FIGURE 2. Spider monkey (*Ateles geoffroyi*).

experiments, spider monkeys are assumed to be the natural hosts and the principal reservoir of *H. ateles* in the wild. This is also consistent with the fact that all *H. ateles* strains were isolated from healthy spider monkeys, and there is no indication that the virus is pathogenic in these animals.

A high incidence of *H. ateles* in spider monkeys of the species *A. paniscus* was documented by serological screening (Falk *et al.*, 1974a). Of 24 monkeys, 14 possessed antibodies against structural viral antigens, detectable by indirect immunofluorescence.

The early seroepidemiology was somewhat confusing, since *H. saimiri* and *H. ateles* can be differentiated by neutralization tests in cell culture, but show some cross-reactivity in immunofluorescence. However, more recent studies on many virus isolates from squirrel monkeys and two spider monkey isolates by discriminatory DNA analyses have clearly indicated that *H. saimiri* and *H. ateles* are different and discrete viruses. Thus, squirrel monkeys are a natural host for *H. saimiri*, and spider monkeys are probably a natural host for *H. ateles*.

III. STRUCTURE, COMPOSITION, AND REPLICATION OF VIRUS PARTICLES

A. Permissive Cell Cultures

The first isolate of *H. saimiri*, S295C (also previously designated SMKI-83, prime strain, or prototype strain), was identified in a primary-kidney-cell culture from a healthy squirrel monkey (Melendez *et al.*, 1968). The initial cytopathic changes were characterized by small clumps of refractile and swollen rounded cells, which slowly progressed to complete destruction of the monolayer within 1 month. Staining with hematoxylin and eosin revealed typical morphological features of herpesvirus cytopathic effect (CPE) with intranuclear acidophilic inclusions of variable size and pyknotic nuclei. Further studies showed that *H. saimiri* has a spectrum of permissive cells that includes a number of epithelial primary cultures and cell lines as well as fibroblastic-cell strains from marmosets, owl monkeys, and other New World primates (Melendez *et al.*, 1969a, 1972d; Wolfe *et al.*, 1971b). Owl monkey kidney (OMK) cells seem the best choice as a productive system for propagation of *H. saimiri* in cell culture. Infectivity titers in the range of $10^{7.0}$ median tissue-culture infective doses ($TCID_{50}$)/ml can be reached in primary OMK-cell cultures and early subpassages from them. A number of permanent cell lines have been established (Daniel *et al.*, 1976; Todaro *et al.*, 1978), and all *H. saimiri* strains grow lytically on OMK cells, resulting in complete destruction of the monolayers within 2–20 days, depending on the multiplicity of infection. Final infectivity titers obtained on later passages of OMK-cell cultures may be lower on average (about 10^5 $TCID_{50}$/ml) than in primary culture or early-passage OMK cells. The number of DNA-containing virus particles in any well-grown and fully infected culture is usually around 10^9/ml; this calculation of the number of DNA-containing viral particles is based on the yields of viral DNA from cell-culture fluids.

Epithelial cells of Old World primates are less susceptible to *H. saimiri*. The virus replicates in primary African green monkey kidney cells and two kidney-cell lines of this species, Vero and CV-1. However, the yield is variable and poor in many cases. Vero cells have been used for microplaque and infectious-focus assays using fluid medium. Vero and rhesus monkey kidney cells have been converted into cultures that produce *H. saimiri* continuously in low titers without visible cytopathic changes (Oie *et al.*, 1973; Didier *et al.*, 1975; Daniel *et al.*, 1980). *Herpesvirus saimiri* does not seem to grow in human epithelial-cell cultures; human embryonic fibroblasts, however, seem to be semipermissive (Ablashi *et al.*, 1971b, 1974a; Daniel *et al.*, 1974b). The virus can be propagated by serial passage on human fibroblasts; the infectivity is low (about 10^3 $TCID_{50}$/ml), but slightly increased by pretreatment with diethylaminoethyl (DEAE)–dextran. Cell cultures of nonprimate origin are usually not permissive. The virus was shown to replicate in mink lung cells that are simultaneously infected with squirrel monkey retrovirus (Smith *et al.*, 1979), but most attempts to grow *H. saimiri* in various cultures from chicken, mice, hamsters, dogs, cats, and rabbits were not successful (Melendez *et al.*, 1969a).

Herpesvirus saimiri infectivity can be quantitated by plaque assays. Daniel *et al.* (1971, 1972) developed the plaquing techniques on OMK, marmoset kidney, squirrel monkey lung, heart, and intestine cells using methylcellulose, starch, or agar to prepare semisolid media. The plaques remain small (0.5–2.5 mm) within 10 days of incubation, and the plaque populations are always heterogeneous in size and shape. Harvests of single plaques consistently result in heterogeneous plaque populations on serial plating.

Herpesvirus ateles has a similar but slightly different spectrum of permissive cell cultures, mostly epithelial and fibroblast cells from owl, marmoset, squirrel, rhesus, and African green monkeys. In general, *H. ateles* grows to lower infectivity titers than *H. saimiri*. There are OMK-cell lines such as line No. 210 on which *H. ateles* grows unsatisfactorily, causing a slowly progressing focal CPE; on the other hand, it grows rapidly on OMK-cell line No. 637, on which it forms large syncytia (Daniel *et al.*, 1974a; B. Fleckenstein, unpublished). In contrast to *H. saimiri*, *H. ateles* does not appear to grow in Vero cells; this difference has been used to distinguish the two viruses (Falk *et al.*, 1974a; Melendez *et al.*, 1972d,e).

B. Virion Deoxyribonucleic Acid Structures

1. *Herpesvirus saimiri*

a. Genome Structure of Wild-Type Strains

The structural analysis of *H. saimiri* and *H. ateles* DNA revealed a genome organization that is very much different from that of all other herpesviruses so far characterized and seems to be unique among all animal viruses. Most strikingly, genomes of *H. saimiri* and *H. ateles* possess manyfold repetitive DNA that amounts to nearly one third of total genome size, and the repetitive DNA is extremely high in guanine plus cytosine (G + C) content in comparison to nonreiterated DNA.

The genomes of all original *H. saimiri* strains (Table I) follow the same structural plan. They are linear double-stranded DNA molecules of slightly variable length. Most intact DNA molecules extracted from purified virus particles are in the range between 96 and 110 megadaltons (Md), which is equivalent to 145–170 kilo-base-pairs (kbp) (Fleckenstein and Wolf, 1974).

The pronounced heterogeneity in base composition becomes obvious from isopyknic centrifugation in salt gradients in the analytical or preparative ultracentrifuge, from determination of the melting temperatures of the DNA (Fleckenstein *et al.*, 1975a), and from analysis of hydrolyzed virion DNA by high-performance liquid chromatography (HPLC) (M. Ehrlich, J. Gehrke, and R. C. Desrosiers, unpublished). Analytical density scannings showed that DNA molecules that are carefully extracted from purified virus particles to avoid mechanical shear band at two densities, 1.729 and 1.705 g/ml. These density values correspond to 70.6% and 45.4% G + C, respectively, in duplex DNA that does not contain unusual bases. The high-density DNA, designated H-DNA, represents about 10% of total virion DNA, while the majority of DNA (M-DNA) is found in the 1.705 g/ml density range. As described below in more detail, M-DNA represents infectious viral DNA, while H-DNA molecules are highly defective genomes. If total virion DNA from *H. saimiri* has been fragmented by mechanical shearing prior to analytical centrifugation, two density classes are again found: H-DNA with 70.6% G + C and low-density DNA (L-DNA) with 35.8% G + C, the latter banding at 1.695 g/ml. In these gradients of sheared DNA, approximately 40% of total DNA is H-DNA and 60% L-DNA. These data are explained by a model that proposes that *H. saimiri* virus particles contain two types of virus genomes: (1) a majority of M-DNA genomes that are composed of 70% L-DNA and 30% H-DNA in covalent linkage and (2) a minority of H-DNA genomes consisting of high-G + C DNA only (Fleckenstein and Wolf, 1974; Fleckenstein and Bornkamm, 1975; Simonds *et al.*, 1975).

Base-composition analysis by HPLC yields values of 45.7% G + C for M-DNA and 70.0% G + C for H-DNA (M. Ehrlich, J. Gehrke, and R.C.

Desrosiers, unpublished). The intramolecular base composition of M-DNA was confirmed by determination of the melting temperature (t_m), since the t_m analysis revealed a characteristic biphasic curve. About 70% of M-DNA sequences were heat-denatured at 75.4°C (in 50 mM Na^+, pH 6.8), while 30% of the DNA had a t_m of 89.6°C (Fleckenstein *et al.*, 1975a). The experimentally determined values of t_m for M- and H-DNA agree within 1°C with calculated expected values as determined from the G + C content and salt concentration (Frank-Kamenetskii, 1971).

The H-DNA of *H. saimiri* is manyfold repetitive. This is obvious from reassociation kinetics in liquid-phase hybridization and cleavage with a number of restriction endonucleases. Furthermore, these experiments show that the H-DNA sequences in H-genomes and M-genomes are indistinguishable, in all likelihood identical. Many restriction endonucleases cleave H-DNA into a limited number of distinct equimolar classes of fragments with a total molecular weight of 830,000 (Table II) (Bornkamm *et al.*, 1976). This indicates that H-DNA sequences of *H. saimiri* are chains of identical repeat units in strict tandem orientation, and each repeat unit has 1.4 kbp. Thus, H-genomes have about 120 identical repeat units in head-to-tail arrangement, and necessarily, H-genomes are a form of defective genomes because of their low genetic complexity. M-genomes possess approximately 36 H-DNA repeat units, equivalent to about 30% of the full genome length.

The arrangement of H- and L-DNA within M-genomes was analyzed first by partial denaturation in the electron microscope and computer alignment of the denaturation maps (Bornkamm *et al.*, 1976). The L-DNA in M-genomes is present as one continuous nonpermuted DNA segment of 71.6 Md. The L-DNA segment is inserted between two H-DNA stretches. The lengths of these terminal H-DNA stretches are highly variable, with dimensions between 1 and about 21 μm. In general, molecules with a long H-DNA stretch at one end have a short H-DNA stretch at the opposite end, resulting in a limited size heterogeneity of the total M-DNA molecule. Alignment of M-DNA molecules by the denaturation pattern of their L-DNA segment shows that the longer H-DNA terminus may be located at either end of the M-genomes. The H-DNA segments of M-DNA differ in length by integral numbers of repeat units; this becomes obvious when the sizes of terminal fragments are analyzed after cleavage with a restriction enzyme (Group II) that cuts in the outer portion of the L-region without cleaving in H-DNA (Fleckenstein and Mulder, 1980). The H-DNA sequences at both ends of M-genomes are oriented in the same direction; this follows from the observation that intact single strands of *H. saimiri* M-DNA do not have a tendency to form any of the foldback structures that are easily found in the single strands of herpesvirus genomes with inverted repeats (P. Sheldrick, N. Berthelot, and B. Fleckenstein, unpublished).

The peculiar structure of M-genomes with components of widely different base composition and highly repetitive segments has facilitated

TABLE II. Cleavage Specifity of Some Restriction Endonucleases in Unique (L) and Repetitive (H) Deoxyribonucleic Acid of Representative *Herpesvirus saimiri* and *Herpesvirus ateles* Strains[a]

Restriction endonuclease	Recognition sequence[b] 5′ → 3′	*H. saimiri* prime strain (S295C)		*H. saimiri* strain 11		*H. ateles* strain 73	
		L-DNA[c]	H-DNA[d]	L-DNA[c]	H-DNA[d]	L-DNA[c]	H-DNA[d]
Group I							
*Bgl*I	GCCNNNN ↓ NGGC	nt[e]	4	15	5	15	3
*Hae*III	GG ↓ CC	nt	nt	>40	10	nt	nt
*Hinf*I	G ↓ ANTC	nt	nt	>50	7	nt	2
*Hpa*II (*Msp*I)	CCGG	16	nt	>17	8	nt	nt
*Pst*I	CTGCA ↓ G	nt	nt	>25	4	18	5
*Pvu*II	CAG ↓ CTG	nt	nt	20	3	nt	nt
*Sac*I	GAGCT ↓ C	nt	nt	16	2	Group II	Group II
*Sac*II	CCGC ↓ GG	nt	nt	1	3	7	3
*Sma*I (*Xma*I)	CCC ↓ GGG	Group III	Group III	Group III	Group III	3	5
*Taq*I	T ↓ CGA	nt	nt	>50	1	nt	2

Group II							
*Aos*I	TGC ↓ GCA	7	0	7	0	nt	nt
*Bam*HI	G ↓ GATCC	5	0	5	0	4	0
*Cla*I	AT ↓ CGAT	3	0	2	0	nt	nt
*Eca*I	G ↓ GTNACC	3	0	2	0	nt	nt
*Eco*RI	G ↓ AATTC	14	0	13	0	18	0
*Hind*II	GTPy ↓ PuAC	nt	nt	>40	0	nt	nt
*Hind*III	A ↓ AGCTT	nt	nt	>40	0	nt	nt
*Hpa*I	GTT ↓ AAC	nt	nt	15	0	nt	nt
*Kpn*I	GGTAC ↓ G	3	0	6	0	10	0
*Pvu*I	CGATCG	1	0	1	0	nt	nt
*Sac*I	GAGCT ↓ C	nt	nt	Group I	Group I	14	0
*Sal*I	G ↓ TCGAC	Group IV	Group IV	2	0	2	0
*Xba*I	T ↓ CTAGA	29	0	26	0	36	0
*Xho*I	C ↓ TCGAG	9	0	8	0	9	0
Group III							
Sma (*Xma*I)	CCCGGG	0	3	0	4	Group I	Group I
Group IV							
*Sal*I	G ↓ TCGAC	0	0	Group II	Group II	Group II	Group II

[a] From Fleckenstein and Mulder (1980) and C. Mulder, M. Koomey, G. Keil, and B. Fleckenstein (unpublished). [b] According to Roberts (1980).
[c] Number of cleavage sites in total L-region. [d] Number of cleavage sites in each H-DNA repeat unit. [e] nt (not tested).

physical mapping by restriction enzymes. The enzyme *Sma*I, which recognizes the sequence CCCGGG, has proven to be very useful for the initial restriction-enzyme analyses, since it does not cleave within the entire length of L-DNA of all *H. saimiri* strains tested so far. It does cleave three or four times within each H-DNA repeat unit depending on the strain of *H. saimiri* (Table II). Other endonucleases (group II enzymes) are found to cleave the L-DNA of *H. saimiri* into a small number of fragments without cleavage of H-DNA. The enzymes *Aos*I, *Bam*HI, *Cla*I, *Eca*I, *Kpn*I, *Pvu*I, and *Xho*I cleave the L-region into fewer than ten fragments (Table II). The location of H-DNA segments with variable length at the termini of M-genomes allows easy mapping of terminal L-DNA fragments. Restriction endonuclease (*endo*R) *Sma*I cleaves H-DNA, but not L-DNA; group II enzymes cleave L-DNA but not H-DNA. A comparison of the fragment pattern of *H. saimiri* M-DNA cleaved by a group II enzyme alone with the pattern produced by the group II enzyme plus *endo*R *Sma*I allows identification of L-DNA fragments located at the H–L border. Each time *endo*R *Sma*I is present, series of high-molecular-weight bands disappear and two new discrete bands appear. These two discrete fragments represent the terminal fragments of L-DNA that are not seen when a group II enzyme cleaves the DNA alone, since in that case they are linked to a variable number of H-DNA repeat units. Further mapping of the L-region of *H. saimiri* strain 11 has been achieved by preparative isolation of the fragments generated by one endonuclease from agarose gels and recleaving with one or several of the other enzymes (Fig. 3) (Fleckenstein and Mulder, 1980; C. Mulder, M. Koomey, G. Keil, and B. Fleckenstein, unpublished).

Physical maps have also been developed for the repetitive H-DNA of *H. saimiri* strain No. 11. The order of the four *endo*R *Sma*I H-DNA fragments was deduced from the sizes of DNA fragments after partial digestion. Additional maps for *endo*R *Hinf*I, *Hpa*II, *Pst*I, *Pvu*II, *Sac*II, and *Taq*I have resulted from multiple cleavages of isolated H-DNA fragments in various combinations (Bornkamm *et al.*, 1976; Fleckenstein, 1979; Fleckenstein and Mulder, 1980).

Herpesvirus saimiri strain 11 was the first wild-type strain used for structural analysis. The three other strains characterized so far, strains S295C, SMHI, and OMI (HOT), possess the same basic structure of M- and H-genomes, and there is a high degree of sequence homology among all strains. Cross-hybridization experiments using CoT analysis did not detect significant base-pair divergence among the strains. This is in accordance with the results of the comparative restriction enzyme cleavages of L- and H-DNA of these *H. saimiri* strains whereby distinct but related patterns were found consistently. A more quantitative determination of the degree of base-pair homology was obtained by measuring t_m depression of heteroduplex molecules. In these experiments, we found less than 3°C for Δt_m between *H. saimiri* L-DNA homoduplex and heteroduplex molecules, indicating at most 2% average base-pair divergence over most of the length of the L-region (Keil *et al.*, 1980).

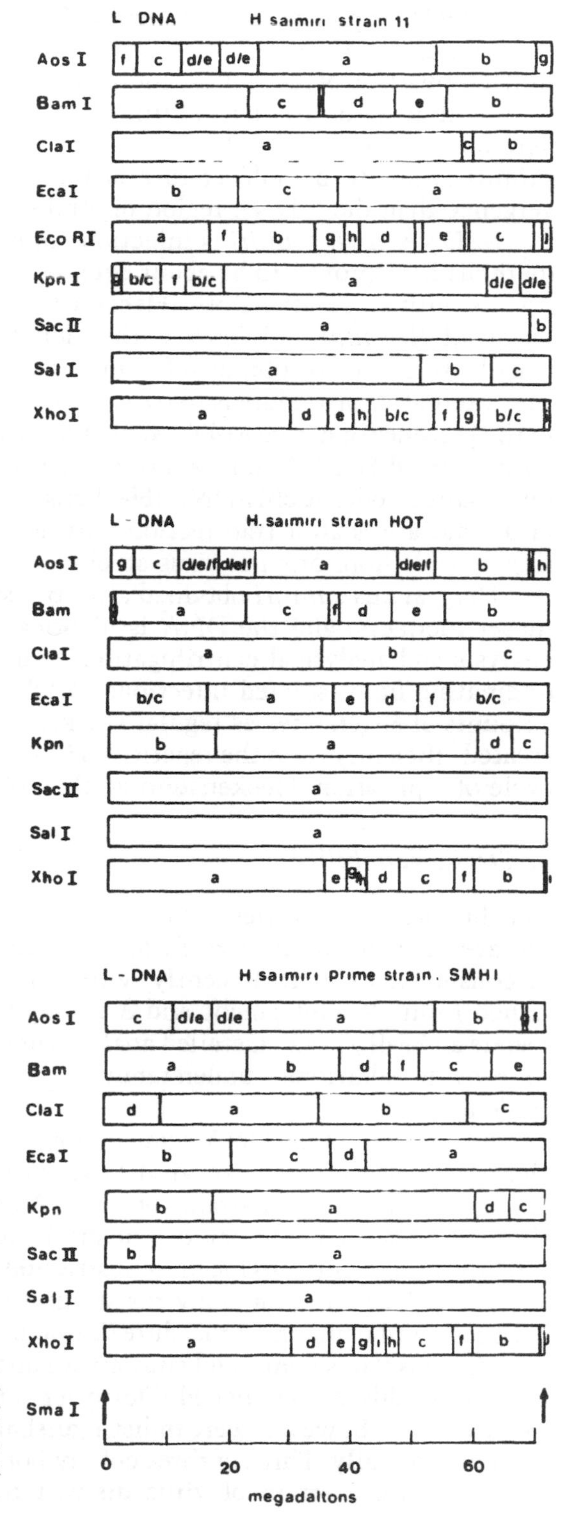

FIGURE 3. Cleavage maps of the L-DNA regions from three *H. saimiri* strains. From C. Mulder, M. Koomey, G. Keil, and B. Fleckenstein (unpublished).

H-DNA sequences from different *H. saimiri* strains show more divergence than the average of L-DNA sequences. This becomes obvious from comparison of restriction-endonuclease cleavage patterns of H-DNA from various virus strains and from measuring the t_m values of H-DNA heteroduplex molecules (Keil *et al.*, 1980). The repetitive DNA of *H. saimiri* seems to be subject to a faster evolutionary base-sequence divergence than the L-DNA region of viral genomes.

M-DNA of *H. saimiri* is infectious in permissive cell cultures. Two methods have proven to be about equally efficient in transfection experiments with *H. saimiri* DNA (Fleckenstein *et al.*, 1975a). After precipitation of *H. saimiri* M-DNA with calcium phosphate, 10 ng DNA is usually sufficient to obtain infection of OMK cells within 2–7 weeks. The alternative procedure is to treat permissive cells with DEAE–dextran in the presence of native M-DNA. No productive infection can be induced with isolated H-DNA, nor is any to be expected from the low genetic-information content of H-DNA (Fleckenstein and Bornkamm, 1975). The virus that arises after transfection with M-DNA was found to be indistinguishable from original virus stocks.

Herpesvirus saimiri obtained after transfection is oncogenic in marmoset monkeys, and the DNA is identical in restriction endonuclease analysis and analytical centrifugation. Remarkably, *H. saimiri* particles originating from isolated infectious M-DNA contain the same relative amounts of H-genomes as regular stocks, indicating that H-genomes are created efficiently from the repetitive H-DNA ends of M-genomes in each cycle of replication (Fleckenstein *et al.*, 1975a).

b. Strain Variability

The biological properties, genome structure, and restriction-fragment linkage maps of the four *H. saimiri* strains described above are known in considerable detail. Recently, virus and virion DNAs have been obtained from 19 wild-caught and 3 colony-born squirrel monkeys (Desrosiers and Falk, 1982). Detailed analyses of these 22 new isolates promise to provide additional understanding of *H. saimiri* systematics and epidemiology.

Except for actual DNA sequencing, analysis of restriction-endonuclease cleavage products of viral DNA is the most sensitive means of distinguishing different strains of virus. Any two fresh isolates of herpes simplex virus (HSV) unrelated epidemiologically can be distinguished by the DNA fragments produced by restriction endonucleases (Buchman *et al.*, 1978). Transmission of a virus strain from one person to another has been followed using this procedure (Linnemann *et al.*, 1978). Although all *H. saimiri* isolates contained similarities and common features, 19 of the 22 were readily distinguished (Desrosiers and Falk, 1982, Fig. 4). Three of the isolates, however, were indistinguishable and probably were related epidemiologically. Three of three colony-born squirrel monkeys that were tested yielded a strain of virus distinct from that obtained from the mother. In separate experiments, two of three animals chosen randomly yielded a strain of virus different from that originally obtained 16 and 22 months previously; only one of the three animals examined yielded the same strain of virus 22 months after the original isolation. The degree of restriction endonuclease fragment variability among *H. saimiri* strains appeared to be greater than previously observed for herpes simplex virus (Buchman et al., 1978) and *Herpesvirus tamarinus* (Desrosiers and Falk, 1981).

In the case of HSV, the differences in restriction-endonuclease cleavage patterns can usually be explained by a simple gain or loss of a recognition site by point mutation. The variability in *H. saimiri* strains is so great that either many such gains or losses occur in one strain vs. another or other factors such as rearrangements, deletions, insertions, or hypervariability in discrete sequences contribute to the variability. This can be simply seen by comparing the *Bam*HI cleavage sites in L-DNA of strains OMI (HOT) and SMHI (see Fig. 3). Although *Bam*HI cleaves only five and six times, respectively, in these DNAs, five gains or losses of recognition sites would be required to explain the differences, if this were the only factor contributing to the differences in *Bam*HI cleavage maps. Similarly, six gains or losses would be required to explain the differences between No. 11 and SMHI. Measurement of Δt_m between homoduplex and heteroduplex DNA molecules in reassociation experiments has indicated less than 2% average base-pair divergence in L-DNA among laboratory strains of *H. saimiri* (Keil *et al.*, 1980). A 2% random base-pair divergence would result in 88.6% of six nucleotide recognition sequences remaining unaltered. The cleavage sites of *Eco*RI appear to be more highly conserved than those of other enzymes, but the variability in cleavage sites with most restriction endonucleases is much greater than would be expected from a simple 2% base-pair divergence. Thus, the causative factors that underlie restriction-endonuclease cleavage-site variability may be complicated and are not fully understood.

One region of L-DNA that has apparently changed by means other than point mutations is the leftmost 6% (7 kbp). The genomes of five of these new isolates have been compared to the DNA from four laboratory strains, 11, OMI, SMHI, and S295C, by restriction-endonuclease cleavage, blotting to filter paper, and hybridization. All nine strains showed extensive homology over most of the length of the genome. On the basis of homology of the leftmost 7 kbp of L-DNA, however, they could be grouped into two classes. The leftmost 7-kbp region from one group was not detectably homologous with DNA from the other group (Medveczky *et al.*, 1980). This lack of homology is particularly interesting in that two independently derived attenuated, nononcogenic strains of *H. saimiri* have deletions in this same region (see Section VI.D). Viruses from both classes can produce tumors in experimental animals, but there are possible differences in the degree of oncogenicity, i.e., percentage of animals that die or survival time.

Most herpesviruses are not easily spread from person to person

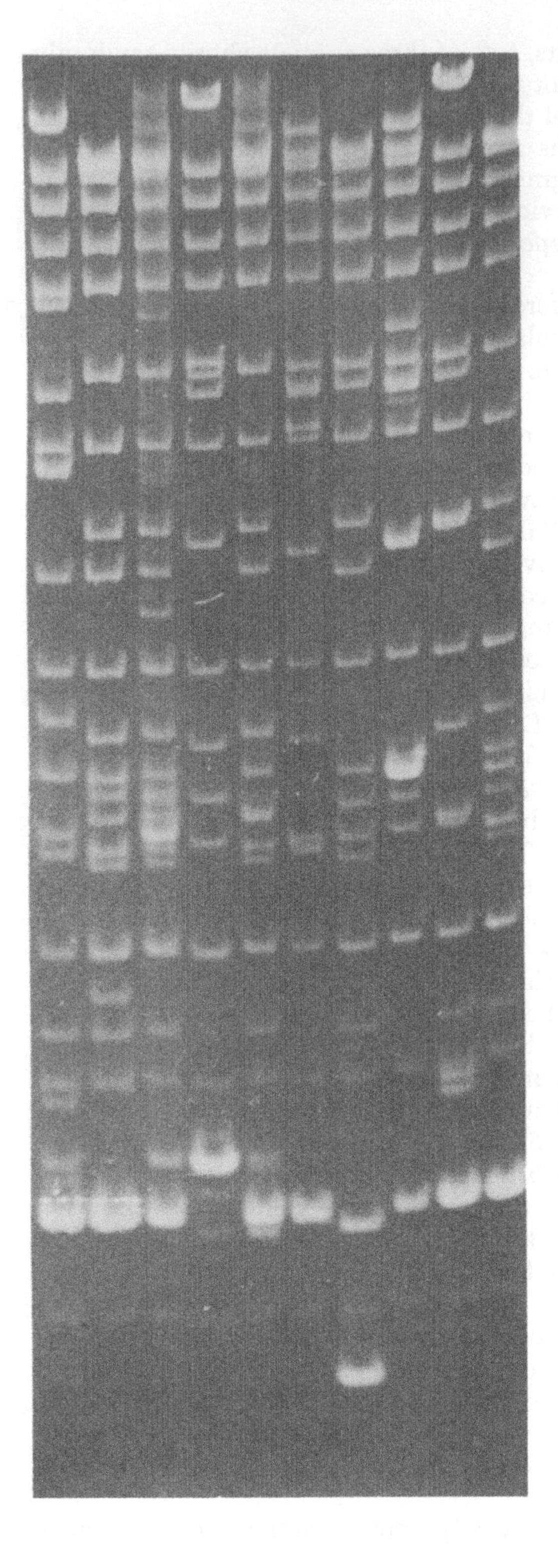

FIGURE 4. *Herpesvirus saimiri* strain variability as determined by restriction-endonuclease digestion of virion DNA. *Herpesvirus saimiri* was isolated from ten feral, imported squirrel monkeys by cocultivation of peripheral lymphocytes with permissive OMK cells. Stock virus was prepared from this initial cocultivation and used for the preparation of sufficient virus for isolation of DNA. DNA was prepared from purified virions as previously described (Bornkamm *et al.*, 1976). Virion M-DNA was digested with *endo*R *Sac*I and electrophoresed through a 0.5% agarose gel. DNA fragments were stained with ethidium bromide and photographed under UV light. DNA from the isolate shown in slot 2 was from a Peruvian squirrel monkey imported from Iquitos, Peru. The other isolates were from Colombian squirrel monkeys imported from Leticea, Colombia. Animal markings and in some cases karyotyping were used to confirm the origin of the animals. The supermolar fragments toward the bottom of the gel are derived from cleavage within repetitive H-DNA. Additional multimolar fragments in slot 8 are probably due to the presence of defective DNA. Data from R. Desrosiers and L. Falk (unpublished).

through aerosols. It has been hypothesized that horizontal spread of herpesviruses within a given community or region is limited, compared with that of respiratory viruses, and may frequently be restricted to family clusters (Roizman and Buchman, 1979). Geographic clustering of strain types would have important implications for the epidemiology of human diseases. The geographic clustering of EBV-related tumor diseases (Burkitt lymphoma and nasopharyngeal carcinoma) is well documented (de Thé *et al.*, 1978; Judson *et al.*, 1977). It is conceivable that particular strain types are prevalent in a particular geographic region and contribute to the observed differences in disease patterns. *Herpesvirus saimiri* provides an ideal model system to study this since (1) squirrel monkey subgroups coming from different regions of South and Central America have been identified; (2) virus can be easily isolated by cocultivation of peripheral lymphocytes with permissive OMK cells; and (3) viral DNA can easily be prepared in sufficient quantities from permissive infection for restriction-endonuclease analysis. Further analyses of these 22 plus other isolates may reveal characteristic features peculiar to an individual geographic region.

c. Recombinant Strains

For most viruses, the isolation of conditionally lethal mutants is the primary approach to their genetics. For *H. saimiri*, difficulties have been encountered with the isolation of temperature-sensitive mutants, since the virus grows slowly, and the mutants tended to leakiness and reversion (Schaffer *et al.*, 1975; P. Schaffer, personal communication). To facilitate a genetic analysis, a method has been developed to generate recombinants between different *H. saimiri* strains from large restriction fragments of their DNAs (Keil *et al.*, 1980). Permissive monolayers were infected with long, overlapping fragments of *H. saimiri* DNA. Complete M-genomes could arise only by recombination of DNA segments from different virus strains, imposing a strong selection for recombinants between different strains of *H. saimiri*. Restriction-endonuclease analysis of virion DNA arising from such mixed transfections is used to determine the nature of the recombination events.

In a first approach, *Bam*HI DNA digests of one *H. saimiri* strain and *Sma*I-cleaved DNA from a second strain were used for transfection of OMK cells. *Endo* R *Sma*I yields a complete *H. saimiri* L-DNA region, while *Bam*HI leaves intact H-stretches with long adjacent L-DNA segments. From these cultures, we could rescue five isolates, designated RF 1, RF 2, RF 4, RF 6, and RK 2. The viruses were plaque-purified, and the DNA was purified from each strain and cleaved with appropriate restriction endonucleases. For each of the isolates, it became obvious that the H-DNA sequences were derived from one strain and most of the L-DNA was derived from the other parental strain. Construction of cleavage maps for restriction endonucleases *Aos*I, *Bam*HI, *Kpn*I, *Sac*II, *Sma*I, and *Xho*I for each of the recombinants allowed us to determine which parts of the recombinant M-genome originated from either of the parents (Fig. 5). The maps of four strains suggested that the recombination occurred by crossing over in accurately corresponding L-sequence loci of parental M-genomes. One recombinant strain had an insertion of 1.3 Md close to the right crossing-over point and in parallel a deletion of 1.1 Md on the left side (Keil *et al.*, 1980).

Cleavage of L-DNA from all *H. saimiri* strains with *endo*R *Kpn*I yields a long A-fragment of 42.5 Md representing the interior part of the L-DNA region. The ends of the *Kpn*A fragment overlap with the terminal

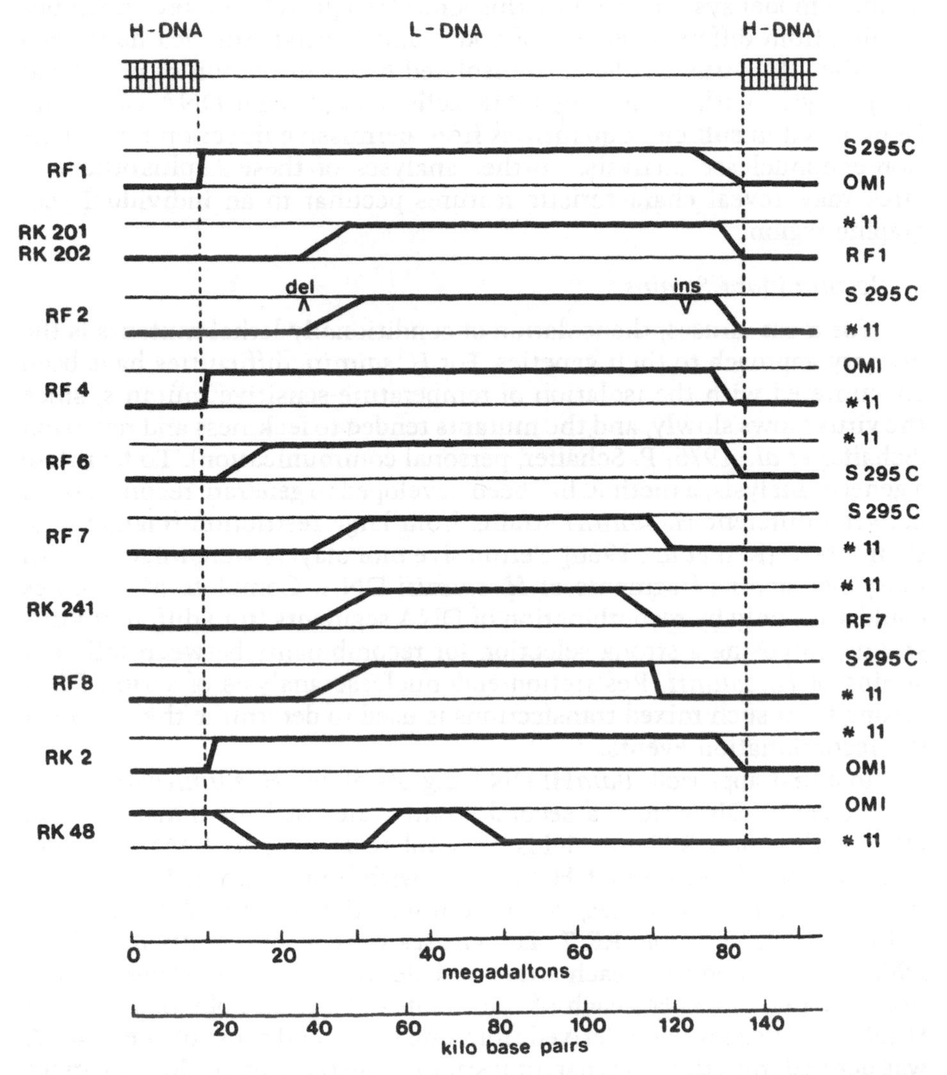

FIGURE 5. Schematic representation of *H. saimiri* recombinants. The recombinant strains were isolated after transfection of long, overlapping restriction fragments from two different *H. saimiri* strains in each experiment. The parental DNA sequences represented in each recombinant strain are indicated by a bold line (Keil *et al.*, 1980).

*Bam*HI A- and B-fragments of strain 11 L-DNA. Thus, cotransfection of *Kpn*I-cleaved DNA of strain S295C with *Bam*HI digests of strain 11 M-DNA resulted in the development of two recombinant viruses, designated RF 7 and RF 8. Necessarily, the sites of crossing over were at a greater distance from the H-DNA junctions than in the first series of recombinants. As was to be expected, recombinant virus RF 8 possessed one *Sal*I cleavage site in L-DNA only. Surprisingly, M-DNA of RF 7 could not be cleaved at all by *Sal*I; this could be due to an additional double crossover with a short sequence exchange or by loss of the cleavage site unrelated to recombination. With the use of an analogous approach, a virus (RK 241) was isolated after cotransfection of the *Kpn*I A fragment of strain 11 with the *Bam*HI-cleaved M-DNA of recombinant strain RF 7, and three further recombinants (RK 48, RK 201, and RK 202) were obtained from the large *Kpn*I A fragment of *H. saimiri* strain 11 M-DNA with a *Cla*I digest of OMI DNA or with a mixture of strain 11 M-DNA *Xho*I fragments and strain RF 1 *Bam*HI fragments, respectively (Fig. 5).

Two further recombinant viruses were obtained after cotransfection with *Bam*HI fragments of strain OMI, an *Xho*I digest of strain 11 and the internal L-region from strain S295C. Cleavage maps are not completed as yet; however, assuming a mode of generation by two crossing-over events akin to the mechanism described before, these two recombinants could possess different H-DNA sequences at the two ends.

As described in more detail in Section VI.D, attenuated *H. saimiri* (strain 11^{att}) has a 1.1 Md deletion at the left terminus of L-DNA. In a first series of experiments, recombination between the attenuated *H. saimiri* and the highly oncogenic wild-type strain OMI resulted in an isolate that preserved the L-DNA deletion (Fig. 6). Analysis of this type of recombinant may help in the future to identify the functional role of the left end of L-DNA for oncogenicity.

The generation of recombinants between different *H. saimiri* strains by cotransfection with large, overlapping restriction fragments is a convenient and effective procedure in comparison to other possible methods, since conditionally lethal mutants of the viruses are not needed and recombinants are necessarily selected for. Thus, the peculiarities of the *H. saimiri* genome structure and the availability of restriction endonucleases that cleave very infrequently may allow a first useful approach toward molecular genetics of this virus. Analyses of the biological and biochemical features of *H. saimiri* recombinants may provide map locations for some functions in future studies.

As mentioned above, one recombinant strain (RF 8) has one *Sal*I cleavage site within the entire M-genome. Further experiments aimed at deleting the other *Sal*I site of strain 11 by a different recombination are in progress. If a single *Sal*I recognition site in recombinant strains of *H. saimiri* were not part of essential coding sequences, a recombinant genome with one *Sal*I site could be regarded as a large eukaryotic cloning vector that may be able to incorporate a long segment of foreign DNA into the unique *Sal*I site without loss of viral genetic information.

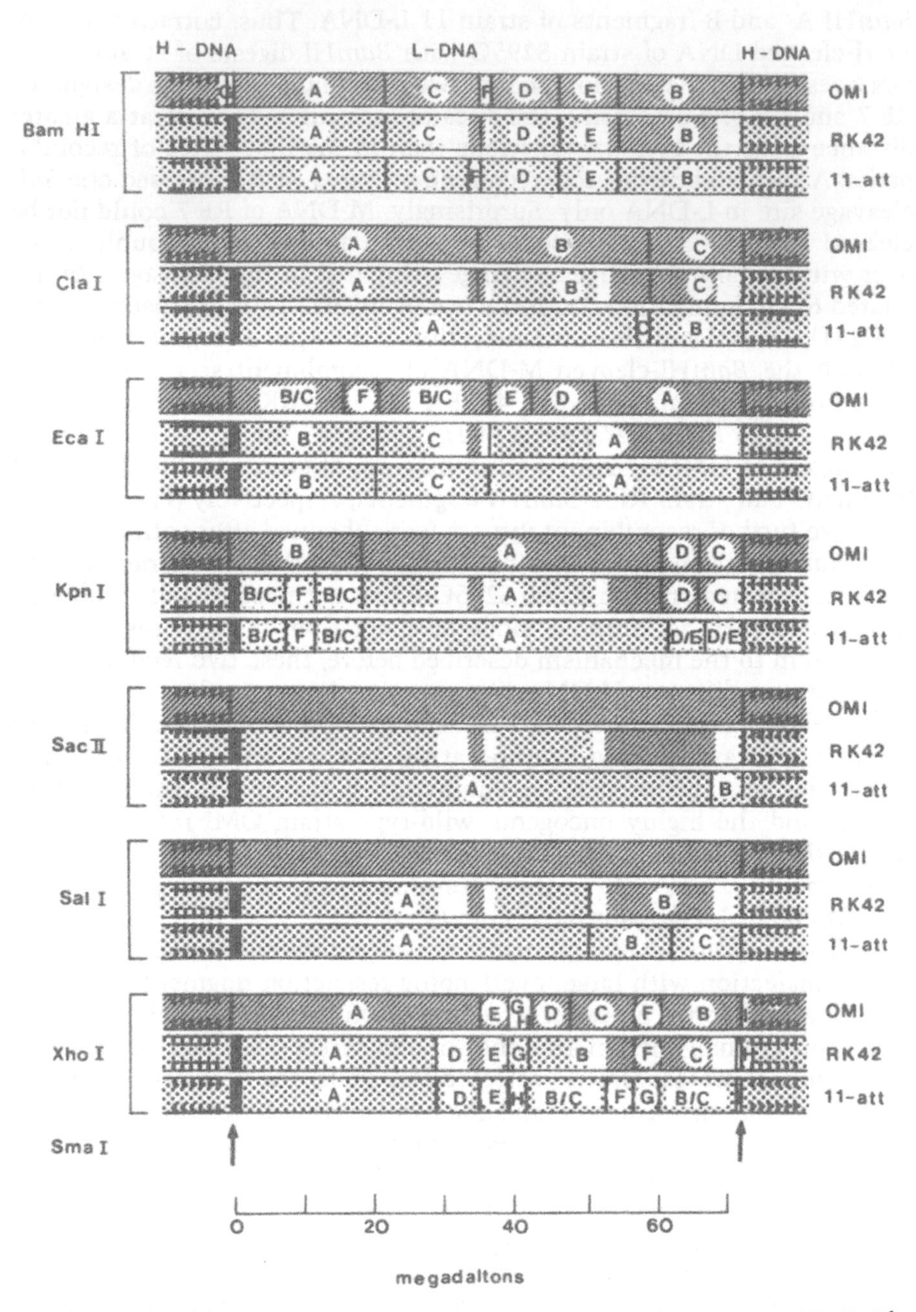

FIGURE 6. DNA sequence representation in *H. saimiri* recombinant strain RK 42. The recombinant was obtained from transfection with overlapping restriction fragments of the attenuated *H. saimiri* strain 11^{att} and the highly oncogenic strain OMI. The left-end 1.1-Md deletion of strain 11^{att} is preserved in the recombinant virus. From G. Keil and B. Fleckenstein (unpublished).

2. *Herpesvirus ateles*

Herpesvirus ateles and *H. saimiri* are similar in the structural organization of their genomes (Fleckenstein *et al.*, 1978a): their DNAs share the most characteristic features, an extreme heterogeneity in base composition and the presence of highly repetitive DNA in complete and defective genomes.

Similar to virions of *H. saimiri*, purified *H. ateles* virions contain repetitive H-DNA of very high G + C (75%) and nonreiterated L-DNA of low G + C (38%). As shown by partial denaturation in the electron microscope, the two types of DNA sequences are assembled in at least two types of double-stranded molecules, M-genomes and H-genomes, both with molecular weights of approximately 90×10^6. Defective H-genomes consist of H-DNA sequences only that are approximately 100-fold reiterated per molecule. M-genomes contain L-DNA segments that represent 74% of the molecule length inserted between terminal H-DNA sequences. Similar to *H. saimiri* DNA, the sizes of H-DNA termini in M-genomes of *H. ateles* are highly variable, differing in length by integral numbers of repeat units, and all H-sequences in one molecule are oriented in the same direction. Usually, an M-genome with a long H-terminus at one end has a short H-sequence at the other end, the longer end being located at either side of the molecule.

Despite these similarities, differences are found between *H. ateles* and *H. saimiri* DNA: (1) H-DNA, as well as L-DNA, of *H. ateles* has a significantly higher G + C content than the corresponding sequences of *H. saimiri*; (2) the relative amount of H-DNA in the M-genome of *H. ateles* is lower than the percentage of H-DNA in the M-genome of *H. saimiri*; and (3) the repeat units of *H. ateles* H-DNA (about 1.5 kbp) are slightly longer than the *H. saimiri* H-DNA repeats (1.4 kbp).

A number of restriction endonucleases cleave infrequently in *H. ateles* DNA. Thus, physical gene maps for the L-region in the M-genome of *H. ateles* strain 73 have been developed using *endo*R *Sac*II, *Sal*I, and *Sma*I (Fleckenstein and Mulder, 1980). Comparative restriction analysis of virus strains 810 and 73 indicates a higher degree of variability between the two *H. ateles* isolates than among the *H. saimiri* group of virus isolates. This is consistent with the observation that an appreciable base-sequence divergence is found between the DNAs of these *H. ateles* strains by analysis of heteroduplex hybrids (Fleckenstein *et al.*, 1978a).

C. Virion Proteins

Sodium dodecyl sulfate–polyacrylamide gel electrophoresis (SDS-PAGE) of purified *H. saimiri* virions has revealed major and minor proteins in numbers similar to HSV (Roizman and Furlong, 1974) and EBV

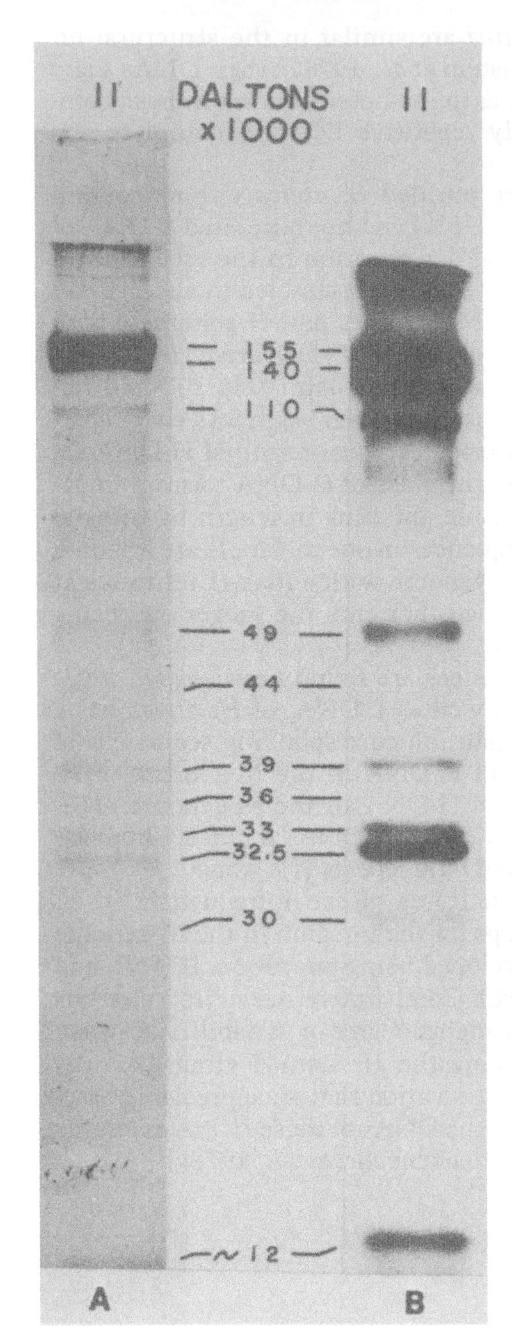

FIGURE 7. SDS-PAGE of *H. saimiri* virion proteins. OMK cells were infected with strain 11 at a multiplicity of infection (Moi) of 0.5–3.0. At 18 hr postinfection, the medium was removed and replaced with medium containing [^{35}S]methionine. At 44 hr postinfection, virus was purified from the cell-free supernatant by standard procedures (Bornkamm *et al.*, 1976). Purified virus was analyzed by electrophoresis in 10% polyacrylamide gels using bisacrylamide for cross-linking. The molecular weights are relative to ^{14}C standards of New England Nuclear. (A) Coomassie brilliant blue stain pattern; (B) fluorograph (using sodium salicylate). Data from S. Tracy, and R. Desrosiers (unpublished).

(Dolyniuk *et al.*, 1976) (Fig. 7) (Keil *et al.*, 1980; Randall and Honess, 1980; S. Modrow and H. Wolf, personal communication; W. Bodemer and G. Keil, unpublished). Although the designated molecular-weight values of these groups vary, they almost certainly have detected the same major virion polypeptides. Figure 7 shows a profile of ^{35}S-labeled *H. saimiri* strain 11 purified virion polypeptides; Fig. 7A shows the pattern with Coomassie brilliant blue strain and Fig. 7B shows the exposed film following fluorography. Modrow and Wolf found no differences in the molecular weights of polypeptides from purified virions of strain 11, the attenuated strain derived from 11 (11^{att}), and virus recovered from a marmoset inoculated with 11^{att}. Following SDS-PAGE, the polypeptide profiles of the two *H. ateles* strains, 73 and 810, were similar but distinguishable from the *H. saimiri* profile, and the profiles of 73 and 810 even showed differences in some of the major polypeptides (S. Modrow and H. Wolf, personal communication). Keil *et al.* (1980) found differences among various *H. saimiri* strains in the molecular weights of some of the minor virion polypeptides, especially in the 12–16 kd range.

D. Mechanism of Virus Replication

1. DNA Replication

The structure of replicating DNA from *H. saimiri* or *H. ateles* has not yet been visualized. A hypothesis on the mode of DNA replication has to take into account the peculiar genome structures and the dynamics of their formation. Terminal H-DNA stretches are of discrete sizes that differ by multiples of the length of a repeat unit; thus, it is likely that all molecules end at each terminus with one distinct terminal sequence that is present once only in each repeat unit. This could easily be explained by assuming that circular or concatameric forms are intermediates of *H. saimiri* DNA replication and that these intermediates are cleaved into monomers by an enzyme activity present in the nuclei of lytically infected cells. The arrangement of tandem repeats with variable lengths at both ends would be consistent with this hypothesis.

De novo creation of H-genomes from M-genomes after transfection with M-DNA could involve a mechanism akin to gene amplification in other eukaryotic systems. A rolling-circle model of DNA replication could account for all structural phenomena of *H. saimiri* DNA observed so far. The peculiar structure of *H. saimiri* DNA may help in the future to elucidate possible replication mechanisms of herpesvirus DNA.

2. Transcription

Very little information is available at present on characterization of the RNA and proteins coded by *H. saimiri* and *H. ateles*. Experiments mapping viral polypeptides to particular regions of the genome have not yet been performed.

The only experiments regarding RNA synthesis have involved determining transcription from H- vs. L-sequences. Fleckenstein *et al.* (1976) have reported that total RNA from infected OMK and from lymphoid-tumor cells can hybridize significantly to a major portion of H-DNA. Tracy and Desrosiers (1980) have found that less than 1%, if any, of viral-specific cytoplasmic RNA labeled late in infection is derived from H-sequences. Immediate early transcripts of HSV (Anderson *et al.*, 1980) and vaccinia virus (Wittek *et al.*, 1980) are derived from reiterated regions of DNA. Transcription from repeated H-DNA sequences very early in infection could possibly explain the aforementioned results, but this has not been demonstrated.

3. Protein Synthesis

Early work on identifying *H. saimiri*-specific proteins concentrated on characterizing the antigens of permissively infected cells. Late antigens (LAs) are predominately cytoplasmic antigens synthesized late in the viral replicative cycle. Antibodies to LAs persist for life in naturally or experimentally infected squirrel monkeys and can frequently be found in experimentally infected owl or marmoset monkeys (Pearson *et al.*, 1973). Early antigens (EAs), by definition, are detected before the onset of viral DNA replication and primarily are not major structural proteins (Klein *et al.*, 1973). Antibodies against EAs appear in sera from animals with active *H. saimiri* infections. EAs seem to be located primarily in the nucleus of infected cells when assayed by indirect-immunofluorescent-antibody tests, and two different types of staining patterns have been observed. Qualtière and Pearson (1978) have solubilized *H. saimiri* membrane antigens (MAs) by limited papain digestion. The solubilized MA preparation was active in inhibiting antibody-dependent lymphocyte cytotoxicity. Ohno *et al.* (1979) have detected a DNA-binding *H. ateles* nuclear antigen (HATNA) using an acid-fixed nuclear binding technique. HATNA is probably analogous to the nuclear antigen of EBV. The detectability of HATNA appeared to correlate with the number of *H. ateles* genome equivalents per cell (Johnson *et al.*, 1980) (see Table IV).

Herpesvirus saimiri-specific neutralizing sera have little or no ability to cross-neutralize *H. ateles* and vice versa. Infectivity-neutralization tests thus can be used to distinguish *H. saimiri* from *H. ateles* or *H. saimiri*-specific sera from *H. ateles*-specific sera. Antigenic cross-reactivity (EA and LA) can be detected, however, by indirect-immunofluorescence tests (Falk *et al.*, 1974a; L. Falk and M.D. Daniel, unpublished).

Randall and Honess (1980) have found that two of the major virion proteins of approximately 160 and 28 kd are preferentially and specifically released into culture fluid during infection of permissive OMK cells.

These proteins were also specifically found in a membrane fraction. Immunization of rabbits against these polypeptides gave rise to neutralizing antibodies.

Work is just beginning in several laboratories on biochemical characterizations of the structural and nonstructural proteins of *H. saimiri*. One approach that has been successfully used with HSV is to examine labeled viral polypeptides in infected cells against a declining background of host-cell polypeptide synthesis. This procedure allows for numerous characterizations, such as varying the label to detect glyco- and phosphoproteins, subcellular fractionation to determine the location of various polypeptides, and varying the time of labeling for detection of immediate early, early, and late viral polypeptides. Although this approach will undoubtedly be useful, its success in the *H. saimiri* system is hampered by a number of factors: First, the highest titers of infectious *H. saimiri* are usually around $10^{6.5}$. To guarantee initial infection of 100% of the cells, either large volumes or concentrated virus preparations are required. The particle/plaque-forming unit ratio for *H. saimiri* is not usually known, and it is not known how many virus particles are actually entering the cells under these conditions. Second, the shutoff of host protein synthesis is slow and occurs late. Third, the infectious cycle of *H. saimiri* is considerably longer than the cycles of other viruses, and infection may not proceed synchronously. A single cycle is longer than 36 hr (Morgan and Epstein, 1971). The tumor promoter 12-*O*-tetradecanoylphorbol-13 acetate (TPA) has been found to accelerate and enhance virus polypeptide production and to shut down host protein synthesis more rapidly (Ablashi *et al.*, 1980a; S. Modrow and H. Wolf, personal communication); a similar stimulatory effect on lytic replication of *H. saimiri* was described for 5-tungsto-2-antimonate (Ablashi *et al.*, 1977b). This may help in the analysis of *H. saimiri*-specified protein synthesis.

It is possible that the temporal classification of *H. saimiri* polypeptide synthesis may not be analogous to the well-characterized HSV system (Roizman and Furlong, 1974). P. O'Hare, R.W. Honess, and R.E. Randall (unpublished) have found that only a subset of early *H. saimiri* polypeptide synthesis proceeded normally in the presence of the antiherpes agent phosphonoacetic acid (PAA) and that host protein synthesis was not shut off in infected cells at a concentration of PAA sufficient to inhibit virus yield 99%. This contrasts with the action of PAA during HSV replication, when only late protein synthesis is selectively inhibited and host protein synthesis is effectively shut off.

4. Inhibitors of Virus Replication

A number of virostatic drugs have been tested to determine their inhibitory effects on the lytic replication cycle of *H. saimiri*. The examination of antiviral compounds was directed mainly at finding a substance that could efficiently suppress the primary infection of *H. saimiri* in experimentally infected marmoset or owl monkeys, thus preventing the development of tumor disease. PAA and phosphonoformic acid (PFA) suppress growth of *H. saimiri* in cell culture at a concentration (50 μg/ml) that does not exert obvious toxic effects on OMK cells (Barahona *et al.*, 1977; Ablashi *et al.*, 1977a; Daniel *et al.*, 1980; Honess *et al*, 1982). The replication is still inhibited if PAA is applied at an advanced stage of cytopathic change, and the inhibitory effect is reversible by removing the drug. As is to be expected for any inhibitor of DNA replication, synthesis of structural polypeptides in the intranuclear and intracytoplasmic LA complex and MAs is arrested (Morgan and Epstein, 1977).

Two carcinogenic polycyclic aromatic hydrocarbons, benzo[a]pyrene and 3-methylcholanthrene, inhibit the infectious cycle of *H. saimiri* by a mechanism different from that of PAA. In the presence of these tumor initiators, EAs and intranuclear structural viral proteins are expressed, but MAs do not appear. These results indicate that some but not all late events of the replication cycle are affected. The inhibitory effects of benzo[a]pyrene and 3-methylcholanthrene on virus replication are not readily reversible (Pearson and Beneke, 1977). The inhibition of virus replication by these carcinogens stands in contrast to the enhancement of viral polypeptide synthesis by the tumor-promoting agent TPA (see Section III.D.3).

Replication of *H. saimiri* is also inhibited by cytosine arabinoside (Klein *et al.*, 1973; Ablashi *et al.*, 1974b), adenine arabinoside (ARA-A), and tilorone (Adamson *et al.*, 1972, 1974). Acycloguanosine (Daniel *et al.*, 1980; Honess *et al.*, 1982) and rifamycin (Adamson *et al.*, 1972) are relatively ineffective. Human leukocyte, lymphoblast, and fibroblast interferons suppress replication of *H. saimiri* and *H. ateles* in OMK cells at concentrations of 500 IU/ml (Laufs *et al.*, 1974a; Daniel *et al.*, 1979b, 1980, 1981; Lvovsky *et al.*, 1981). Interferon inducers such as polyriboinosinic:polyribocytidylic acid [poly(I:C)] have also been used to inhibit *H. saimiri* growth in OMK cells, and this effect is enhanced by DEAE–dextran (Barahona and Melendez, 1971; Lvovsky *et al.*, 1981).

E. Morphogenesis

The ultrastructure of *H. saimiri* and *H. ateles* in OMK cells has been described in several morphological studies (Morgan *et al.*, 1970, 1973, 1976; Heine *et al.*, 1971; Heine and Ablashi, 1974; King *et al.*, 1972; Friedman *et al.*, 1976; Banfield *et al.*, 1977; Tralka *et al.*, 1977; Luetzeler *et al.*, 1979). The morphogenesis follows the same general principles as that of other herpes-group viruses. Nucleocapsids are first formed in the nuclei of infected cells in large amounts. The nucleocapsids are approximately 100 nm in diameter, possess ring-shaped or dense nucleoids, and are intermixed with empty capsids. Enveloped virus particles are formed by budding from the nuclear matrix into the perinuclear cisterna. Thus, the virus envelope is derived from the internal lamella of the nuclear membrane. Most mature virus particles, 150–200 nm in diameter, are found in intracytoplasmic vacuoles of the endoplasmic reticulum or the Golgi apparatus. Virions in the perinuclear space have a regular, uniform circular symmetry, while cytoplasmic and extracellular particles appear to be elliptical, eccentric, or ragged. The changes in morphology during intracytoplasmic maturation steps were explained by the action of lysosomal enzymes in lytically infected monkey kidney-cell cultures (Morgan *et al.*, 1976).

The budding process of *H. saimiri* and *H. ateles* particles may also take place through the cytoplasmic membrane during relatively early phases of *H. saimiri* replication, often resulting in virions with eccentric nucleocapsids and crescentic, thickened outer envelopes (Morgan *et al.*, 1976; Smith *et al.*, 1979; Leutzeler *et al.*, 1979). Other reports described accumulations of naked nucleocapsids in the stroma of cytoplasm, mainly at advanced stages of cytopathic changes (Friedmann *et al.*, 1976; Tralka *et al.*, 1977). This may be due to aberrant processing of nucleocapsids, e.g., extension of nucleocapsids through nucleopores, partial intracytoplasmic disaggregation of enveloped particles, or fusion of viral envelope membranes with membrane of the endoplasmic reticulum. Also, nucleocapsids may be assembled in the cytoplasm; however, immunofluorescence studies with specific antisera against capsid protein have shown that the bulk of nucleocapsids must be formed at intranuclear sites (Morgan and Epstein, 1977). Another electron-microscopic study suggested that mature *H. saimiri* particles may also be formed by budding into intranuclear vacuoles (Heine *et al.*, 1971). However, it should be cautioned that deepenings in the inner lamella of the nucleolemma may be mistaken for intranuclear vacuolization in ultrathin-section pictures (King *et al.*, 1972; Morgan *et al.*, 1976).

Herpesvirus saimiri-infected monkey kidney-cell cultures contain "bizarre" aggregations of intranuclear material. The structures seem to be equivalents of the eosinophilic intranuclear inclusion bodies that are usually found in the nuclei of cells that are productively infected with herpesviruses. In monkey kidney cells that produce *H. saimiri*, the "bizarre" aggregations appear as bundles of fibrils, granular filaments, tubuli, or laminated stacks (Heine *et al.*, 1971; Morgan *et al.*, 1973, 1976; Friedmann *et al.*, 1976; Tralka *et al.*, 1977). Lamellae of the stacked forms have striations, periodicity, and thickness equal to those of the capsid of herpesvirus particles, and tubular and stacked complexes seem to be made up of the same basic structure. The unanimous interpretation is that all these aggregations are the result of aberrant assembly of viral nucleocapsid proteins.

IV. CLASSIFICATION OF ONCOGENIC PRIMATE HERPESVIRUSES

A. Phylogenetic Relationship between *Herpesvirus saimiri* and *Herpesvirus ateles*

As discussed in Section III, *H. saimiri* and *H. ateles* share a number of biological and molecular properties. Fleckenstein *et al.* (1978a) have further investigated the relatedness of these viruses by more detailed analyses of their DNAs. Under the annealing conditions used, 35% of *H. ateles* low-density DNA (L-DNA) was found to cross-hybridize with *H. saimiri* L-DNA. In contrast, L-DNA from *H. ateles* strain 73 cross-hybridized with strain 810 L-DNA to greater than 90%. Similarly, the extent of L-DNA cross-hybridization among *H. saimiri* strains was greater than 90%. The degree of mismatching in each of these heterologous hybrids was estimated by measuring the change in melting temperature (Δt_m) of isolated hybrids. According to Ullman and McCarthy (1973), each 1.5° Δt_m corresponds to approximately 1% mismatching. The average degree of mismatching between L-DNAs of *H. ateles* strains 73 and 810 was 2.3%; it was less than 2% among L-DNAs of *H. saimiri* strains and was 9% for *H. saimiri–H. ateles* heterologous L-DNA hybrids.

Similar analyses have been performed with viral repetitive high-density DNA (H-DNA). Under the hybridization conditions used, low cross-hybridization was detected between H-DNAs of *H. saimiri* and *H. ateles*—about 10%. Cross-hybridization of *H. ateles* strain 810 H-DNA with strain 73 H-DNA was greater than 90%, as was the cross-hybridization among H-DNAs of *H. saimiri* strains. The percentage mismatching in strains 810 and 73 H-DNA was only about 3.3%, and among *H. saimiri* strains, about 2.5%. The small amount of hybrid formed between *H. saimiri* and *H. ateles* exhibited about 13% mismatch.

These DNA sequence-homology results and the variation in restriction-endonuclease cleavage sites parallel the differences described for virion proteins—the two characterized *H. ateles* isolates exhibit greater differences than any of the *H. saimiri* isolates. It should be noted that all *H. saimiri* isolates have been obtained from the South American *Saimiri sciureus*, while *H. ateles* strain 810 was obtained from *Ateles geoffroyi* and *H ateles* strain 73 from *A. paniscus*.

These results coupled with the restriction-endonuclease analyses described in Sections III.B.1 and III.B.2 allow the following conclusions: First, *H. saimiri* and *H. ateles* are distinct viruses closely related in evolution. Second, the individual isolates of *H. saimiri* are strains of the same virus that are most easily distinguished by the sensitive technique of restriction-endonuclease analysis. The same holds for strains 810 and 73 of *H. ateles*. Third, sequences in H-DNA appear to have diverged at a faster rate than those in L-DNA. This is apparent both from analysis of various *H. saimiri* strains (Keil *et al.*, 1980) and from *H. saimiri–H. ateles* heteroduplex DNA molecules (Fleckenstein *et al.*, 1978a).

Another virus, *Herpesvirus aotus* type 2, shares in common with *H. saimiri* and *H. ateles* the overall genomic structure: a unique L-DNA and highly repeated H-DNA. No cross-hybridization has been detected, however, between DNA of *H. aotus* type 2 and *H. saimiri* or *H. ateles* (R. Rüger and B. Fleckenstein, unpublished).

B. Position of Oncogenic Herpesviruses from Nonhuman Primates within the Herpes Group

The herpesviruses are defined by a common morphology. Most published schemes for herpesvirus classification and the preliminary recommendation of the International Committee on Taxonomy of Viruses recognize three classes, termed alpha, beta, and gamma (see Chapter 1). The human herpes simplex viruses (HSVs) are the prototypes of the alpha class. This class also includes viruses similar to HSV from other species (e.g., pseudorabies of pigs, *Herpesvirus tamarinus* of New World primates, bovine mammillitis virus, equine herpes type I). The human cytomegalovirus is the prototype of the beta class, which includes similar members from other species (e.g., Old World primates, New World primates, and mice). The human Epstein–Barr virus (EBV) is the prototype of the gamma or lymphotropic class, which contains lymphotropic viruses of other species, including Old World primate viruses, *H. saimiri* and *H. ateles* of New World primates, and Marek's disease virus (MDV) of chickens.

The basis of this tentative classification scheme is mostly biological with a smattering of molecular relationships. Members of the alpha class have a tendency to remain latent in nerve cells *in vivo*. They exhibit rapid lytic growth in cell culture with broad host range. Cross-hybridization has been detected among many members of the alpha class (Ludwig *et al.*, 1972; Sterz *et al.*, 1973–1974; Desrosiers and Falk, 1981). Also, virion DNA of some members has been shown to exhibit an unusual two- or four-isomer arrangement with inversion in certain regions of the genome (for review, see Roizman, 1979). Most members of the beta class grow more slowly in cell culture with limited host range, and their genomes are considerably larger than members of the alpha class (usually about 150 vs. 100 Md). Members of the gamma class tend to remain latent in cells of the lymphatic series *in vivo*. Unlike members of the alpha and beta class, members of the gamma class produce tumors in experimental animals. This class, however, probably exhibits the greatest heterogeneity: (1) DNA of EBV cross-hybridizes significantly with DNA from the similar Old World primate viruses (Gerber *et al.*, 1976; Rabin *et al.*, 1978), but significant cross-hybridization with *H. saimiri* DNA has not been detected (B. Fleckenstein, unpublished). Even MDV and herpesvirus of turkeys, which cross-react antigenically, exhibit minimal (<4%) cross-hybridization in their DNAs (Lee *et al.*, 1979; Kaschka-Dierich *et al.*, 1979). (2) Permissive cell-culture systems do not exist for EBV or the Old

World primate gamma-class viruses, but *H. saimiri*, *H. ateles*, and MDV do grow permissively in monolayer cultures with narrow host range. (3) EBV and the gamma-class viruses of Old World primates are B-cell tropic, while *H. saimiri*, *H. ateles*, and MDV are T-cell tropic.

There are at least two problems with this and perhaps any herpesvirus classification scheme. First, biological properties of viruses are frequently overlapping and misleading. It is probably best that any classification scheme ultimately depend on carefully defined molecular properties. Second, the herpesviruses are a very diverse group, and simple classification schemes are bound to result in many exceptions. The evolutionary relationship of *H. saimiri* and *H. ateles* with other herpesviruses and definite classification should await more detailed molecular characterization.

Cross-hybridization under stringent conditions probably underestimates base-sequence homology (Yang and Wu, 1979). A minimal base-sequence homology is probably necessary to detect cross-hybridization under the usual rather stringent conditions. Hybridizations are usually performed 25–30°C below the t_m, since this gives optimal rates of reannealing (Wetmur and Davidson, 1968). A randomly distributed sequence divergence of 30% (70% sequence homology) would probably yield very little cross-hybridization at 25–30°C below the t_m of homologous hybrid (the usual conditions), since this temperature is above the t_m of heterologous hybrids. Thus, the absence or low level of cross-hybridization does not rule out base-sequence homologies, even as high as 70–80%. The possible relatedness of *H. saimiri* and EBV DNAs and that of other herpesvirus DNAs will be determined only when hybridization protocols are designed that allow for a lower sequence homology.

The genetic relatedness (by cross-hybridization) of baboon herpesvirus SA8 (35%), New World primate *H. tamarinus* (15%), and pseudorabies virus (7%) with the human HSV suggests a classic evolutionary lineage in parallel with the host species (Desrosiers and Falk, 1981). It is possible that such an evolutionary relatedness is not readily detected in the gamma class of herpesviruses because of a much faster rate of sequence divergence. The high degree of strain variability in individual *H. saimiri* isolates and low or absent cross-hybridization of DNAs, even from similar viruses of closely related species, are consistent with a fast rate of sequence divergence in the gamma class of herpesviruses.

V. PATHOLOGY OF NEOPLASTIC DISEASES CAUSED BY *Herpesvirus saimiri* AND *Herpesvirus ateles*

A. Lymphomas in Marmoset Monkeys

Tamarin marmosets (*Saguinus* spp.) are the most susceptible animals for the induction of lymphatic tumors by *H. saimiri* and *H. ateles*. All monkeys of the genus *Saguinus* [cottontop marmosets (Fig. 8A), white-lip marmosets, white-moustache marmosets] that were infected with wild-type *H. saimiri* or *H. ateles* have died from a rapidly progressing neoplasia of the lymphatic system, irrespective of the age of the animal or the amount of inoculum (see Table III). The survival time depends to some extent on the marmoset species and the infecting virus dose. Cottontop marmosets that are inoculated intramuscularly with 10^5 infectious units usually die between 3 and 4 weeks thereafter. Lower virus doses can result in a more prolonged course (up to 7 weeks), and the longest known survivor reached the 18th week. In white-lip marmosets, slower-progressing forms of lymphoma with fatal outcome later than 4 weeks after inoculation are more common.

The symptoms of tumor disease and pathology of the lesions in cottontop marmosets have been described comprehensively by Hunt *et al.* (1970). The animals develop hepatosplenomegaly and generalized enlargement of lymph nodes as early as 10 days after infection (Fig. 8B). The terminal stage is characterized by inactivity, food refusal, weakness, subnormal body temperature, lethargy, and progressive lymphadenopathy. On autopsy, the most striking macroscopic alterations are found in organs

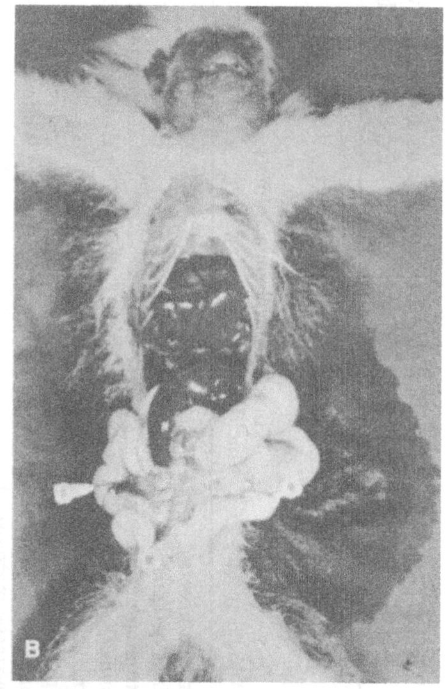

FIGURE 8. (A) Cottontop marmoset (*Saguinus oedipus*). (B) Cottontop marmoset, dying from malignant lymphoma 27 days after experimental infection with *H. saimiri*.

TABLE III. Oncogenicity of *Herpesvirus saimiri* and *Herpesvirus ateles* in New World Monkeys and Rabbits after Experimental Infection

	Herpesvirus saimiri			*Herpesvirus ateles*		
Animal species	Tumor incidence[a]	Survival in tumor-developing animals (days)	References	Tumor incidence[a]	Survival in tumor-developing animals (days)	References
Tamarin marmosets						
Saguinus oedipus (cottontop marmosets)	+	18–126	Hunt *et al.* (1970), Deinhardt *et al.* (1974)	+	14–40	Melendez *et al.* (1972a), Falk *et al.* (1974a)
Saguinus fuscicollis, S. nigricollis (white-lip marmosets)	+	19–116	Wolfe *et al.* (1971a)	+	26–28	Falk *et al.* (1974a)
Saguinus mystax (white-moustache marmosets)	+	70	Wolfe *et al.* (1971a)	NT	—	—
Common marmosets (*Callithrix jacchus*)	±	22–42	Wright *et al.* (1977), Melendez *et al.* (1971)	+	36–104	Laufs and Melendez (1973b)
Owl monkeys (*Aotus trivirgatus*)	±	13–330	Cicmanec *et al.* (1974)	+	21–28	Hunt *et al.* (1978)
Spider monkeys (*Ateles geoffroyii*)	±	129–179	Hunt *et al.* (1972b)	0	—	
Capuchin monkeys (*Cebus albifrons*)	±	18–20	Melendez *et al.* (1970a)	NT	—	—
Howler monkeys (*Alouatta caraya*)	+	36–52	Rangan *et al.* (1977)	NT	—	—
Squirrel monkeys (*Saimiri sciureus*)	0	—		0	—	—
Rabbits (*Oryctalogus cuniculus*)	±	17–196	Daniel *et al.* (1974a)	±	20–100	Daniel *et al.* (1977)

[a] (+) Malignant lymphoproliferative neoplasias developing consistently after infection; (±) variable tumor incidence; (0) no tumors.

of the reticuloendothelial system, spleen, lymph nodes, and thymus, as well as in parenchymatous organs, mainly liver, kidneys, and adrenals. Peripheral and visceral lymph nodes are 2–4 times enlarged, often coalescent, with hemorrhagic areas. Similarly, in young animals, the swollen thymus shows hemorrhagic foci. The spleen, deep red or black, is enlarged from 2 to 10 times the normal size; an extraordinary splenomegaly is quite characteristic for lymphoma in cottontop marmosets. The liver, often light red-brown, shows a prominent gray and red reticular pattern. Kidneys are swollen, with focal grayish mottling of the cortex and hemorrhages. Often, enlarged adrenals show extensive hemorrhages and focal grayish mottling of the cortex, and cortex and medulla are not discernible. Lesions are also found macroscopically in bone marrow, tonsils, lung, prostate, testicle, epididymis, and choroid plexus, and in the connective tissues and muscles surrounding the inoculation site.

Microscopic examination reveals extensive cellular infiltrations and organ replacement as the basic features. The infiltrates are dense accumulations of lymphoreticular cells of relatively uniform morphology, mostly in solid sheets or clumps. The infiltrating cells are classified as reticulum cells or immature cells of the lymphatic series, on the basis of histochemical and ultrastructural characterization (King and Melendez, 1972; King *et al.*, 1972). The cells possess large oval leptochromic nuclei with pronounced nucleoli; mitotic figures are frequent; viral structures were never found in these cells by either light or electron microscopy.

The infiltration of lymphatic and parenchymatous organs in cottontop marmosets varies from animal to animal. The most extensive infiltrations of the liver are usually seen in the periportal area, extending into the sinusoids of the hepatic lobules. Widespread necrosis of hepatocytes may accompany the picture. Similarly, kidney tubuli are often necrotic if the organ is heavily infiltrated and hemorrhagic. Adrenals may show extensive infiltration and replacement with a relative sparing of the zona fasciculata of the cortex; alternatively, extensive necrosis and hemorrhage with moderate infiltrations may dominate. In the lymph nodes, the cytoarchitecture may be completely obscured. Specific tissues are entirely replaced by solid sheets of reticulum cells that also infiltrate into capsule and perinodal tissues. Sometimes, lymph-node infiltrations are confined to perifollicular areas, and moderate amounts of reticulum cells are found in medullary and cortical sinuses. The bulk of splenic tissues is usually replaced by infiltrating cells, and follicles are necrotic. In the thymus, cortex and medulla are no longer discernible; foci of necrotic cells are found interspersed between solid sheets of infiltrating reticulum cells.

A leukemic reaction is common in the final stage of disease in cottontop marmosets, with a total white blood cell count between 16,000 and 100,000/mm^3. The differential blood picture shows from 10 to more than 50% immature cells that are described as prolymphocytes, lymphoblasts, or reticulum cells. Occasionally, low lymphocyte counts in the blood with pronounced neutropenia are found in cottontop marmosets. Anemia occurs in the last days antemortem. Necrosis of parenchymatous and lymphatic cells is found to be a consistent feature in organs of fatally diseased cottontop marmosets. This cannot be explained by compression or multiple thrombosis alone, since the lesions are usually not infarctlike and are often more pronounced in the less infiltrated organs. Presumably, toxic effects, immune reactions, or direct virus effects are responsible for the necrosis. The presence of giant cells that are seen occasionally in kidneys may also be reminiscent of productive virus replication. In general, monkeys inoculated with higher virus doses show more necrosis and less infiltration than animals that received less virus.

The neoplastic disease caused by *H. saimiri* in cottontop marmosets is classified as malignant lymphoma of the reticulum-cell or poorly differentiated lymphocytic type (Hunt *et al.*, 1970; King and Melendez, 1972; King *et al.*, 1972). However, it should be noted that this tumor form is different from any other type of lymphatic neoplasia described before. It has unique features like extensive necrosis of parenchymas and an extremely rapid fatal outcome.

The neoplastic disease caused by *H. saimiri* in white-lip marmosets shows a wider variability in the clinical and pathological picture than the relatively uniform malignant lymphoma of cottontop marmosets (Wolfe *et al.*, 1971a). White-lip marmosets may also die after a very short time from an acute lymphoma of the reticulum-cell type, usually without splenomegaly. In other cases, the course is prolonged and the pathology is different: The neoplastic cell type is a medium-sized lymphoblast or a rather well-differentiated lymphocyte. Liver and spleen are hardly enlarged, much less infiltrated, and there is little destruction of cytoarchitecture in lymphatic organs. Bone marrow is heavily infiltrated, and leukemia is more pronounced than in cottontop marmoset tumor tissues. White blood cell counts up to 340,000/mm^3 have been reached.

Common marmosets (*Callithrix jacchus*) have reacted variably in different studies when inoculated with *H. saimiri*. Laufs and Fleckenstein (1973) found that nine adult animals infected with *H. saimiri* (S295C) remained healthy, but were persistently infected and developed neutralizing antibodies. The virus could be isolated from peripheral white blood cells (Laufs and Melendez, 1973a; Laufs *et al.*, 1974b). However, a common marmoset inoculated with the same virus strain at the age of 32 days came down 42 days later with an acute malignant lymphoproliferative disease characterized by massive reticulum-cell infiltrations of spleen and lymph nodes (Laufs and Fleckenstein, 1973). Ablashi *et al.* (1978) also made the observation that adult common marmosets can be refractory to induction by *H. saimiri* (strain S295C). The inoculated animals developed antibodies against viral early antigens (EAs) and late antigens (LAs) without tumor response.

Wright *et al.* (1977) have found that *H. saimiri* strain 11 produced malignant lymphoproliferative disease in five of five adult common marmosets. The susceptibility of common marmosets to the S295C strain in these studies has not yet been definitively determined (Wright *et al.*, 1977; R. Desrosiers and L. Falk, unpublished). The attenuated variant of strain 11 (see Section VI.D) produced a persistent infection of common marmosets without lymphoma and death (Wright *et al.*, 1977).

Herpesvirus saimiri strain OMI was found to be highly oncogenic in the *C. jacchus* colony of the New England Primate Research Center (R. Hunt and M.D. Daniel, personal communication). The tumors resemble very closely the malignant lymphomas of the reticulum-cell type that are usually found in cottontop marmosets. It thus appears that the sensitivity of common marmosets is highly strain-dependent, and it is possible that other factors such as virus-passage history and age of the animal could also play a role.

Persistent infection with *H. saimiri* attenuated strain 11 does protect common marmosets from challenge by wild-type oncogenic virus (Wright *et al.*, 1980) (see Section VI.D). It is not known whether persistent infection of common marmosets with strain S295C can protect against challenge by strain 11.

Herpesvirus ateles is similar to *H. saimiri* in its oncogenic properties for tamarin marmosets. All *H. ateles*-infected animals die within 2–5 weeks. Symptomatology and pathological anatomy show many parallels, but also a few distinct differences (Hunt *et al.*, 1972a; Falk *et al.*, 1974a). The neoplastic disease caused by *H. ateles* in cottontop marmosets was classified as an acute lymphomatic leukemia or malignant lymphoma of the poorly differentiated type (Hunt *et al.*, 1972a). *Herpesvirus ateles* lymphoma in white-lip marmosets is very similar or identical (Falk *et al.*, 1974a). Most monkeys die with the full picture of advanced lymphoma; the course is shortened in some animals by concurrent pulmonary complications. Gross pathology shows a rather uniform pattern in all monkeys. General lymphadenopathy with lymph nodes enlarged 2–5 times, splenomegaly, and swollen thymus with lymph nodes enlarged 2–5 times are the most predominant features. A slightly enlarged tan-yellow liver with reticular pattern on capsular and cut surfaces and hemorrhages into adrenals, colon mucosa, and the skin are common findings.

The microscopic picture of *H. ateles* lymphoma in tamarins is dominated by a rather uniform type of infiltrating cells, presumably lymphoblasts. Dense sheets of invasive cells obscure the normal structure of thymus and lymph nodes. The distribution of infiltrations in spleen, kidneys, adrenals (with preference for the corticomedullary junction), and liver (mostly periportal areas) resembles a similar pattern in the *H. saimiri*-induced lymphoma of cottontop marmosets. Further invasion occurs in heart muscle, lungs, salivary glands, pancreas, bone marrow, and choroid plexus. Lymph nodes of some animals show a "starry-sky picture" in histology due to large mononuclear cells (macrophages) with basophilic cytoplasmic inclusions scattered among the dark background of densely packed neoplastic lymphoblasts. All animals have shown sharply demarcated foci of coagulation necrosis that are presumably due to infarctions by multiple thromboses of surrounding blood vessels. Most animals have a moderate leukemic reaction with total white blood cell counts between 14,000 and 25,000/mm^3. Sometimes, a few neoplastic blood cells appear in the periphery; sometimes up to half the peripheral white blood cells are lymphoblasts or atypical lymphocytic cells.

In summary, the comparison of *H. ateles*-induced tumors in tamarin marmosets with *H. saimiri* lymphomas reveals slight differences only. The clinical picture is more uniform in *H. ateles* lymphoma. In histology, the neoplastic cell type is a poorly differentiated lymphoblast, more uniform in comparison to some variability in *H. saimiri* tumors. In general, *H. ateles*-infected marmosets have a leukemic reaction, but with lower cell counts than *H. saimiri*-tumor animals. Multiple infarctions due to thrombosis seem to be somewhat restricted to *H. ateles* neoplasms. However, it should be noted that from the limited number of studies, we do not know what the range of variability between the two viruses will be, to what extent the clinical and pathological picture can overlap, and what factors contribute to the variability of tumor response.

Common marmosets react with tumor disease on experimental infection with *H. ateles* strain 810 (Laufs and Melendez, 1973a,b). When *H. saimiri* strain S295C persistently infected common marmosets were superinfected with virulent *H. ateles*, they developed malignant lymphomas (Laufs and Melendez, 1973a). Apparently, the *H. saimiri* carrier state does not convey protection against the tumor-inducing potential of *H. ateles*.

B. Leukemia and Lymphoma in Owl Monkeys

Herpesvirus saimiri causes a neoplastic disease in owl monkeys that progresses to death within a few weeks or months (Fig. 9). However, the tumor response upon experimental infection is variable. Tumor susceptibility, course of disease, and histopathology were different in several reports. This variability may depend on a combination of any or all of the following factors: virus strain, passage history of the virus, infectivity titer of the inoculum, route of inoculation, age of the animal, and the amazing intraspecies variation of owl monkeys (Ma *et al.*, 1976; Hunt *et al.*, 1976).

The first reports on *H. saimiri* infections in owl monkeys indicated that this species can be highly susceptible to early-passage *H. saimiri* prime strain (S295C) (Melendez *et al.*, 1969b, 1970c; Hunt *et al.*, 1970). All of ten monkeys died within 4 weeks after experimental infection. In subsequent studies, usually 50–80% of owl monkeys died from lympho-

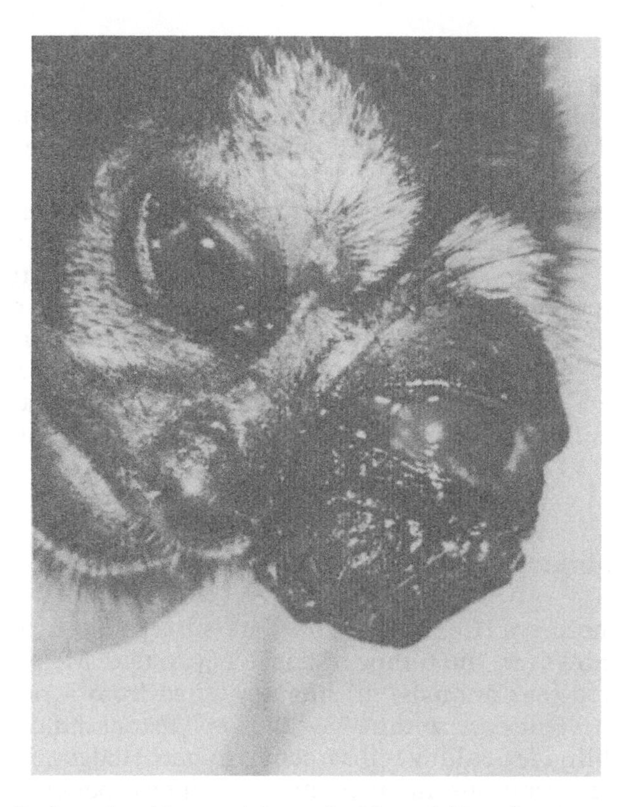

FIGURE 9. Owl monkey (*Aotus trivirgatus*) with partial lymphosarcoma after spontaneous infection with *H. saimiri*. Courtesy of M. D. Daniel.

proliferative disease after a single inoculation with *H. saimiri* (Melendez *et al.*, 1971; Ablashi *et al.*, 1971a; Cicmanec *et al.*, 1974). In most of these experiments, the average survival period in tumor-developing owl monkeys was several months long, in some cases up to 12 months.

Lymphomatous owl monkeys may have enlarged peripheral and mesenteric lymph nodes and hepatosplenomegaly of variable degree, histology being characterized by reticulum-cell infiltration similar to the neoplastic disease of tamarin marmosets (Hunt *et al.*, 1970). In other cases, extensive infiltrations can be confined to liver and kidney (Melendez *et al.*, 1970c). Often, infiltrations are rather mature cells of the lymphatic series. The cytoarchitecture of lymph nodes is destroyed by focal or diffuse infiltrations, and extensive infiltration is seen in perinodal tissues, spleen, and small intestine. There is less destruction of parenchymatous organs, with moderate infiltrations in kidneys, adrenals, heart, lungs, pancreas, liver, and other organs. Hepatic sinusoids are excessively filled with lymphocytes (Cicmanec *et al.*, 1974).

About 75% of the tumor-developing animals acquire leukemia. Lymphoblasts and prolymphocytes are predominant in the white blood picture, and the total white blood cell count reaches up to 175,000/mm^3. Some animals have leukemoid reactions with alternative periods of disease and remission. It is a characteristic feature of leukemias in owl monkeys that the absolute leukocyte counts drop in the last few days *ante mortem*.

Owl monkeys are also susceptible to *H. ateles*, developing tumors within 3–4 weeks. Pathology shows a significantly different picture in comparison to *H. saimiri* tumors. The distribution of infiltrations is unique. Invasion is most extensive in the kidneys, where the cells form expanding nodules in the cortex and medulla; it is minimal or absent in adrenals, lungs, and liver. Most important, no abnormalities have been noted in peripheral blood of any monkey. The *H. ateles* neoplasia in owl monkeys is classified as lymphoblastic or stem-cell lymphoma (Hunt *et al.*, 1978).

C. Lymphoproliferative Diseases in Other Primates

Capuchin monkeys (*Cebus albifrons*) are susceptible to *H. saimiri* (strain S295C); however, the tumor response is variable. Melendez *et al.* (1970a) inoculated four animals; all monkeys died from a rapidly fatal lymphoproliferative disease within 18–20 days. Histopathology showed a picture resembling reticulum-cell sarcoma, in part Hodgkin's sarcoma, with extensive infiltrations in liver, lungs, kidneys, pancreas, spleen, thymus, and lymph nodes. The four monkeys had been infected with an apathogenic cebus monkey adenovirus several months before. Rabin *et al.* (1975b) also inoculated five capuchin monkeys with strain S295C; all animals seroconverted to significant anti-LA and anti-EA titers and developed persistent infection, but no tumor disease.

Six spider monkeys (*Ateles geoffroyi*) were used in a study by Hunt *et al.* (1972b). Two of the animals developed malignant lymphomas. Macroscopic and histological examination of one animal resembled closely the lymphoblastic type of lymphoma of *H. saimiri*-infected marmoset and owl monkeys. Necropsy of the other animal revealed a large solid tumor. The esophagus was surrounded by the solid masses of a lymphosarcoma that infiltrated into the stomach and neighboring organs. The observation of a malignant *H. saimiri*-induced neoplasia in spider monkeys is of particular interest, since these animals are the natural hosts of *H. ateles*. *Herpesvirus ateles*-associated tumors have been observed neither in squirrel monkeys nor in spider monkeys (Melendez *et al.*, 1972e; Falk *et al.*, 1974a).

Howler monkeys (*Alouatta caraya*) are another species in which *H. saimiri* exerts oncogenic effects. Rangan *et al.* (1977) described a malignant lymphoproliferative disease with terminal leukemia. All animals in a group of four succumbed within 5–8 weeks.

Old World primates are usually refractory to *H. saimiri*-induced disease. Melendez *et al.* (1970b) found a neoplasia in one of two African green monkeys (*Cercopithecus aethiops*) that developed as a reticuloproliferative disease similar to that observed in capuchin monkeys after *H. saimiri* infection. In other studies, African green monkeys were not susceptible, not even an antibody response being detected after *H. saimiri* inoculation (Laufs and Fleckenstein, 1973; Rabin, 1971). No other cases of *H. saimiri* tumors were observed in virus-inoculated macaques, baboons, or chimpanzees (Melendez *et al.*, 1970b, 1972d).

Humans are more closely related to Old World monkeys and apes than to the South American primates. It must be considered whether *H. saimiri* and *H. ateles* could eventually be oncogenic in the human organism. The number of persons who might have been exposed to the virus is not small. It includes zoo personnel, animal dealers, persons living in close contact with squirrel monkeys, experimenters working with squirrel and spider monkeys, and laboratory personnel handling infected cell cultures. Sporadic accidental self-inoculations have occurred, and bites by these monkeys are not rare. So far, antibodies against *H. saimiri* have never been detected in the sera of these contact persons. So far, there is no known human case of a malignant lymphatic tumor that could be traced to an *H. saimiri* or *H. ateles* infection. Perhaps the viruses are not able to replicate in the human organism; permissive cells or target cells for transformation may lack appropriate receptors for *H. saimiri* infection. Nevertheless, the odds of a human infection, possibly under conditions of immunosuppression, must not be ignored, since infection could be a statistically infrequent event. Immune or tumor response could be dose-dependent. The oncogenic herpesviruses of nonhuman primates are regarded as moderate-risk tumor viruses, and the appropriate safety measures should be strictly observed by laboratory workers and personnel handling squirrel and spider monkeys.

D. Neoplasias in Rabbits

Both *H. saimiri* and *H. ateles* are able to induce malignant tumors in rabbits, extending the spectrum of susceptible animals beyond the order Primates. The incidence of neoplastic disease from *H. saimiri* is between about 20 and 100%. Lymphoproliferative diseases were induced in New Zealand white rabbits and in the inbred strains ACCRB and III/J by inoculation with *H. saimiri* strain S295c or SMHI (Daniel *et al.*, 1974a, 1975; Hunt *et al.*, 1975a; Rangan *et al.*, 1976; Ablashi *et al.*, 1980b). Among many other possible factors, tumor incidence seems to be dependent on the rabbits used for inoculation experiments (M.D. Daniel, personal communication). Moreover, the chance of tumor induction is significantly influenced by the route of inoculation. Intraveneous infections result in a significantly higher tumor rate than subcutaneous or intramuscular inoculations (Daniel *et al.*, 1975; Rangan *et al.*, 1976).

Rabbits bearing *H. saimiri* lymphomas have succumbed 17–196 days after experimental inoculation. Few symptoms are seen in the rabbits that die early; the animals with extended clinical course contract a severe conjunctivitis, nasal discharge, and dyspnea resulting from lymphocytic invasion of ocular and nasal tissues. This picture has not been observed in lymphomatous monkeys. Pathology shows a disease pattern similar to that of the primate tumors, characterized by massive diffuse infiltration of most organs and tissues with lymphocytic or lymphoblastic elements. A slight infiltration of the brain is a unique feature in rabbit lymphomas, since it has never been noted in primates. A number of rabbits show a terminal leukemic reaction with up to 150,000 leukocytes/mm^3; most of these cells are mature small lymphocytes (Hunt *et al.*, 1975a; Rangan *eg al.*, 1976).

The incidence of *H. ateles* lymphomas apparently depends on the rabbit strain. New Zealand white rabbits are relatively resistant to tumor induction; the ACCRB inbred strain is highly susceptible (Daniel *et al.*, 1977). ACCRB rabbits usually die within 3–14 weeks with a lymphoblastic type of lymphoma. Diffuse invasion of lymphatic organs, liver, and skin predominates; however, minor infiltrations are found in other tissues, and leukemic reactions have never occurred.

The rabbit system may allow experimenters to conduct tumor studies with *H. saimiri* and *H. ateles* at considerably less expense. This is of particular importance, since many New World primates, including all tamarins, are on the list of endangered species. Importation of these animals is strictly limited, and they are available only in minimal numbers for experimental purposes.

VI. MECHANISMS OF PATHOGENESIS IN ONCOGENIC TRANSFORMATION AND TUMOR INDUCTION

A. Permissive and Transforming Infection in Tumor-Developing Primates

It is not yet known in what types of cells, tissues, or organs *H. saimiri* and *H. ateles* replication occurs. There are two possibilities: (1) Virus infection and replication are confined to cells of the lymphatic system; in this case, a certain lymphocyte population would be the sole target for a limited productive infection and for transformation. (2) Alternatively, other types of cells, epithelial or mesenchymal, may be permissive for a productive virus replication, thus amplifying the amount of infectious virus far above the original inoculum. This would imply that in parallel to the events of oncogenic transformation and tumor development, a concomitant lytic virus replication would keep the remaining lymphoid target cells under a permanent shower of infectious virus.

No experiments have been performed to decide between these two possibilities. The first hypothesis may be supported by the fact that all attempts to isolate cell-free virus from serum and oral swabs of tumor-developing owl monkeys and marmosets remained unsuccessful (Deinhardt *et al.*, 1973b, 1974; Neubauer *et al.*, 1974c). *Herpesvirus saimiri*-infected marmosets and owl monkeys do not excrete appreciable amounts of infectious virus. Transmission of *H. saimiri* between infected marmosets or owl monkeys and their cagemates was not observed (Wolfe *et al.*, 1971a; Ablashi *et al.*, 1976). Similarly, infected rabbits are not contagious for their cagemates (M. D. Daniel, personal communication). Only one case is documented wherein a cottontop marmoset acquired *H. ateles* infection and tumor disease from an experimentally infected marmoset that was kept in the same cage. However, the failures of attempts to isolate cell-free virus do not exclude either an early productive infection in some organs or transient viremia.

The hypothesis that a productive virus replication takes place during the incubation period of the tumor disease is supported by a number of arguments: (1) The extraordinarily high susceptibility of marmosets and owl monkeys to *H. saimiri* becomes plausible assuming that the virus is amplified in a productive cycle after entrance into the host organism. Minimal amounts of infectivity about [1 plaque-forming unit (PFU)] are sufficient to induce tumors and immune response in owl monkeys (Ablashi *et al.*, 1973) and cottontop marmosets (Laufs and Steinke, 1975a). The high susceptibility is also documented by the observation that animals acquire infection and tumor disease after close contact with squirrel monkeys (Hunt *et al.*, 1973; Barahona *et al.*, 1975; Rabin *et al.*, 1975a). (2) *Herpesvirus saimiri* replicates in a relatively wide spectrum of permissive cell cultures. Epithelial cells from owl monkeys and marmosets are efficient virus producers in tissue culture. In view of the situation with other herpesviruses, it seems unlikely that the host range of *H. saimiri in vitro* would be so much broader than *in vivo*. (3) the pattern of humoral antibody response in tumor-developing marmosets and, especially, owl monkeys is most easily explained by the notion that the virus grows at early stages in cells other than the lymphoid cells that are the target for transformation (see below). High titers of antibodies against structural viral antigens (anti-LA, anti-MA) are independent of the number of transformed cells in the organism and do not reflect whether a quick or delayed tumor response will follow or whether the animals are refractory to tumor induction by the virus (Pearson *et al.*, 1974). (4) There was no significant difference in the usual humoral immune response whether animals were inoculated with virus particles or with infectious DNA (M–DNA) (Fleckenstein *et al.*, 1978b).

Pathogenesis in rabbits may be different; a lytic replication may not be involved. Epithelial rabbit cells in cell culture are not permissive for *H. saimiri*. Rabbits do not develop neutralizing antibodies against *H. saimiri* until shortly before the tumor appears, and virus isolations from peripheral lymphocytes remain negative until that time, irrespective of the duration of the latency period (Daniel *et al.*, 1974a).

Thymus-dependent lymphocytes (T lymphocytes) seem to be the primary and sole target cells for transformation by *H. saimiri* and, similarly, *H. ateles*. When B and T cells of infected marmosets were separated, viruses was rescued from T-cell fractions only (Wright *et al.*, 1976). Lymphocyte surface markers have been determined in virus-transformed lymphoid cells from owl monkeys and marmosets, including infiltrating cells from lymphatic organs, peripheral lymphoblasts, and tumor-cell lines. All cells have consistently shown T-cell-specific membrane markers of primate cells. They form nonimmune rosettes with sheep erythrocytes, do not synthesize substantial amounts of immunoglobulins, and do not have complement receptors (Wallen *et al.*, 1973). The consistency in the T-cell phenotype of all *H. saimiri*- and *H. ateles*-carrying lymphocytes in marmosets and owl monkeys may indicate a strong general T-cell tropism of the viruses, rather than a merely incidental expression of T-cell properties as a result of oncogenic transformation in nonpredetermined lymphocyte precursors. Epstein–Barr virus (EBV) infects only lymphocytes with B-cell properties, irrespective of the primate species from which the lymphoid cells are derived (Robinson *et al.*, 1977). Cell-membrane receptors for EBV and complement (C3) receptors are closely associated, like parts of one single membrane structure. This supports the assumption that the membrane receptor for EBV is present in B cells but not in T cells (Yefenof *et al.*, 1976). A similar receptor specificity might be responsible for the T-cell preference of *H. saimiri* (Johnson *et al.*, 1980). However, it may also be possible that lymphoid elements other than T cells may be infected to a limited extent without transformation, amplification, and outgrowth as tumor cells (Falk *et al.*, 1979; Daniel *et al.*, 1979c).

The different cell tropisms of EBV and the oncogenic herpesviruses of nonhuman primates imply basic differences in pathogenesis between human EBV-associated lymphoma and *H. saimiri*-induced neoplasia. EBV-transformed B cells seem to be mostly eliminated by T-cell mediated cellular immune mechanisms within a few weeks or months after primary infection of humans. Development of a clone of EBV-transformed cells into tumor is a rare event; development of Burkitt lymphoma occurs long after EBV primary infection; when the tumor appears, cell-free transforming and infectious virus is presumably not involved. This is consistent with the fact that Burkitt lymphoma is a monoclonal tumor. In contrast, *H. saimiri* and *H. ateles* directly affect the cellular immune systems of tumor-developing animals. Primary infection is accompanied by a massive and synchronous infection and transformation of T lymphocytes, immediately followed by foudroyant neoplastic growth. Tumor development is an event of high likelihood, and tumors are probably polyclonal, at least in tamarins (Chu and Rabson, 1972; Marczynska *et al.*, 1973). This emphasizes clearly that the analogy between Burkitt lymphoma and the "primate models" *H. saimiri* or *H. ateles* lymphoma should not be taken too far.

Tumorous organs of *H. saimiri*-infected marmosets contain considerable amounts of viral DNA (see Section VI.F.2). The presence of multiple DNA copies in tumor cells stands in remarkable contrast to the low expression of viral genes. Cell-free virus has never been isolated from cell-free extracts direct from tumor tissue or cells, and virus particles or structural viral components cannot be detected in fresh tumor cells by morphological or immunological approaches (Deinhardt *et al.*, 1973b). This might be due in part to insensitive methodology; however, it is plausible to assume that the persisting virus in the tumor cells is in a state of repression (Deinhardt *et al.*, 1974). This becomes evident if tumor cells are explanted into cell culture. At first, lymphoid cells in culture do not produce virus. After a brief period, they seem to gradually lose the suppressive influence that constrains virus expression *in vivo*. A few hours or days after initiation of the cell culture, transformed lymphoid cells start to synthesize viral structural antigens and infectious virus (Laufs and Fleckenstein, 1972; Falk *et al.*, 1972c; Pearson *et al.*, 1972; Giddens, 1975).

We do not know how strictly the production of infectious virus by tumor cells is suppressed in lymphoma-bearing animals. Some leakiness is possible, at least eventually under certain conditions. Marczynska *et al.* (1973) transplanted tumor tissues between marmosets of different karyotypes; the recipient animals developed lymphomatous tumors of their own karyotype, indicating that the lymphomas were not the result of outgrowth of donor cells. Small amounts of infectious virus could have been transferred with the tumor-cell inoculum, or the graft cells could have produced cell-free virus in the seronegative host organism, resulting in oncogenic transformation of recipient cells.

Common marmoset, owl, and capuchin monkeys become infected when inoculated with *H. saimiri*. As outlined in Sections V.A–C, some animals remain persistently infected with *H. saimiri* without symptoms of malignant lymphoma or other disease. The virus can easily be isolated by cocultivation of their lymphocytes with permissive monolayers (Laufs and Fleckenstein, 1973; Laufs *et al.*, 1974b; Rabin *et al.*, 1975b; Armstrong *et al.*, 1976). Persistently infected common marmosets may continuously have 10^4–10^5 virus-carrying lymphocytes/ml in the blood. The cells can be detected and quantitated by explantation and cocultivation with permissive cells in infectious-center assays. This carrier state may be analagous to the persistence of *H. saimiri* in the leukocytes of the natural host, which does not seem to develop *H. saimiri* lymphoma. A number of possible explanations can be given for the failure of persistently infected common marmoset, owl, capuchin, and squirrel monkeys to develop malignant lymphomas. The constraints of the cellular immune system and immunosurveillance may prevent outgrowth of these cells as a tumor. However, it has not been possible so far to demonstrate unambiguously a strong cell-dependent immune response against *H. saimiri* in infected monkeys (Pearson and Davis, 1976; Neubauer *et al.*, 1980b). It therefore seems possible as well that the state of persistent viral DNA in lymphoid cells in squirrel monkeys and in the tumor-resistant common marmosets and capuchin monkeys is not equivalent to the virus genome–host cell interaction in oncogenic transformation.

The isolation of numerous oncornaviruses from Old and New World monkeys and apes in recent years raises the question whether RNA tumor viruses could be involved in the etiology of *H. saimiri*- and *H. ateles*-induced tumors. Neubauer *et al.* (1974a,b) searched intensively for oncornavirus particles and reverse transcriptase activity in tumor cells and transformed cell cultures; however, these workers could not find any traces of RNA tumor viruses. Attempts to demonstrate oncornavirus particles with a simultaneous detection test (Schlom and Spiegelman, 1971) for 70 S RNA complex and reverse transcriptase also remained unsuccessful (Laufs and Steinke, 1975b). Nucleic acid hybridizations also gave negative results (Neubauer *et al.*, 1974a). This, however, does not exclude the presence of oncornaviruses, since the putative RNA tumor viruses of New World primate tumors do not necessarily possess nucleic acid sequences that are homologous to the radioactive probes that were used in these studies and, similarly, negative radioimmunoprecipitations are of limited value to disprove that oncornavirus particles or structural proteins are synthesized in *H. saimiri*-induced tumors. There is evidence that tumor induction by *H. saimiri* does not depend on oncornaviruses that may be transmitted as contaminants of *H. saimiri* inocula: *Herpesvirus saimiri* purified in the ultracentrifuge by consecutive density runs (Laufs and Fleckenstein, 1972) and virus obtained through transfection with isolated M-DNA (Fleckenstein and Bornkamm, 1975) are also oncogenic. Furthermore, we have found that it is possible to induce tumors by injecting purified *H. saimiri* DNA into cottontop marmoset monkeys (see Section VI.E) (Fleckenstein *et al.*, 1975b, 1978b). However, Rangan (1976) found oncornavirus particles of C-type morphology in a tumor-cell line from a howler monkey (*Alouatta caraya*) and we isolated an endogenous owl monkey C-type virus (OMC-1) from owl monkey kidney cells (Todaro *et al.*, 1978; Barbacid *et al.*, 1980). At present, it is purely a matter of speculation whether persisting viral genomes of such retroviruses could be activated in *H. saimiri*-transformed cells or whether proviral sequences could even be integrated into the genomes of oncogenic primate herpesviruses. So far, there is no evidence that oncornaviruses play a role in *H. saimiri*- and *H. ateles*-induced tumors.

B. Immune Reactions in Neoplastic Diseases

1. Humoral Immunity

Most studies on antibody reactions against *H. saimiri* have been conducted in owl monkeys, since the survival period in tamarin marmosets is usually too short to allow a sufficient humoral immune response. Antibodies against structural viral antigens, late antigens (LAs) and membrane antigens (MAs), usually become detectable within 2 months after infection of owl monkeys, sometimes as early as 2 weeks postinfection, but sometimes delayed until the 4th month (Klein *et al.*, 1973; Pearson *et al.*, 1973; Prevost *et al.*, 1976). The anti-LA titers range from 1:40 to 1:640 by indirect immunofluorescence. Anti-LA and anti-MA develop in parallel, but it is not known whether the antibodies are identical or different, since the LA and MA complexes have not yet been characterized. There is also a good correlation in the time course between the antibody titers against structural viral antigens detected by neutralization and indirect immunofluorescence tests. Titers of neutralizing antibodies and anti-LA and anti-MA titers measured by immunofluorescence are independent of tumor development; owl monkeys with progressing lymphomas and leukemias and tumor-resistant animals produce these antibodies equally well (Pearson *et al.*, 1973, 1974). Antibodies against LAs, and MAs of *H. saimiri* detected in owl monkey sera do not exert appreciable complement-dependent cytotoxic effects against cells expressing viral antigens (Neubauer and Rabin, 1979).

Prevost *et al.* (1975, 1976) developed as assay to detect cytotoxic antibodies in the presence of normal peripheral-blood lymphocytes from rhesus monkeys. This type of cytotoxic reaction, designated as antibody-dependent lymphocyte cytotoxicity (ADLC), is mediated by the interaction of antigen–antibody complexes with lymphoid cells bearing constant-fragment (Fc) receptors. Sera from tumor-bearing animals caused the rapid destruction of lytically infected cells *in vitro* by normal peripheral lymphocytes from rhesus monkeys. ADLC titers in general increased along with anti-MA determined by immunofluorescence, indicating that antibodies detected by immunofluorescence could also be involved in the ADLC reaction. However, in a number of the tumor-developing animals, the sera were discordant in the two assays, suggesting that other factors may be relevant in mediating ADLC. The two reactions may in part detect different antibody populations directed against different antigenic groups in the membranes of *H. saimiri*-infected and -transformed cells. ADLC titers increase in tumor-developing animals. Tumor-resistant owl monkeys remain at relatively low or moderate levels in ADLC assays, but remarkably high titers (up to almost 1:500,000) are found in the sera of tumor-developing animals. In owl monkeys with intermittent leukemoid reaction, a decrease in ADLC titers is observed during remissions (Wallen *et al.*, 1976). So far, it has not been possible to find virus-instructed MAs in fresh tumor cells. Thus, it is not known whether reactions of the ADLC type are of any importance in the *in vivo* defense of the host organism against *H. saimiri*-transformed cells.

Titers of antibodies against early antigens (anti-EA) detected by immunofluorescence also correlate with course and progression of neoplastic disease in owl monkeys (Klein *et al.*, 1973). Anti-EA appears 2–4 weeks after anti-LA and anti-MA. A sharp increase of anti-EA up to 1:320–640 precedes or coincides with the development of malignant tumors. The titers seem to correlate with the burden of tumor cells or the number of virus-carrying lymphoid cells in the blood (Pearson *et al.*, 1974). Owl monkeys that do not develop malignant disease usually do not show anti-EA titers. In a few of many cases examined, owl monkeys refractive to tumor induction have developed a long-lasting low-titered anti-EA response. The correlation between tumor load and anti-EA antibodies shows a striking similarity to the serology of EBV in Burkitt lymphoma. In the human condition, rising anti-EA titers are an unfavorable sign for the prognosis, since this indicates neoplastic proliferation, and the anti-EA titers increase in parallel with the burden of tumor cells (Henle *et al.*, 1973).

The immune response of tamarin marmosets in *H. saimiri*-caused neoplasia is relatively rapid. Anti-LA and anti-EA become detectable simultaneously, usually during the 3rd week postinoculation (Klein *et al.*, 1973). At the same time, neutralizing antibodies can be found. If the clinical course of lymphoma is very rapid and the animals die before the end of the 3rd week, neutralizing antibodies may not become detectable. Similarly, the titers of virus antibodies in *H. ateles*-infected tamarin marmosets usually remain low (Falk *et al.*, 1974a). The time course of malignancies in marmosets is too short to allow specific observations on a relationship between anti-EA titers and tumor growth. Complement-fixing assays with marmoset sera are reportedly difficult, since the sera often contain anticomplementary activities. Occasionally, virus-infected marmosets have heterophil antibodies, the significance of which is not known (Deinhardt *et al.*, 1974).

Capuchin monkeys (*Cebus* spp.) have a seroconversion against LAs of *H. saimiri* between 1 and 3 months after infection (Rabin *et al.*, 1975b). Anti-EA appears around the same time or is delayed up to 3 months later. Stable anti-EA titers were found over several months in persistently infected *Cebus* monkeys that did not show signs of tumor disease. This is remarkable, since the anti-EA in owl monkeys is essentially correlated with the load of tumor cells in the organism.

The humoral immune response of rabbits against *H. saimiri* and *H. ateles* is significantly delayed in comparison to tumors that develop in New World primates. Neutralizing antibodies always appear late, about 1 week before clinical signs of malignant lymphoma become manifest. This is independent of the length of the incubation period, which may be as long as 13 months (Daniel *et al.*, 1974a). This could mean that in the rabbit, virus replication is confined to the population of virus-transformed proliferating lymphocytes without productive replication in extralymphatic tissues.

2. Cell-Mediated Immunity

Studies on cellular immunity have concentrated on *H. saimiri*-infected owl monkeys. This system is particularly interesting, since tumor-bearing animals succumb after a varying, usually extended, time course. A major question still unanswered is that of what factors influence tumor susceptibility or resistence. Attempts to detect specific active cellular immune reactions in uninfected, persistently infected, or tumor-bearing owl monkeys have usually failed, at most indicating low responsiveness. While infected squirrel monkeys exhibit long-lasting virus-specific cell-mediated immunity, such reactivity could not be demonstrated in freshly or persistently infected owl monkeys by a proliferative assay with peripheral lymphocytes (Pearson and Davis, 1976; Neubauer *et al.*, 1980a). Yet cells reactive to *H. saimiri* could be demonstrated in lymph nodes of tumorous owl monkeys. The failure of owl monkeys to react with a strong general cellular immune response to *H. saimiri* antigens is not due to a basic immunological defect. It may explain in part the tumor susceptibility of owl monkeys in comparison to the natural host, but these parameters do not seem to explain why some owl monkeys remain persistently infected while others develop tumors. Also, disease patterns could not be attributed to subspecies differences of owl monkeys. Seven karyotypes of owl monkeys have been described (Ma *et al.*, 1976); none of them is overrepresented among tumor-resistant monkeys, indicating that chromosome polymorphism cannot be correlated with tumor susceptibility or resistance (Neubauer *et al.*, 1980).

While T cells generally may play an important role in the defense against tumor cells, *H. saimiri* and *H. ateles* neoplasias are characterized by the oncogenic transformation of T lymphocytes themselves. A multitude of abnormal immune reactions become possible, dependent on which functional T-cell populations are stimulated or paralyzed in their specific function by a virus carrier state or oncogenic transformation. A very high proportion of the blood lymphocytes become infected before tumors arise; as much as 80% of all peripheral leukocytes may produce virus in an infectious-center assay. Preliminary results indicate that several or all populations of T cells, e.g., T cells with and without receptors for the Fc portion of immunoglobulin G (IgG), are virus-infected and proportionally increased in number (Neubauer *et al.*, 1981). Tumor-developing animals show severe impairment of T-cell functions (Pearson *et al.*, 1974; Wallen *et al.*, 1974, 1975a). This is manifested by a loss of responsiveness to T-cell mitogens, i.e., phytohemagglutinin and concanavalin A. This loss of mitotic reactivity is disease-related. In animals with rapidly progressing tumors, hyporesponsiveness is usually noticed at the time when virus-positive cells appear in the peripheral blood. Inapparently infected monkeys with persisting virus in the peripheral lymphocytes do not have an obvious impairment of T-cell stimulation, since only the subpopulation of IgG-Fc-receptor-positive T cells are hyporesponsive (Neubauer *et al.*, 1981). Abolished T-cell responsiveness of tumor-developing owl monkeys could not be restored by levamisole (Neubauer and Rabin, 1978). Wallen *et al.* (1975b) found evidence that the inhibition of T-lymphocyte responsiveness is mediated by a lymphocyte subpopulation with suppressor function. When the lymphocytes of tumor-developing owl monkeys lose the ability to react to T-cell mitogens, they can also stimulate tumor cells and tumor-derived cell lines to produce a soluble antiproliferation factor (APF) that inhibits the mitogen response of isogeneic and allogeneic peripheral T lymphocytes (Neubauer *et al.*, 1975, 1976). APF has been partially characterized with respect to its physiochemical properties (Neubauer and Rabin, 1979; Neubauer *et al.*, 1975). APF activity could be demonstrated in the sera of tumor-bearing monkeys (Neubauer and Rabin, 1979). Further studies on the role of suppressor cells in *H. saimiri* lymphomas of owl monkeys may provide highly significant insights into the mechanism of initiation of tumor growth. Not all T-cell populations are changed in the sense of a paralysis or constituitive overfunction in tumor-developing owl monkeys; helper T cells seem not to be affected by *H. saimiri* infection. Also, the levels of nonspecific cytotoxicity of lymphocytes from infected animals are as high as those from uninfected monkeys, and this activity is maintained at a normal level during disease (Neubauer *et al.*, 1981).

C. Vaccination and Chemotherapy against Tumor Disease

Attempts to prevent *H. saimiri*-induced malignant lymphoma with chemotherapeutic agents have been largely unsuccessful. Several groups have used high doses of human interferon with at most only a very slight delay to the time of death (Laufs *et al.*, 1974a; Rabin *et al.*, 1976; L. Falk, M. D. Daniel, R. D. Hunt, and G. Bekesi, unpublished), even though human interferon is effective in preventing growth of *H. saimiri in vitro* in owl monkey kidney (OMK) cells (Daniel *et al.*, 1980). Phosphonoacetic acid (PAA), a potent inhibitor of many herpesviruses that is relatively specific for the herpesvirus-induced DNA polymerase, is also a potent inhibitor of *H. saimiri* growth *in vitro* (Daniel *et al.*, 1980). This compound, however, has not been useful in animal studies with New World primates. Difficulties were encountered in maintaining effective levels of PAA in the blood, and PAA solutions were toxic to these animals (R. D. Hunt and M. D. Daniel, unpublished).

Greater success has been achieved using live attenuated or killed virus vaccines. *Herpesvirus saimiri* inactivated with formaldehyde at 56°C was immunogenic and was able to protect animals against challenge by wild-type oncogenic virus (Laufs and Steinke, 1976; Laufs, 1974). High challenge doses did result in delayed lymphoma development. Attempts to rescue virus from circulating lymphocytes were unsuccessful, indicating that circulating neutralizing antibodies had prevented infection.

Another approach to the development of noninfectious vaccine has been the preparation of plasma vesicles from lytically infected OMK cells (Pearson and Scott, 1977). These vesicles were free of virus and viral DNA, but did contain viral MAs. MAs usually contain the viral proteins responsible for neutralization by antisera (Randall and Honess, 1980). Cottontop marmosets inoculated with these membrane vesicles seroconverted when assayed by neutralization and immunofluorescence tests. Such animals were partially protected against lymphoma induction on challenge by oncogenic *H. saimiri* (Pearson and Scott, 1978). Two of three animals survived challenge greater than 300 days.

Live attenuated strains of *H. saimiri* have also been successfully used for active immunization. These are discussed in the next section.

D. Attenuation of Virus Strains

Two independent attempts to attenuate *H. saimiri* have been successful. The development of these attenuated strains not only has been useful for studying the process of active immunization by live vaccine, but also has opened the door for studying viral functions responsible for cell transformation and tumorigenesis.

Attenuation of *H. saimiri* strain 11 was achieved by serial propagation in Vero cells at 39°C (Schaffer *et al.*, 1975). When cottontop marmosets were inoculated with this attenuated strain, designated 11^{att} (previously A-HVS), they did not develop lymphoma even though they were infected. Infection was confirmed by antibody response and by recovery of virus after cocultivation of circulating lymphocytes with permissive OMK cells. Virus recovered as long as 4 years after inoculation was confirmed as the original strain 11^{att} by restriction-endonuclease analysis of virion DNA (Falk *et al.*, 1980). Two cottontop marmosets latently infected with 11^{att} were challenged with 770 PFU of strain 11, resulting in malignant lymphoma and death 75 and 80 days postinoculation (Falk *et al.*, 1976). This time represented more than 3 times longer to death than control marmosets inoculated with strain 11. Wright *et al.* (1980), however, have found that common marmosets latently infected with strain 11^{att} resisted challenge with 100–825 PFU of oncogenic strain 11 virus. Virus recovered from circulating lymphocytes of one animal 2 years after challenge was analyzed by restriction-endonuclease digestion of DNA;

only 11^{att} could be detected (Falk *et al.*, 1980). Thus, *H. saimiri* strain 11^{att} does not produce lymphoma in cottontop or common marmosets, and it can protect common marmosets from challenge by wild-type oncogenic virus. The original attenuation of strain 11 may have been somewhat of a chance occurrence; attempts to obtain other attenuated strains of S295C and 11 by the same procedure have been unsuccessful (Falk *et al.*, 1976).

Analysis of virion DNA with restriction endonucleases has revealed that 1.1 Md from the left terminus of low-density DNA (L-DNA) has been deleted from strain 11 in the generation of 11^{att}. Thus, the *Aos*I F and *Eco*RI A fragments were 1.1 Md shorter in *H. saimiri* 11^{att} DNA. In agreement with this, strain 11^{att} DNA totally lacked the *Kpn*I G fragment of strain 11. Although the adjacent *Kpn*I C fragment was apparently unchanged in size, it had become the left L-terminal *Kpn*I fragment located at the H–L border, since the terminal high-density DNA (H-DNA) was attached to this fragment. No other genome alterations have been detected in 11^{att} by these procedures. This genome alteration is a stable property of the virus, whether examined after passage *in vitro* or after infection of marmosets (Mulder *et al.*, 1982).

Attenuation of *H. saimiri* strain SMHI was achieved by persistent infection of semipermissive Vero cells (Daniel *et al.*, 1979a). These persistently infected Vero cells have been subpassaged in culture for over 5 years and continuously produce an attenuated strain, designated $SMHI^{att}$. Although strain $SMHI^{att}$ grows to normal titers in OMK cells in culture, $SMHI^{att}$ does not produce disease in cottontop marmoset or owl monkeys, even after repeated inoculations or inoculation of high doses. Although inoculated animals develop antibodies to LAs, attempts to demonstrate infection by virus recovery have repeatedly failed. In one study, two cottontop marmosets were inoculated 10 times over a period of 211 days; both animals developed neutralizing antibodies. Both animals were then challenged with 100 median tissue-culture infectious doses ($TCID_{50}$) of oncogenic SMHI; one animal developed malignant lymphoma, while the other survived for 3 years. The surviving animal then succumbed to challenge by a very high dose of virus (3×10^6 $TCID_{50}$). In other experiments, cottontop marmosets inoculated with $SMHI^{att}$ have not been protected from challenge by wild-type oncogenic virus (Daniel *et al.*, 1979a).

$SMHI^{att}$ virion DNA contains genomic alterations when compared to SMHI. Use of restriction endonucleases (*endo*Rs) *Aos*I, *Kpn*I, *Xho*I, *Eco*RI, and *Bam*HI has shown that 3.5 Md from the left terminus of L-DNA has been deleted from strain SMHI in the generation of $SMHI^{att}$. The deletion appears analogous to the deletion found in 11^{att}. In addition, $SMHI^{att}$ DNA has a 7.2-Md fragment of L-DNA inserted in the repetitive H-DNA of the virus. This segment represents an inverted repetition of a 3.6-Md region of the right end of L-DNA. $SMHI^{att}$ DNA preparations also have variable amounts of defective DNA molecules, consisting of a tandemly repeated portion of 8.2 Md of L-DNA plus 0.8 Md of H-DNA. This sort of defective molecule was also found in some other *H. saimiri* DNA preparations, especially after high-multiplicity serial passage (Mulder *et al.*, 1982).

It is not known at present by what mechanism these attenuated strains are no longer oncogenic. Several possibilities seem likely: These attenuated strains may have lost the ability to transform or immortalize lymphocytes through a change or loss of a gene coding for a hypothetical transforming protein. Alternatively, an attenuated strain may express in a different fashion a viral antigen that allows the host animal to prevent malignant outgrowth through normal immune mechanisms. It also seems possible that a viral product that functions in suppressing normal immune reaction could be defective. Unfortunately, *H. saimiri* wild-type strains have functioned very poorly in *in vitro* transformation assays, so it is not yet possible to directly test the *in vitro* transforming potential of the attenuated strains.

It is remarkable that both attenuated strains have developed a deletion in the same region of the genome, but there is as yet no direct evidence that these deletions are responsible for the attenuation. The absence of an *in vitro* transformation assay makes it impossible to test the effect on transformation of alterations in this region relative to the effects of alterations in other regions. Nevertheless, results from various lines of investigation point to the importance of the leftmost region of the L-DNA in transformation: (1) Sequences are deleted from this region of the genome in the attenuated strains. (2) Viral DNA sequences from this region fall into two groups that appear to be structurally unrelated (see Section III.B.1.b). This is possibly analogous to the similar but structurally unrelated *src* genes responsible for growth transformation in various strains of avian sarcoma virus (Yoshida *et al.*, 1980). Also, it is known that the *src* genes can be readily lost in some cases by continued passage of the virus. (3) At least some continuous cell lines derived from *H. saimiri*-induced tumors contain this region of the genome in a molarity double that of other regions (Werner *et al.*, 1978; Desrosiers, 1981). (4) Although most L-DNA sequences are methylated in tumor cells, this region of the genome contains specifically unmethylated sites (see Section VI.F.4).

Further work is necessary to determine the basis for the lack of oncogenicity of attenuated strains and to determine what role, if any, the left-terminal L-DNA sequences play in transformation and attenuation.

E. Infectivity of Viral Deoxyribonucleic Acid

Purified DNA of *H. saimiri* and *H. ateles* is capable of inducing malignant lymphoma in tamarin marmosets (Fleckenstein *et al.*, 1975b; 1978b). In the course of several experiments, 14 cottontop marmosets were injected with 50 ng to 10 μg calcium-phosphate-precipitated *H. saimiri* M-DNA (strain 11) by the intravenous, intraperitoneal, and intramuscular routes. Three animals that received 1 μg DNA or more developed fatal tumor disease within 60–85 days. Infectious virus was isolated from each of the animals. The virus was identified as *H. saimiri* by serology or by cleavage patterns of its DNA with restriction endonucleases. Viral DNA was found in tumor autopsy materials by CoT hybridization. Two animals had neutralizing antibodies and anti-LA detectable by immunofluorescence. So far, we cannot decide whether the DNA-induced neoplastic disease in marmosets is mediated by transfection of permissive cells involving a primary productive infection or, alternatively, whether transformable lymphoid T cells are the direct targets for infectious DNA.

Herpesvirus ateles DNA was used in one study on tumor induction by DNA (Fleckenstein *et al.*, 1978b). Four cottontop marmosets were injected with precipitated *H. ateles* DNA, each animal receiving 20 μg M-DNA. One of these monkeys died with the clinical and pathological features of malignant lymphoma. However, there was a striking difference from the *H. saimiri* DNA inoculations described above; in the case of the *H. ateles*-DNA-induced tumor, we were not able to isolate infectious virus or to find humoral antibodies to *H. ateles*. However, *H. ateles* DNA sequences were found in the tumor tissues in multiple copies per cell, as determined by reassociation kinetics. The extent of hybridization suggested that a portion of the *H. ateles* genome was lacking in the tumor cells. This could explain the failure to demonstrate infectious particles and antibody response, since lymphoid target cells could be transformed by a fragment of genomic DNA without productive replication taking place. However, this possibility needs further experimental investigation. The ability of herpesvirus DNA to induce tumors indicates that DNA of large size can be incorporated into the cells of the living organism.

F. Persistence and Expression of Viral Genes in Tumor Cells and Cell Lines

1. Establishment of Cell Lines

Tumor cells from circulating lymphocytes, lymph nodes, spleen, or thymus may be cultured *in vitro* following *H. saimiri*-induced malignant lymphoma. Tumor cells from cottontop, white-lip, owl, and black howler monkeys have been cultured *in vitro* for several months to over 8 years (Rabson *et al.*, 1971; Falk *et al.*, 1972d; Rangan, 1976; Fleckenstein *et al.*, 1977). Rabson *et al.* (1971) described the first lymphoma cell line, designated MLC-1, which was derived from an infiltrated lymph node of a cottontop marmoset. Similar cell lines have also been established from

H. ateles-induced lymphomas (Falk *et al.*, 1974b). *Herpesvirus saimiri*- and *H. ateles*-induced tumor-cell lines grow mainly in suspension as small clumps or as larger cell aggregates of a hundred cells or more. Tumor-cell lines continue to preserve T-cell-specific markers—rosette formation with sheep erythrocytes, the lack of membrane and intracellular immunoglobulins, and membrane Fc and complement receptors (Wallen *et al.*, 1974; Falk, 1980b).

Marmosets are hematopoietic chimeras. Marmosets that are born after a twin pregnancy with a fraternal twin of the opposite sex will have male and female hematopoietic cells. Freshly established tumor-cell lines from *H. saimiri*-infected marmosets do still have this characteristic (Marcyzynska *et al.*, 1973). Even after 10 months of continuous culture, an *H. saimiri*-induced tumor-cell line remained chimeric, with the proportion of male to female cells essentially unchanged (Chu and Rabson, 1972). The 45% female/55% male ratio after 10 months was considered within the random selection range. This indicates that the tumor cells were polyclonal and that there was no selective growth advantage for either cell type *in vitro*.

Recently established cell cultures generally produce detectable levels of virus within the first few months of growth *in vitro*, with 1–10% of the cells able to produce virus when cocultivated with permissive OMK cells. After a prolonged period of cultivation *in vitro*, tumor-cell cultures generally diminish in their ability to produce virus. Some cell lines lose their ability to produce virus entirely, and others have maintained a low level of ability to produce virus over several years.

A summary of *H. saimiri* and *H. ateles* continuous cell lines is shown in Table IV.

2. Quantities of Viral DNA in Tumor Cells

Cell DNA has been isolated from various organs of animals bearing *H. saimiri*-induced tumors and has been used to demonstrate and quantitate viral DNA sequences. Both types of viral sequences, L-DNA and H-DNA, have been detected by measurement of reassociation kinetics using radioactively labeled viral DNA as probe (Fleckenstein *et al.*, 1977). Autopsy material showed from 0.014 to 0.075% of total DNA as L-DNA sequences and from 0.01 to 0.11% as H-DNA sequences. This amount of L-DNA is equivalent to 7–38 viral M-genome equivalents per diploid cell. The tissues used for extraction of DNA of course contained a complex mix of tumor and nontumor cells and perhaps even lytically infected cells. Thus, it was not possible to accurately estimate the amount of viral DNA per neoplastic cell. The presence of multiple viral genome copies in neoplastic cells appears analogous to similar findings with Marek's disease, Burkitt lymphoma, and nasopharyngeal carcinoma.

Viral DNA has also been quantitated in a similar manner in *H. saimiri* and *H. ateles* continuous cell lines (Fleckenstein *et al.*, 1977;

TABLE IV. *Herpesvirus saimiri*- and *Herpesvirus ateles*-Transformed Cell Lines

Cell-line designation	Virus and strain	Source of cells	Ability to produce virus[a,b]	Genome equivalents per diploid cell[b]	HATNA[b,c]	Reference	Comments
1670	*H. saimiri* S295C	Spleen of male white lip marmoset tumor	Neg.	211[d]		Marczynska *et al.* (1973)	Analyses of DNA have shown strain 11 virus (Desrosiers, 1981).
70N2	*H. saimiri* 11	Thymus of female cottontop marmoset tumor	Neg.	274[d]		Falk *et al.* (1972d)	—
77/5	*H. saimiri* S295C	Lymph node of cottontop marmoset tumor	Pos.	230[d]		Fleckenstein *et al.* (1977)	—
1926	*H. saimiri* 11	Lymph node of female white-lip marmoset tumor	Pos.	NT		Falk *et al.* (1972d)	—
MLC-1	*H. saimiri* S295C	Lymph node of female cottontop marmoset tumor	Pos.	83[d]		Rabson *et al.* (1971)	—
1591	*H. saimiri* OMI	*In vitro* transformation of cottontop marmoset lymphocytes	Pos. & Neg.[e]	38[c]		B. Fleckenstein (unpublished)	—
1022	*H. ateles* 73	Male cottontop marmoset tumor	Pos.	4, 96[c]	− and +	Falk *et al.* (1974b)	1022 cells passed in Erlangen vs. Stockholm have different properties.
22CM37	*H. ateles* 73	*In vitro* transformation of cottontop marmoset lymphocytes	Pos.	8, 10[c]	−	Falk *et al.* (1978)	Transformed by virus shed from X-irradiated 1022 cells

A651	*H. ateles* 73	*In vitro* transformation of cottontop marmoset lymphocytes	Pos.	110[c]	–	B. Fleckenstein (unpublished)	—
A661	*H. ateles* 73	*In vitro* transformation of cottontop marmoset lymphocytes	Pos.	103[c]	+	B. Fleckenstein (unpublished)	—
A1601	*H. ateles* 73	*In vitro* transformation of cottontop marmoset lymphocytes	Pos.	240[c]	+	B. Fleckenstein (unpublished)	—
22CM61	*H. ateles* 73	*In vitro* transformation of cottontop marmoset lymphocytes	NT	95[c]	+	Falk *et al.* (1978)	Transformed by virus shed from X-irradiated 1022 cells.
461	*H. ateles* 73	*In vitro* transformation of cottontop marmoset lymphocytes	NT	72[c]	–	Falk *et al.* (1978)	—
4086	*H. ateles* 73	*In vitro* transformation of cottontop marmoset lymphocytes	Pos.	39[c]	–	Falk *et al.* (1978)	—
4156	*H. ateles* 73	*In vitro* transformation of cottontop marmoset lymphocytes	Neg.	177[c]	+	Falk *et al.* (1978)	—
70KI-1HVA	*H. ateles* 73	*In vitro* transformation of cottontop marmoset lymphocytes	Neg.	33[c]	–	Falk *et al.* (1978)	—
A2241	*H. ateles* 73	*In vitro* transformation of cottontop marmoset lymphocytes	NT	NT	NT	B. Fleckenstein (unpublished)	—
A2543	*H. ateles* 73	*In vitro* transformation of cottontop marmoset lymphocytes	NT	NT	NT	B. Fleckenstein (unpublished)	—
A2585	*H. ateles* 73	*In vitro* transformation of white-lip marmoset lymphocytes	NT	NT	NT	B. Fleckenstein (unpublished)	—

(*Continued*)

TABLE IV. (*Continued*)

Cell-line designation	Virus and strain	Source of cells	Ability to produce virus[a,b]	Genome equivalents per diploid cell[b]	HATNA[b,c]	Reference	Comments
A2549	*H. ateles* 73	*In vitro* transformation of red-belly marmoset lymphocytes	NT	NT	NT	B. Fleckenstein (unpublished)	—
RLA	*H. ateles* 810	ACCRB rabbit tumor	Pos.	343[c]	+	M. D. Daniel (unpublished)	—
RLB	*H. ateles* 810	ACCRB rabbit tumor	Pos.	NT	NT	M. D. Daniel (unpublished)	ACCRB rabbit inoculated with RLA cells.
RLC-1	*H. ateles* 810	ACCRB rabbit tumor	Pos.	326[c]	+	M. D. Daniel (unpublished)	ACCRB rabbit inoculated with RLB cells.
RLC-2	*H. ateles* 810	ACCRB rabbit tumor	Pos.	NT	NT	M. D. Daniel (unpublished)	ACCRB rabbit inoculated with RLB cells.

[a] Measured by the ability of 10^6 cells to produce virus when cocultivated with permissive OMK cells.
[b] (NT) Not tested. [c] Johnson *et al*. (1980). [d] Fleckenstein *et al*. (1977).
[e] Early and late subcultures are available that are positive and negative for the ability to produce virus.

Johnson *et al.*, 1981). The findings are summarized in Table IV. Various cell lines show from as few as 4 to over 300 genome equivalents per diploid cell. The *H. ateles* rabbit tumor-cell lines appear to have the highest amount of viral DNA. There seems to be no correlation with the number of genome equivalents per cell and the ability of cells to produce virus. There does, however, appear to be a correlation with the number of genome equivalents and the ability to detect the *H. ateles* nuclear antigen (HATNA) (Johnson *et al.*, 1981).

3. DNA Circularization and Sequence Rearrangements in Tumor Cells

In situ hybridizations with metaphase chromosomes of a nonproducing cell line suggested that much of the viral DNA persisting in these cells is somehow associated with cellular chromosomes (Fleckenstein *et al.*, 1976). Similar conclusions were drawn from cytogenetic studies (Periman *et al.*, 1972). Since much of viral DNA in the 1670 and other cell lines is present as covalently closed circular DNA not covalently joined to host-cell DNA sequences (see below), the nature of this association with the host chromosomes remains unclear.

The many similarities between *H. saimiri* and EBV prompted investigations into whether viral DNA in *H. saimiri*-transformed cells is present as covalently closed circular DNA. That much of EBV DNA is present in such supercoiled structures is well documented (for review, see Adams, 1979). Werner *et al.* (1977) found that *H. saimiri* DNA in the nonproducer cell line 1670 is present in circular form. Covalently closed circles were demonstrated by equilibrium centrifugation in CsCl–ethidium bromide and by direct visualization of circles in the electron microscope. Length measurements, however, revealed an unusual feature of viral circular DNA molecules in 1670 cells—they were considerably larger than virion DNA, 131 vs. 100 Md.

The nature of the circular DNA in 1670 cells has been further probed using the technique of partial denaturation mapping (Werner *et al.*, 1977, 1978). In this procedure, partially purified viral circular DNA molecules were partially denatured with formamide and then examined by electron microscopy. Regions low in G + C content tend to denature and regions very high in G + C content (e.g., H-DNA) invariably remain double-stranded. Examination of many such partially denatured DNA molecules showed that two L-DNA regions (L1 = 57.8 Md; L2 = 33.7 Md) were separated by two H-DNA regions (H1 = 22.8 Md; H2 = 17.2 Md). A partially denatured circular DNA molecule from 1670 cells is shown in Fig. 10. Fine-detailed analysis of 1670 DNA molecules with partially denatured L-DNA regions indicated that L2 sequences were a subset of L1 sequences that represented the leftmost 33.7 Md of virion L-DNA. Furthermore, the common sequences of L2 and L1 were oriented in the same direction. These results thus indicated that approximately 13.8 Md of viral DNA information was absent in these circular DNA molecules

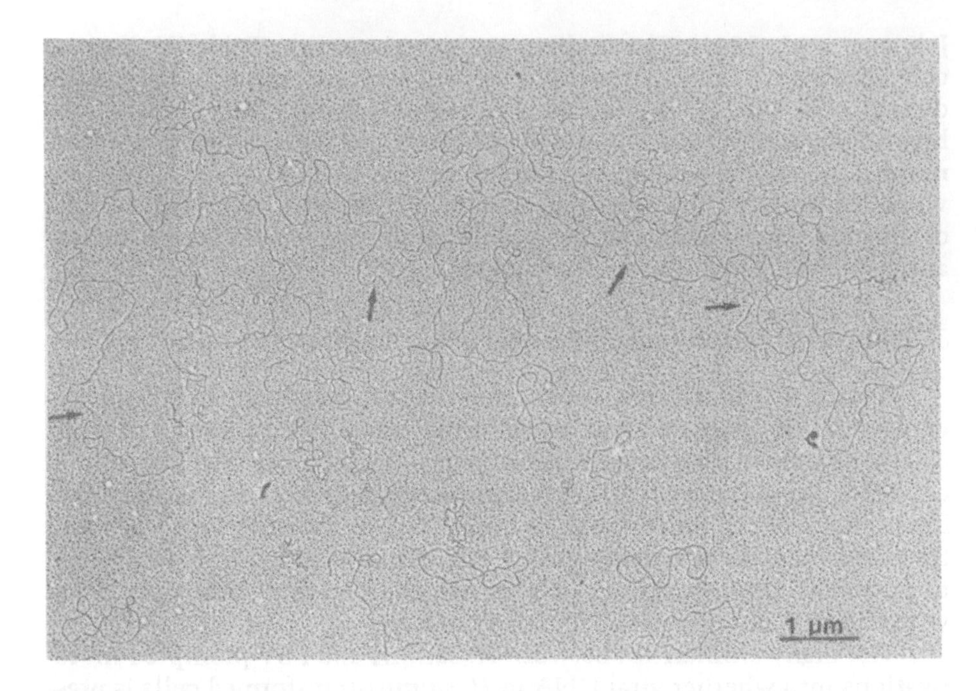

FIGURE 10. Electron micrograph of a partially denatured episome from *H. saimiri*-transformed marmoset tumor cell line 1670. The arrows indicate the transitions between H- and L-sequences. The small circles are PM2-DNA. Courtesy of F.-J. Werner.

from 1670 cells. Due to the limitations of partial denaturation mapping, the location of the 13.8-Md deletion within L-DNA could not be determined with confidence; the rightmost 24 Md of L1 DNA could not be aligned with virion DNA unambiguously.

The blotting technique of Southern (1975) has confirmed and extended these findings on *H. saimiri* DNA in 1670 cells (Desrosiers, 1981). DNA from 1670 cells was digested with restriction endonucleases, electrophoresed through agarose gels, transferred to nitrocellulose filters, and hybridized to high-specific-activity ^{32}P-labeled virion L-DNA. The viral DNA fragments *Eco*RI G, H, D, and I, *Kpn*I A, and *Bam*HI D and E were not detected in Southern transfers of DNA from the nonproducing 1670 cell line. For each restriction endonuclease, a new fragment appeared consistent with a 13.0-Md deletion of viral DNA sequences. These results unambiguously placed the deletion in L1 between 35 and 48 ± 0.6 Md from the left end of L-DNA. The size of the deletion determined by this procedure agrees quite well with the size determined by difference measurements with the electron microscope.

The partial denaturation mapping and the hybridization data are consistent with the sequence arrangement of viral DNA in 1670 cells shown schematically in Fig. 11. This model represents the major population of *H. saimiri* DNA sequences in 1670 cells, but there are still deficiencies in our understanding of the viral DNA in this cell line. For example, it is not known whether any viral DNA is integrated with host-cell sequences. Also, new discrete fragments that were mostly confined to or highly enriched in partially purified linear or episomal DNA fractions were produced with some restriction endonucleases (Kaschka-Dierich *et al.*, 1981; Desrosiers, 1981); these fragments did not comigrate with any virion DNA fragments. Most of these unexplained fragments were present in three of the four cell lines examined, so it seems likely that they have arisen through some common mechanism. They could

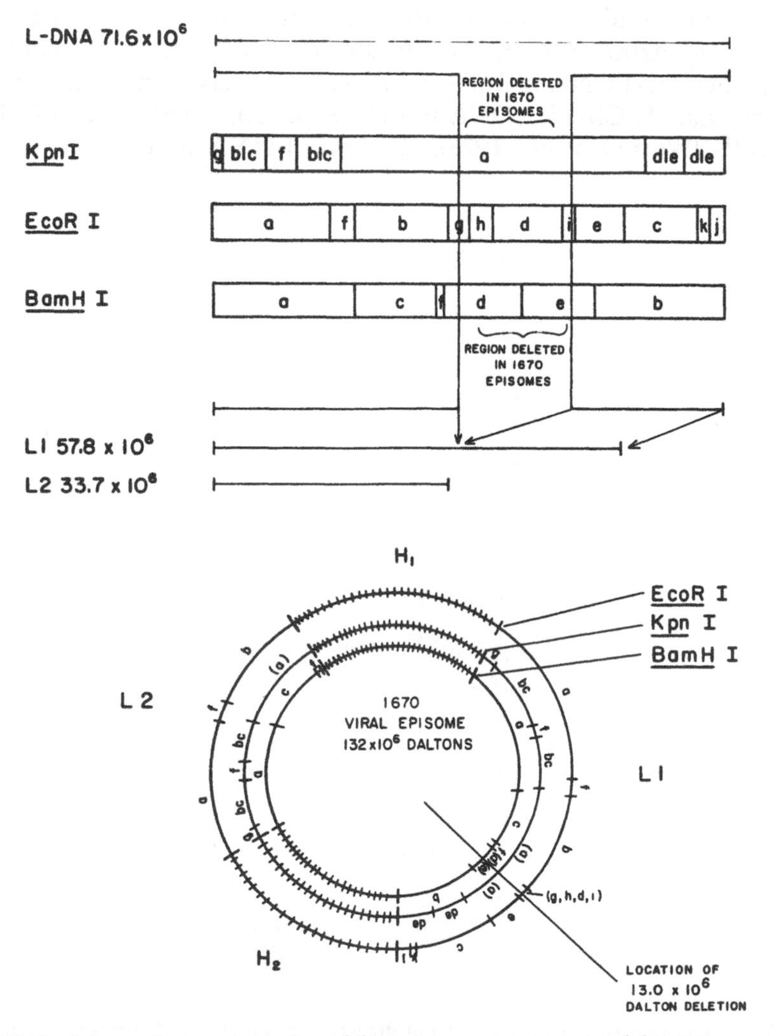

FIGURE 11. Schematic representation of the sequence arrangement of the major population of *H. saimiri* DNA in 1670 cells. From Desrosiers (1981).

possibly represent highly defective DNA molecules of very low sequence complexity. Also, the structure of *H. saimiri* DNA in 1670 cells shown in Fig. 11 may not include finer sequence arrangement changes that would not have been detected with the endonucleases used. Such changes could include small deletions or insertions of L-DNA sequences into H-DNA.

The question arises whether the unusual sequence arrangement described above is peculiar to 1670 cells. The viral DNA sequences deleted within *Eco*RI fragments G, H, D, and I in 1670 cells are also deleted in 70N2 cells (Desrosiers, 1981). Figure 12 shows the missing *Eco*RI fragments D and G in 1670 and 70N2 cells (fragments H and I ran off this gel), and it shows that the new fragment from fusion of sequences with *Eco*RI G and I is identical or nearly identical in size in 1670 and 70N2 cells. This amazing similarity is not simply due to cell cross-contamination, since viral DNA in 1670 and 70N2 cells may be distinguished by three criteria: (1) Circular DNA is 131 Md in 1670 cells and 119 Md in 70N2 cells (Werner *et al.*, 1978). (2) Viral H-DNA is more extensively

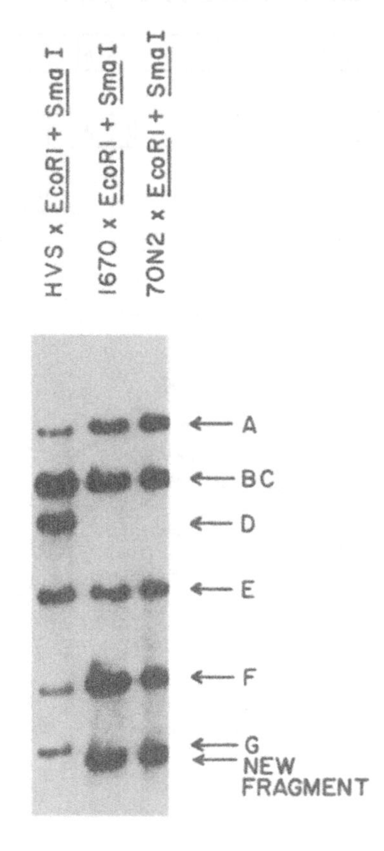

FIGURE 12. Comparison of *Eco*RI + *Sma*I digestion products of *H. saimiri* (HVS) DNA from virions, 1670 cells, and 70N2 cells. Total cell DNA (3 μg) or strain 11 virion DNA (6 ng) was digested with *Sma*I followed by *Eco*RI, electrophoresed through a 0.5% agarose gel, transferred to a nitrocellulose filter, hybridized to [^{32}P]L-DNA, and autoradiographed.

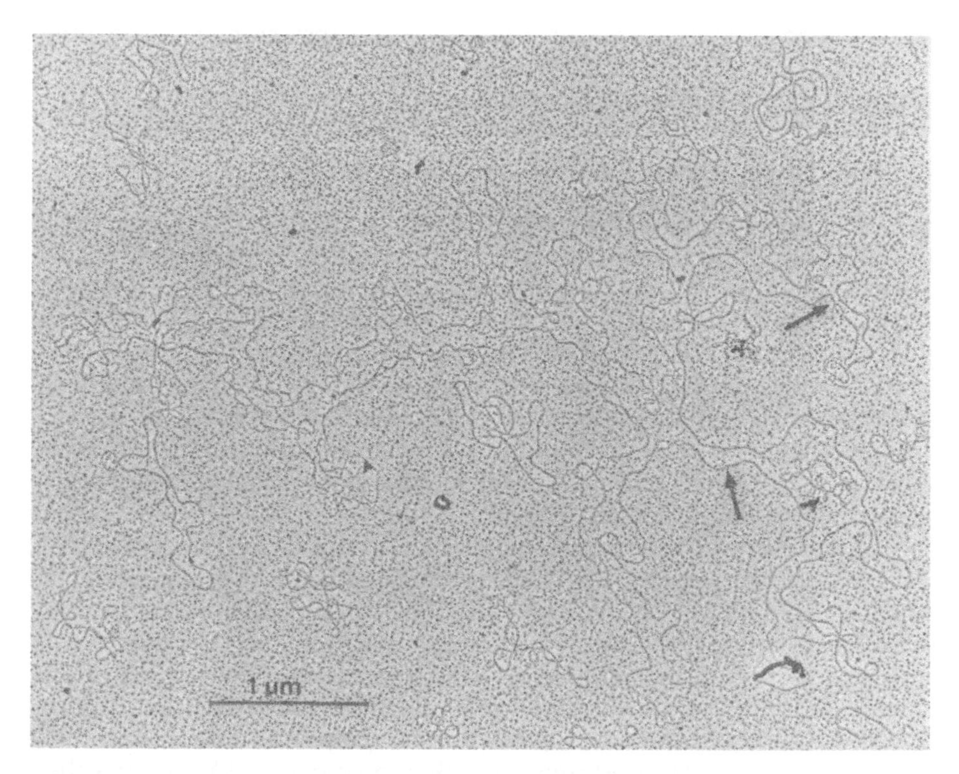

FIGURE 13. Electron micrograph of a partially denatured episome from *H. ateles*-transformed rabbit tumor cell line. The arrows indicate the transitions between H- and L-sequences. The small circles are PM2 DNA. From C. Kaschka-Dierich *et al.* (1982).

methylated in 1670 cells than in 70N2 cells (Desrosiers *et al.*, 1979). (3) DNA from these lines contains nonidentical L2 regions—33.7 Md in 1670 cells and approximately 18 Md in 70N2 cells (Kaschka-Dierich *et al.*, 1982; Desrosiers, 1981). These three characteristics are stable properties of the 1670 and 70N2 cell lines. Thus, surprisingly similar deletions and rearrangements appear to have occurred independently in the 1670 and 70N2 cell lines.

Fewer changes have occurred in cell lines still able to produce virus. Viral DNA from two *H. saimiri*-producing transformed cell lines (1926 and 77/5) contained all the viral DNA fragments present in purified virion DNA. DNA from tumor cells taken directly from a tumor-bearing animal contained viral DNA indistinguishable from the parental virion DNA by the Southern blotting analysis (Desrosiers, 1982). Partial denaturation of a circular viral DNA molecule from the virus-producing RLC cells indicated one L- and one H-DNA region of the same lengths found in virion DNA (Kaschka-Dierich *et al.*, 1982) (Fig. 13).

These results indicate that considerable viral DNA sequence rearrangements can occur on passage of tumor cells in culture and that excision of DNA sequences from the viral genome may play a role in establishing the nonproducing state of some tumor-cell lines. The decreased viral sequence complexity and limited gene expression in some nonproducing lines may aid in identifying viral gene products responsible for maintaining the transformed phenotype.

4. Methylation of Viral DNA in Transformed Cells

DNA methylation in mammalian cells occurs exclusively at cytosine, forming 5-methylcytosine. Furthermore, 5-methylcytosine occurs predominately in the dinucleotide CG (Cedar *et al.*, 1979; Bird, 1980), and the dinucleotide CG is by far the least common of the 16 dinucleotides (Nussinov, 1980). The *endo*R isoschizomers *Msp*I and *Hpa*II (recognition sequence CCGG) have been very useful for methylation analyses in mammalian cells, since *Hpa*II does not cleave when the C of the CG dinucleotide in its recognition sequence is methylated and *Msp*I cleaves whether or not the C of the CG dinucleotide in its recognition sequence is methylated (Waalwijk and Flavell, 1978).

Three lines of evidence indicate that *H. saimiri* virion DNA produced from lytic infection of OMK cells is not significantly methylated: (1) Acken *et al.* (1979) labeled virions with [^{3}H]DNA precursors and analyzed hydrolyzed virion DNA by thin-layer chromatography. They found less than one methyl group per genome-length molecule. (2) Analysis of unlabeled virion DNA (R. Desrosiers, M. Ehrlich, and J. Gehrke, unpublished) by high-performance liquid chromatography could not detect 5-methylcytosine (at least 30 times less than host-cell DNA) or any other modified nucleosides. (3) *Hpa*II and *Msp*I cleave purified virion DNA identically (Desrosiers *et al.*, 1979).

During the course of restriction-endonuclease analysis of DNA in *H. saimiri* tumor-cell lines, unusual behavior of viral H-DNA was observed with some restriction endonucleases (Desrosiers *et al.*, 1979). For example, *Sma*I, which cleaves four times in each repeat unit of H-DNA and does not cleave in L-DNA, cannot cleave H-DNA in 1670 cells to completion (Fig. 14). Fewer than 10% of the *Sma*I cleavage sites (CCCGGG) are actually able to be cleaved in H-DNA of 1670 cells. Methylation of C of the CG dinucleotide apparently blocks the action of *Sma*I (Gautier *et al.*, 1977). *Msp*I cleaves viral H-DNA in 1670 cells indistinguishable from viron H-DNA, but fewer than 10% of the same sites in 1670 H-DNA are cleaved by *Hpa*II (Desrosiers *et al.*, 1979). This is convincing evidence for extensive methylations of the CG dinucleotide in H-DNA in 1670 cells. Youssoufian and Mulder (1981) have found that *Xma*I, an isoschizomer of *Sma*I, cleaves H-DNA in 1670 cells indistinguishable from virion DNA. This provides further proof that H-DNA resistance to *Sma*I digestion in 1670 cells is due to extensive methylation and indicates that *Sma*I and *Xma*I are another isoschizomer pair with different sensitivities to methylation. Similarly, H-DNA in 1670 cells is refractory to cleavage by *Sac*II (CCGCGG), but not to cleavage by *Sac*I (GAGCTC), *Pvu*II (CAGCTC), or *Pst*I (CTGCAG) (Desrosiers *et al.*, 1979).

Viral H-DNA in the nonproducing 70N2 cells was also found to be

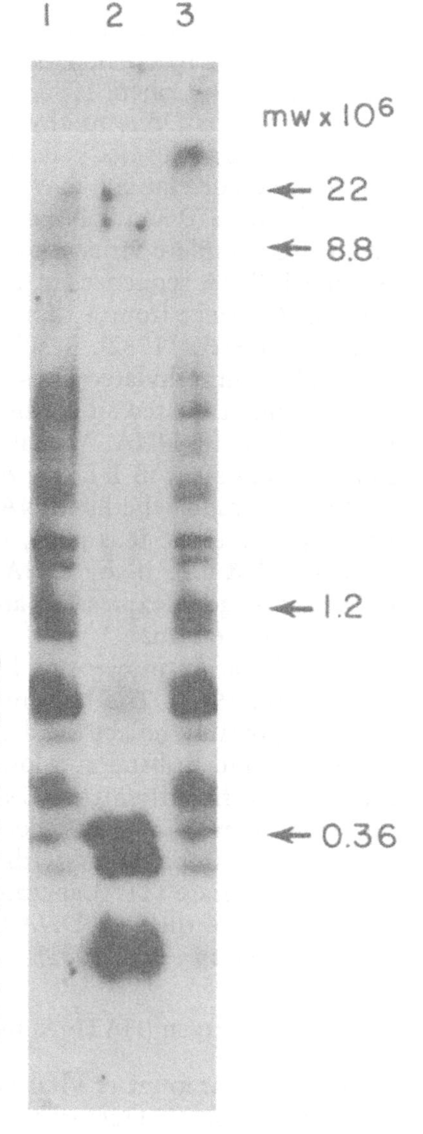

FIGURE 14. Methylation of H-DNA in 1670 cells—resistance to digestion by *Sma*I. Total cell DNA (3 μg) or strain 11 virion DNA (6 ng) was digested with restriction endonucleases, electrophoresed through a 0.7% agarose gel, transferred to a nitrocellulose filter, hybridized to [^{32}P]-M-DNA, and autoradiographed. In slot 2, virion DNA was digested with *Bam*HI + *Sma*I. *Sma*I digestion products of H-DNA (0.36, 0.30, and 0.14 Md) are clearly evident, and the *Bam*HI cleavage products of L-DNA (8.8–22 Md) are indicated in the top portion of the gel. In slot 3, 1670 DNA was digested with *Sma*I alone in excess; only a small portion of the H-DNA is cleaved to the same size as in virion DNA. Cleavage of 1670 DNA with *Bam*HI + *Sma*I, shown in slot 1, results in several L-DNA products in the 6–22 Md range. From Desrosiers *et al.* (1979).

methylated, but in this case only 50–90% of the CCCGGG sites were methylated. Viral H-DNAs in the producing cell lines 1926, 77/5, and 1591 were not detectably methylated (Desrosiers *et al.*, 1979). Consistent with the ability to produce virus when cocultivated with permissive OMK cells, viral H-DNA in tumor cells direct from a tumor-bearing animal was not detectably methylated (R. Desrosiers, unpublished).

This raised the possibility that DNA methylation somehow plays a role in the prevention of complete expression of *H. saimiri* in nonproducing lymphoid-cell lines. Since this initial report on *H. saimiri* DNA methylation, a number of other reports have appeared with other virus systems or specific cellular genes in which DNA methylation and gene expression apparently correlated inversely (van der Ploeg and Flavell, 1980; Sutter and Doerfler, 1980; McGhee and Ginder, 1979; Mandel and Chambon, 1979; Cohen, 1980; Guntaka *et al.*, 1980).

Any role for DNA methylation in the control of *H. saimiri* gene expression in transformed cells will most certainly be more complex than these simple correlations may indicate. The deletion of 13.0 Md, 18%, of L-DNA in 1670 and 70N2 cells more than likely must play a role in preventing complete virus expression. Furthermore, recent results indicate that L-DNA sequences in 1670, 70N2, the producing line 1926, and tumor cells direct from a tumor-bearing animal are extensively methylated (Desrosiers, 1982). Results to date indicate that there are a few specifically unmethylated sites in L-DNA of tumor cells and that specifically unmethylated sites cluster within the *Eco*RI A fragment in the leftmost 20% of L-DNA. Work is continuing to precisely map specifically unmethylated sites in L-DNA with as many endonucleases as possible and to determine whether RNA sequences originate from specifically unmethylated sites. It is clear, however, that much work is needed to elucidate what role, if any, DNA methylation contributes to the control of *H. saimiri* gene expression and the ability of transformed lymphoid cells to produce virus.

5-Methylation of cytosine lowers the density of DNA in CsCl gradients (Kirk, 1967). The buoyant density of viral DNA in transformed cells may possibly reflect the degree of methylation. Viral DNA of 1670 cells was found to band at a lower density in CsCl than cellular DNA, despite a significantly higher G + C content in viral DNA. Viral DNA in 70N2 cells showed a slighter density shift. Treatment of 1670 cells with methylation inhibitors or with the phorbol ester 12-*O*-tetradecanoyl phorbol-13 acetate (TPA) appeared to increase the density of at least a major portion of the viral DNA to that expected from the G + C content (Kaschka-Dierich *et al.*, 1981).

5. Nuclear Antigen (HATNA) in Tumor-Cell Lines

The large amount of viral DNA in *H. saimiri*- and *H. ateles*-transformed cells stands in remarkable contrast to the difficulties encountered in detecting virus-specific gene expression in these cells. Until recently, no viral proteins had been detected in tumor cells and cell lines by immunofluorescence studies, radioimmunoprecipitations, and many other attempts. This is in remarkable contrast to other DNA tumor virus systems, since papovavirus- or adenovirus-transformed cells have readily detectable early viral proteins and EBV-transformed lymphoblastoid cells and tumor cells contain EBV nuclear antigen (EBNA), a viral DNA-binding protein.

Ohno *et al.* (1979) detected a viral protein in some *H. ateles*-transformed lymphoid-cell lines, using the acid-fixed nuclear-binding (AFNB) technique in combination with anticomplement immunofluorescence.

In this procedure, concentrated salt-extracted soluble antigens from lymphoid-cell lines are incubated with methanol/acetic-acid-fixed amphibian erythrocytes; after being stained with a live-virus-boostered anti-*H. saimiri*-positive squirrel monkey serum, the erythrocyte nuclei appear in brilliant fluorescence. The DNA-binding protein (HATNA), in all likelihood a nuclear protein of virus-transformed cells, can be demonstrated in some marmoset and rabbit tumor-cell lines from *H. ateles*-induced tumors and in lymphoid primate cell lines transformed *in vitro* by *H. ateles*. Permissive cells productively infected with *H. ateles* also contain HATNA. There is no apparent correlation between the expression or absence of HATNA in the transformed cells and a virus-producer or -non-producer status, respectively. HATNA-positive cell lines have at least 95 persisting *H. ateles* genome copies per cell, while the majority of cell lines that remain negative in the AFNB test contain significantly less viral DNA. This suggests a quantitative correlation between the amount of viral DNA and viral DNA-binding nuclear antigen in *H. ateles*-transformed cells (Johnson *et al.*, 1981). It is still a matter of speculation whether the HATNA of *H. ateles*-transformed T cells is functionally related to the EBNA of EBV-transformed cells or to T antigen of papova- and adenoviruses.

G. *In Vitro* Transforming Potential of *Herpesvirus ateles* and *Herpesvirus saimiri*

Herpesvirus ateles is capable of transforming lymphocytes from marmoset monkeys *in vitro*. First, Falk *et al.* (1974b) showed that peripheral and splenic lymphocytes from marmoset monkeys can be immortalized by cocultivation with lethally X-irradiated virus-producing lymphoid cells. Later, it was shown that cell-free *H. ateles* immortalizes marmoset lymphocytes (Falk *et al.*, 1975, 1978; B. Fleckenstein, unpublished). Lymphoid-cell lines were obtained from tamarin marmosets (*Saguinus oedipus, S. fuscicollis, S. labiatus*) and common marmosets (*Callithrix jacchus*). Transforming assays are usually done by infecting a lymphocyte suspension culture of approximately 5×10^6 cells that are isolated in a Ficoll–Hypaque density gradient. The first signs of virus-induced proliferation usually become discernible after 3–4 weeks. It is difficult to quantitate the number of transformed foci in a culture. Transformed cell lines have T-cell markers, e.g., rosetting with sheep erythrocytes, reaction with an anti-marmoset T-lymphocyte serum that is detectable by cell lysis in ^{51}Cr microcytotoxicity tests, and lack of appreciable complement receptor and IgG (Falk *et al.*, 1978). All *in vitro*-transformed cell lines produce infectious virus on cocultivation with permissive cells, at least during the first months of cultivation; however, the number of cell-free virus particles in culture fluids of producer cell lines is low (<10 PFU/ml) (Falk, 1980b). Attempts to transform marmoset lymphocytes by transfection with *H. ateles* DNA in the presence of DEAE–dextran or after precipitation with calcium phosphate have proven unsuccessful so far (B. Fleckenstein, unpublished).

Attempts to transform lymphocytes *in vitro* by *H. saimiri* have been surprisingly unsuccessful. Several groups have experienced difficulties in trying to achieve immortalization of peripheral leukocytes by *H. saimiri* in cell culture. In one of numerous experiments, we obtained a cottontop marmoset cell line (H 1591) from infecting blood lymphocytes with *H. saimiri* strain OMI, employing squirrel monkey fibroblasts as a feeder layer (B. Fleckenstein and co-workers, unpublished). The sharp contrast between high oncogenicity in animals and poor transforming potential *in vitro* is remarkable. On the other hand, it should also be considered that the extreme oncogenicity in tamarins and other monkeys may be more reflective of a state of paralysis in the cellular immune system of the tumor-developing host than of the transforming power of the virus. It is possible that the primary target for transformation is a more immature stem cell rather than a well-differentiated peripheral lymphocyte or that some unknown factor that contributes to transformation *in vivo* is absent in the *in vitro* transformation.

VII. CONCLUDING REMARKS

The primary attention focused on the oncogenic herpesviruses of nonhuman primates arose from the obvious similarities to Epstein–Barr virus (EBV). The natural history of *H. saimiri* in squirrel monkeys and, as far as is known, of *H. ateles* in spider monkeys is reminiscent of the epidemiology of EBV in humans. All three viruses are able to infect lymphocytes, to persist in these cells, and to immortalize lymphoid cells in culture. The neoplasias associated with virus infection are tumors of the lymphatic system in most cases. The patterns of humoral-antibody reactions are similar in tumor-bearing organisms, and transformed tissue-culture cells and tumor cells contain multiple copies of viral genomes that persist, at least in part, in a circular form.

On the other hand, it should also be emphasized that many differences exist between EBV and the lymphoma-inducing herpesviruses from New World primates. *Herpesvirus saimiri* and *H. ateles* have a genome structure that is very much different from EBV DNA. EBV DNA has an internal stretch of repetitive DNA in tandem orientation with a constant length, and a few short terminal repeats. The New World primate virus DNA does not possess internal reiterated DNA, but does have enormous terminal redundancies. EBV DNA is high in G + C content; *H. saimiri* and *H. ateles* have coding sequences of low G + C. EBV DNA molecules are far less heterogeneous with regard to G + C content. *Herpesvirus saimiri* and *H. ateles* are extremely heterogeneous in the base composition of genome segments. EBV and oncogenic herpesviruses of New World primates do not share common antigenic determinants of their structural polypeptides, and there is no other molecular parameter that would hint at a particular evolutionary relatedness. *Herpesvirus saimiri* and *H. ateles* are consistently T-cell tropic in the natural host and in tumor-developing animals, while EBV infects B lymphocytes only. This results in substantial differences in the pathogenesis of *H. saimiri*- and *H. ateles*-caused experimental monkey lymphomas vs. human EBV-associated tumors. In human EBV infection, the T-cell system plays a role in elimination of transformed B lymphocytes. *Herpesvirus saimiri* and *H. ateles* tumors are characterized by the total breakdown of the T-cell system, since the majority of all T cells are infected and transformed by the viruses. *Herpesvirus saimiri* and *H. ateles* lymphomas, probably always polyclonal, usually arise shortly after experimental infection, with a time course shorter than that of other human or experimental animal tumor diseases.

The chain of events in the etiopathology of the described monkey tumors and the pathological anatomy of these diseases are unique in many regards and distinct from those of any other tumor virus system described before. The ability to induce reproducibly fast-growing T-cell lymphomas is a valuable advantage for certain experiments in tumor immunology and studies on chemotherapy. The variety of phenomena involved in the stimulation and paralysis of different T-cell subpopulations provides many avenues for immunological investigation. The viruses combine high oncogenicity and transforming potential with the possibility of applying the full spectrum of genetic and molecular methodology in tumor virus research. Thus, the biological and molecular pecularities of *H. saimiri* and *H. ateles* promise further new perspectives on the pathogenesis of malignant diseases in primates and the molecular biology of oncogenic herpesvirus transformation.

ACKNOWLEDGMENTS. Work described in this chapter was supported by Stiftung Volkswagenwerk, Wilhelm-Sander-Stiftung, Deutsche Forschungsgemeinschaft (SFB 118), and by grants RR00168 to the New England Regional Primate Research Center from the Division of Research Resources (NIH), 1R01 CA 31363–01 from the National Institutes of Health, and 1511-C-1 from the American Cancer Society, Massachusetts Division. R.C. Desrosiers was a fellow of the Medical Foundation, Inc., of Boston, Massachusetts. We thank Ingrid Altjohann for typing the manuscript.

REFERENCES

Ablashi, D.V., Loeb, W.F., Valerio, M.G., Adamson, R.H., Armstrong, G.R., Bennett, D.G., and Heine, U., 1971a, Malignant lymphoma with lymphatic leukemia in owl monkeys induced by *Herpesvirus saimiri*, *J. Natl. Cancer Inst.* **47**:837.

Ablashi, D.V., Armstrong, G.R., Heine, U., and Manaker, R.A., 1971b, Propagation of *Herpesvirus saimiri* in human cells, *J. Natl. Cancer Inst.* **47**:241.

Ablashi, D.V., Loeb, W.F., Pearson, G., Valerio, M.G., Armstrong, G.R., Rabin, H., Kingsbury, E.W., and Heine, U., 1973, Induction of lymphoma in owl monkeys with heated, noncytopathogenetic *Herpesvirus saimiri, Nature (London) New Biol.* **242**:28.

Ablashi, D.V., Armstrong, G.R., Heine, U., and Manaker, R.A., 1974a, Propagation of *Herpesvirus saimiri* in human cells, in: *Herpesvirus I: Recent Studies* (J.A. Bellanti, H. Kabuta, and G. L. Rosenberg, eds.), pp. 219–225, MSS Information, New York.

Ablashi, D.V., Armstrong, G.R., Easton, J.M., Chopra, H.C., and Adamson, R.H., 1974b, Evaluation of effects of cystosine arabinoside on oncogenic *Herpesvirus saimiri* in owl monkey kidney cells, *Biomedicine* **21**:57.

Ablashi, D.V., Armstrong, G.R., and Easton, J.M., 1976, Absence of horizontal transmission of *Herpesvirus saimiri* between experimentally infected and noninfected owl monkeys, *Cancer Lett.* **2**:35.

Ablashi, D.V., Armstrong, G., Fellows, C., Easton, J., Pearson, G., and Twardzik, D., 1977a, Evaluation of the effects of phosphonoacetic acid and 2-deoxy-*d*-glucose on *Herpesvirus saimiri* and Epstein–Barr virus, in: *Prevention and Detection of Cancer* (H.E. Nieburgs, ed.), pp. 245–261, Marcel Dekker, New York.

Ablashi, D.V., Twardzik, D., Easton, J.M., Armstrong, G.R., Luetzeler, J., Jasmin, C., and Chermann, J.-C., 1977b, Effects of 5-tungsto-2-antimoniate in oncogenic DNA and RNA virus-cell systems, *Eur. J. Cancer* **13**:713.

Ablashi, D.V., Pearson, G., Rabin, H., Armstrong, G., Easton, J., Valerio, M., and Cicmanec, J., 1978, Experimental infection of *Callithrix jacchus* marmosets with *Herpesvirus ateles, Herpesvirus saimiri* and Epstein–Barr virus, *Biomedicine* **29**:7.

Ablashi, D.V., Bengali, Z.H., Eichelberger, M.A., Sundar, K.S., Armstrong, G.R., Daniel, M., and Levine, P.H., 1980a, Increased infectivity of oncogenic herpes viruses of primates with tumor promoter 12-*O*-tetradecanoylphorbol-13-acetate (40901), *Proc. Soc. Exp. Biol. Med.* **164**:485.

Ablashi, D.V., Sundar, K.S., Armstrong, G., Golway, P., Valerio, M., Bengali, Z., Lemp, J., and Fox, R.R., 1980b, *Herpesvirus saimiri*-induced malignant lymphoma in inbred strain III/J rabbits (*Oryctolagus cuniculus*), *J. Cancer Res. Clin. Oncol.* **III**:1.

Acken, U.V. Simon, D., Grunert, F., Döring, H.-P., and Kröger, H., 1979, Methylation of viral DNA *in vivo* and *in vitro, Virology* **99**:152.

Adams, A., 1979, The state of the virus genome in transformed cells and its relationship to host cell DNA, in: *The Epstein–Barr Virus* (M.A. Epstein and B.G. Achong, eds.), pp. 156–183, Springer-Verlag, New York.

Adamson, R.H., Ablashi, D.V., Armstrong, G.R., and Ellmore, N.W., 1972, Effect of cytosine arabinoside, adenine arabinoside, tilorone, and rifamycin SV on multiplication of *Herpesvirus saimiri in vitro, Antimicrob. Agents Chemother.* **1**:82.

Adamson, R.H., Ablashi, D.V., Rabin, H., Cicmanec, J., and Dalgard, D.W., 1974, Chemotherapy of *Herpesvirus saimiri* induced lymphoma and leukemia in owl monkey, *J. Med. Primatol.* **3**:68.

Anderson, K.P., Costa, R.H., Holland, L.E., and Wagner, E.K. 1980, Characterization of herpes simplex virus type 1 RNA present in the absence of *de novo* protein synthesis, *J. Virol.* **34**:9.

Anzil, A.P., Bowden, D.M., and Ploog, D., 1977, Malignant lymphoma in a squirrel monkey (*Saimiri sciureus*), *J. Med. Primatol.* **6**:251.

Armstrong, G.R., Orr, T., Ablashi, D.V., Pearson, G.R., Rabin, H., Luetzeler, J., Loeb, W.F., and Valerio, M.G., 1976, Brief communication: Chronic *Herpesvirus saimiri* infection in an owl monkey, *J. Natl. Cancer Inst.* **56**:1069.

Banfield, W.G., Lee, C.W., Tralka, T.S., and Rabson, A.S., 1977, Lamella–particle complexes in nuclei of owl monkey kidney cells infected with *Herpesvirus saimiri, J. Natl. Cancer Inst.* **58**:1421.

Barahona, H.H., and Melendez, L.V., 1971, *Herpesvirus saimiri*: *In vitro* sensitivity to virus-induced interferon and to polyriboinosimic acid: polyribocytidylic acid, *Proc. Soc. Exp. Biol. Med.* **136**(4):1163.

Barahona, H., Melendez, L.V., Hunt, R.D., Forbes, M., Fraser, C.E.O., and Daniel, M.D.,

1975, Experimental horizontal transmission of *Herpesvirus saimiri* from squirrel monkeys to an owl monkey, *J. Infect. Dis.* **132**:694.

Barahona, H., Daniel, M.D., Bekesi, J.G., Fraser, C.E.O., King, N.W., Hunt, R.D., Ingalls, J.K., and Jones, T.C., 1977, *In vitro* suppression of *Herpesvirus saimiri* replication by phosphonoacetic acid (39687), *Proc. Soc. Exp. Biol. Med.* **154**:431.

Barbacid, M., Daniel, M.D., and Aaronson, S.A., 1980, Immunological relationship of OMC-1, an endogenous virus of owl monkeys, with mammalian and avian type C viruses, *J. Virol.* **33**:561.

Bird, A.P., 1980, DNA methylation and the frequency of CpG in animal DNA, *Nucleic Acids Res.* **8**:1499.

Bornkamm, G.W., Delius, H., Fleckenstein, B., Werner, F.-J., and Mulder, C., 1976, The structure of *Herpesvirus saimiri* genomes: Arrangement of heavy and light sequences within the M genome, *J. Virol.* **19**:154.

Bornkamm, G.W., Delius, H., Zimber, U., Hudewentz, J., and Epstein, M.A., 1980, Comparison of Epstein–Barr virus strains of different origin by analysis of the viral DNAs, *J. Virol.* **35**:603.

Buchman, T.G., Roizman, B., Adams, G., and Stover, B.H., 1978, Restriction endonuclease fingerprinting of herpes simplex virus DNA: A novel epidemiological tool applied to a nosocomial outbreak, *J. Infect. Dis.* **138**:488.

Cedar, R., Solage, A., Glaser, G., and Razin, A., 1979, Direct detection of methylated cytosine in DNA by use of the restriction enzyme *Msp* I, *Nucleic Acids Res.* **6**:2125.

Chu, E.W., and Rabson, A.S., 1972, Chimerism in lymphoid cell culture line derived from lymph node of marmoset infected with *Herpesvirus saimiri, J. Natl. Cancer Inst.* **48**:771.

Cicmanec, J.L., Loeb, W.F., and Valerio, M.G., 1974, Lymphoma in owl monkeys (*Aotus trivirgatus*) inoculated with *Herpesvirus saimiri:* Clinical, hematological and pathologic findings, *J. Med. Primatol.* **3**:8.

Cohen, J.C., 1980, Methylation of milk-borne and genetically transmitted mouse mammary tumor virus proviral DNA, *Cell* **19**:653.

Daniel, M.D., Rabin, H., Barahona, H., and Melendez, L.V., 1971, *Herpesvirus saimiri.* III. Plaque formation under multi agar, methyl cellulose, and starch overlays, *Proc. Soc. Exp. Biol. Med.* **136**:1192.

Daniel, M.D., Meléndez, L.V., and Barahona, H.H., 1972, Plaque characterization of viruses from South American nonhuman primates, *J. Natl. Cancer Inst.* **49**:239.

Daniel, M.D., Melendez, L.V., Hunt, R.D., King, N.W., Anver, M., Fraser, C.E.O., Barahona, H., and Baggs, R.B., 1974a, *Herpesvirus saimiri.* VII. Induction of malignant lymphoma in New Zealand white rabbits, *J. Natl. Cancer Inst.* **53**:1803.

Daniel, M.D., Melendez, L.V., and Silva, D., 1974b, Multiplication, cytopathogenicity and latent infection of human cell cultures by *Herpesvirus saimiri* strains, *In Vitro* **9**:354.

Daniel, M.D., Hunt, R.D., DuBose, D., Silva, D., and Melendez, L.V., 1975, Induction of *Herpesvirus saimiri* lymphoma in New Zealand white rabbits inoculated intravenously, in: *Oncogenesis and Herpesviruses II*, Part 2 (G. de Thé, M.A. Epstein, H. zur Hausen, and W. Davis, eds.), pp. 205–208, IARC Press, Lyon.

Daniel, M.D., Silva, D., and Ma, N., 1976, Establishment of owl monkey kidney 210 cell line for virological studies, *In Vitro* **12**:290.

Daniel, M.D., Hunt, R.D., King, N.W., and Ingalls, J.K., 1977, The oncogenicity of *Herpesvirus ateles* in rabbits, in: Abstracts of The Third International Symposium on Oncogenesis and Herpesviruses, Cambridge, Massachusetts, p. 213.

Daniel, M.D., Silva, D., Koomey, J.M., Mulder, C., Fleckenstein, B., Tamulevich, R., King, N.W., Hunt, R.D., Seghal, P., and Falk, L.A., 1979a, *Herpesvirus saimiri*: Strain SMHI modification of oncogenicity, in: *Advances in Comparative Leukemia Research* (D.S. Yohn, B.A. Lapin, and J.R. Blakeslee, eds.), pp. 395–396, Elsevier/North-Holland, New York, Amsterdam, Oxford.

Daniel, M.D., Tamulevich, R., Silva, D., Bekesi, G., Holland, J., King, N.W., and Falk, L.A., 1979b, Inhibitory effects of acycloguanosine and human lymphoblastoid interferon on *Herpesvirus saimiri* and *Herpesvirus ateles, Proc. Am. Assoc. Cancer Res.*, p. 233.

Daniel, M.D., Falk, L., Tamulevich, R., and King, N.W., 1979c, Interaction of *Herpesvirus saimiri* (HVS) and *Herpesvirus ateles* (HVA) with human and primate lymphoblastoid cells, *Bacteriol. Proc.*, p. 289.

Daniel, M.D., Falk, L.A., King, N.W., Tamulevich, R., Holland, J.S., and Bekesi, G., 1980, Comparative studies of interferon and three antiviral agents on neurotropic and oncogenic herpesviruses, *Antimicrob. Agents Chemother.* **18**:622.

Daniel, M.D., Tamulevich, R., Bekesi, J.G., King, N.W., Falk, L.A., Silva, D., and Holland, J.F., 1981, Selective antiviral activity of human interferons on primate oncogenic and neurotropic herpesviruses: Selective viral inhibition by interferon, *Int. J. Cancer,* **27**:113.

Deinhardt, F., Wolfe, L., Northrop, R., Marczynska, B., Ogden, J., McDonald, R., Falk, L., Shramek, G., Smith, R., and Deinhardt, J., 1972, Induction of neoplasms by viruses in marmoset monkeys, *J. Med. Primatol.* **1**:29–50.

Deinhardt, F., Falk, L., and Wolfe, L., 1973a, Simian herpesviruses, *Cancer Res.* **33**:1424.

Deinhardt, F., Falk, L., Marczynska, B., Shramek, G., and Wolfe, L., 1973b, *Herpesvirus saimiri*: A simian counterpart of Epstein–Barr virus of man?, in: *Bibliotheca Haematologica* (R.M. Dutcher and L. Chieco-Bianchi, eds.), pp. 416–427, S. Karger, Basel.

Deinhardt, F.W., Falk, L.A., and Wolfe, L.G., 1974, Simian herpesviruses and neoplasia, in: *Advances in Cancer Research* (G. Klein and S. Weinhouse, eds.), pp. 167–205, Academic Press, New York.

Desrosiers, R. C., 1981, *Herpesvirus saimiri* DNA in tumor cells — deleted sequences and sequence rearrangements, *J. Virol.* **39**:497.

Desrosiers, R. C., 1982, Specifically unmethylated CG sites in *Herpesvirus saimiri* DNA in tumor cells. *J. Virol.* **43**:in press.

Desrosiers, R. C., and Falk, L. A., 1981, *Herpesvirus tamarinus* and its relation to herpes simplex virus, *J. Gen. Virol.* **56**:119.

Desrosiers, R. C., and Falk, L. A., 1982, *Herpesvirus saimiri* strain variability, *J. Virol.* **43**:000.

Desrosiers, R.C., Mulder, C., and Fleckenstein, B., 1979, Methylation of *Herpesvirus saimiri* DNA in lymphoid tumor cell lines, *Proc. Natl. Acad. Sci. U.S.A.* **76**(8):3839.

de Thé, G., Geser, A., and Day, N.E., 1978, Epidemiological evidence for causal relationship between Epstein–Barr virus and Burkitt's lymphoma from Ugandan prospective study, *Nature (London)* **274**:756.

Didier, M.L., Ablashi, D.V., Oie, H.K., Armstrong, G.R., Easton, J.M., Chu, E.W., and Rabson, A.S., 1975, Some biological properties of *Herpesvirus saimiri* from chronically infected monolayer and suspension cultures, in: *Oncogenesis and Herpesviruses II*, Part I (G. de Thé, H. zur Hausen, and A.M. Epstein, eds.), p. 491, IARC Press, Lyon.

Dolyniuk, M., Pritchett, R., and Kieff, E., 1976, Proteins of Epstein–Barr virus, *J. Virol.* **17**:935.

Falk, L.A., Jr., 1980a, Biology of *Herpesvirus saimiri* and *Herpesvirus ateles*, in: *Viral Oncology* (G. Klein, ed.), pp. 813–832, Raven Press, New York.

Falk, L.A., Jr., 1980b, Simian herpesviruses and their oncogenic properties, in: *Oncogenic Herpesviruses* (F. Rapp, ed.), pp. 145–173, CRC Press, Boca Raton, Florida.

Falk, L.A., Wolfe, L.G., and Deinhardt, F., 1972a, Isolation of *Herpesvirus saimiri* from blood of squirrel monkeys (*Saimiri sciureus*), *J. Natl. Cancer Inst.* **48**:1499.

Falk, L., Wolfe, L., and Deinhardt, F., 1972b, Epidemiology of *Herpesvirus saimiri* infection in squirrel monkeys, in: *Medical Primatology III* (E.I. Goldsmith and J. Moor-Jankowski, eds.), pp. 151–158, S. Karger, Basel.

Falk, L., Wolfe, L.G., Hoekstra, J., and Deinhardt, F., 1972c, Demonstration of *Herpesvirus saimiri*-associated antigens in peripheral lymphocytes from infected marmosets during *in vitro* cultivation, *J. Natl. Cancer Inst.* **48**:523.

Falk, L.A., Wolfe, L.G., and Marczynska, F., 1972d, Characterization of lymphoid cell lines established from *Herpesvirus saimiri* (HVS)-infected marmosets, *Bacteriol. Proc.* **38**:191.

Falk, L.A., Wolfe, L.G., and Deinhardt, F., 1973a, *Herpesvirus saimiri*: Experimental infection of squirrel monkeys (*Saimiri sciureus*), *J. Natl. Cancer Inst.* **51**(1):165.

Falk, L.A., Nigida, S., Deinhardt, F., Cooper, R.W., and Hernandez-Camacho, J.I., 1973b, Brief communication: Oral excretion of *Herpesvirus saimiri* in captive squirrel monkeys and incidence of infection in feral squirrel monkeys, *J. Natl. Cancer Inst.* **51**(6):1987.

Falk, L.A., Nigida, S.M., Deinhardt, F., Wolfe, G., Cooper, R.W., and Hernandez-Camacho, J.I., 1974a, *Herpesvirus ateles*: Properties of an oncogenic herpesvirus isolated from circulating lymphocytes of spider monkeys (*Ateles sp.*), *Int. J. Cancer* **14**:473.

Falk, L., Wright, J., Wolfe, L., and Deinhardt, F., 1974b, *Herpesvirus ateles*: Transformation *in vitro* of marmoset splenic lymphocytes, *Int. J. Cancer* **14**:244.

Falk, L.A., Wolfe, L., and Deinhardt, F., 1975, Transformation *in vitro* with *Herpesvirus ateles*, in: *Oncogenesis and Herpesviruses II*, Part I (G. de Thé, M.A. Epstein, and H. zur Hausen, eds.), pp. 379–384, IARC Press, Lyon.

Falk, L., Wright, J., Deinhardt, F., Wolfe, L., Schaffer, P., and Benyesh-Melnick, M., 1976, Experimental infection of squirrel and marmoset monkeys with attenuated *Herpesvirus saimiri*, *Cancer Res.* **36**:707.

Falk, L., Johnson, D., and Deinhardt, F., 1978, Transformation of marmoset lymphocytes *in vitro* with *Herpesvirus ateles*, *Int. J. Cancer* **21**:652.

Falk, L., Daniel, M.D., and King, N., 1979, Interaction of herpesviruses with lymphoblastoid cells, *In Vitro*, p. 170.

Falk, L., Desrosiers, R., and Hunt, R.D., 1980, *Herpesvirus saimiri* infection in squirrel and marmoset monkeys, in: *Viruses in Naturally Occurring Cancers* (G. Todaro, M. Essex, and H. zur Hausen, eds.), pp. 137–143, Cold Spring Harbor Press, Cold Spring Harbor, New York.

Fischer, R.G., Falk, L.A., Rytter, A., Burton, G.J., Luecke, D.H., and Deinhardt, F., 1974, *Herpesvirus saimiri*: Viability in four species of hematophagous insects and attempted insect transmission to marmosets, *J. Natl. Cancer Inst.* **52**:1477.

Fleckenstein, B., 1979, Oncogenic herpesviruses of non-human primates, *Biochim. Biophys. Acta* **560**:301.

Fleckenstein, B., and Bornkamm, G.W., 1975, Structure and function of *H. saimiri* DNA, in: *Oncogenesis and Herpesviruses II*, Part 1 (G. de Thé, H. zur Hausen, and M.A. Epstein, eds.), pp. 145–150, IARC Press, Lyon.

Fleckenstein, B., and Mulder, C., 1980, Molecular biological aspects of *Herpesvirus saimiri* and *Herpesvirus ateles*, in: *Viral Oncology* (G. Klein, ed.), pp. 799–812, Raven Press, New York.

Fleckenstein, B., and Wolf, H., 1974, Purification and properties of *H. saimiri* DNA, *Virology* **58**:55.

Fleckenstein, B., Bornkamm, G.W., and Ludwig, H., 1975a, Repetitive sequences in complete and defective genomes of *Herpesvirus saimiri*, *J. Virol.* **15**:398.

Fleckenstein, B., Werner, J., Bornkamm, G.W., and zur Hausen, H., 1975b, Induction of a malignant lymphoma by *Herpesvirus saimiri* DNA, in: *Epstein–Barr Virus Production, Concentration, and Purification* (G. de Thé and D.V. Ablashi, eds.), pp. 159–164, IARC Press, Lyon.

Fleckenstein, B., Bornkamm, G.W., and Werner, F.-J. 1976, The role of *Herpesvirus saimiri* genomes in oncogenic transformation of primate cells, in: *Bibliotheca Haematologica*, No. 43 (J. Clemmesen and D.S. Yohn, eds.), pp. 308–312, S. Karger, Basel.

Fleckenstein, B., Müller, I., and Werner, F.-J. 1977, The presence of *Herpesvirus saimiri* genomes in virus-transformed cells, *Int. J. Cancer* **19**:546.

Fleckenstein, B., Bornkamm, G.W., Mulder, C., Werner, F.-J., Daniel, M.D., Falk, L.A., and Delius, H., 1978a, *Herpesvirus ateles* DNA and its homology with *Herpesvirus saimiri* nucleic acid, *J. Virol.* **25**:361.

Fleckenstein, B., Daniel, M.D., Hunt, R.D., Werner, F.-J., Falk, L.A., and Mulder, C., 1978b, Tumor induction with DNA of oncogenic primate herpesviruses, *Nature (London)* **274**:57.

Frank-Kamenetskii, M.D., 1971, Simplification of the empirial relationship between melting temperature of DNA, its GC content and concentration of sodium ions in solution, *Biopolymers* **10**:2623.

Friedmann, A., Coward, J.E., and Morgan, C., 1976, Electron microscopic study of the development of *Herpesvirus saimiri*, *Virology* **69:**810.

Gautier, F., Büneman, H., and Grotjahn, F., 1977, Analysis of calf-thymus satellite DNA: Evidence for specific methylation of cytosine in C-G sequences, *Eur. J. Biochem.* **80:**175.

Gerber, P., Pritchett, R.F., and Kieff, E.D., 1976, Antigens and DNA of a chimpanzee agent related to Epstein–Barr virus. *J. Virol.* **19:**1090.

Giddens, W.E., 1975, Replication of *Herpesvirus saimiri* in cultured lymphocytes of infected owl monkeys (*Aotus trivirgatus*): An electron microscopic and immunofluorescent study, *Lab. Invest.* **32**(4):492.

Guntaka, R.V., Rao, P.Y., Mitsialis, A., and Katz, R., 1980, Modification of avian sarcoma proviral DNA sequences in non-permissive XC cells but not in permissive chicken cells, *J. Virol.* **34:**569.

Heine, U., and Ablashi, D.V., 1974, Morphological studies of *Herpesvirus saimiri* in primate cell cultures, *J. Med. Primatol.* **3**(1):18.

Heine, U., Ablashi, D.V., and Armstrong, G.R., 1971, Morphological studies on *Herpesvirus saimiri* in subhuman and human cell cultures, *Cancer Res.* **31:**1019.

Henle, W., Henle, G., Gunvén, P., Klein, G., Clifford, P., and Singh, S., 1973, Patterns of antibodies to Epstein–Barr virus-induced early antigens in Burkitt's lymphoma: Comparison of dying patients with long term survivors, *J. Natl. Cancer Inst.* **50:**1163.

Honess, R.W., O'Hare, P., and Young, D., 1982, Comparison of thymidine kinase activities induced in cells productively infected with *Herpesvirus saimiri* and Herpes simplex virus, *J. Gen. Virol.* **58:**237.

Hunt, R.D., Melendez, L.V., King, N.W., Gilmore, C.E., Daniel, M.D., Williamson, M.E., and Jones, T.C., 1970, Morphology of a disease with features of malignant lymphoma in marmosets and owl monkeys inoculated with *Herpesvirus saimiri*, *J. Natl. Cancer Inst.* **44:**447.

Hunt, R.D., Melendez, L.V., Garcia, F.G., and Trum, B.F., 1972a, Pathologic features of *Herpesvirus ateles* lymphoma in cotton-topped marmosets (*Saguinus oedipus*), *J. Natl. Cancer Inst.* **49:**1631.

Hunt, R.D., Melendez, L.V., King, N.W., and Garcia, F.G., 1972b, *Herpesvirus saimiri*: Malignant lymphoma in spider monkeys: A new susceptible host, *J. Med. Primatol.* **1:**114.

Hunt, R.D., Garcia, F.G., Barahona, H.H., King, N.W., Fraser, C.E.O., and Melendez, L.V., 1973, Spontaneous *Herpesvirus saimiri* lymphoma in an owl monkey, *J. Infect. Dis.* **127:**723.

Hunt, R.D., Daniel, M.D., Baggs, R.B., Blake, B.J., Silva, D., DuBose, D., and Melendez L.V., 1975a, Clinico-pathologic characterization of *Herpesvirus saimiri* malignant lymphoma in New Zealand White rabbits, *J. Natl. Cancer Inst.* **54:**(6):1401.

Hunt, R.D., Barahona, H.H., King, N.W., Fraser, C.E.O., Garcia, F.G., and Melendez, L.V., 1975b, Spontaneous *Herpesvirus saimiri* lymphoma in owl monkeys, in: *Comparative Leukemia Research 1973* (Y. Ito and R.M. Dutcher, eds.), *Bibliotheca Haematologica*, No. 40, pp. 351–355, S. Karger, Basel.

Hunt, R.D., Blake, J.B., and Daniel, M.D., 1976, *Herpesvirus saimiri* lymphoma in owl monkeys (*Aotus trivirgatus*): Susceptibility, latent period, hematologic picture and course, *Theriogenology*, 6:139.

Hunt, R.D., Barahona, H., and Daniel, M.D., 1978, *Herpesvirus ateles* malignant lymphoma in owl monkeys: A new susceptible primate, in: *Advances in Comparative Leukemia Research* (D.S. Yohn and P. Bentvelzen, eds.), pp. 198–200, Elsevier/North-Holland, Amsterdam.

Johnson, D.R., Klein, G., and Falk, L., 1980, Interaction of *Herpesvirus ateles* and *Herpesvirus saimiri* with primate lymphocytes. I. Selective adsorption of virus by lymphoid cells, *Intervirology* **13:**21.

Johnson, D.R., Ohno, S., Kaschka-Dierich, C., Fleckenstein, B., and Klein, G., 1981, Relationship between *Herpesvirus ateles* associated nuclear antigen (HATNA) and the number of viral genome equivalents in HVA-carrying lymphoid lines, *J. Gen. Virol.* **52**:221.

Judson, S.C., Henle, W., and Henle, G., 1977, A cluster of Epstein–Barr virus-associated American Burkitt's lymphoma, *N. Engl. J. Med.* **297**:468.

Kaschka-Dierich, C., Bornkamm, G.W., and Thomssen, R., 1979, No homology detectable between Marek's disease virus (MDV) and herpesvirus of the turkey (HVT) DNA, *Med. Microbiol. Immunol.* **165**:223.

Kaschka-Dierich, C., Bauer, I., Fleckenstein, B., and Desrosiers, R.C., 1981, Episomal and non-episomal herpesvirus DNA in lymphoid tumor cell lines, in: *Modern Trends in Human Leukemia* (R. Neth, ed.), Springer-Verlag, Heidelberg, p. 197.

Kaschka-Dierich, C., Bauer, I., Schirm, S., and Fleckenstein, B., 1982, Structure of non-integrated covalently closed circular viral DNA in *Herpesvirus saimiri* and *Herpesvirus ateles* transformed cells (submitted).

Keil, G., Müller, I., Fleckenstein, B., Koomey, J. M., and Mulder, C., 1980, Generation of recombinants between different strains of *Herpesvirus saimiri*, in: *Viruses in Naturally Occurring Cancer* (H. zur Hausen, G. Todaro, and M. Essex, eds.), pp. 145–161, Cold Spring Harbor Press, Cold Spring Harbor, New York.

King, N.W., and Melendez, L.V., 1972, The ultrastructure of *Herpesvirus saimiri*-induced lymphoma in cotton-top marmosets (*Saguinus oedipus*), *Lab. Invest.* **26**:682.

King, N.W., Daniel, M.D., Barahona, H.H., and Melendez, L.V., 1972, Viruses from South American monkeys: Ultrastructural studies, *J. Natl. Cancer Inst.* **49**:273.

Kirk, J.T.O., 1967, Effect of methylation of cytosine residues on the buoyant density of DNA in caesium chloride solution, *J. Mol. Biol.* **28**:171.

Klein, G., Pearson, G., Rabson, A., Ablashi, D.V., Falk, L., Wolfe, L., Deinhardt, F., and Rabin, H., 1973, Antibody reactions to *Herpesvirus saimiri* (HVS)-induced early and late antigens (EA and LA) in HVS-infected squirrel marmoset and owl monkeys, *Int. J. Cancer* **12**:270.

Laufs, R., 1974, Immunization of marmoset monkeys with a killed oncogenic herpesvirus, *Nature (London)* **249**:571.

Laufs, R., and Fleckenstein, B., 1972, Malignant lymphoma induced by partially purified *Herpesvirus saimiri* and recovery of infectious virus from tumorous lymph nodes, *Med. Microbiol. Immunol.* **158**:135.

Laufs, R., and Fleckenstein, B., 1973, Susceptibility to *Herpesvirus saimiri* and antibody development in Old and New World monkeys, *Med. Microbiol. Immunol.* **158**:227.

Laufs, R., and Melendez, L.V., 1973a, Latent infection of monkeys with oncogenic herpes-virus, *Med. Microbiol. Immunol.* **158**:299.

Laufs, R., and Melendez, L.V., 1973b, Oncogenicity of *Herpesvirus ateles* in monkeys, *J. Natl. Cancer Inst.* **51**:599.

Laufs, R., and Steinke, H., 1975a, Vaccination of non-human primates against malignant lymphoma, *Nature (London)* **253**:71.

Laufs, R., and Steinke, H., 1975b, No evidence for particles encapsulating RNA-instructed DNA polymerase and high molecular weight virus-related RNA in herpesvirus induced tumors of non-human primates, *J. Gen. Virol.* **27**:239.

Laufs, R., and Steinke, H., 1976, Vaccination of non-human primates with killed oncogenic herpesviruses, *Cancer Res.* **36**:704.

Laufs, R., Steinke, H., Jacobs, C., Hilfenhaus, J., and Karges, H., 1974a, Influence of interferon on the replication of oncogenic herpesviruses in tissue cultures and in non-human primates, *Med. Microbiol. Immunol.* **160**:285.

Laufs, R., Steinke, H., Steinke, G., and Petzold, D., 1974b, Latent infection and malignant lymphoma in marmosets (*Callithrix jacchus*) after infection with two oncogenic her-pesviruses from primates, *J. Natl. Cancer Inst.* **53**:195.

Lee, Y., Tanaka, A., Silver, S., Smith M., and Nonoyama, M., 1979, Minor DNA homology between herpesvirus of turkey and Marek's disease virus?, *Virology* **93**:277.

Linnemann, C.C., Light, I.J., Buchman, T.G., and Ballard, J.L., 1978, Transmission of herpes-simplex virus type 1 in a nursery for the newborn: Identification of viral isolates by DNA "fingerprinting," *Lancet* **1**:964.

Lvovsky, E., Levine, P.H., Fuccillo, D., Ablashi, D.V., Bengali, Z.H., Armstrong, G.R., and Levy, H.B., 1981, Epstein–Barr virus and *Herpesvirus saimiri*: Sensitivity to interferons and interferon inducers, *J. Natl. Cancer Inst.* **66**:1013.

Ludwig, H., Biswal, N., and Benyesh-Melnick, M., 1972, Studies on the relatedness of herpesviruses through DNA–DNA hybridization, *Virology* **49**:95.

Luetzeler, J., Heine, U., Wendel, E., Prasad, U., and Ablashi, D.V., 1979, Ultrastructural studies on the replication of *Herpesvirus ateles*-73 in owl monkey kidney cells, *Arch. Virol.* **60**:59.

Ma, N.S.F., Jones, T.C., Miller, A.C., Morgan, L.M., and Adams, E.A., 1976, Chromosome polymorphism and banding patterns in the owl monkey (*Aotus*), *Lab. Anim. Sci.* **26**(6):1022.

Mandel, J.L., and Chambon, P., 1979, DNA methylation: Organ specific variations in the methylation pattern within and around ovalbumin and other chicken genes, *Nucleic Acids Res.* 7(8):2081.

Marczynska, B., Falk, L., Wolfe, L., and Deinhardt, F., 1973, Transplantation and cytogenetic studies of *Herpesvirus saimiri*-induced disease in marmoset monkeys, *J. Natl. Cancer Inst.* **50**:331.

Martin, L.N., and Allen, W.P., 1975, Response to primary infection with *Herpesvirus saimiri* in immunosuppressed juvenile and newborn squirrel monkeys, *Infect. Immun.* **12**:526.

McGhee, J.D., and Ginder, G.D., 1979, Specific DNA methylation sites in the vicinity of the chicken beta-globin genes, *Nature* (*London*) **280**:419.

Medveczky, P., Desrosiers, R.C., Falk, L.A., Fleckenstein, B., and Mulder, C., 1980, Structural differences in the genomes of 14 strains of *Herpesvirus saimiri*, Abstracts of the Fifth Cold Spring Harbor Meeting on Herpesviruses, August 26–31, 1980.

Melendez, L.V., Daniel, M.D., Hunt, R.D., and Garcia, F.G., 1968, An apparently new herpesvirus from primary kidney cultures of the squirrel monkey (*Saimiri sciureus*), *Lab. Anim. Care* **18**(3):374.

Melendez, L.V., Daniel, M.D., Garcia, F.G., Fraser, C.E.O., Hunt, R.D., and King, N.W., 1969a, *Herpesvirus saimiri*. I. Further characterization studies of a new virus from the squirrel monkey, *Lab. Anim. Care* **19**:372.

Melendez, L.V., Hunt, R.D., Daniel, M.D., Garcia, F.G., and Fraser, C.E.O., 1969b, *Herpesvirus saimiri*. II. An experimentally induced malignant lymphoma in primates, *Lab. Anim. Care* **19**:378.

Melendez, L.V., Hunt, R.D., Daniel, M.D., Fraser, C.E.O., Garcia, F.G., and Williamson, M.E., 1970a, Lethal reticuloproliferative disease induced in *Cebus albifrons* monkeys by *Herpesvirus saimiri*, *Int. J. Cancer* **6**:431.

Melendez, L.V., Hunt, R.D., Daniel, M.D., and Trum, B.F., 1970b, New World monkeys, herpesviruses and cancer, in: *Infections and Immunosuppression in Sub-human Primates* (H. Balner and W.J.B. Beveridge, eds.), pp. 111–117, Munksgaard, Copenhagen.

Melendez, L.V., Daniel, M.D., Hunt, R.D., Fraser, C.E.O., Garcia, F.G., King, N.W., and Williamson, M.E., 1970c, *Herpesvirus saimiri*. V. Further evidence to consider this virus as the etiological agent of a lethal disease in primates which resembles a malignant lymphoma, *J. Natl. Cancer Inst.* **44**:1175.

Melendez, L.V., Hunt, R.D., Daniel, M.D., Blake, B.J., and Garcia, F.G., 1971, Acute lymphocytic leukemia in owl monkeys inoculated with *Herpesvirus saimiri*, *Science* **171**:1161.

Melendez, L.V., Hunt, R.D., King, N.W., Barahona, H.H., Daniel, M.D., Fraser, C.E.O., and Garcia, F.G., 1972a, *Herpesvirus ateles*, a new lymphoma virus of monkeys, *Nature* (*London*) *New Biol.* **235**:182.

Melendez, L.V., Castellano, H., Barahona, H.H., Daniel, M.D., Hunt, R.D., Fraser, C.E.O., Garcia, F.G., and King, N.W., 1972b, Two new herpesviruses from spider monkeys (*Ateles geoffroyi*), *J. Natl. Cancer Inst.* **49**:233.

Melendez, L.V., Barahona, H.H., Daniel, M.D., Hunt, R.D., Fraser, C.E.O., Garcia, F.G., King, N.W., and Castellanos, H., 1972c, Two new herpesviruses from spider monkeys (*Ateles geoffroyi*), *Proceedings of the Second International Symposium on Health*

Aspects of the International Movement of Animals, (Pan American Health Organization Scientific Publication No. 235, p. 145.

Melendez, L.V., Hunt, R.D., Daniel, M.D., Fraser, C.E.O., Barahona, H.H., Garcia, F.G., and King, N.W., 1972d, Lymphoma viruses of monkeys: *Herpesvirus saimiri* and *Herpesvirus ateles*, the first oncogenic herpesviruses of primates—a review, in: *Oncogenesis and Herpesviruses* (P.M. Biggs, G. de Thé, and L.N. Payne, eds.), pp. 451–461, IARC Press, Lyon.

Melendez, L.V., Hunt, R.D., Daniel, M.D., Fraser, C.E.O., Barahona, H.H., King, N.W., and Garcia, F.G., 1972e, *Herpesvirus saimiri* and *ateles*—their role in malignant lymphomas of monkeys, *Fed. Proc. Fed. Am. Soc. Exp. Biol.* **31:**1643.

Morgan, D.G., and Epstein, M.A., 1977, Sequential immunofluorescence and infectivity studies on the replication of *Herpesvirus saimiri* on owl monkey kidney cells, *J. Gen. Virol.* **34:**61.

Morgan, D.G., Epstein, M.A., Achong, B.G., and Melendez, L.V., 1970, Morphological confirmation of the herpes nature of a carcinogenetic virus of primates (*Herpes saimiri*), *Nature (London)* **228:**170.

Morgan, D.G., Achong, B.G., and Epstein, M.A., 1973, Unusual intranuclear tubular structures associated with the maturation of *Herpesvirus saimiri* in monkey kidney cell cultures, *Br. J. Cancer* **27:**434.

Morgan, D.G., Achong, B.G., and Epstein, M.A., 1976, Morphological observation on the replication of *Herpesvirus saimiri* in monkey kidney cell cultures, *J. Gen. Virol.* **32:**461.

Mulder, C., Koomey, J.M., Daniel, M.D., Desrosiers, R.C., Fleckenstein, B., and Falk, L., 1982, Structural alterations in the virion DNA of non-oncogenic variants of *Herpesvirus saimiri*, to be published.

Neubauer, R.H., and Rabin, H., 1978, *In vitro* effect of levamisole on the mitogenic response of owl monkey peripheral blood lymphocytes with the development of *Herpesvirus saimiri*-induced lymphoma, in: *Comparative Leukemia Research* (D.S. Yohn and P. Bentvelzen, eds.), pp. 406–407, Elsevier, Amsterdam.

Neubauer, R.H., and Rabin, H., 1979, Herpes-induced lymphomas: Immunodepression and disease, in: *Naturally Occurring Biological Immunosuppressor Factors and the Relationship to Disease* (R.H. Neubauer, ed.), pp. 203–231, CRC Press, West Palm Beach, Florida.

Neubauer, R.H., Wallen, W.C., Parks, W.P., Rabin, H., and Cicmanec, J.L., 1974a, Attempts to demonstrate type-C virus in normal and neoplastic tissues of non-human primate origin, *Lab. Anim. Sci.* **24:**235.

Neubauer, R.H., Wallen, W.C., and Rabin, H., 1974b, Stimulation of *Herpesvirus saimiri* expression in the absence of evidence for type C virus activation in a marmoset lymphoid cell line, *J. Virol.* **14:**745.

Neubauer, R.H., Wallen, W.C., Rabin, H., Pearson, G.R., and Ablashi, D.V., 1974c, Virological investigations of *Herpesvirus saimiri*-infected owl monkeys, *J. Med. Primatol.* **3:**27.

Neubauer, R.H., Wallen, W.C., and Rabin, H., 1975, Inhibition of the mitogenic response of normal peripheral lymphocytes by extracts or supernatant fluids of a *Herpesvirus saimiri* lymphoid tumor cell line, *Infect. Immun.* **12:**1021.

Neubauer, R.H., Wallen, W.C., and Rabin, H., 1976, Antilymphocytic factor derived from a *Herpesvirus saimiri* lymphoid tumor cell line, in: *Comparative Leukemia Research* (J. Clemmensen and D.S. Yohn, eds.), *Bibliotheca Haematologica*, pp. 388–395, S. Karger, Basel.

Neubauer, R.H., Rabin, H., and Dunn, F.E., 1980, Antiviral cell-mediated immune responses and effect of chromosome polymorphism in *Herpesvirus saimiri*-infected monkeys, *Infect. Immun.* **27:**549.

Neubauer, R.H., Dunn, F.E., and Rabin, H., 1981, Infection of multiple T cell subsets and changes in lymphocyte functions associated with *Herpesvirus saimiri* infection of owl monkeys, *Infect. Immun.* **32:**698.

Nussinov, R., 1980, Some rules in the ordering of nucleotides in the DNA, *Nucleic Acids Res.* **8**(19):4545.
Ohno, S., Luka, J., Klein, G., and Daniel, M.D., 1979, Detection of a nuclear antigen in *Herpesvirus ateles*-carrying marmoset lines by the acid-fixed nuclear binding technique, *Proc. Natl. Acad. Sci. U.S.A.* **76**:2042.
Oie, H.K., Ablashi, D.V., Armstrong, G.R., Pearson, G.R., Orr, T., and Heine, U., 1973, A continuous *in vitro* source of *Herpesvirus saimiri*, *J. Natl. Cancer Inst.* **51**:1077.
Pearson, G.R., and Beneke, J.S., 1977, Inhibition of *Herpesvirus saimiri* replication by phosphonoacetic acid, benzo(a)pyrene, and methylcholanthrene, *Cancer Res.* **37**:42.
Pearson, G.R., and Davis, S., 1976, Immune response of monkeys to lymphotropic herpesvirus antigens, *Cancer Res.* **36**:688.
Pearson, G.R., and Scott, R.E., 1977, Isolation of virus-free *Herpesvirus saimiri* antigen-positive plasma membrane vesicles, *Proc. Natl. Acad. Sci. U.S.A.* **14**:2546.
Pearson, G.R., and Scott, R.E., 1978, Potential for immunization against herpesvirus infection with plasma membrane vesicles, in: *Oncogenesis and Herpesviruses III*, Part 2 (G. de Thé, W. Henle, and F. Rapp, eds.), pp. 1019–1025, IARC Press, Lyon.
Pearson, G., Ablashi, D., Orr, T., Rabin, H., and Armstrong, G., 1972, Intracellular and membrane immunofluorescence investigations on cells infected with *Herpesvirus saimiri*, *J. Natl. Cancer Inst.* **49**:1417.
Pearson, G.R., Orr, T., Rabin, H., Cicmanec, J., Ablashi, D., and Armstrong, G., 1973, Antibody patterns to *Herpesvirus saimiri*-induced antigens in owl monkeys, *J. Natl. Cancer Inst.* **51**:1939.
Pearson, G.R., Rabin, H., Wallen, W.C., Neubauer, R.H., Orr, T.W., and Cicmanec, J.L., 1974, Immunological and virological investigations on owl monkeys infected with *Herpesvirus saimiri*, *J. Med. Primatol.* **3**:54.
Periman, P., Tyrrell, S., and Rabson, A.S., 1972, *Herpesvirus saimiri* in marmoset-mouse hybrid cell lines, *J. Natl. Cancer Inst.*, **49**:387.
Prevost, J.M., Orr, T.W., and Pearson, G.R., 1975, Augmentation of lymphocyte cytotoxicity by antibody to *Herpesvirus saimiri* associated antigens, *Proc. Natl. Acad. Sci. U.S.A.* **72**:1671.
Prevost, J.M., Pearson, G.R., Wallen, W.C., Rabin, H., and Qualtière, L., 1976, Antibody responses to membrane antigens in monkeys infected with *Herpesvirus saimiri*, *Int. J. Cancer* **18**:679.
Qualtiere, L.F., and Pearson, G.R., 1978, Solubilization and characterization of *Herpesvirus saimiri* induced membrane antigens, *J. Virol.* **25**:852.
Rabin, H., 1971, Assay and pathogenesis of oncogenic viruses in non-human primates, *Lab. Anim. Sci.* **21**:1032.
Rabin, H., Pearson, G., Klein, G., Ablashi, D., Wallen, W., and Cicmanec, J., 1973, *Herpesvirus saimiri* antigens and virus recovery from cultured cells and antibody levels and virus isolations from squirrel monkeys, *Am. J. Phys. Anthropol.* **38**:491.
Rabin, H., Neubauer, R.H., Pearson, G.R., Cicmanec, J.L., Wallen, W.C., Loeb, W.F., and Valerio, M.G., 1975a, Brief communication: Spontaneous lymphoma associated with *Herpesvirus saimiri* in owl monkeys, *J. Natl. Cancer Inst.* **54**:499.
Rabin, H., Pearson, G.R., Wallen, W.C., Neubauer, R.H., Cicmanec, J.L., and Orr, T.W., 1975b, Infection of capuchin monkeys (*Cebus albifrons*) with *Herpesvirus saimiri*, *J. Natl. Cancer Inst.* **54**:673.
Rabin, H., Adamson, R.H., Neubauer, R.H., Cicmanec, J.L., and Wallen, W.C., 1976, Pilot studies with human interferon in *Herpesvirus saimiri*-induced lymphoma in owl monkeys, *Cancer Res.* **36**:715.
Rabin, H., Neubauer, R.H., Hopkins, R.F., and Nonoyama, M., 1978, Further characterization of a herpesvirus-positive orangutan cell line and comparative aspects of *in vitro* transformation with lymphotropic Old World primate herpesviruses, *Int. J. Cancer* **21**:762.
Rabson, A.S., O'Connor, G.T., Lorenz, D.E., Kirschstein, R.L., Legallais, F.Y., and Tralka, T.S., 1971, Lymphoid cell-culture line derived from lymph node of marmoset infected with *Herpesvirus saimiri*-preliminary report, *J. Natl. Cancer Inst.* **46**:1099.

Randall, R.E., and Honess, R.W., 1980, Proteins of *Herpesvirus saimiri*: Identification of two virus polypeptides released into the culture medium of productively infected cells, *J. Gen. Virol.* **51**:445.

Rangan, S.R.S., 1976, Brief communication: Oncornavirus particles in lymphoid cultures from a howler monkey with *Herpesvirus saimiri*-induced disease, *J. Natl. Cancer Inst.* **57**(4):951.

Rangan, S.R.S., Martin, L.N., Enright, F.M., and Allen, W.P. 1976, *Herpesvirus saimiri*-induced malignant lymphoma in rabbits, *J. Natl. Cancer Inst.* **57**:151.

Rangan, R.S., Martin, L.N., Enright, F., and Abee, C., 1977, *Herpesvirus saimiri* induced lymphoproliferative disease in howler monkeys, *J. Natl. Cancer Inst.* **59**:165.

Roberts, R., 1980, Restriction and modification enzymes and their recognition sequences, *Gene* **8**:329.

Robinson, J.E., Andiman, W.A., Henderson, E., and Miller, G., 1977, Host-determined differences in expression of surface marker characteristics on human and simian lymphoblastoid cell lines transformed by Epstein–Barr virus, *Proc. Natl. Acad. Sci. U.S.A.* **74**:749.

Roizman, B., 1979, The organization of the herpes simplex virus genomes, in: *Annual Review of Genetics*, Vol. 13 (H.L. Roman, A. Campbell, and L.M. Sandler, eds.), pp. 25–57, Annual Reviews, Palo Alto, California.

Roizman, B., and Buchman, T., 1979, The molecular epidemiology of herpes simplex virus, *Hosp. Pract.*, Jan. 1979, p. 95.

Roizman, B., and Furlong, D., 1974, The replication of herpes-viruses, in: *Comprehensive Virology*, Vol. 3 (H. Fraenkel-Conrat and R.R. Wagner, eds.), pp. 229–403, Plenum Press, New York.

Schaffer, P.A., 1975, Genetics of herpesviruses—A review, in: *Oncogenesis and Herpesviruses II*, Part I (G. de Thé, M.A. Epstein, and H. zur Hausen, eds.), pp. 195–217, IARC Press, Lyon.

Schaffer, P.A., Falk, L.A., and Deinhardt, F., 1975, Brief communication: Attenuation of *Herpesvirus saimiri* for marmosets after successive passage in cell culture at 39°C, *J. Natl. Cancer Inst.* **55**:1243.

Schlom, J., and Spiegelman, S., 1971, Simultaneous detection of reverse transcriptase and high molecular weight RNA unique to oncogenic RNA viruses, *Science* **174**:840.

Schlom, J., and Spiegelman, S., 1974, Breast cancer: Molecular basis for virus etiology, *N. Y. State J. Med.* **74**:1373.

Simonds, J.A., Rabey, W.G., Graham, B.J., Oie, H., and Vande Woude, G.F., 1975, Purification of *Herpesvirus saimiri* and properties of the viral DNA, *Arch. Virol.* **49**:249.

Smith, C., Heberling, R.L., Barker, S.T., and Kalter, S.S., 1979, Squirrel monkey retrovirus and *Herpesvirus saimiri*: Observation in the same cell following isolation, *J. Natl. Cancer Inst.* **63**:983.

Southern, E.M., 1975, Detection of specific sequences among DNA fragments separated by gel electrophoresis, *J. Mol. Biol.* **98**:503.

Sterz, H., Ludwig, H., and Rott, R., 1973–1974, Immunologic and genetic relationship between herpes simplex virus and bovine herpes mammillitis virus, *Intervirology* **2**:1.

Sutter, D., and Doerfler, W., 1980, Methylation of integrated adenovirus type 12 DNA sequences in transformed cells is inversely correlated with viral gene expression, *Proc. Natl. Acad. Sci. U.S.A.* **77**:253.

Todaro, G.J., Scherr, C.J., Sen, A., King, N., Daniel, M.D., and Fleckenstein, B., 1978, Endogenous New World primate type C viruses isolated from owl monkey (*Aotus trivirgatus*) kidney cell line, *Proc. Natl. Acad. Sci. U.S.A.* **75**:1004.

Tracy, S., and Desrosiers, R.C., 1980, RNA from unique and repetitive DNA sequences of *Herpesvirus saimiri*, *Virology* **100**:204.

Tralka, T.S., Costa, J., and Rabson, A., 1977, Electron microscopic study of *Herpesvirus saimiri*, *Virology* **80**:158.

Ullman, J.S., and McCarthy, B.J., 1973, The relationship between mismatched base pairs and the thermal stability of DNA duplexes. II. Effects of deamination of cytosine, *Biochim. Biophys. Acta* **294**:416.

Van der Ploeg, L.H.T., and Flavell, R.A., 1980, DNA methylation in the human γδβ-globin locus in erythroid and non-erythroid tissues, *Cell* **19**:947.

Waalwijk, C., and Flavell, R.A., 1978, *Msp* I, an isoschizomer of *Hpa* II which cleaves both unmethylated and methylated *Hpa* II sites, *Nucleic Acids Res.* **5**(9):323.

Wallen, W.C., Neubauer, R.H., Rabin, H., and Cicmanec, J.L., 1973, Nonimmune rosette formation by lymphoma and leukemia cells from *Herpesvirus saimiri*-infected owl monkeys, *J. Natl. Cancer Inst.* **51**:967.

Wallen, W.C., Neubauer, R.H., and Rabin, H., 1974, *In vitro* immunological characteristics of lymphoid cells derived from owl monkeys infected with *Herpesvirus saimiri, J. Med. Primatol.* **3**:41.

Wallen, W.C., Rabin, H., Neubauer, R.H., and Cicmanec, J.L., 1975a, Depression in lymphocyte response to general mitogens by owl monkeys infected with *Herpesvirus saimiri, J. Natl. Cancer Inst.* **54**:679.

Wallen, W.C., Neubauer, R.H., and Rabin, H., 1975b, Evidence for suppressor cell activity associated with induction of *Herpesvirus saimiri*-induced lymphoma, *Clin. Exp. Immunol.* **22**:468.

Wallen, W.C., Pearson, G.R., Prevost, J.M., Neubauer, R.H., and Rabin, H., 1976, Correspondence between T cell suppression and ADLC titers during *Herpesvirus saimiri* (HVS) infection in owl monkeys, *Bibl. Haematol.* **43**:339.

Werner, F.-J., Bornkamm, G.W., and Fleckenstein, B., 1977, Episomal viral DNA in a *Herpesvirus saimiri*-transformed lymphoid cell line, *J. Virol.* **22**:794.

Werner, F.-J., Desrosiers, R.C., Mulder, C., Bornkamm, G.W., and Fleckenstein, B., 1978, Physical mapping of viral episomes in *Herpesvirus saimiri* transformed lymphoid cells, in: *Persistent Viruses*, Vol. XI (J.G. Stevens, G.J. Todaro, and C.F. Fox, eds.), pp. 189–200, Academic Press, New York.

Wetmur, J.G., and Davidson, N., 1968, Kinetics of renaturation of DNA, *J. Mol. Biol.* **31**:349.

Wittek, R., Cooper, J.A., Barbosa, E., and Moss, B., 1980, Expression of the vaccinia virus genome: Analysis and mapping of mRNAs encoded within the inverted terminal repetition, *Cell* **21**:487.

Wolfe, L.G., Falk, L.A., and Deinhardt, F., 1971a, Oncogenicity of *Herpesvirus saimiri* in marmoset monkeys, *J. Natl. Cancer Inst.* **47**:1145.

Wolfe, L.G., Marczynska, B., Rabin, H., Smith, R., Tischendorf, T., Gavitt, F., and Deinhardt, F., 1971b, Viral oncogenesis in nonhuman primates, in: *Medical Primatology* (E.I. Goldsmith and J. Moor-Jankowski, eds.), pp. 671–682, S. Karger, Basel.

Wright, J., Falk, L.A., Collins, D., and Deinhardt, F., 1976, Brief communication: Mononuclear cell fraction carrying *Herpesvirus saimiri* in persistently infected squirrel monkeys, *J. Natl. Cancer Inst.* **57**:959.

Wright, J., Falk, L.A., Wolfe, L.G., Ogden, J., and Deinhardt, F., 1977, Susceptibility of common marmosets (*Callithrix jacchus*) to oncogenic and attenuated strains of *Herpesvirus saimiri, J. Natl. Cancer Inst.* **59**:1475.

Wright, J., Falk, L.A., Wolfe, L.G., and Deinhardt, F., 1980, *Herpesvirus saimiri*: Protective effect of attenuated strain against lymphoma induction, *Int. J. Cancer* **26**:477.

Yang, R.C., and Wu, R., 1979, BK virus DNA sequence: Extent of homology with Sv40 DNA, *Proc. Natl. Acad. Sci. U.S.A.* **76**:1179.

Yefenof, E., Klein, G., Jondal, M., and Oldstone, M.B.A., 1976, Surface markers on human B- and T-lymphocytes in two-color immunofluorescence studies on the association between EBV receptors and complement receptors on the surface of lymphoid cell lines, *Int. J. Cancer* **17**:693.

Youssoufian, H., and Mulder, C., 1981, Detection of methylated sequences in eukaryotic DNA with the restriction endonucleases SmaI and XmaI, *J. Mol. Biol.* **150**:133.

Yoshida, M., Kawai, S., and Toyoshima, K., 1980, Uninfected avian cells contain structurally unrelated progenitors of viral sarcoma genes, *Nature (London)* **287**:653.

CHAPTER 7

The Molecular Biology of Marek's Disease Herpesvirus

MEIHAN NONOYAMA, Ph.D.

I. INTRODUCTION

Marek's disease (MD) is a lymphoproliferative disease in chickens that was originally described by Marek (1907). A herpesvirus was isolated in cell cultures derived from tumors and from peripheral-blood lymphocytes (Churchill and Biggs, 1967; Solomon *et al.*, 1968; Nazerian *et al.*, 1968; Nazerian and Burmester, 1968), which implied that the virus is the causative agent of this disease (Churchill and Biggs, 1968; Witter *et al.*, 1969). Although Marek's disease virus (MDV) is generally found in the cell-associated form in tissue culture, cell-free virus is produced in the feather follicles of infected birds and released into the air (Calneck *et al.*, 1970; Nazerian and Witter, 1970). Both cell-associated virus in infected cells and cell-free virus, extracted from the feather follicle, can induce MD tumor in chickens by inoculation.

Herpesvirus of turkeys (HVT) was isolated from the turkey and found to be immunologically related to MDV (Kanamura *et al.*, 1969; Witter *et al.*, 1970). This virus is apathogenic for chickens as well as for the turkey and is now commonly used for vaccination against MD induction by MDV infection (Okazaki *et al.*, 1970; Purchase, 1973) [see Chapter 8 (Section VII.D)].

MDV is important in tumor virology, since it is one of the herpesviruses that are involved in the induction of tumors in their natural hosts. The similarity of this system to that of Epstein–Barr virus (EBV) in relation to oncogenicity makes the study of MDV more attractive in molecular virology.

MEIHAN NONOYAMA, Ph.D. • Department of Virology, Showa University Research Institute for Biomedicine in Florida, 5180 113th Avenue North, Clearwater, Florida 33520.

II. MAREK'S DISEASE VIRUS DEOXYRIBONUCLEIC ACID

A. Characterization

MDV DNA is a linear double-stranded DNA of considerably high molecular weight. The sedimentation coefficient of this viral DNA was determined as 56 S by comparing it with the sedimentation of T4 phage DNA and herpes simplex virus (HSV) DNA through a sucrose gradient by centrifugation (L.F. Lee *et al.*, 1971). This corresponds to 1.2×10^8 daltons, which is slightly larger than HSV DNA. In an alkaline gradient, MDV DNA shows a highly heterogeneous population, with the fastest-sedimenting peak of 70 S corresponding to 6×10^7 daltons. This implies the presence of nicks or gaps within the DNA strands (L.F. Lee *et al.*, 1971). The density of MDV DNA in CsCl was once measured as 1.715–1.716 g/cm^3 (L.F. Lee *et al.*, 1969), but was later reported as 1.705 g/cm^3 (L.F. Lee *et al.*, 1971) corresponding to 46 mole % guanine plus cytosine. In a comparative study of MDV DNA and HVT DNA, it was concluded that HVT DNA has a sedimentation coefficient similar to that of MDV DNA, but a slightly higher density in CsCl, 1.707 g/cm^3 (Hirai *et al.*, 1979).

The total sum of the molecular weights of MDV DNA fragments produced by digestion with *Eco*RI, *Hin*dIII, and *Sal*I was calculated as $1.0–1.1 \times 10^8$ daltons, which is slightly less than the value obtained from the sedimentation coefficient (Hirai *et al.*, 1979; Lee, unpublished). The electron-microscopic picture of self-annealed, single-stranded viral DNA showed large and small loops, indicating that the viral DNA molecules contain inverted repeat sequences at the end of the molecules and within the molecules (Cebrian *et al.*, 1982). This represents a situation similar to the HSV DNA molecule, which has four different forms produced by the combination of the two different orientations of the two unique large and small sequences, as discussed in Chapter 3. However, the restriction-enzyme patterns of MDV DNA are reasonably homogeneous and do not reveal any quarter-molar ratios, suggesting that at least the majority of the molecules contain a fixed orientation of the two unique sequences. The physical map of MDV DNA is being constructed confirming the presence of inverted repeats at the end and within the molecule.

A hetrogeneous MDV DNA population has been reported recently that consists of viral DNA with densities of 1.705 and 1.700 g/cm^3 in CsCl equilibrium centrifugation (Tanaka *et al.*, 1980). The viral DNA obtained from a plaque-purified virus preparation is homogeneous, showing only the higher density of 1.705 g/cm^3. The lower-density DNA accumulates during the repeated passages of cell-associated virus through chick embryo fibroblasts (CEF). The lower-density DNA coincides with the density of cellular DNA. However, hybridization with complementary RNA (cRNA) specific to the viral DNA indicated that the DNA with the lower density is of viral origin. By the 20th passage, at a ratio of 5 uninfected CEF to 1 infected CEF, 90% of the viral DNA becomes the lower-density type. Since a high titer of cell-free infectious virus is difficult to obtain in this system, the defective nature of the virus with the lower-density DNA has not been demonstrated.

The infectious DNA of MDV has been reported by Kaaden (1978). His data showed that the specific infectivity is 10 plaques/μg MDV DNA on CEF, and the latency period of viral replication is approximately 2 weeks. The viral DNA is also capable of inducing tumors in chicken within 6 weeks by intraabdominal inoculation.

B. Homology between Marek's Disease Virus and Herpesvirus of Turkeys Deoxyribonucleic Acid

HVT is known to be antigenically related to MDV (Kanamura *et al.*, 1969; Witter *et al.*, 1970), but is apathogenic for both the turkey and the chicken. HVT is successfully used as a vaccine to protect chickens against the development of MDV-induced lymphoma (Okazaki *et al.*, 1970; Purchase, 1973). It is important to know the degree of relatedness between these two viruses for further studies on the mechanism of tumor induction and vaccination. Surprisingly, neither cRNA hybridization nor DNA–DNA reassociation kinetics revealed any sequence homology between HVT and MDV DNA (Y.S. Lee *et al.*, 1979; Kaschka-Dierich *et al.*, 1979a) or significant hybridization (Hirai *et al.*, 1979). Southern-blot hybridization using a radioactive MDV DNA probe and total HVT DNA digested by *Eco*RI on nitrocellulose filter, however, showed the hybridization of three fragments, 5.6, 4.9, and 4.7 $\times 10^6$ daltons, when exposed for 4 days (Y.S. Lee *et al.*, 1979) (Fig. 1, column 3). The reciprocal experiment also showed the hybridization of four MDV DNA fragments, 7.1, 4.5, 3.9, and 3.5 $\times 10^6$ daltons (Fig. 1, column 5). It was roughly estimated from densitometer tracings that the homologous sequences between HVT and MDV DNA are within the range of 1–4%. The minor homology of these two viral DNAs is rather unexpected considering the close antigenic relationship of the two viruses and the vaccine capability of HVT. It would be interesting to determine whether the homologous sequences present in HVT DNA belong to the genes for the protective function of HVT against tumor development induced by MDV infection.

C. Viral Genome in Tranformed Cells and Tumors

The presence of multiple copies of a herpesvirus genome in tumors and transformed cells was first demonstrated in biopsies from Burkitt-lymphoma patients (zur Hausen *et al.*, 1970; Nonoyama *et al.*, 1973) and in the lymphoblastoid cells derived from Burkitt-lymphoma patients (Nonoyama and Pagano, 1971; zur Hausen and Schulte-Holthausen,

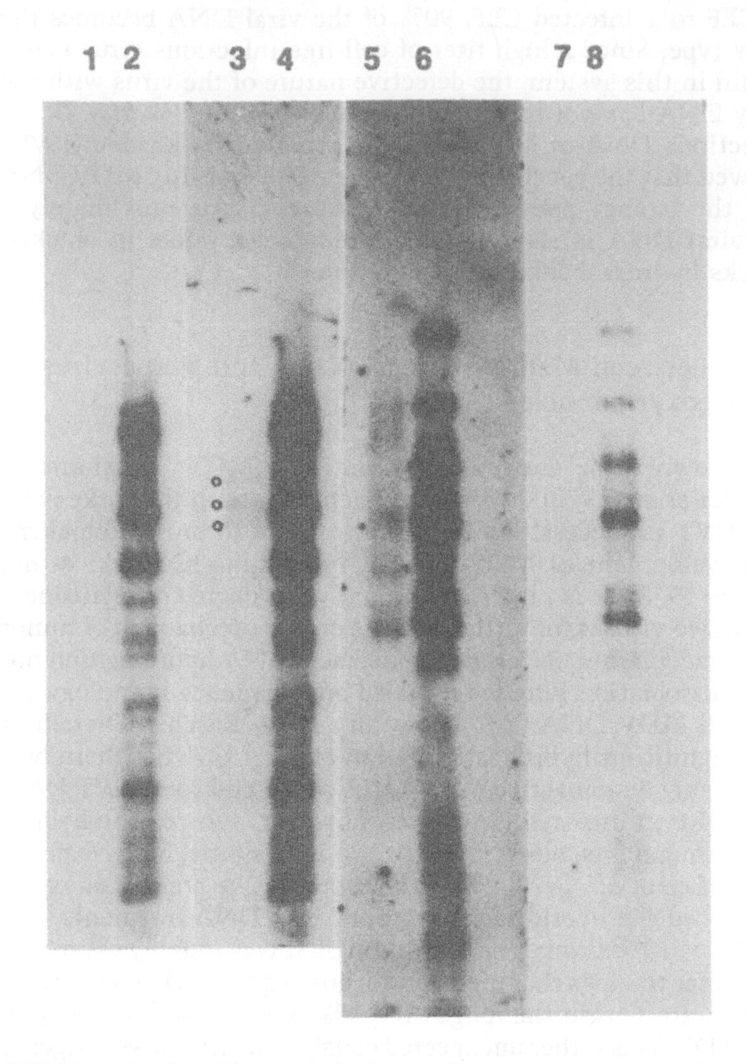

FIGURE 1. Southern-blotting hybridization between MDV DNA and HVT DNA: autoradiographs of agarose slab gel electrophoresis of MDV DNA and HVT DNA digested with *Eco*RI endonuclease. MDV DNA (0.75 μg) or HVT DNA (0.25 μg) was digested with *Eco*RI and electrophoresed in 0.5% agarose gel at 80 V for 16 hr. The DNA fragments were transferred to a nitrocellulose filter paper. The filters were backed at 80° for 2 hr under vacuum and hybridized with ^{32}P-labeled viral DNA (1×10^7 cpm). Hybridized fragments were exposed to autoradiography film (RP/R2) with intensifying screens (DuPont Cronex). Columns: 1, ^{32}P-labeled MDV DNA + 0.25 μg HVT DNA exposed for 0.5 day; (2) ^{32}P-labeled MDV DNA + 0.75 μg MDV DNA exposed for 0.5 day; (3) ^{32}P-labeled MDV DNA + 0.25 μg HVT DNA exposed for 4 days; (4) ^{32}P-labeled MDV DNA + 0.75 μg MDV DNA exposed for 4 days; (5) ^{32}P-labeled HVT DNA + 0.75 μg MDV DNA exposed for 4 days; (6) ^{32}P-labeled HVT DNA + 0.25 μg HVT DNA exposed for 4 days; (7) ^{32}P-labeled HVT DNA + 0.75 μg MDV DNA exposed for 0.5 day; (8) ^{32}P-labeled HVT DNA + 0.25 μg HVT DNA exposed for 0.5 day. From Y. S. Lee *et al.* (1979).

1970). The presence of viral DNA in tumors induced by MDV infection in chickens was first demonstrated by Nazerian *et al.* (1973) by cRNA hybridization specific to MDV DNA. Five tumors from different organs, liver, testis, and ovary of different chickens were tested and found to contain viral DNA ranging from 3 to 80 genomes/cell (Nazerian *et al.*, 1973; L.F. Lee *et al.*, 1975). These tumors were negative for viral antigens by an immunofluorescence test, thus indicating that the viral DNA found in these tumors must be present in the latent form. Lymphoblastoid-cell lines established from tumors that developed in MDV-infected chickens also contained viral DNA. The MSB-1 cell line, first established by Akiyama *et al.* (1973), was tested by cRNA hybridization test (Nazerian and Lee, 1974), and 60–90 genomes/cell were obtained. Since this cell line is a productive cell line and a certain percentage of the cells support viral replication, the number of genomes reported may not accurately represent the number of latent viral genomes per cell. The MKT-1 cell line reported by Tanaka *et al.* (1978) is a nonproductive cell line, and the 15 viral genomes/cell obtained by DNA–DNA reassociation kinetics is the true number of latent viral genomes for this cell line. Silver *et al.* (1979a) also reported the number of viral genomes in MD tumors in various organs of MDV-infected chickens. All seven tumors tested, including liver, kidney, spleen, and ovary, contained from 3 to 12 viral genomes/cell. The relatively small number of viral genomes per cell in tumors may be due to the mixture of normal cells and tumor cells in the samples tested. The consistent presence of viral DNA in MD tumors presents close analogy to the relationship of Burkitt lymphoma and EBV DNA (zur Hausen *et al.*, 1970; Nonoyama *et al.*, 1973). Although HVT-transformed cells have not been obtained yet, a lymphoblastoid-cell line cotransformed by HVT and MDV has been established, and the coexistence of both viral genomes has been shown by DNA–DNA reassociation kinetics (Hirai, unpublished data).

Since MDV genomes are present in nonproductively infected tumors and cells, the next obvious question is whether these viral genomes are integrated into cellular DNA or exist as plasmid DNA such as EBV DNA in Raji cells (Tanaka and Nonoyama, 1974; Lindahl *et al.*, 1975). The status of viral DNA in MKT-1 cells was examined by a 5–10% neutral glycerol gradient (Tanaka *et al.*, 1978). The majority of viral DNA was detected in two peaks around the 100 and 60 S regions by cRNA hybridization, whereas the high-molecular-weight celullar DNA was sedimented to the bottom of the gradient. The 100 S DNA sedimented as 200 S DNA in alkaline-gradient centrifugation through 10–30% glycerol gradient and banded in ethidium bromide–CsCl equilibrium centrifugation at the position expected for closed circular molecules. These characteristics of the viral DNA are compatible with the physical properties of closed (100 S) and open (60 S) circular molecules of herpes viral DNA of 100 million daltons, thus indicating that the majority of MDV DNA present in MKT-1 cells is not integrated into cellular DNA and exists as circular plasmid viral DNA. These results, however, do not answer the question whether or not some viral DNA may be integrated in MKT-1 cells. Contradictory results to this earlier report have been presented by Kaschka-Dierich *et al.* (1979b) using two different cell lines, MSB-1 (low-producer cell line) and HPRS-1 (nonproducer cell line), which were established from ovarian and splenic lymphomas, respectively (Akiyama *et al.*, 1973; Powell *et al.*, 1974). MDV DNA (1.705 g/cm^3) in HPRS-1 did not show any separation from cellular DNA (1.700 g/cm^3) by equilibrium centrifugation in neutral CsCl, whereas viral DNA in MSB-1 cells was partly separated from cellular DNA. After HPRS-1 DNA was sheared to a molecular weight of 8×10^6 daltons, some MDV DNA with a free-viral DNA density was detected, but a large portion of the viral DNA still remained associated with the cellular DNA. This association of MDV DNA with HPRS-1 cellular DNA was shown to be alkaline-stable. These experiments suggested that the viral DNA in established lymphoblastoid cells of the chicken is integrated into cellular DNA. Of particular interest is the viral DNA in HPRS-1 cells that appeared to be integrated in fragments because the mechanical shearing did not release free viral DNA. The conflicting results obtained in different laboratories may be due to the types of cell lines used. However, the use of CsCl centrifugation analysis to answer the viral DNA integration problem leaves various technical questions and ambiguous explanations. The final evidence of integration of viral DNA into cellular DNA must await the results from Southern-blot analysis and the cloning of viral and cellular DNA joint-fragments into a bacterial plasmid.

Regardless of the state of viral DNA in transformed cells, the following two experiments indicated that these resident viral DNAs are well regulated by cellular functions. Phosphonoacetate specifically inhibits MDV DNA replication, and the activity of the isolated viral-specific DNA polymerase was inhibited by the drug (L.F. Lee *et al.*, 1976). However, the number of viral genomes per cell remained constant during the cultivation of transformed cells in the presence of phosphonoacetate (Nazerian and Lee, 1976), indicating that the replication of the resident viral DNA is not affected by the drug. This may indicate that viral DNA polymerase is not used for the resident viral DNA replication in transformed cells. The second experiment indicated that the resident viral DNA replicates only once during the cell cycle in the early S phase before active synthesis of cellular DNA occurs. MKT-1 cells were synchronized by double thymidine blocks, and the number of viral genomes per cell was measured at various times in the cell cycle. The number of viral genomes per cell increased 2-fold just before significant cellular DNA synthesis was detected, and quickly dropped to the normal level (Lau and Nonoyama, 1980). Thus, the resident viral DNA duplicates only once during the cell cycle at the early S phase and is not allowed to replicate extensively as in the lytically infected cells. The number of viral genomes per cell can be maintained constant in this manner. These two experiments indicate that the viral DNA, once established within the transformed cells, is regarded as a member of cellular DNA and controlled by the cellular regulatory mechanism.

The recent report of Hughes *et al.* (1980) has shown the chromosomal location of the resident MDV DNA in transformed cells. Metaphase chromosomes from an MDV-induced lymphoblastoid-cell line, MSB-1, were isolated and partially fractionated by a sucrose density-gradient centrifugation. DNA from each fraction was digested by a restriction enzyme, and Southern-blot hybridization was conducted for the DNA in each fraction with an MDV DNA probe. The result indicated that at least two distinct size classes of chromosomes contained the major resident viral DNA. These results are analogous to the situation of Raji cells in which all the EBV DNA was found in chromosomes despite the nonintegrated status of the viral DNA (Nonoyama and Pagano, 1971).

III. PROTEINS

Various immunologically defined antigens have been identified in animals bearing MDV-induced tumors, in transformed cells, and in virus-infected cells. These include membrane antigens, gel precipitin antigens, early and late antigens, and Marek's disease tumor-associated surface antigen (MATSA), which is discussed in Chapter 8. In this section, only biochemically defined proteins will be discussed.

A. A and B Antigens

Three antigens have been detected in MDV-infected cell extracts and culture supernatants by an immunodiffusion method using sera from infected animals (Chubb and Churchill, 1968). A antigens are found both in culture supernatants and in infected-cell extracts, whereas B and C antigens are predominantly located in the extracts of virus-infected cells. There are conflicting observations concerning the simultaneous loss of pathogenicity and A antigens (Churchill *et al.*, 1969; Bülow, 1971; Purchase *et al.*, 1971). This antigen is a glycoprotein and stable at pH 2 (Settnes, 1972; Ross and Biggs, 1973; Onuma, *et al.*, 1974; Long *et al.*, 1975a). The apparent molecular weight is estimated to range from 33,000 (Onuma *et al.*, 1974,) and 44,850 (Long *et al.*, 1975a) to 80,000 (Ross and Biggs, 1973). The most purified antigen preparation has an isoelectric point of 6.5 (Long *et al.*, 1975b). The protein tends to aggregate, and the aggregated proteins appears to have a lower isoelectric point. B antigen is also a glycoprotein and stable at pH 2 (Velicer *et al.*, 1978). B antigen is resistant to trypsin treatment under conditions in which A antigen is sensitive to the treatment. The molecular weight of B antigen is slightly higher, 58,250 daltons, than that of A antigen, 44,850 as reported by the same investigators (Long *et al.*, 1975a; Velicer *et al.*, 1978). The isolectric point of B antigen is 4.54 (Velicer *et al.*, 1978). Both A and B antigens are common antigens between MDV and HVT (Velicer *et al.*, 1978).

B. Deoxyribonucleic Acid Polymerase

MDV-induced DNA polymerase has been partially purified by phosphocellulose chromatography and characterized (Boezi *et al.*, 1974). The virus-induced DNA polymerase had a sedimentation coefficient of 5.9 S in the presence of 0.25 M KCl, whereas DNA polymerase isolated from the uninfected nuclear fraction showed 7.3 and 3.1S. From this sedimentation value, the molecular weight of MDV-induced enzyme was estimated to be 100,000 daltons. MDV DNA polymerase could not use either poly(dA) · oligo(dT) or poly(dC) · oligo(dG), whereas the cellular DNA polymerase used these synthetic oligonucleotides. The addition of ammonium sulfate to the reaction mixture markedly inhibited the reaction of MDV DNA polymerase.

Phosphonoacetate and phophonoformate are inhibitors of MDV-induced DNA polymerase (L.F. Lee *et al.*, 1976; Leinbach *et al.*, 1976; Reno *et al.*, 1978) as well as of viral replication (L.F. Lee *et al.*, 1976; Nazerian and Lee, 1976). The apparent inhibition constants of these drugs are 1–3 μM (Reno *et al.*, 1978). Phosphonoacetate and phosphonoformate are mutually exclusive inhibitors, indicating that the two drugs bind DNA polymerase at the same site on the enzyme. Leinbach *et al.* (1976) reported that phosphonoacetate interacts with virus-induced DNA polymerase at the pyrophosphate-binding site. Phosphonoformate was not an effective inhibitor of the phosphonoacetate-resistant mutant virus or of its induced DNA polymerase (Reno *et al.*, 1978). The isolation of virus mutants with an altered property of DNA polymerase that are resistant to these drugs suggests that the virus-induced DNA polymerase is indeed virus-gene-coded.

C. Thymidine Kinase

HVT-induced thymidine kinase has been characterized by Kit *et al.* (1973–1974, 1974). HVT-induced thymidine kinase was found in the cytosol fraction of infected chicken cells, whereas cellular thymidine kinase was found in both the mitochondrial and cytosol fractions of uninfected cells. The sedimentation coefficients for viral, cellular cytosol, and mitochondrial enzymes are 5.3, 4.9, and 4.5 S, respectively. The molecular weight of the viral-induced enzyme is 87,700 daltons calculated from these sedimentation values. The isoelectric point of the viral thymidine kinase is 6.1, whereas those of the cytosol and mitochondria enzymes in uninfected cells are 9.7 and 5.7, respectively. Both ATP and CTP can be phosphate donors for the HVT-induced thymidine kinase as well as for the mitochondrial enzyme, whereas the cellular cytosol kinase can use only ATP as a phosphate donor. HVT-induced thymidine kinase and HSV-1 thymidine kinase share common properties, such as a similar isoelectric point, a similar sedimentation coefficient, and the type of phosphate donors; however, the antibody to HSV-1 kinase does not neutralize the activity of HVT-induced thymidine kinase.

IV. MAREK'S DISEASE VIRUS GENOME EXPRESSION

A. Transformed Cells

The previous section described the establishment of lymphoblastoid-cell lines with T-cell characteristics in a number of laboratories. Most of these cell lines are virus-productive with viral antigens expressed, but some of them are virus-nonproductive and do not show any evidence of viral antigens by immunological methods. These nonproductive cells, however, contain multiple copies of the viral genome in the form of non-integrated plasmid viral DNA or possibly some as viral DNA covalently linked to the cellular DNA. The absence of immunologically reactive antigens in the nonproductive cells does not necessarily indicate the absence of viral gene expression. At present, since there is no way of recognizing viral proteins in transformed cells except immunologically, the evidence of viral gene expression is best studied by viral genome transcription. In the EBV system, nonproductive cells showed a limited transcription of viral genomes; 10–25% of viral DNA is transcribed, and not all the viral RNA is found in polyribosomes (Hayward and Kieff, 1976; Tanaka *et al.*, 1977), whereas nearly 50% of viral DNA or all the viral genes, if asymmetric transcription is assumed, are transcribed in the productive cells as expected. DNA–RNA hybridization kinetics were conducted by using *in vitro*-labeled viral DNA and RNA extracted from nonproductive lymphoblastoid cells, MKT-1 (Silver *et al.*, 1979b). The result indicated that 12–14% of the viral DNA template was transcribed in these nonproductive cells, whereas nearly 50% of the viral DNA was hybridized to viral RNA in a productive cell system. This showed the restrictive transcription of viral genome in nonproductively infected tumor cells. Furthermore, only 60–70% of viral RNA species among the viral RNA synthesized in these nonproductive cells was found in the polyribosome fraction. Thus, these results revealed two different types of viral gene regulation, one at the transcriptional level and the other at the viral messenger RNA transfer process to polyribosomes. The viral RNA transcribed but not found in the polyribosomes is probably not due to the loss of these sequences by splicing, since nearly all the viral RNA sequences are found in the polyribosomes in productively infected cells. The treatment of MKT-1 cells with 5-iodo-2-deoxyuridine (IUdR) induced viral antigens, but viral DNA replications remained restricted. However, 45% of the viral DNA template was transcribed by this treatment, showing an increase to a level nearly equal to the transcription of the viral genome in productively infected cells. A hybridization experiment using a mixture of the two types of RNA, one from IUdR-induced MKT-1 cells and the other from productively infected cells, with a viral DNA probe indicated that the two types of RNA are transcribed from the same region of the viral DNA. Again, the polyribosomal fraction of IUdR-treated MKT-1 cells contained only 60–70% of viral RNA species transcribed, indicating the presence of a block of viral RNA transfer. Therefore, failure of the completion of viral replication after the IUdR treatment may also be due to a block at the transfer process. Probably a similar or the same regulatory mechanism is operating in MKT-1 cells to control viral replication regardless of IUdR induction of viral replication. The absence of some species of viral RNA in the polyribosomal fraction may be ascribed to the RNA processing step, such as RNA splicing. The detailed transcriptional control pattern must await the completion of viral DNA physical mapping. The results so far obtained for the transcriptional pattern of MKT-1 cells follow a pattern similar to EBV genome transcription as presented in Chapter 3.

B. Marek's-Disease-Virus-Induced Tumors

It is of interest to see whether the transcriptional pattern of the resident viral genomes in MDV-induced tumors is the same as that in MKT-1 cells, since MKT-1 cells were originally established from an MD lymphoid tumor. RNA was extracted from MDV-induced tumors in the liver, kidney, spleen, and ovary of diseased chickens, and DNA–RNA hybridization kinetics were conducted with an *in vitro*-labeled viral DNA probe (Silver *et al.*, (1979a). The results indicated that all these tumors contained viral RNA transcripts from a limited region of the viral genome, ranging from 5 to 15% of the viral DNA. Again, the polyribosomal fraction contained only about 60% of the RNA species found in the total viral RNA transcripts. A mixed RNA hybridization experiment showed that viral RNA transcribed in MKT-1 cells and in tumors are derived from the same region of the viral genome. Thus, these experiments demonstrated that resident viral DNA in MDV-induced tumor is controlled in a similar manner as in MKT-1 cells, one at the transcription step and the other at the RNA transfer stage or possibly RNA processing stage. This warrants the study of MKT-1 cells to understand viral gene control in MDV-induced tumors.

TABLE I. Comparison of MKT-1 and Raji Cells

Characteristic	Raji	MKT-1
Cell type	B lymphoblastoid	T lymphoblastoid
VCA, EA, MA	Not detected	Not detected
Virus-specific nuclear antigen	EBNA	Not detected
Viral genome	Multiple; 50/cell	Multiple; 15/cell
Viral DNA status	Circular plasmid (integration?)	Circular plasmid (integration?)
Viral genome transcription	Restricted	Restricted
Total	25%	15%
Polyribosomal	10%	7%

V. COMPARISON OF LYMPHOBLASTOID CELLS TRANSFORMED BY MAREK'S DISEASE VIRUS WITH THOSE TRANSFORMED BY EPSTEIN–BARR VIRUS

It would be useful to compare the characteristics of a typical lymphoblastoid-cell line established from MD tumor, MKT-1, with a cell line derived from African Burkitt lymphoma, Raji, in order to reemphasize the similarity of these two virus-transformed cell lines. Thus, the importance of the study of MDV may be realized not only for the understanding of MDV itself but also as a model for the study of EBV to aid in developing knowledge of the role of EBV in human tumors. Table I summarizes the characteristics of Raji and MKT-1 cells that are comparable. A virus-specific nuclear antigen EBNA, is detected in EBV-transformed cells (Reedman and Klein, 1973), but similar nulcear antigens have not been found so far either in established lymphoblastoid cells or in tumors induced by MDV infection. In addition to the characteristics summarized in Table I, the pattern of IUdR induction is comparable in both systems. Both Raji and MKT-1 cells showed virus antigens but no viral DNA replication after IUdR treatment (Silver *et al.*, 1979b). Most of viral genomes were transcribed, but limited species of viral RNA were found in the polyribosomal fraction. Thus, the block of viral gene expression in these two systems appears to be similar. The similarities in the status of viral genomes in transformed cells suggest that MDV and EBV may share common properties in the mechanism of cellular transformation.

REFERENCES

Akiyama Y., Kato, S., and Iwa, N., 1973, Continuous cell culture from lymphoma of Marek's diseae, *Biken J.* **16**:177.

Boezi, J.A., Lee, L.F., Blakesley, R.W., Koenig, M., and Towle, H.C., 1974, Marek's desease herpesvirus-induced DNA polymerase, *J. Virol.* **14**:1209.

Bülow, V. von, 1971, Diagnosis and certain biological properties of the virus of Marek's disease, *Am. J. Vet. Res.* **32**:1275.

Calnek, B.W., Adldinger, H.K., and Kahn, D.E., 1970, Feather follicle epithelium: A source of enveloped and infectious cell-free herpesvirus from Marek's disease, *Avian Dis.* **14**:219.

Cebrian, J., Kaschka-Dierich, C., Berthelot, N., and Sheldrick, P., 1982, Inverted repeat nucleotide sequences in the genomes of Marek's disease virus and the herpesvirus of the turkey, *Proc. Natl. Acad. Sci. U.S.A.* **79**:555.

Chubb, R.C., and Churchill, A.E., 1968, Precipitating antibodies associated with Marek's disease, *Vet. Res.* **83**:4.

Churchill, A.E., and Biggs, P.M., 1967, Agent of Marek's disease in tissue culture, *Nature (London)* **215**:528.

Churchill, A.E., and Biggs, P.M., 1968, Herpes-type virus isolated in cell culture from tumors of chickens with Marek's disease. II. Studies *in vivo*, *J. Natl. Cancer Inst.* **41**:951.

Churchill, A.E., Chubb, R.C., and Baxendale, W., 1969, The attenuation, with loss of antigenicity of the herpes-type virus of Marek's disease (strain HPRS-16) on passage in cell culture, *J. Gen. Virol.* **4**:557.

Hayward, S.D., and Kieff, E.D., 1976, Epstein Barr virus specific RNA. I. Analysis of viral RNA in cellular extracts and in the polyribosomal fraction of permissive and nonpermissive lymphoblastoid cell lines, *J. Virol.* **18**:518.

Hirai, K., Ikuta, K., and Kato, S., 1979, Comparative studies on Marek's disease virus and herpesvirus of turkey DNAs, *J. Gen. Virol.* **45**:119.

Hughes, S.H., Stubblefield, E., Nazerian, K., and Varmus, H.E., 1980, DNA of a chicken herpesvirus is associated with at least two chromosomes in a chick lymphoblastoid cell line, *Virology* **105**:234.

Kaaden, O.R., 1978, Transfection studies *in vitro* and *in vivo* with isolated Marek's disease virus DNA, in: *Oncogenesis and Herpesviruses III* (G. de Thé, W. Henle, F. Rapp, eds.), IARC Scientific Publication, Part 2, p. 267, Lyon.

Kanamura, H., King, D.J., and Anderson, D.P., 1969, A herpesvirus isolated from kidney cell culture of normal turkeys, *Avian Dis.* **13**:853.

Kaschka-Dierich, C., Bornkamm, G.W., and Thomssen, R., 1979a, No homology detectable between Marek's disease virus (MDV) DNA and herpesvirus of the turkey (HVT) DNA, *Med. Microbiol. Immunol.* **165**:223.

Kaschka-Dierich, C., Nazerian, K., and Thomssen, R., 1979b, Intracellular state of Marek's disease virus DNA in two tumour-derived chicken cell lines, *J. Gen. Virol.* **44**:271.

Kit, S., Jorgensen, G.N., Leung, W.C., Irkula, D., and Dubbs, D.R., 1973–1974, Thymidine kinases induced by avian and human herpesviruses, *Intervirology* **2**:299.

Kit, S., Leung, W.C., Jorgensen, G.N., and Dubbs, D.R., 1974, Distinctive properties of thymidine kinase isozymes induced by human and avian herpesviruses, *Int. J. Cancer* **14**:598.

Lau, R.Y., and Nonoyama, M., 1980, Repliction of the resident Marek's disease virus genome in synchronized nonproducer MKT-1 cells, *J. Virol*, **44**:912.

Lee, L.F., Roizman, B., Spear, P.G., Kieff, E.D., Burmester, B.R., and Nazerian, K., 1969, Marek's disease herpesvirus: A cytomegalovirus?, *Proc. Natl. Acad. Sci. U.S.A.* **64**:951.

Lee, L.F., Kieff, E.D., Bachenheimer, S.L., Roizman, B., Spear, P.G., Burmester, B.R., and Nazerian, K., 1971, Size and composition of Marek's disease virus deoxyribonucleic acid, *J. Virol.* **7**:289.

Lee, L.F., Nazerian, K., and Boezi, J.A., 1975, Marek's disease virus DNA in a chicken lymphoblastoid cell line (MSB-1) and in virus-induced tumours, in: *Oncogenesis and Herpesviruses II* (G. de Thé, W. Henle, F. Rapp, eds.), IARC Scientific Publication, Part 2, p. 199, Lyon.

Lee, L.F., Nazerian, K., Leinbach, S.S., Reno, J.M., and Boezi, J.A., 1976, Effect of phosphonoacetate on Marek's disease virus replication, *J. Natl. Cancer Inst.* **56**:823.

Lee, Y.S., Tanaka, A., Silver, S., Smith, M., and Nonoyama, M., 1979, Minor DNA homology between herpesvirus of turkey and Marek's disease virus?, *Virology* **93**:277.

Leinbach, S.S., Reno, J.M., Lee, L.F., Isbell, A.F., and Boezi, J.A., 1976, Mechanism of phosphonoacetate inhibition of herpesvirus-induced DNA polymerase, *Biochemistry* **15**:426.

Lindahl, T., Adams, A., Bjursell, G., Bornkamm, G.W., Kaschaka, C., and Jehn, U., 1976, Covalently closed circular duplex DNA of Epstein Barr virus in a human lymphoid cell line, *J. Mol. Biol.* **102**:511.

Long, P.A., Clark, J.L., and Velicer, L.F., 1975a, Marek's disease herpesviruses. II. Purification and further characterization of Marek's disease herpes-virus A antigen, *J. Virol.* **15**:1192.

Long, P.A., Kaveh-Yamini, P., and Velicer, L.F., 1975b, Marek's disease herpesviruses. I. Production and preliminary characterization of Marek's disease herpesvirus A antigen, *J. Virol.* **15**:1182.

Marek, J., 1907, Multiple Nervenentzündung (Polyneuritis) bei Hühnern, *Dtsch. Tieraerztl. Wochenschr.* **15**:417.

Nazerian, K., and Burmester, B.R., 1968, Electron microscopy of a herpes-virus associated with the agent of Marek's disease in cell culture, *Cancer Res.* **28**:2454.

Nazerian, K., and Lee, L.F., 1974, Deoxyribonucleic acid of Marek's disease virus in a lymphoblastoid cell line from Marek's disease tumors, *J. Gen. Virol.* **25**:317.

Nazerian, K., and Lee, L.F., 1976, Selective inhibition by phosphono-acetic acid of MDV DNA replication in a lymphoblastoid cell line, *Virology* **74**:188.

Nazerian, K., and Witter, R.L., 1970, Cell-free transmission and *in vivo* replication of Marek's disease virus, *J. Virol.* **5**:388.

Nazerian, K., Solomon, J.R., Witter, R.L., and Burmester, B.R., 1968, Studies on the etiology of Marek's disease. II. Finding of a herpes-virus in cell culture, *Proc. Soc. Exp. Biol. Med.* **127**:177.

Nazerian, K., Lindahl, T., Klein, G., and Lee, L.F., 1973, Deoxyribonucleic acid of Marek's disease virus in virus-induced tumors, *J. Virol.* **12**:841.

Nonoyama, M., and Pagano, J.S., 1971, Detection of Epstein–Barr viral genome in nonproductive cells, *Nature (London) New Biol.* **233**:103.

Nonoyama, M., Huang, C.H., Pagano, J.S., Klein, G., and Singh, S., 1973, DNA of Epstein–Barr virus in tissue of Burkitt's lymphoma and nasopharyngeal carcinoma, *Proc. Natl. Acad. Sci. U.S.A.* **70**:3265.

Okazaki, W., Purchase, H.G., and Burmester, B.R., 1970, Protection against Marek's disease by vaccination with a herpesvirus of turkeys (HVT), *Avian Dis.* **14**:413.

Onuma, M., Mikami, T., and Hayashi, T., 1974, Properties of the common antigen associated with Marek's disease herpesvirus and turkey herpesvirus infections, *J. Natl. Cancer Inst.* **52**:805.

Powell, P.C., Payne, L.N., Frazier, J.A., and Rennie, M.T., 1974, Lymphoblastoid cell lines from Marek's disease lymphomas, *Nature (London)* **251**:79.

Purchase, H.G., 1973, Control of Marek's disease by vaccination, *Wild Poult. Sci. J.* **29**:238.

Purchase, H.G., Burmester, B.R., and Cunningham, C.H., 1971, Pathogenicity and antigenicity of clones from strains of Marek's disease virus and the herpes-virus of turkeys, *Infect. Immun.* **3**:295.

Reedman, B., and Klein, G., 1973, Cellular localization of an Epstein–Barr associated complement-fixing antigen in producer and nonproducer lymphoblastoid cell lines, *Int. J. Cancer* **11**:499.

Reno, J.M., Lee, L.F., and Boezi, J.A., 1978, Inhibition of herpesvirus replication and herpesvirus-induced dexoxyribonucleic acid polymerase by phosphonoformate, *Antimicrob. Agents Chemother.* **13**:188.

Ross, L.J.N., and Biggs, P.M., 1973, Purification and properties of the "A" antigen associated with Marek's disease virus infections, *J. Gen. Virol.* **18**:291.

Settnes, O.P., 1972, Some characteristics of the A antigen in Marek's disease virus-infected cell cultures, *Pathol. Microbiol. Scand.* **80**:817.
Silver, S., Smith, M., and Nonoyama, M., 1979a, Transcription of the Marek's disease virus genome in virus-induced tumors, *J. Virol.* **30**:84.
Silver, S., Tanaka, A., and Nonoyama, M., 1979b, Transcription of the Marek's disease virus genome in a nonproductive chicken lymphoblastoid cell line, *Virology* **93**:127.
Solomon, J.J., Witter, R.L., Nazerian, K., and Burmester, B.R., 1968, Studies on the etiology of Marek's disease: Propagation of the agent in cell culture, *Proc. Soc. Exp. Biol. Med.* **127**:173.
Tanaka, A., and Nonoyama, M., 1974, Latent DNA of Epstein–Barr virus: Separation from high molecular weight cell DNA in neutral glycerol gradient, *Proc. Natl. Acad. Sci. U.S.A.* **74**:4658.
Tanaka, A., Nonoyama, M., and Glaser, R., 1977, Transcription of latent Epstein–Barr virus genomes in human epithelial Burkitt hybrid cells, *Virology* **82**:63.
Tanaka, A., Silver, S., and Nonoyama, M., 1978, Biochemical evidence of the nonintegrated status of Marek's disease virus DNA in virus-transformed lymphoblastoid cells of chicken, *Virology* **88**:19.
Tanaka, A., Lee, Y.S., and Nonoyama, M., 1980, Heterogeneous population of virus DNA in serially passaged Marek's disease virus preparation, *Virology* **103**:510.
Velicer, L.F., Yager, D.R., and Clark, J.L., 1978, Marek's disease herpesviruses. III. Purification and characterization of Marek's disease herpesvirus B antigen, *J. Virol.* **27**:205.
Witter, R.L., Burgoyne, G.H., and Solomon, J.J., 1969, Evidence for a herpesvirus as an etiologic agent of Marek's disease, *Avian Dis.* **13**:171.
Witter, R.L., Nazerian, K., Purchase, H.G., and Burgoyne, G.H., 1970, Isolation from turkeys of a cell associated herpesvirus antigenically related to Marek's disease virus, *Am. J. Vet. Res.* **31**:525.
Zur Hausen, H., and Schulte-Holthausen, H., 1970, Presence of EB virus nucleic acid homology in a virus free line of Burkitt tumor cells, *Nature (London)* **227**:245.
Zur Hausen, H., Schulte-Holthausen, H., Klein, G., Henle, W., Henle, G., Clifford, P., and Samtessen, L., 1970, EBV DNA in biopsies of Burkitt tumors and anaplastic carcinomas of the nasopharynx, *Nature (London)* **228**:1056.

CHAPTER 8

Biology of Marek's Disease Virus and the Herpesvirus of Turkeys

LAURENCE NOEL PAYNE, Ph.D.

I. INTRODUCTION

Marek's disease (MD) is a lymphoproliferative disorder of the domestic chicken caused by a herpesvirus and is characterized principally by T-cell lymphoma formation and a demyelinative peripheral neuropathy. It is named after Marek (1907), who first described the neural form of the disease; an associated visceral lymphomatosis was reported later by Pappenheimer *et al.* (1926, 1929a,b).

MD has attracted attention for two main reasons. First, it has been recognized since the 1920s as an important cause of economic loss to the poultry industry. For many years, unsuccessful efforts were made to understand the nature of, and to control, MD. The problem was greatly exacerbated by the appearance in the 1960s of a more virulent form of the disease, responsible for MD becoming the major cause of poultry mortality in many countries. This stimulated increased research on MD, with the consequence that by 1970 the cause of the disease was known and vaccines were available commercially for its control.

Apart from its importance in veterinary medicine, MD also attracted the notice of biomedical research workers as the first neoplastic disease shown to be caused by a herpesvirus and the first common, naturally occurring, lymphomatous disorder to be controlled by vaccination. This interest was heightened by a number of common features between the

LAURENCE NOEL PAYNE, Ph.D. • Houghton Poultry Research Station, Houghton, Huntingdon, Cambridgeshire PE17 2DA, England.

MD herpesvirus (MDV) and the Epstein–Barr herpesvirus, discovered in 1964, associated with Burkitt's lymphoma of African children and considered to be a candidate for the first recognized tumor virus of man (see Epstein and Achong, 1979). Subsequently, MD has been increasingly used as an experimental model in viral oncology; more recently, because of other pathological features of the disease, its possible value as a model in neurological and vascular research has been emphasized (Payne, 1973, 1979).

In this chapter, the biology of MDV and the related herpesvirus of turkeys (HVT) is reviewed; biochemical and molecular aspects of these viruses are discussed in Chapter 7. The reader is also referred to other recent reviews on general aspects of MD (Nazerian *et al.*, 1976; Calnek and Witter, 1978; Nazerian, 1979, 1980; Calnek, 1980) and on the more specialized topics of immunity (Witter, 1976; Sharma, 1978; Theis, 1979; Biggs, 1980; Powell, 1981), pathogenesis (Payne *et al.*, 1976), HVT (Prasad, 1979), and vaccination and other control methods (Biggs, 1975; Purchase, 1975, 1976).

II. HISTORICAL CONSIDERATIONS

The modern era of research on Marek's disease (MD) dates from 1960 and stemmed from the adoption of the newer techniques of experimental virology and the application of ideas then evolving in tumor virology. An extensive literature on MD exists from before that time (see L.G. Chubb and Gordon, 1957; Biggs, 1968) dealing particularly with the pathology and transmissibility of MD, but this is now mainly of historical interest only. There was confusion about whether the disease was infectious and transmissible and whether the neural lesions and associated visceral lymphomatous lesions were related, etiologically and pathologically, to other leukotic diseases of the fowl. Experimental-transmission studies designed to answer these questions produced equivocal results because, it is now known, of the almost ubiquitous presence in domestic chickens of both Marek's disease virus (MDV) and the unrelated retroviruses that also cause lymphoid and other neoplasms. Furthermore, the retroviruses are vertically transmitted and so were no doubt often present even in young chicks, whereas MDV is persistent and contagious, and was difficult to exclude except by isolation of experimental birds. Consequently, transmission experiments frequently resulted in manifestations of both MD and the leukoses, and these occurred in both inoculated and uninoculated control birds.

The confusion was reflected in the terminology used to describe these conditions, the all-embracing term "avian leukosis complex" being widely used (L.G. Chubb and Gordon, 1957). Various pathologically based classifications under this term were used (Anonymous, 1941; Cottral, 1952; L.G. Chubb and Gordon, 1957; Biggs, 1961; J.G. Campbell, 1954,

1961), a widely used one being that which included the terms "neural and visceral lymphomatosis" (Anonymous, 1941). Subsequent experimental studies (Biggs and Payne, 1964) confirmed the view of J.G. Campbell (1954, 1956) that visceral lymphomatosis included two pathologically distinct lymphoid neoplasms, one (MD lymphoma) now known to be caused by a herpesvirus and the other (lymphoid leukosis) by an retrovirus.

The term "Marek's disease" was adopted in 1961 (Biggs, 1961; J.G. Campbell, 1961) to designate the disease first described by Marek (1907) and known previously by a variety of names. The popular names "fowl paralysis" and "range paralysis" referred to the commonly observed clinical signs of the disease. Other terms reflected differing opinions about the pathological nature of the neural lesion, and included "neurolymphomatosis gallinarum" (Pappenheimer *et al.*, 1926, 1929a,b), indicating a neoplastic condition, and "neurogranulomatosis infectiosa gallinarum (Marek)" (Lerche and Fritzsche, 1934), denoting an inflammatory disorder. Pappenheimer regarded the associated lymphoproliferative lesions in the ovary and elsewhere as visceral lymphomas, whereas J.G. Campbell (1956) considered them to be "lymphogranulomas."

Thus, in 1960, there was uncertainty about whether MD was infectious, how it was related to other leukotic diseases, and whether it was an inflammatory or a neoplastic disorder. By 1970, these questions had been answered and commercially available vaccines were in use to control MD. The main events during this period, reviewed by Witter (1971), were: (1) in 1962, the experimental transmission of MD by inoculation of susceptible young chicks with blood or lymphoma cells from birds with MD (Sevoian *et al.*, 1962; Biggs and Payne, 1963); (2) in 1967, growth of the agent in cell culture and its recognition as a herpesvirus (Churchill and Biggs, 1967; Solomon *et al.*, 1968; Nazerian *et al.*, 1968); (3) in 1969, production of live virus vaccines based either on tissue-culture-attenuated MDV (Churchill *et al.*, 1969a) or on the antigenically related herpesvirus of turkeys (Okazaki *et al.*, 1970).

III. NATURAL OCCURRENCE

A. Clinical Signs

Three forms of Marek's disease (MD) are recognized: the classic form, the acute form, and transient paralysis. The *classic form* refers to the neural form of the disease (fowl paralysis) described by Marek (1907) and early workers. Up to 30% of a flock may be affected, and signs occur usually when birds are 2–12 months old. The signs are neurological and vary with the peripheral nerves affected; commonly, there is paresis or paralysis affecting the legs and wings (Fig. 1), occasionally torticollis, and respiratory and alimentary disorders, accompanied by loss of weight. The

FIGURE 1. Paralysis of the legs of a 2-month-old chicken infected with the HPRS-16 strain of MDV.

acute form (Biggs *et al.*, 1965) describes the more virulent type of MD first reported by Benton and Cover (1957) as "acute leukosis," in which lymphomatosis of various visceral organs and tissues is common, and mortality may reach 80% of a flock. Birds may be affected from 6 weeks of age, and losses commonly occur between 3 and 6 months of age. Neural involvement and signs may not be evident, particularly in older birds. Marked lymphomatous enlargement of the liver and other organs is not uncommon, and in older birds such cases are easily confused grossly with lymphoid leukosis. Lymphomatous involvement of the skin ("skin leukosis") may be evident, and iritis may occur in both classic and acute forms. *Transient paralysis* is an uncommon acute encephalitic expression of infection by Marek's disease virus (MDV) (Zander, 1959; Kenzy *et al.*, 1973), occurring in chickens 5–18 weeks of age. Apparently healthy birds suddenly develop various locomotory disorders, such as paralysis of legs, wings, and neck, or incoordination. Mortality is low (less than 5%), and symptoms usually disappear within 24–48 hr. Experimentally, transient paralysis occurs 8–9 days after infection with MDV, but its occurrence depends on strain of bird and virus.

B. Host Range and Prevalence

Under natural conditions, infection by MDV occurs almost entirely in domestic chickens and is ubiquitous among poultry populations throughout the world. In commercial chicken flocks, virtually all birds become infected, commonly within the first few weeks of life (Witter *et al.*, 1971; Biggs *et al.*, 1972). The virus persists in infected birds throughout life. The incidence of clinical MD varies greatly among flocks, poultry houses, and pens within a house (Biggs *et al.*, 1972); factors responsible for this variability are discussed in Sections VIII.B and VIII.C.

There are a number of pathological reports of MD-like lesions in other species of birds, including the budgerigar, canary, duck, goose, owl, partridge, pheasant, quail, and swan (see Purchase, 1972; Cho and Kenzy, 1975). However, it should be noted that MD-like lesions can also be caused by reticuloendotheliosis virus (Witter *et al.*, 1970c). MDV or MDV antibodies have been found naturally in chickens, including feral jungle fowl (Weiss and Biggs, 1972), turkeys (Witter and Solomon, 1971), and quails (Kenzy and Cho, 1969). Cho and Kenzy (1975) examined 111 zoo birds, representing 49 species from 14 orders, for MDV infection and detected it only in birds of the genus *Gallus*; they concluded that the natural hosts of MDV are Galliformes birds. However, infection without clinical disease can occur in ducks (Baxendale, 1969). Experimentally, MD has been induced in turkeys with MDV of turkey and chicken origin (Witter *et al.*, 1974 Elmubarak *et al.*, 1981) and in quails with chicken MDV (Khare *et al.*, 1975; Mikami *et al.*, 1975).

Herpesvirus of turkeys (HVT) is a widely distributed, naturally occurring, apathogenic virus of turkeys. It was discovered by Witter *et al.* (1970b) while they were investigating the cause of lymphomas in turkeys, and was similar to a herpesvirus previously isolated from turkeys by Kawamura *et al.* (1969). HVT does not occur naturally in chickens, but is now widely used to vaccinate chickens against MD (see Section VII.D).

Mammalian hosts and cell cultures are considered to be resistant to infection by MDV and HVT, although infectivity may persist transiently in mammalian cells inoculated with avian cell-associated virus (Hložánek and Sovova, 1974; Witter and Sharma, 1974).

C. Public-Health Significance

Many people are heavily exposed in their occupations to MDV or HVT, but no direct or epidemiological evidence exists that these viruses present any hazard to man. Naito *et al.* (1971) found higher antibody levels to HVT and Epstein–Barr virus (EBV) in a greater proportion of sera from poultry workers than from office workers, possibly related to exposure to MDV or HVT antigens, and to antigens common to these viruses and EBV (Ono *et al.*, 1970; Evans *et al.*, 1973). Sharma *et al.* (1973a), on the other hand, found HVT or MDV antibody in only 8% of human sera, with no association with exposure to virus. Monkeys inoculated with MDV or HVT and observed for 1 year or longer developed no viremia or clinical disease (Sharma *et al.*, 1972, 1973a). Priester and Mason (1974) reported excessive deaths from cervical and ovarian cancer, and from multiple myeloma, in high compared with low poultry population areas. However, only the incidence of cervical cancer was significantly higher than that for the total United States, for which social differences were held likely to be responsible.

D. Economic Loss to the Poultry Industry

During the 1960s, MD in commercial poultry increased to such an extent that it became the most important cause of economic loss to the industry in many countries. These losses resulted from mortality, carcass condemnation, reduced feed conversion, and reduced egg production, detailed estimates of which have been given for the United States by Purchase (1977). In 1970, the last year before HVT vaccine was extensively used in the field, condemnations of broilers due to leukosis (MD) by meat inspectors in the United States reached a peak of nearly 1.6%, and 2.9% in Georgia, the worst-affected state. Following vaccination, these losses were reduced by 80% or more (Purchase, 1975). In the United States, annual losses from MD by 1970 were estimated to be in excess of $200 million, and annual benefit from vaccination in 1974 was estimated to be $168 million (Purchase, 1977).

IV. VIROLOGY

A. Virus Morphology and Morphogenesis

Marek's disease virus (MDV) is a typical herpesvirus (Fig. 2). It has been designated *Gallid herpesvirus* 1 and provisionally placed, with the herpesvirus of turkeys (HVT) (*Gallid herpesvirus* 2), in the subfamily Gammaherpesvirinae of the family Herpesviridae (Matthews, 1979). The nucleocapsid measures about 100 nm and has 162 hollow tubular capsomeres (6 × 9 nm) arranged in icosahedral symmetry. Enveloped particles in the nucleus of infected cells measure 150–180 nm in diameter; those from the cytoplasm measure 250–280 nm (Nazerian, 1971; Nazerian *et al.*, 1971). The nucleoid measures 50–60 nm, contains the DNA, and has a toroidal morphology responsible for the cross-shaped appearance of the virion in thin sections (Nazerian, 1974) (Fig. 3). A cohelical arrangement of the MDV DNA has been described (K. Okado *et al.*, 1980). The morphology of HVT resembles that of MDV (Nazerian *et al.*, 1971; K. Okada *et al.*, 1972).

Morphological changes during replication of MDV have been described by many workers (Nazerian and Burmester, 1968; Epstein *et al.*, 1968; K. Okada *et al.*, 1972; Hamdy *et al.*, 1974). At 8 hr postinfection, small hexagonal particles, 35–45 nm in diameter, believed to be precur-

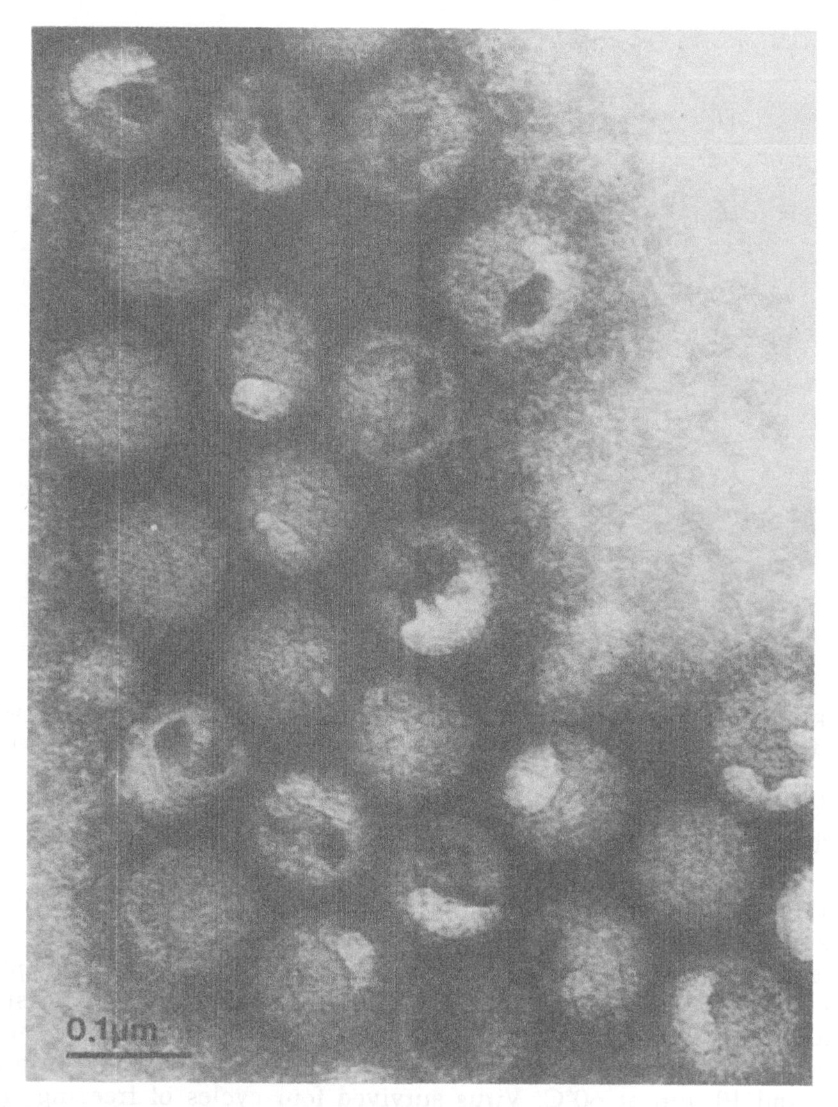

FIGURE 2. Electron micrograph of negatively stained naked MD virions showing typical herpesvirus morphology. Kindly provided by Dr. Judith Frazier.

sors of virion cores, appear in the nucleus. Nucleocapsids appear at about 10 hr and envelopment of nucleocapsids at the nuclear membrane occurs at about 18 hr (Fig. 4). Enveloped virions are also seen in cytoplasmic inclusion bodies (Nazerian and Witter, 1970) (Fig. 5). For a more detailed review of MDV replication, see Payne *et al.* (1976).

Replication of HVT is qualitatively similar to that of MDV, although the 35- to 45-nm nuclear particles are more numerous and form crystalline arrays more frequently, and a larger proportion of naked virions have a cruciform appearance (K. Okada *et al.*, 1972).

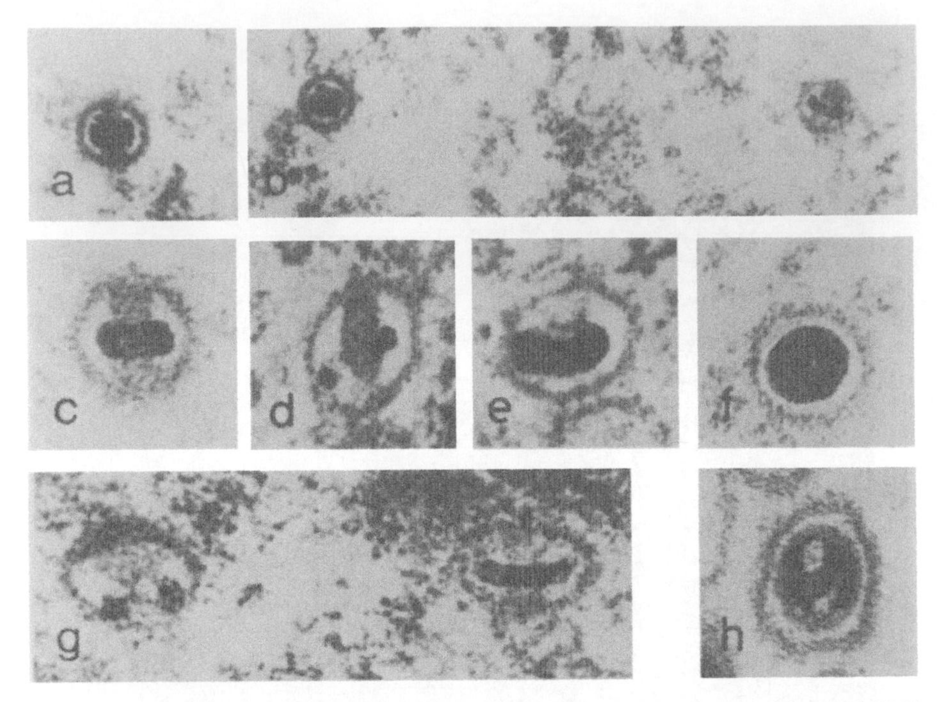

FIGURE 3. Electron micrographs of MDV, showing toroidal arrangement of the viral DNA around a less dense central filamentous structure. From Nazerian (1974) by kind permission.

B. Physical and Chemical Properties

Properties of crude cell-free preparations of MDV obtained from chicken skin were described by Calnek and Adldinger (1971): MDV was stable at pH 7 but lost infectivity after 10 min above and below this pH, with complete loss at pH 3 and 11. Virus titer was stable for at least 7 months at −65°C, but virus titer was lost at −20°C. All infectivity was lost after 2 weeks at 4°C, 4 days at 22–25°C, 18 hr at 37°C, 30 min at 56°C, and 10 min at 60°C. Virus survived four cycles of freezing and thawing, and short periods of sonic vibration. The virus passed a filter of average pore diameter of 300 nm but not 220 nm. Cell-free MDV from cell culture was sensitive to ether and formalin (Nazerian, 1972). Virus-infected feather materials and poultry dust retained infectivity at room temperature for many months (Hložánek *et al.*, 1973; Calnek and Hitchner, 1973). Cell-free MDV and HVT can be lyophilized without loss of titer in the presence of sucrose, phosphate, glutamate, albumin (SPGA) stabilizer and the chelating agent EDTA (Calnek *et al.*, 1970b).

HVT passes filters of 220 nm, and its stability to freezing and thawing, sonication, and temperature is increased in the presence of SPGA (see Prasad, 1979). Lyophilized HVT, sealed under vacuum and stored at 4–6°C, is highly stable (Zanella *et al.*, 1974; Siegmann *et al.*, 1980).

C. Isolation and Cultivation

1. Sources of MDV and HVT

MDV is present in a cell-associated form in many tissues of MDV-infected chickens, as is HVT in HVT-infected turkeys and chickens (Phillips and Biggs, 1972; Witter *et al.*, 1972). Commonly favored for isolation of MDV or HVT in cell culture are viable whole blood cells, buffy coat cells, kidney cells, or [in Marek's disease (MD)] lymphoma cells (Witter *et al.*, 1969); almost all infectivity is lost from such tissues by treatments that disrupt cells, i.e., freezing and thawing, sonication, and lyophilization. Infected cells can be stored at −196°C in the presence of dimethylsulfoxide (Spencer and Calnek, 1967).

Cell-free MDV or HVT is present in feather-follicle epithelium of

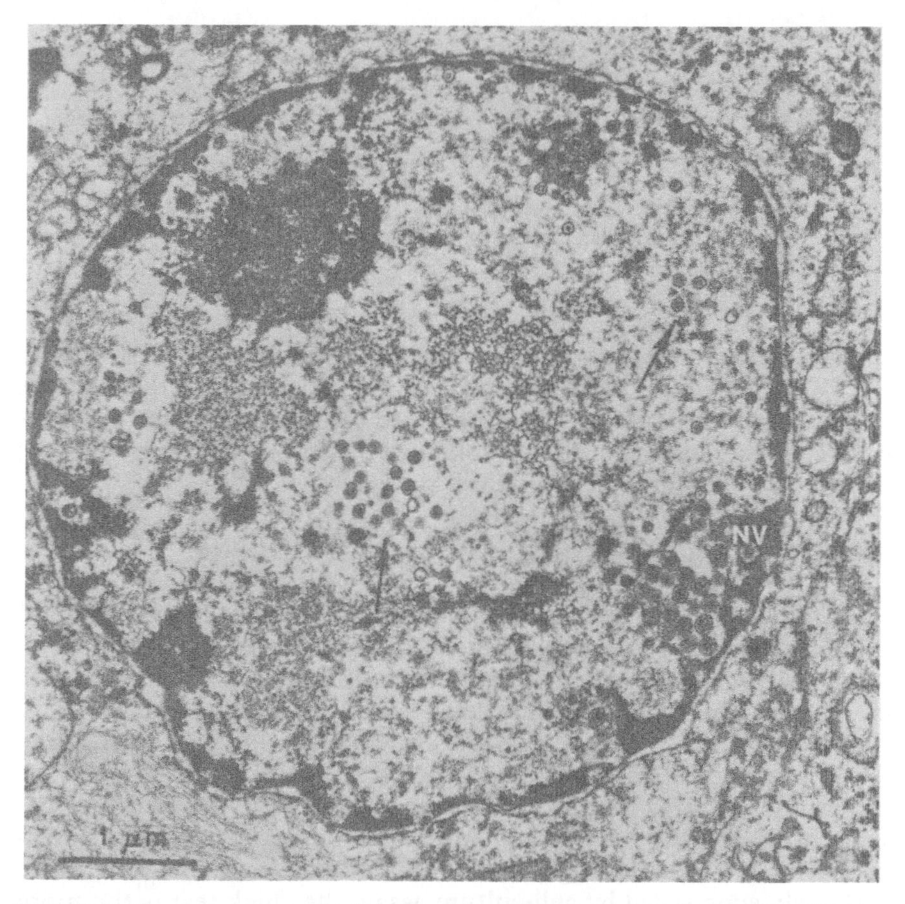

FIGURE 4. Electron micrograph of chick embryo fibroblast infected with MDV. Enveloped virions are seen in a nuclear vesicle (NV), and naked virions are present eleswhere in the nucleoplasm (arrows). Kindly provided by Dr. Judith Frazier.

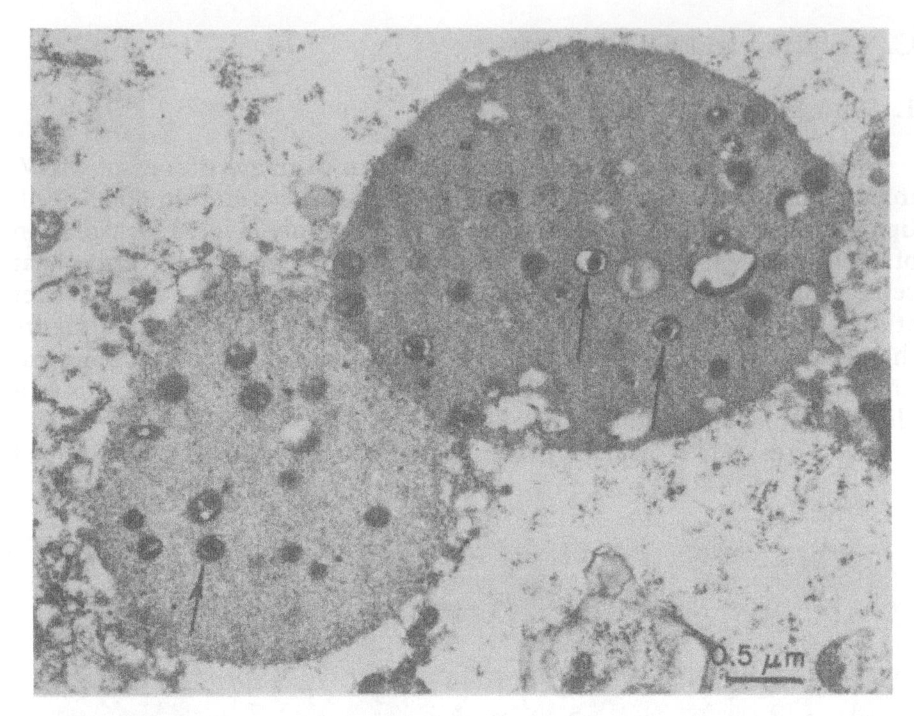

FIGURE 5. Enveloped MD virions (arrows) in cytoplasmic inclusion bodies in feather-follicle epithelium from chicken infected with MDV. From Payne *et al.* (1976) by kind permission of the publishers.

infected chickens or turkeys, respectively (Calnek *et al.*, 1970a; Witter *et al.*, 1972). HVT replicates little, if at all, in feather follicles of chickens. Cell-free virus preparations may be obtained by homogenization and sonication of skin or feather quill tips, followed by slow centrifugation (650g) and filtration (450-nm pore diameter).

2. Chick Inoculation

MDV or HVT can be detected in materials by their intraabdominal inoculation into newly hatched chicks of a susceptible genetic strain. Virulent MDV induces microscopic lesions in nerves and other tissues from 2 weeks postinoculation and gross neural involvement and visceral lymphomas from 3 weeks (Biggs and Payne, 1967; Purchase and Biggs, 1967; Witter *et al.*, 1969). Nonpathogenic MDV, and HVT, do not produce gross lesions, but may be detected by presence of cell-associated viremia, precipitating or virus-neutralizing antibody, and antigen in feather tips (see Sharma, 1975). Although chick-inoculation assays for MDV have been largely superseded by cell-culture assay, the chick test is the more sensitive method of detection (Witter *et al.*, 1969).

3. Cell Culture

Cell-associated and cell-free MDV and HVT produce cytopathic plaques characteristic of herpesviruses within a few days when inoculated onto tissue-culture monolayers of chick kidney cells (CKC) (Churchill, 1968), duck embryo fibroblasts (DEF) (Solomon *et al.*, 1968), chick embryo fibroblasts (CEF) (Nazerian, 1970), quail embryo fibroblasts (Onoda *et al.*, 1970; Colwell *et al.*, 1974), and other avian cells (Purchase *et al.*, 1971a) (Fig. 6). MDV does not produce plaques as efficiently on CEF on primary isolation as on CKC or DEF, but does so after adaptation to cell culture. Plaques consist of foci of refractile rounded or fusiform cells, and polykaryocytes have Cowdry type A intranuclear inclusion bodies (Churchill, 1968; Calnek and Madin, 1969) (Fig. 7). Plaques vary in size and rate of development, depending on the strain of virus. On the basis of growth in CKC, Biggs and Milne (1972) classified isolates of MDV and HVT into those that formed small plaques (naturally occurring mainly apathogenic strains in chicks), medium plaques (pathogenic), or large plaques (tissue-culture-attenuated MDV and HVT). These differences were due to differences in rate of cell-to-cell spread of virus. In CEF,

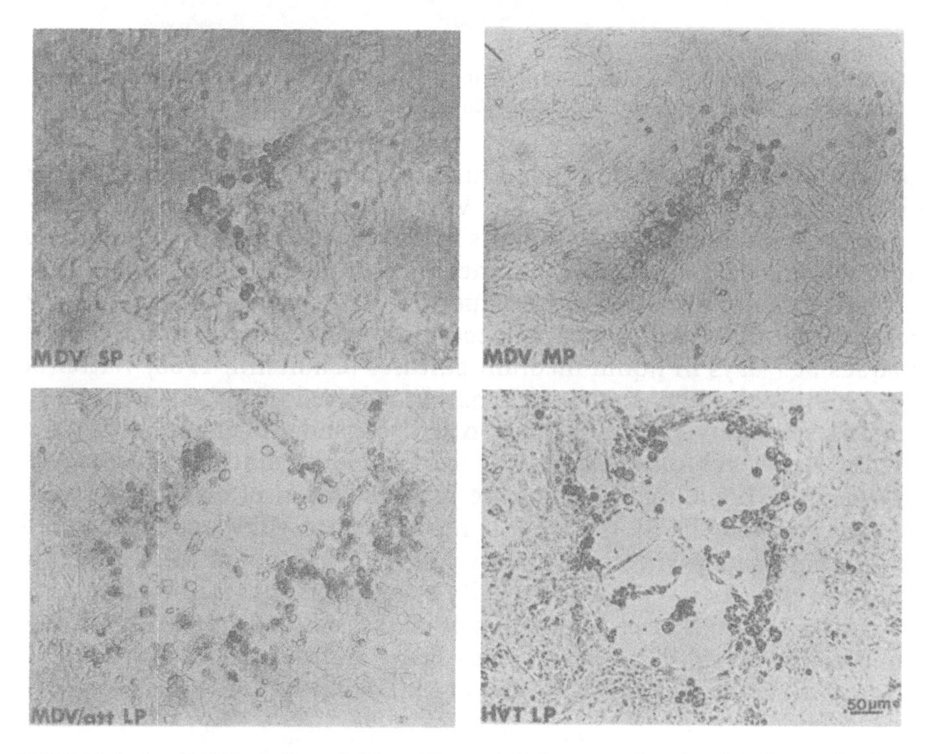

FIGURE 6. Small (SP), medium (MP), and large (LP) plaques induced in CKC by apathogenic, pathogenic, and attenuated MDV, respectively, and HVT. From Biggs and Milne (1972) by kind permission.

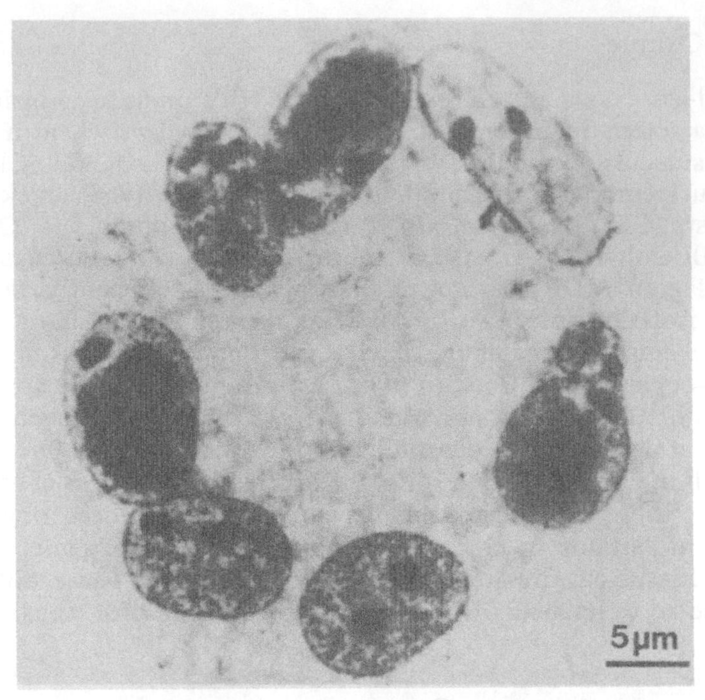

FIGURE 7. Inclusion bodies present in several nuclei of a multinucleated giant cell present in an MDV-induced plaque in CKC. From Churchill (1968) by kind permission.

however, Cho (1976a) found that virulent strains of MDV formed smaller plaques than avirulent strains. The virus in plaque cells is closely cell-associated, although small amounts of cell-free virus can be harvested from disrupted cells, especially in medium containing SPGA and EDTA. Release of cell-free virus and production of secondary plaques do not occur with MDV, which can be accurately assayed by enumeration of plaques at 7 days in liquid-medium cultures (Churchill, 1968; Witter *et al.*, 1969). With HVT, some virus may be released into the supernatant, causing secondary plaque formation (Eidson and Anderson, 1971); the effects of this on virus assay may be overcome by counting HVT plaques at 4 days in liquid-medium culture or by addition of an agar-medium overlay (Withell, 1973). Cho (1981) reported that MDV and HVT could be differentiated *in vitro* by the ability of HVT, but not of MDV, to grow and form plaques in QT35 chemically transformed quail fibroblasts.

Initial absorption of both cell-associated and cell-free MDV of cultured cells is rapid, 50% of the virus being absorbed in 30 min at 37°C (Churchill and Biggs, 1967; Sharma *et al.*, 1969).

4. Embryonated Eggs

MDV and HVT produce pocks on the chorioallantoic membrane (CAM) of the chick embryo, and splenomegaly, when inoculated onto the CAM or into the yolk sac (Bülow, 1969; Biggs and Milne, 1971; Witter *et al.*, 1970b) or intravenously (Longenecker *et al.*, 1975) (Fig. 8). These assays are of similar, or slightly greater, sensitivity compared with cell-culture assays, although Sharma *et al.* (1976) reported that cell-culture adaptation of MDV decreased its pock-forming ability. However, *in ovo* techniques have not been widely used for the assay of MDV or HVT because of the greater convenience of cell-culture assay. Furthermore, the value of the embryo assay may be limited when inocula (e.g., blood, lymphoma cells) contain immunocompetent cells, because of the production of nonspecific pocks caused by a graft-vs.-host reaction.

D. Virus Strains

1. Pathotypes

Many isolates (strains) of MDV have been obtained from naturally infected chickens; some of the better studied of these are listed in Table I. They may be classified into three groups depending on their virulence and pathogenicity in susceptible chicks: (1) highly pathogenic (acute) strains, causing a high incidence of visceral and neural lymphomatosis; (2) moderately pathogenic (classic) strains, causing mainly neural MD, often at a lower incidence; (3) mildly pathogenic or nonpathogenic strains, causing no gross lesions. Usually, the isolated strains are not cloned and

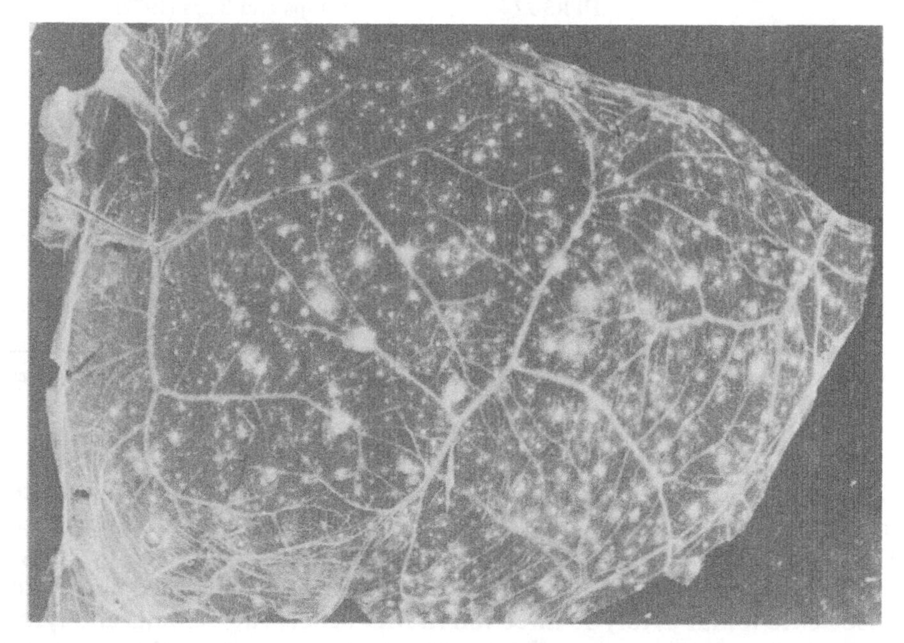

FIGURE 8. Pocks on the CAM of an 18-day-old chick embryo after inoculation of MDV into the yolk sac at 4 days of incubation. From Biggs and Milne (1971) by kind permission.

TABLE I. Some Commonly Studied Isolates of Marek's Disease Virus and Herpesvirus of Turkeys

Type of virus	Strain	References
Highly pathogenic (acute) MDV	JM	Sevoian *et al.* (1962)
	HPRS-16	Purchase and Biggs (1967)
	GA	Eidson and Schmittle (1968a)
	Cal-1	Bankowski *et al.* (1969)
	Conn-B	Jakowski *et al.* (1970)
	Biken C	Kato *et al.* (1970)
	Id-1	Sharma *et al.* (1970)
	RPL 39	Purchase *et al.* (1971a)
	BC-1[a]	Spencer *et al.* (1972)
	HN-1	Kenzy *et al.* (1973)
	Oldenburg	Bülow and Biggs (1975a)
	Md/5, Md/11	Witter *et al.* (1980)
Moderately pathogenic (classic) MDV	HPRS-B14	Biggs and Payne (1963)
	HPRS-17	Purchase and Biggs (1967)
	Conn-A	Chomiak *et al.* (1967)
	WSU-GF	Sharma *et al.* (1969)
	VC	Bülow (1971)
Mildly pathogenic or nonpathogenic MDV	CVI 988	Rispens *et al.* (1972a)
	HN	Cho and Kenzy (1972)
	HPRS-24	Biggs and Milne (1972)
	Beckenham	Blaxland *et al.* (1972)
	C-3	Kenzy *et al.* (1973)
	S-11	M. W. Smith and Calnek (1974)
	Cu-1, -2	M. W. Smith and Calnek (1974)
	HPRS-27	Phillips and Biggs (1972)
	SB-1	Schat and Calnek (1978a)
Artificially attenuated MDV	HPRS-16/att	Churchill *et al.* (1969a)
	JM/att	Nazerian (1970)
	GA/att	Eidson and Anderson (1971)
	BC-1/att	Spencer *et al.* (1972)
HVT	WTHV-1 (= WHG)	Kawamura *et al.* (1969), Eidson and Anderson (1971)
	FC126	Witter *et al.* (1970b)
	HPRS-26	Biggs and Milne (1972)
	PB1	Churchill *et al.* (1973)
	O1	Ono *et al.* (1974)
	TAM-6	Prasad and Spradbrow (1977)

[a] Another pathogenic MDV strain designated BC-1 is associated with the MSB-1 lymphoblastoid-cell line (Akiyama *et al.*, 1973; Murthy and Calnek, 1978).

may therefore consist of mixtures of virus. Some strains of MDV induce transient paralysis (Kenzy *et al.*, 1973); however, the relative importance of virus and host in this syndrome has not been clearly defined. Recently, Witter *et al.* (1980) described variants of MDV with increased ability to cause acute lethal cytolytic infection in MD-susceptible chicks and lymphoma formation in genetically resistant or vaccinated chickens.

As described in Section IV.C.3, the three categories of pathogenicity were associated with differences in plaque type and growth rate in CKC (Biggs and Milne, 1972). The attributes responsible for differences in pathogenicity are not known. Nonpathogenic strains of MDV grow to lower levels in chickens (Phillips and Biggs, 1972) and cause less damage to the immune system than do pathogenic strains (Calnek *et al.*, 1979), suggesting that virus-induced immunosuppression may determine oncogenic expression. In support of this, some strains of low virulence induce lymphomas in immunosuppressed hosts (Calnek *et al.*, 1977). However, other nononcogenic strains (e.g., SB-1) are not lymphomagenic even in immunologically compromised hosts, suggesting that they lack transforming ability.

The pathogenic strains can be artificially attenuated by repeated passage in cell culture (Churchill *et al.*, 1969a; Nazerian, 1970; Eidson and Anderson, 1971); such attenuation is accompanied by loss of pathogenicity, often loss of the A antigen (A^-) (see Section VII.D.1), increased rate of replication in culture, with formation of larger plaques, and sometimes loss of the ability to spread horizontally in chickens. The nature of attenuation is not known.

Strains of HVT are nonpathogenic in turkeys (Witter *et al.*, 1970b; Witter, 1972) and induce only mild microscopic lesions in chickens (Witter *et al.*, 1976).

2. Serotypes

On the basis of indirect-immunofluorescence (IF) tests and agar–gel precipitin (AGP) tests, strains of MDV and HVT have been classified into three serotypes: *Type 1*: pathogenic strains of MDV and their cell-culture-attenuated A^- variants; *Type 2*: naturally apathogenic MDV; *Type 3*: HVT and its A^- variants (Bülow and Biggs, 1975a,b; Bülow, 1977; Schat and Calnek, 1978a). The three types share a group-specific BC antigen (see Section IV.E.1) and partially identical A antigens, but there appears to be a type-specific BC antigen. No serological distinction has been made between virus pathotypes that cause acute or classic MD.

E. Antigens

1. Virus-Associated Antigens

Cells productively infected with MDV or HVT contain virus-associated antigens that can be detected by IF, AGP, or immunoferritin tests. The relationships among antigens detected in these different ways, structural components of virus particles, and viral polypeptides have not yet been completely elucidated (see Ikuta *et al.*, 1981).

Two classes of antigens have been detected by IF tests in cultured cells productively infected with MDV or HVT:

1. *Membrane antigen (MA)* is present on the surface of live cells infected with MDV or HVT (Chen and Purchase, 1970; Ishikawa *et al.*, 1972). Immunoferritin studies suggest that the MA antigen is related to viral structural proteins (Nazerian and Chen, 1973). MA induced by either virus has been divided into two subclasses: early MA (EMA), which appears before replication of viral genome, and late MA (LMA), which appears to be dependent on a late function of the viral genome (Mikami *et al.*, 1973; Inage *et al.*, 1979). EMA and LMA are antigenically different; antibody to HVT LMA, but not to EMA, neutralizes virus infectivity, suggesting that LMA is closely related to virus envelope antigen (Onuma *et al.*, 1975; Inage *et al.*, 1979; Mikami *et al.*, 1980a).
2. *Intracellular antigen (IA)* is detectable in the nucleus and cytoplasm of infected acetone-fixed cells (Purchase, 1969; Purchase *et al.*, 1971a) and is antigenically distinct from MA (Inage *et al.*, 1979; Inoue *et al.*, 1980). IA is found only in cells that contain virus particles (Nazerian and Purchase, 1970); it may consist of more than one antigen. Nuclear IA was reported to be common to MDV and HVT, whereas the cytoplasmic antigens differed (Purchase *et al.*, 1971a). However, by cross-absorption of sera, Bülow and Biggs (1975a) found that IA, both nuclear and cytoplasmic, included serotype-specific components.

Up to six antigens have been detected in extracts of MDV-infected CKC in AGP tests, of which three are regularly present and have been designated A, B, and C (Churchill *et al.*, 1969a). The A antigen is a large glycoprotein mainly present in supernatant fluids of infected cultures (Ross *et al.*, 1973; P.A. Long *et al.*, 1975). The A antigen is lost during attenuation of MDV by continued passage in cell culture, and it was originally thought that it might be a marker for pathogenicity (Churchill *et al.*, 1969a); however, Purchase *et al.* (1971a) found A^- pathogenic strains of MDV, and A^+ nonpathogenic strains occur. Nazerian (1973) suggested that A antigen was related to MA.

A soluble glycoprotein precipitin antigen common to MDV and HVT, probably A antigen, occurs (Witter *et al.*, 1970b; Onuma *et al.*, 1974). It induces virus-neutralizing antibody (Onuma *et al.*, 1975) and appears to be related to LMA (Onuma *et al.*, 1976). Studies by Bülow and Biggs (1975b), however, showed that the A antigen of serotypes 1, 2, and 3 (see Section IV.D.2) were not completely identical and possibly consisted of group-specific and type-specific determinants (Fig. 9). Wyn-Jones and Kaaden (1979) detected three large polypeptides in glycoprotein from HVT-infected cells, which induced virus-neutralizing and precipitating antibodies.

The B and C precipitin antigens are present in extracts of infected cells and have been less studied than A antigen. Bülow and Biggs (1975b) concluded that the BC complex consists of at least three different antigens associated with each of the three serotypes of MDV and HVT studied.

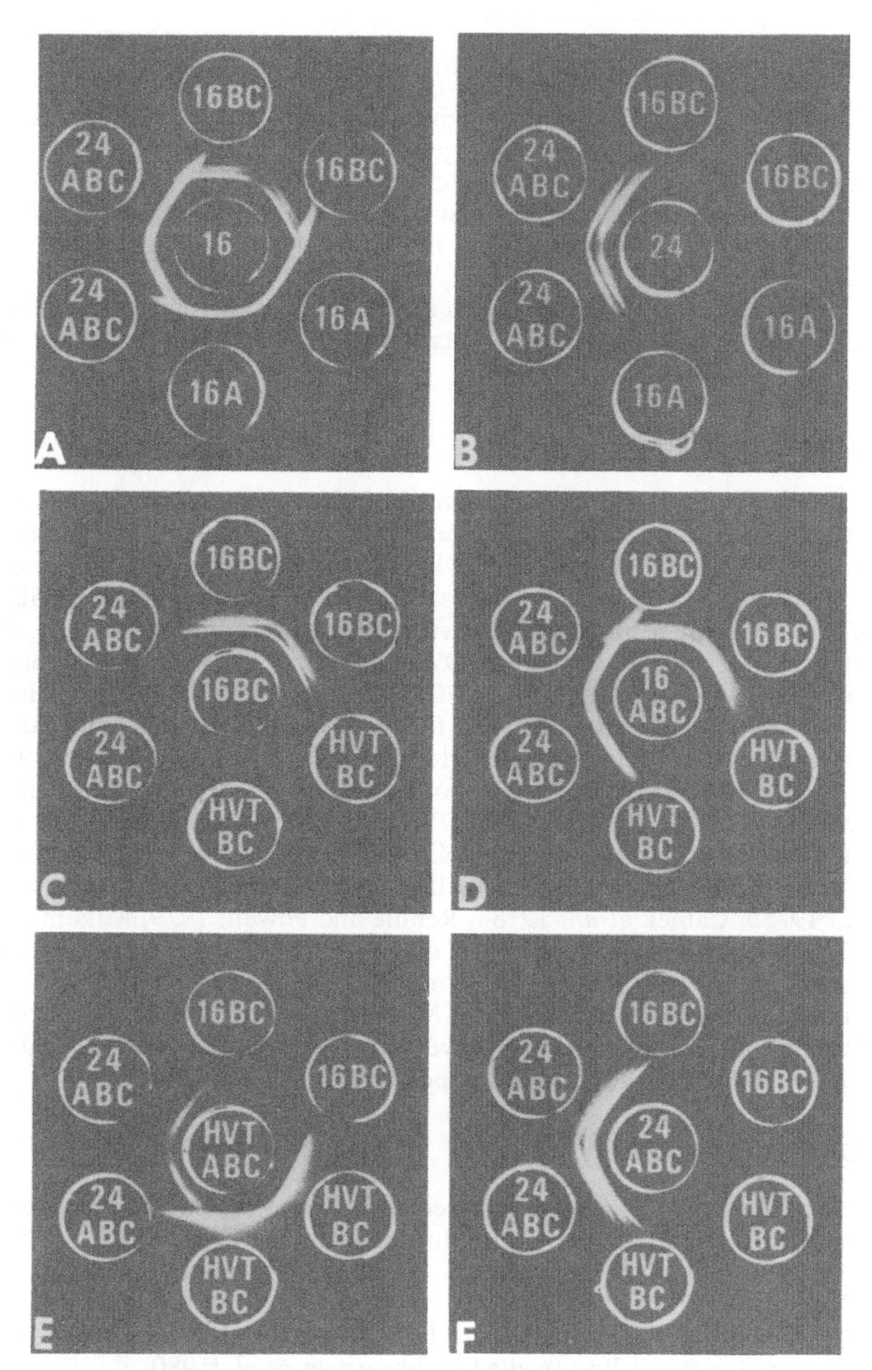

FIGURE 9. Differentiation between A and BC antigens of MDV of serotype 1 (HPRS-16), serotype 2 (HPRS-24), and serotype 3 (HVT). (A) The central well contains HPRS-16 antiserum, and outer wells contain A or BC antigens, or both, from HPRS-16 and HPRS-24 MDV. (B) As in (A), but the central well contains HPRS-24 antiserum. (C) The central well contains attenuated HPRS-16 (BC) antiserum, and outer wells contain ABC or BC antigens from HPRS-16, HPRS-24, and HVT. (D) As in (C), but the central well contains HPRS-16 (ABC) antiserum. (E) As in (C), but the central well contains HVT (ABC) antiserum. (F) As in (C), but the central well contains HPRS-24 (ABC) antiserum. From Bülow and Biggs (1975b) by kind permission.

At least one was group-specific (common to all serotypes) and at least one serotype-specific (Fig. 9). B antigen common to MDV and HVT was characterized as a glycoprotein by Velicer *et al.* (1978).

An MDV-related nuclear antigen comparable to EBV-related nuclear antigen (EBNA) has been sought but not clearly identified (see Nazerian, 1979).

Antigenic relationships between MDV and other herpesviruses from other birds, mammals, and amphibia have been described (Purchase *et al.*, 1972a; Evans *et al.*, 1973).

2. Tumor-Associated Antigen

A cell-membrane antigen designated "Marek's disease tumor-associated surface antigen (MATSA)" is present on a proportion of lymphoma cells and on virtually all cells of MD lymphoma-derived lymphoblastoid-cell lines (Section V.A.2) (Powell *et al.*, 1974; Witter *et al.*, 1975b). MATSA is not present on productively infected, nonlymphoid cells. It is induced not only by virulent MDV but also by attenuated MDV and HVT (Powell and Rennie, 1978, 1980b; Schat and Calnek, 1978d). This fact has been adduced as evidence for ability of attenuated MDV and HVT to transform lymphoid cells, although it has yet to be shown conclusively that MATSA is specific for neoplastically transformed cells. Sharma (1981a) has shown that MDV infection of lymphoid cells does not necessarily result in expression of MATSA. MATSAs on various lymphoblastoid-cell lines are antigenically related but not identical (Witter *et al.*, 1975b; Calnek *et al.*, 1978b; Rennie and Powell, 1979; Mikami *et al.*, 1980b) and are considered to be distinct from chicken fetal antigen (Murthy *et al.*, 1979). Bülow and Schmid (1979), however, cast doubt on the existence of common or related MATSAs on tumor cells. Recently, Coleman and Schierman (1980) have presented evidence for the independence of MATSA and major histocompatibility antigens.

3. Other Antigens

Chicken fetal antigen is expressed on MD lymphoma and lymphoblastoid-line cells, but is not specific for MD tumors (Murthy *et al.*, 1979; Neumann, 1980). Indirect evidence exists for an antigen common to MDV-infected CKC and nonproductive lymphoblastoid-line cells (Powell, 1978). A cell-surface antigen correlated with organ-specific metastasis was detected in a cell line studied by Shearman *et al.* (1980).

F. Role of Retroviruses in Marek's Disease

Interest in the possible cocarcinogenic role of avian retroviruses in induction of MD by the MDV was stimulated by evidence for involvement of RNA tumor viruses in the etiology of Burkitt's lymphoma (Peters *et al.*, 1973). Th roles of both exogenous and endogenous avian leukosis viruses have been considered. Coinfection of chickens by MDV and exogenous leukosis viruses (RAV-1, RAV-7, RAV-50, or REV) resulted in enhanced mortality and tumor development compared with infection by these viruses alone, and in one strain of chicken (LSI-SPF), typical MD was seen only in dually infected birds (Frankel *et al.*, 1974; M.E. Smith *et al.*, 1975). In similar studies by Witter *et al.* (1975a), coinfection with RAV-2 did not alter the character of MD lesions, nor was RAV-2 necessary for oncogenesis by MDV. Chickens free of expressions of endogenous leukosis virus (infectious RAV-O, gs-antigen, and chick helper factor) were susceptible to induction of MD by MDV (Witter *et al.*, 1975a), and occurrence of MD in several strains of fowl did not correlate with, or alter, the natural expressions of endogenous virus (Calnek and Payne, 1976). More recently, however, W.F. Campbell and Frankel (1979) reported enhanced expression of endogenous leukosis virus in MD.

These studies suggest that exogenous leukosis viruses are not essential for oncogenesis in MD, but that they may modify the expression of the disease, possibly by immunosuppressive effects. Whether expression of endogenous viral genome is a prerequisite of oncogenicity of MDV remains to be settled. Absence of such expression in MD lymphoblastoid-cell lines (Nazerian *et al.*, 1978) and evidence for integration of viral DNA with cellular DNA favor a direct transforming role for MDV (see Chapter 7).

V. PATHOLOGY

A. Virus–Cell Relationships

Understanding of the pathology of Marek's disease (MD) is helped by knowledge of the types of virus–cell interactions that occur. These are similar to interactions described for other herpesviruses (Roizman, 1972) and consist of (1) productive infections, in which virions are completely or partially formed, resulting in cell death; and (2) nonproductive infections, in which there is either no, or only limited, expression of viral genome, and in which the infected cell remains viable and is, in some instances, neoplastically transformed.

1. Productive Infection

Productive infection is characterized by synthesis of viral DNA, viral proteins, and virions. *Fully productive infection* by Marek's disease virus (MDV) and herpesvirus of turkeys (HVT) occurs only in the feather-follicle epithelium, from which infectious virus is released (Calnek *et al.*, 1970a; Nazerian and Witter, 1970; Zygraich and Huygelen, 1972; Cho,

1975; Johnson *et al.*, 1975). The production of infectious cell-free virus from this site seems to be associated with the envelopment of virions in cytoplasmic inclusion bodies and the protection afforded to the virion by keratinization of the epithelial cells (see Payne *et al.*, 1976; Calnek, 1980). Recently, Hinshaw and Mora (1980) found mature enveloped virions in cytoplasmic vacuoles of lymphocytes from HVT-viremic turkeys; no stages of virus maturation were seen, and the origin of this virus was not established. Widespread distribution of HVT virions in neural, visceral, and lymphoid tissues, associated with mild degenerative changes, was reported by Hein and Mora (1980) in naturally infected turkeys. *Semiproductive infection* (restrictive or abortive infection) by MDV occurs mainly in the bursa, thymus, and spleen, and to a lesser extent in parenchymatous tissues. These tissues contain infected cells that express viral antigens, naked virions in the nucleus, and limited enveloped virions, mainly associated with the nuclear membrane (Purchase, 1970; Adldinger and Calnek, 1973; Payne and Rennie, 1973). Infection by MDV of *in vitro*-cultured cells is also of the semiproductive type and is closely cell-associated; disruption of the cells usually results in complete loss of infectivity of MDV (Churchill and Biggs, 1967) and almost complete loss of infectivity of HVT (Witter *et al.*, 1970b). Some infectious MDV and increased titers of HVT can be obtained by sonication of infected cells in the presence of a stabilizer (Calnek *et al.*, 1970b).

2. Nonproductive Infection

Nonproductively infected cells contain the viral genome, but this is expressed only to a limited extent. For MDV, this type of infection is typified by lymphoma cells, and the lymphoblastoid-cell lines and transplantable lymphoma derived from them, as discussed below. In these cells, multiple copies of the viral genome are present and in the majority are expressed by immortalization of the cells and the appearance of Marek's disease tumor-associated surface antigen (MATSA). A small proportion of cells may spontaneously express viral antigens and virions, indicative of semiproductive infection, and die.

Blood and spleen lymphocytes in birds infected with MDV and HVT, and some lymphoma cells, are believed to be *latently infected*: they apparently may express no viral or tumor antigens or virions, and yet virus can be rescued by inoculation of cells into birds or onto cultured cells (Calnek, 1980; Calnek *et al.*, 1981a; Sharma, 1981a). However, the possibility that such cells are semiproductively infected, or nonproductively transformed, has not been clearly excluded.

a. Lymphoid-Cell Lines

Lymphoid-cell lines have been established from either primary MDV-induced lymphomas or transplantable lymphomas (Table II). Recently,

TABLE II. Lymphoid-Cell Lines Derived from Marek's Disease Lymphomas[a]

Original designation	New designation[b]	Origin	MDV strain	Host line	Reference
MOB-1	MDCC-MOB1	Ovarian lymphoma	BC-1	—	Akiyama *et al.* (1973)
MSB-1	MDCC-MSB1	Splenic lymphoma	BC-1	—	Akiyama and Kato (1974)
HPRS-1	MDCC-HP1	Ovarian lymphoma	HPRS-16	RIR	Powell *et al.* (1974)
HPRS-2	MDCC-HP2	Lymphoma	HPRS-16	RIR	Powell *et al.* (1974)
MOB-2	MDCC-MOB2	Ovarian lymphoma	BC-1	—	Ikuta *et al.* (1976a)
MOB-3	MDCC-MOB3	Ovarian lymphoma	BC-1	—	Matsuda *et al.* (1976a)
RPL-1	MDCC-RP1	JMV transplant	JM	—[c]	Nazerian *et al.* (1977)
$JM\text{-}VLC_1$	MDCC-SK1	JMV transplant	JM	—	Hahn *et al.* (1977)
$JM\text{-}VLC_8$	MDCC-SK2	JMV transplant	JM	—	Hahn *et al.* (1977)
MOGA-1	MDCC-JP1	Ovarian lymphoma	—	Anthony No. 11 (B^AB^K)	Kawamura *et al.* (1977)
MOGA-2	MDCC-JP2	Ovarian lymphoma	—	Anthony No. 11 (B^AB^K)	Kawamura *et al.* (1977)
JMV-YS-1	MDCC-BP1	JMV transplant	JM	—	Sekiya *et al.* (1977)
GACL-1	MDCC-CU1	MDCT-CU7 transplant	GA	N	Calnek *et al.* (1978b)
CUCL-1	MDCC-CU2	Lymphoma	CU-2	P	Calnek *et al.* (1978b)
JMCL-1	MDCC-CU3	JMCT transplant	JM	PDR	Calnek *et al.* (1978b)
GACL-4	MDCC-CU4	GA/Tr-2 transplant	GA	N	Calnek *et al.* (1978b)
GBCL-1	MDCC-CU5	MDCT-NYMI transplant	Conn-B	G-B1	Calnek *et al.* (1978b)
GACL-6	MDCC-CU6	MD blood lymphocytes	GA	P	Calnek *et al.* (1978b)
GBT	MDCC-SK3	MDCT-NYMI transplant	Conn-B	G-B1	Hahn *et al.* (1978)
MKT-1	MDCC-LS1	Kidney lymphoma	—	—	Tanaka *et al.* (1978)
JMV-1	—	JMV transplant	JM	—	Munch *et al.* (1978)
JM-1	—	MD lymphoma	JM	—	Munch *et al.* (1978)
MDCC-AL1	—	Spleen selected MDCC-RP1 transplant	JM	—	Shearman and Longenecker (1980)
MDCC-AL2	—	Liver selected MDCC-AL1 transplant	JM	—	Shearman and Longenecker (1980)
MDCC-AL3	—	Ovary selected MDCC-AL1 transplant	JM	—	Shearman and Longenecker (1980)

[a] A total of 60 new cell lines from various MDV-hostline combinations were developed by Calnek *et al.* (1981) and Payne *et al.* (1981).
[b] From Witter *et al.* (1979).
[c] Possibly S line (B^1B^1) (see Longenecker *et al.*, 1977).

31 new lymphoblastoid cell lines were developed by Calnek *et al.* (1981b) and 29 new lines by Payne *et al.* (1981), from a variety of chicken lines and MDV strains. Factors helpful for the initiation of lines include culture at 41°C, use of special media, including 2-mercaptoethanol and chicken serum, and derivation from transplantable lymphoma (Kato and Akiyama, 1975; Hahn *et al.*, 1977; Calnek *et al.*, 1978b). The lines consist of lymphoblastoid cells that grow continuously in suspension culture at both 37 and 41°C (Fig. 10). The cells carry T-cell markers and MATSA (Powell *et al.*, 1974; Nazerian and Sharma, 1975; Matsuda *et al.*, 1976b), supporting the T-cell origin of lymphoma cells. Payne *et al.* (1981) developed new lines by initiating them in soft agar medium and showed that most lines were lymphocytoid when first established but changed into lymphoblastoid lines later. This change was associated with increased expression of an embryonic antigen, loss of density dependency, and more rapid growth. Ultrastructural properties of lines have been described (Frazier and Powell, 1975; Akiyama and Kato, 1974). They are essentially diploid, but chromosomal abnormalities may be present (Takagi *et al.*, 1977), none of which appears to be specific for transformation.

Most cells in a line are nonproductively infected, although MDV can be rescued from most lines by inoculating them into chicks or onto cultured cells (Doi *et al.*, 1976 Calnek *et al.*, 1981b; Payne *et al.*, 1981).

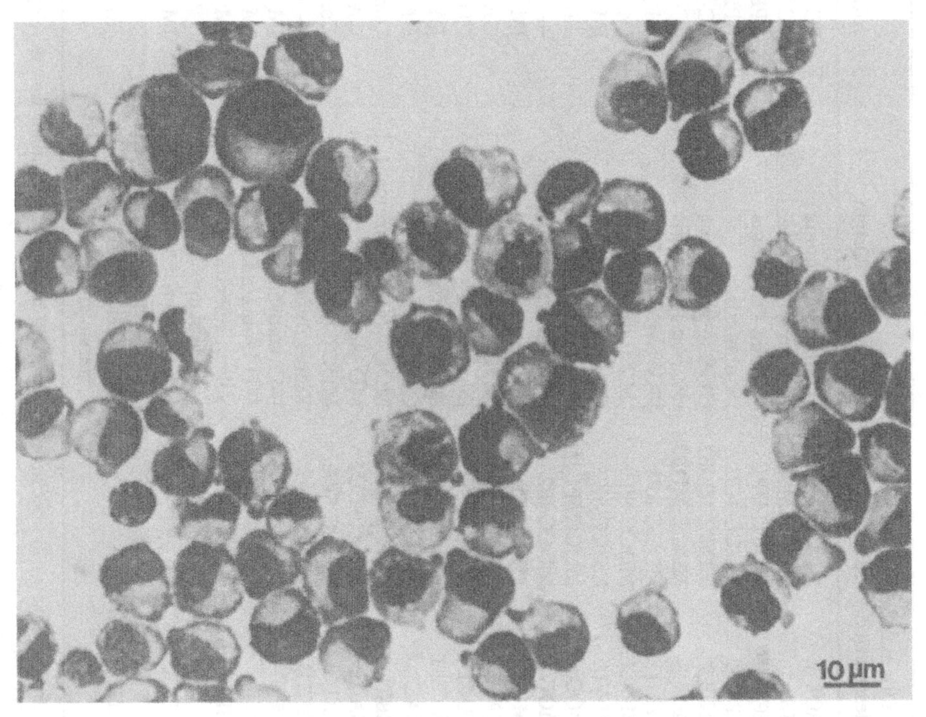

FIGURE 10. Impression smear of MDCC-HPl line of MD lymphoblastoid cells. From Frazier and Powell (1975) by kind permission.

A small proportion of cells spontaneously express viral antigens and immature virions, and this can be increased by treatment with 5-iodo-2-deoxyuridine (IUdR) and 5-bromodeoxyuridine (BUdR) (Nazerian and Lee, 1974; Dunn and Nazerian, 1977; Ross *et al.*, 1977) and decreased with phosphonoacetic acid, an inhibitor of herpesvirus DNA polymerase (Nazerian and Lee, 1976).

The lymphoblastoid cells contain multiple copies of MDV genome: 60–90 genome equivalents/cell were detected in the MSB-1 line (Nazerian and Lee, 1974), although treatment with phosphonoacetic acid reduced the number to about 20, which may be more representative of viral genome in nonproducing cells (Nazerian and Lee, 1976). The MDV DNA has been reported both to be integrated into cellular DNA (Kaschka-Dierich *et al.*, 1979) and to exist mostly in a circular nonintegrated state (Tanaka *et al.*, 1978) (see Chapter 7 for further discussion). Expression of endogenous or exogenous leukosis viruses in lymphoblastoid-line cells is variable and is not a prerequisite for transformation (Ikuta *et al.*, 1976b; Nazerian *et al.*, 1978).

The RPL-1 lymphoblastoid line, derived from the transplantable JMV lymphoma, is a true nonproducer line from which MDV cannot be rescued (Nazerian *et al.*, 1977), as were apparently a number of lines produced by Payne *et al.* (1981).

No lymphoid-cell lines have been established from chickens infected with HVT; whether HVT can transform lymphoid cells is still uncertain. Transformation of lymphoid cells with MDV *in vitro* was reported by Lam and Cho (1981). Recently, Nazerian *et al.* (1982) developed lymphoblastoid cell lines from MDV-induced lymphomas in turkeys; these lines expressed MATSA and carried infectious MDV but, unexpectedly, reacted with antisera against chicken B cells, IgG, and IgM, indicating B cell origin.

b. *Transplantable Lymphomas*

Transplantable lymphomas have been derived from primary MD lymphomas by rapid serial passage of tumor cells in chicks (Table III). The first transplantable tumor so developed, designated JMV, produces an acute lethal lymphoblastoid leukemia that was initially thought to be caused by the presence of a virulent MDV (Sevoian *et al.*, 1964), but was later shown to result from a transplant (Stephens *et al.*, 1976). Some substrains of JMV did apparently carry MDV, but others, including the one studied by Stephens, had no rescuable MDV, although at least part of the MDV genome was present, and the cells expressed MATSA. The JMV lymphoma has also been adapted to grow as a lymphoblastoid-cell line (Nazerian *et al.*, 1977; Munch *et al.*, 1978) and clonal variation associated with metastatic properties has been found (Shearman and Longenecker, 1981).

Several other transplantable lymphomas have been developed (see

TABLE III. Transplantable Lymphomas Derived from Marek's Disease Lymphomas

Original designation	New designation[a]	Origin	MDV strain	Host line	Reference
JMV	—	Lymphoma	JM	S?[b]	Sevoian *et al.* (1964)
JMCT	MDCT-CU8	Lymphoma	JM	PDR	Calnek *et al.* (1969)
MDT-198	MDCT-NYMI	Lymphoma	Conn-B	G-B_1	Theis *et al.* (1974)
JMV-A	—	Attenuated JMV transplant	—	—	Shieh and Sevoian (1974)
Not named	—	Lymphomas	Conn-B	G-B_1	Jakowski *et al.* (1974)
GA/Tr-1	MDCT-CU7	Ovarian lymphoma	GA	N	Calnek *et al.* (1978a)
GA/Tr-2	MDCT-CU9	Ovarian lymphoma	GA	N	Calnek *et al.* (1978b)
MTr-R1	MDCT-RP3	Ovarian lymphomas	GA	7_2	Stephens *et al.* (1980)

[a] From Witter *et al.* (1979).
[b] See Longenecker *et al.* (1977).

Table III). They grow more slowly than JMV in chicks, produce local tumors, are normally dependent on the histocompatability of the recipient for progressive growth, and release MDV. A true nonproducing lymphoblastoid-cell line was developed from GB-1 transplant cells (Hahn *et al.*, 1978).

B. Gross Lesions

The pathognomonic lesions of MD are enlargements of peripheral nerves and lymphoma formation in various organs and tissues; these changes may occur together or alone, depending on factors such as strain of virus, strain of chicken, and age at infection. Other less specific gross changes that may occur are atrophy of the bursa of Fabricius and thymus, muscular wasting, dermatitis, iridocyclitis, and atherosclerosis. Gross lesions are not seen in transient paralysis or in HVT-infected turkeys or chickens. A strain of HVT has been associated with poor semen quality and infertility in turkeys (Thurston *et al.*, 1975).

1. Peripheral Nerves

Nerve enlargement occurs from about 4 weeks after infection. Affected nerves are usually 2–3 times the normal thickness, lose their cross-striations, and show a gray or yellow discoloration, and sometimes edema (Fig. 11). Commonly involved are the celiac and other autonomic nerves, and the sciatic and brachial plexi and trunks (Goodchild, 1969; Sugiyama *et al.*, 1973). The enlargement is caused by lymphomatous or inflammatory infiltrations or both.

2. Lymphoma Formation

Virulent strains of MD induce lymphoma formation, which is usually diffuse, in various organs or tissues, those commonly affected being the ovary, testis, liver, spleen, kidneys, heart, proventriculus, lung, skin, and skeletal muscle (Fujimoto *et al.*, 1971) (Fig. 12). Less virulent strains may induce ovarian lymphomas only. Tumors may become visible 3–4 weeks after infection, or may take many weeks to appear, and their tissue distribution is influenced not only by strain of MDV but also by strain of bird (Morris *et al.*, 1970b) and may depend on selection of antigenically variant cells (Shearman and Longenecker, 1980; Shearman *et al.*, 1980).

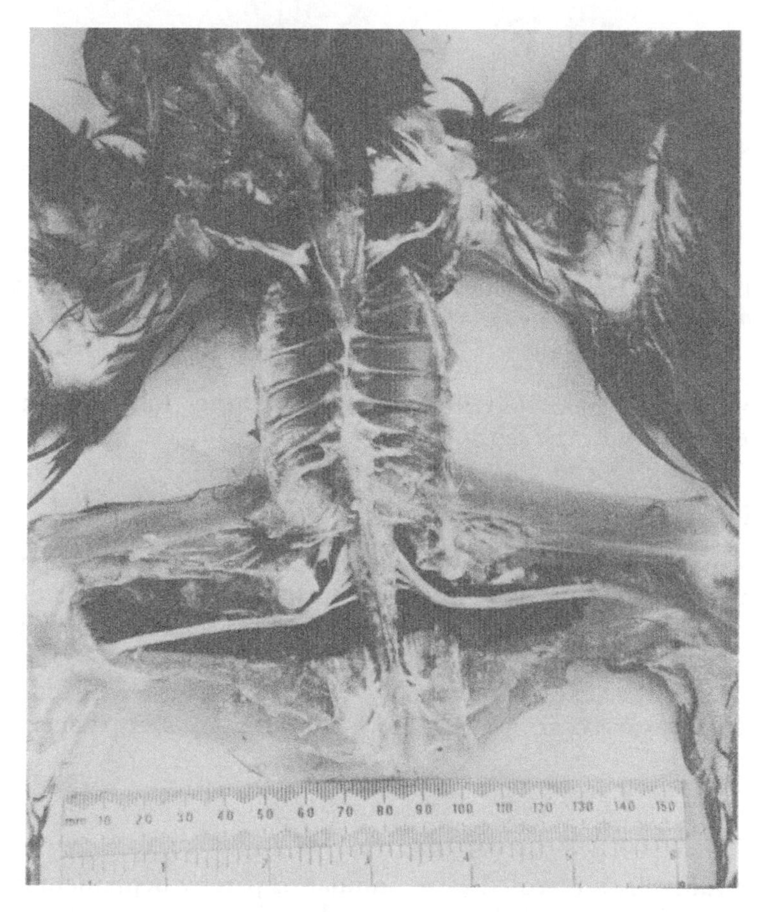

FIGURE 11. Enlarged sciatic and brachial nerves of MD-affected chicken (same case as Fig. 1).

FIGURE 12. Ovarian lymphoma (arrow) in 6-week-old chicken infected with MDV. From Payne *et al.* (1976) by kind permission of the publishers.

Lymphomas in the skin often arise around feather follicles, but may infiltrate diffusely in the dermis.

3. Atrophy of Bursa and Thymus

Loss of bursal and thymus weight caused by lymphoid atrophy may occur as early as 5 days after infection during the acute phase of MDV infection, or several weeks later during the lymphomatous phase (Jakowski *et al.*, 1970; Calnek, 1972b; Payne and Rennie, 1973). Strains of MDV occur that cause death as early as 8–10 days after infection, accompanied by severe atrophy of the bursa and thymus (Witter *et al.*, 1980).

4. Other Gross Lesions

An iridocyclitis revealed as an annular or patchy depigmentation of the iris, and distortion of the pupil, has long been associated with MD (Pappenheimer *et al.*, 1926, 1929a) and has been reproduced experimentally (T.W. Smith *et al.*, 1974). Muscular wasting (Wight, 1966) and dermatitis (Payne, unpublished observations) may occur. Grossly visible occlusive atherosclerotic change, or smaller plaques, have been described recently in large coronary arteries, aorta, and aortic branches of chickens 30 weeks after infection with a strain of MDV of low virulence (Minick *et al.*, 1979).

C. Microscopic Lesions

Extensive, although incomplete, knowledge of the pathogenesis of MD has been acquired largely from studies on the experimentally induced disease. The relationships between the main events that occur when susceptible chicks are infected with virulent MDV are depicted in Fig. 13 and are discussed below, as far as possible in chronological order.

1. Primary Infection

Under natural conditions, MDV infection is usually acquired by inhalation of infective feather debris, dander, and dust in the poultry house. Viral antigen is detectable in the lung at 24 hr postinfection; infectious cell-associated MDV was not found in this site until 5 days, suggesting that the virus is rapidly transported away from the lung, possibly in phagocytic cells. Viral antigen can be detected in peripheral-blood cells as early as 18 hr after infection (Adldinger and Calnek, 1973).

2. Acute Cytolytic Infection of Lymphoid Tissues

An acute cytolytic infection of lymphoid tissues, notably of the bursa, thymus, and spleen, becomes apparent at 3 days after infection, reaches a peak at 5–7 days, and usually resolves by 2 weeks. This phase has been termed an acute lymphoreticulitis (Payne *et al.*, 1976) or the stage of reticuloendothelial proliferation (Witter *et al.*, 1976). Certain strains of MDV can cause mortality during this phase of the disease (Witter *et al.*, 1980). The infection is of the semiproductive type, with abundant viral antigen and cell-associated infectious virus present (Adldinger and Calnek, 1973; Payne and Rennie, 1973; Fujimoto *et al.*, 1974). Histologically, marked cytolysis of lymphocytes and reticulum cells, with numerous intranuclear inclusion bodies, can be seen. Under the electron microscope, immature virions are seen in the nuclei of reticulum cells, which may be the initial target cells for the virus (Frazier, 1974). The virus infection provokes an acute inflammatory response in these organs, with an increase in reticulum cells and macrophages, and granulocytic invasion (Payne and Rennie, 1973). These changes are accompanied by severe regression of thymic cortex and bursal lymphoid follicles, resulting in loss of weight of these organs and probably, at least in part, in the

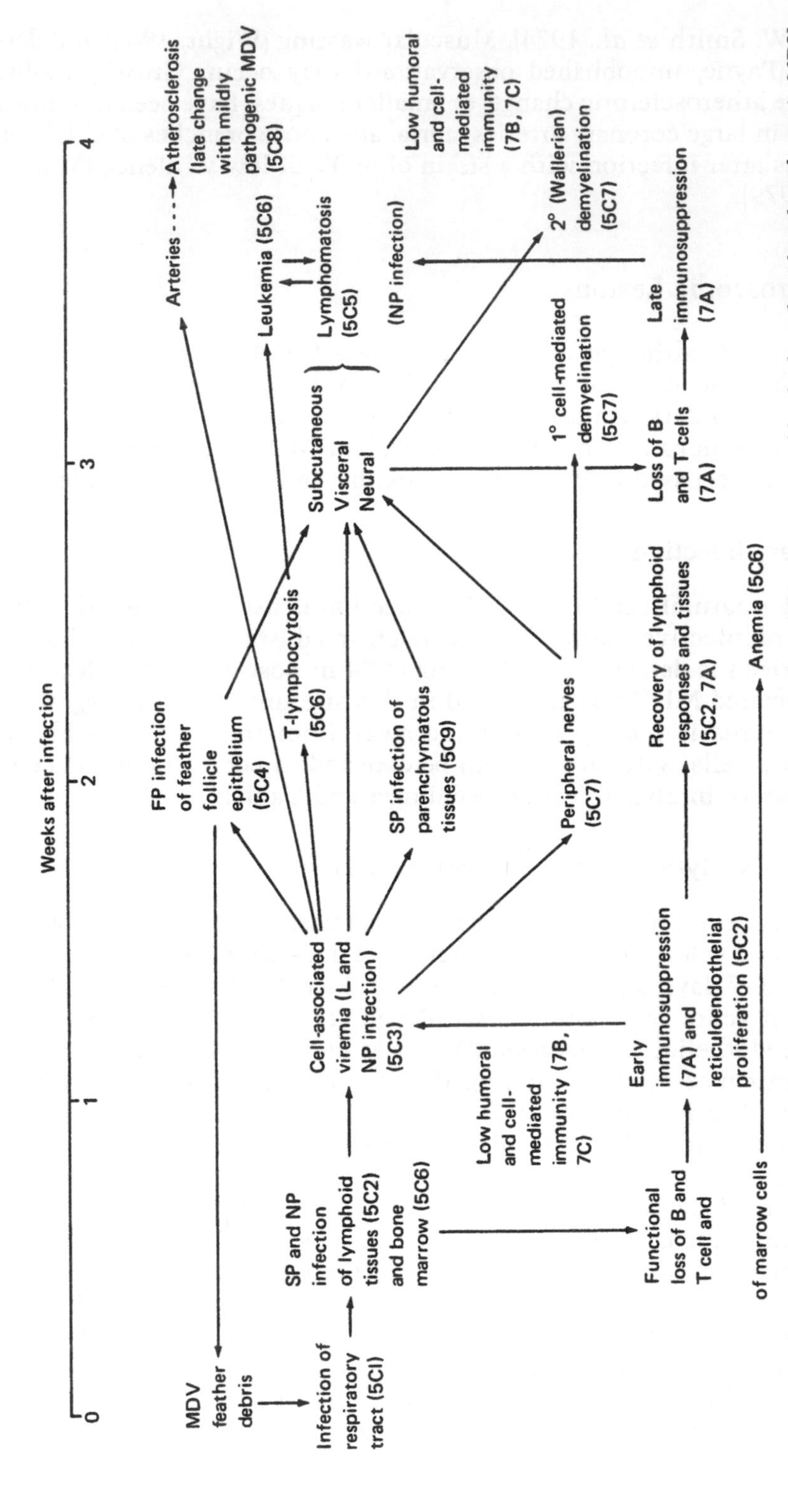

FIGURE 13. Representation of pathogenesis of MD induced in young maternal-antibody-free susceptible chicks infected with pathogenic MDV. Types of infection: (FP) fully productive; (SP) semiproductive; (L) latent; (NP) nonproductive. The numbers in parentheses are those of text sections.

transient immunosuppression that is observed 7 days after infection (Lee *et al.*, 1978a) (Fig. 14). Regressive lymphoid changes do not occur in the spleen, and here the inflammatory changes result in an increase of weight. MATSA-bearing cells first appear in the spleen at 5 days, and in a lower proportion of cells in the thymus and bursa at 7 days (Murthy and Calnek, 1978), suggesting that the target cell for transformation may reside in the spleen and some ameliorative effect of splenectomy on MD has been observed (Schat, 1981). Thymectomy abolished the acute cytolytic infection in the bursa (Goto *et al.*, 1979), as does embryonal bursectomy (Schat *et al.*, 1981), but the mechanisms have not yet been established. After 7 days, the acute inflammatory changes and virus replication disappear, and by 2 weeks the architecture of these organs is considerably restored.

Nononcogenic MDV and HVT similarly localize early after infection in the spleen, bursa, and thymus, and MATSA is expressed. However, they show little or no ability to replicate or induce cytolytic lesions or lymphoid atrophy (Calnek *et al.*, 1979).

3. Viremia

Infection of chickens by MDV or HVT is followed by an early and often persistent cell-associated "viremia" (Phillips and Biggs, 1972). The

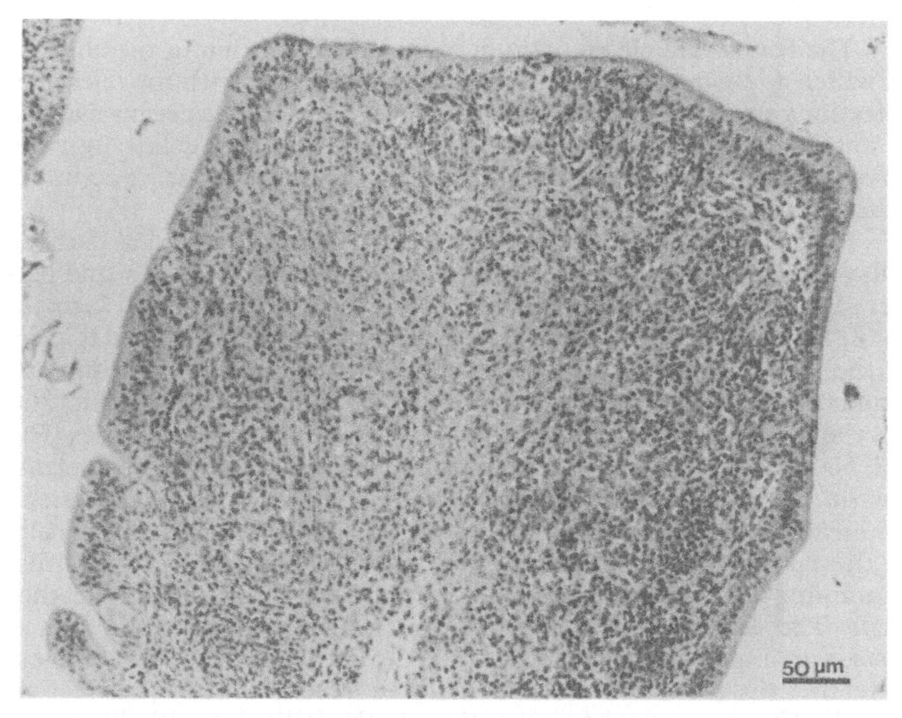

FIGURE 14. Loss of lymphoid follicles, and reticulum-cell hyperplasia, in bursa of 8-day-old chick infected with MDV. From Payne and Rennie (1973).

viruses are present in buffy coat cells (apparently mainly in lymphocytes), but not free in the plasma, and such cells are believed to be the main vehicles for transport of virus to other tissues. The range of cells that carry MDV has not been defined; in studies on HVT-infected turkeys, Hinshaw and Mora (1980) found herpesviruses in lymphocytes, but not in heterophils, eosinophils, or basophils. Calnek *et al.* (1981a) have presented evidence that both B and T lymphocytes from chickens infected with oncogenic or nononcogenic MDV are latently infected; attempts to identify infection of these cells in HVT-infected birds were unsuccessful.

Cell-associated infective MDV was found in susceptible chickens at 5 days postinfection, reached a peak at 8 days, and was followed by a second peak at about 4 weeks (Adldinger and Calnek, 1973), possibly related to the biphasic response observed in lymphoid tissues (Payne and Rennie, 1973). Level of viremia is associated directly with susceptibility of the chicken to MD and virulence of the strain of MDV, and inversely with age at infection (Sharma and Stone, 1972; Calnek, 1973).

A direct relationship between viremia and disease incidence is observed (Witter *et al.*, 1971; Powell *et al.*, 1980a). An early and persistent cell-associated viremia is similarly observed in HVT-infected turkeys (Witter *et al.*, 1972).

4. Feather-Follicle Epithelium

The feather-follicle epithelium is the single location in the chicken in which a fully productive infection by MDV occurs with the release of infectious enveloped virus. The biochemical basis for this permissive site is not known. Virus is shed in feathers and in feather debris into the environment, where it can persist for many months and serve to infect other birds.

After infection with MDV, viral antigen can be detected as early as 5 days later (Purchase, 1970), is extensively present by 14 days, and persists for many weeks (Lapen *et al.*, 1970, 1971; Adldinger and Calnek, 1973; Payne and Rennie, 1973). Cytopathic changes are restricted to cells of the transitional and corneous layers of the epithelium, and include cloudy swelling and hydropic change, margination of chromatin, and development of intranuclear and intracytoplasmic inclusion bodies (Fig. 15). Naked and occasionally enveloped virions are present in the nucleus, and numerous enveloped particles develop in the cytoplasmic inclusions (Calnek *et al.*, 1970a; Nazerian and Witter, 1970). The outermost cells of the epithelium degenerate (as they do normally) and release virus-containing debris into the space between the epithelium and the feather shaft. The detection of viral antigen in feather tips in radial-diffusion tests is an important diagnostic method (Haider *et al.*, 1970), and extracts of feather tips provide a source of cell-free infectious virus (see Section IV.C.1). The appearance of viral antigen in the follicular epithelium (Fig. 16) is followed shortly by local perivascular and perifollicular aggregation

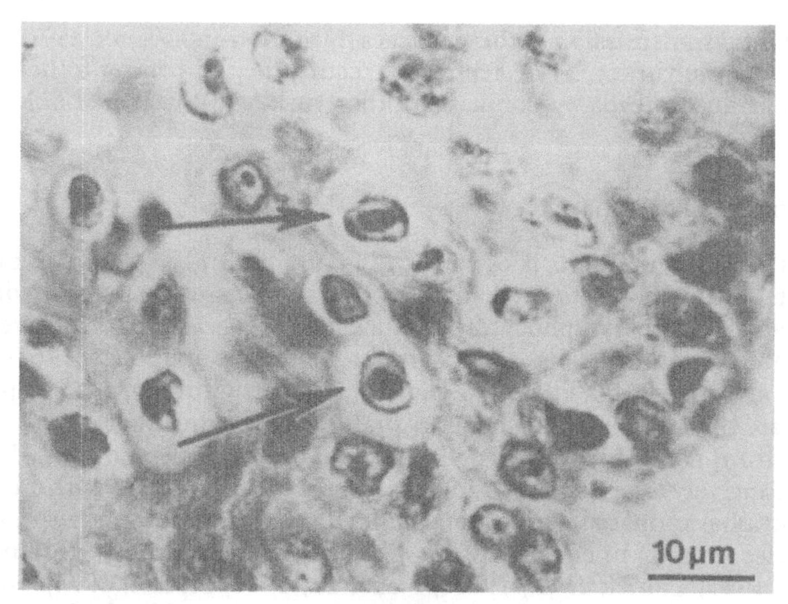

FIGURE 15. Inclusion bodies in feather-follicle epithelial cells (arrows) of MDV-infected chicken. From Lapen *et al.* (1970) by kind permission.

and proliferation of lymphoid cells (Fig. 17); in some birds, these develop into lymphomatous tumors (skin leukosis) (Lapen *et al.*, 1971; Payne and Rennie, 1973).

Attenuated strains of MDV and HVT replicate poorly and only for a week or so in feather follicles of the chicken, accounting for the poor

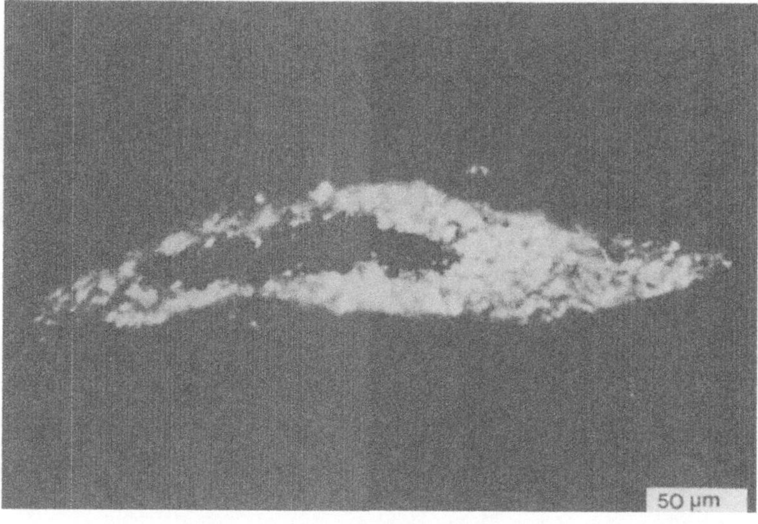

FIGURE 16. Viral antigen demonstrated by immunofluorescence in feather-follicle epithelium of MDV-infected chicken. From Lapen *et al.* (1971) by kind permission.

horizontal transmission of these viruses (Nazerian and Witter, 1970; Cho, 1975). In the turkey, HVT replicates readily in the feather-follicle epithelium and spreads by contact to other turkeys (Witter and Solomon, 1971).

5. Lymphoma Formation

The MD lymphoma is cytologically complex and its nature not fully understood; it consists predominantly of a mixture of small, medium, and large (blastic) lymphocytes, with a small proportion of primitive and activated reticulum cells, and macrophages (Payne and Biggs, 1967; Purchase and Biggs, 1967) (Figs. 18 and 19). Ultrastructurally, the lymphoid cells are pleomorphic and show no qualitative features that differentiate them from normal lymphoid cells (Doak *et al.*, 1973; Payne *et al.*, 1976); quantitatively, they show an increased proportion of cells with nuclear projections compared with normal cells. Rarely, intranuclear herpesvirus particles, mainly unenveloped, are seen in lymphoma cells (Schidlovsky *et al.*, 1969; Calnek *et al.*, 1970c; Nazerian, 1971; Frazier, 1974). Foci of lymphomatous proliferation first appeared in susceptible chicks about 7 days after infection with a pathogenic strain of MDV, particularly in the gonads, liver, proventriculus, and peripheral nerves; they can become

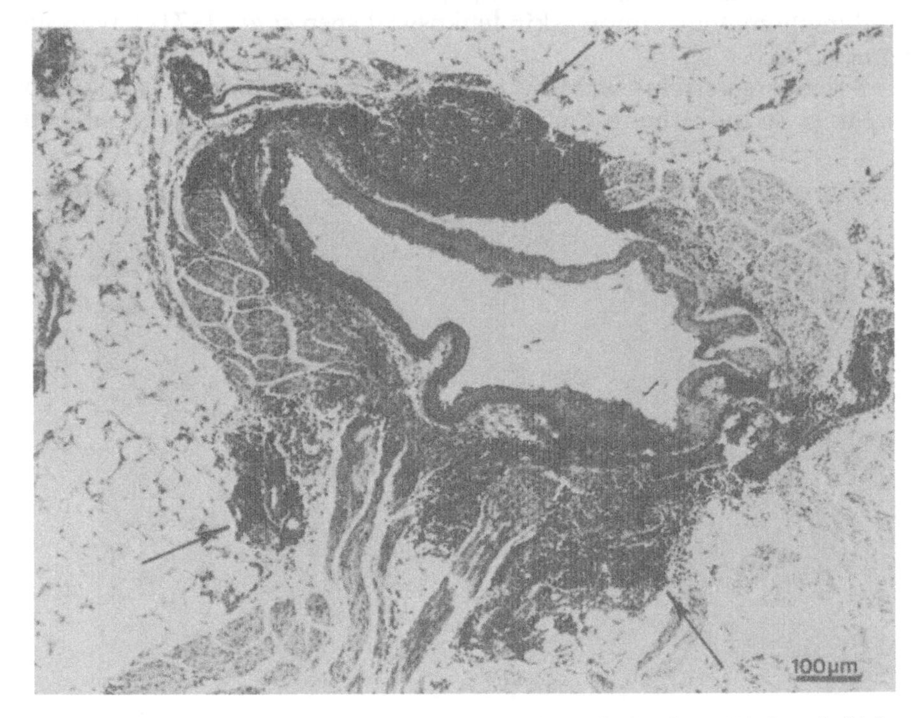

FIGURE 17. Lymphoid tissue (arrows) around feather follicle of MDV-infected chicken. From Payne (1972) by kind permission of the publishers.

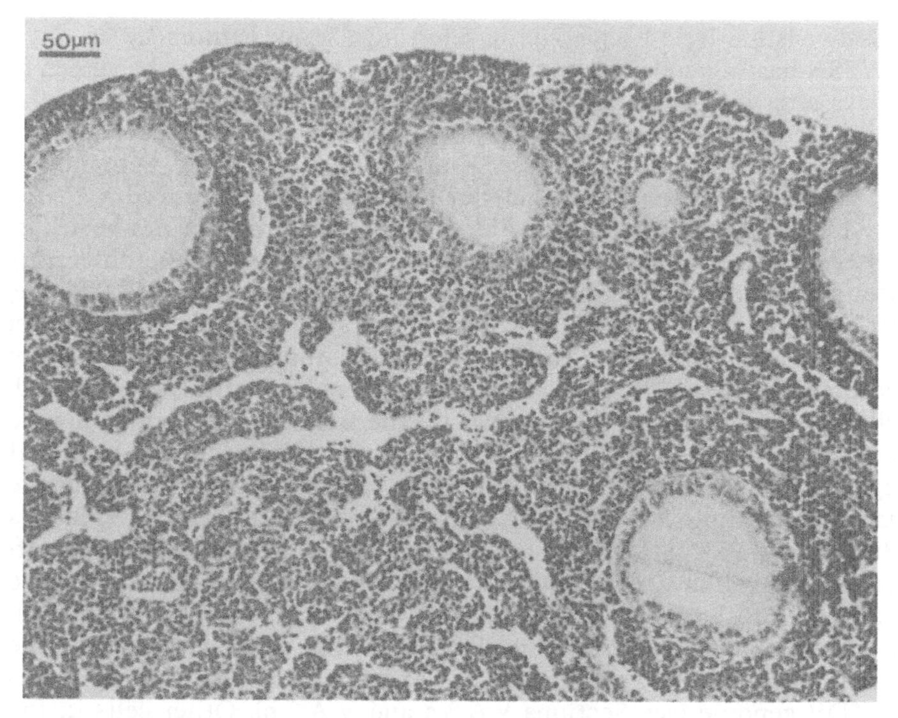

FIGURE 18. Ovarian lymphoma in 4-week-old chicken infected with MDV. From Payne (1972) by kind permission of the publishers.

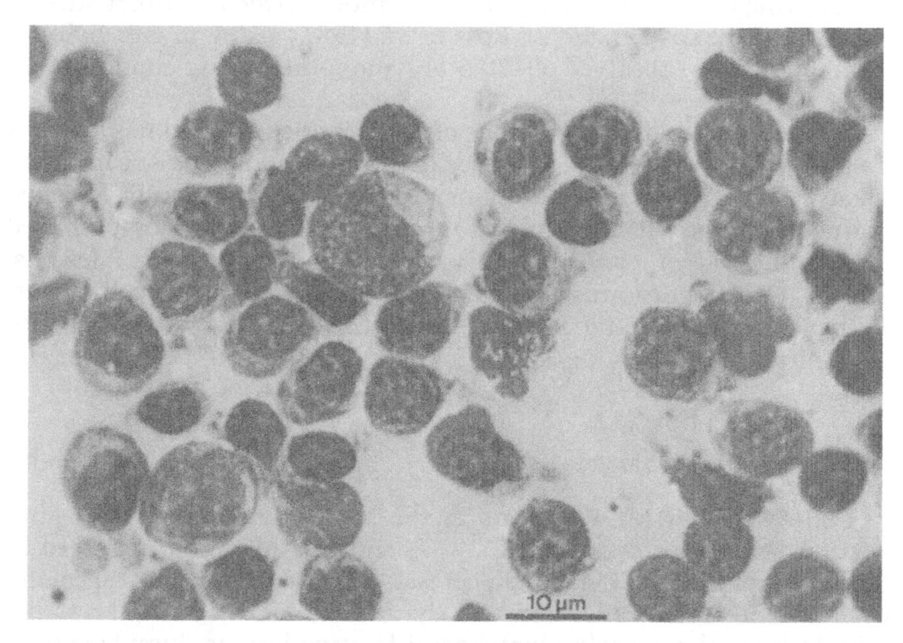

FIGURE 19. Impression smear of ovarian lymphoma cells showing mixed morphology of lymphoid cells.

grossly visible by 14 days and develop into frank tumors by 21 days. MATSA-bearing cells, as mentioned earlier, first appear in the spleen at 5 days after infection, and in other lymphoid organs by 7 days (Murthy and Calnek, 1978), but whether this represents the first appearance of transformed cells is unsettled (see Sections IV.E.2 and V.C.2). Lymphomas appear to be polyclonal in origin, but this has not been proved.

The nature of the stimulus for lymphoma formation has been discussed in terms of extrinsic stimulation, from outside the proliferating lymphoid cells, and intrinsic stimulation, lying inside the cells (Payne, 1972; Payne *et al.*, 1976). Current evidence suggests that both types of stimuli may contribute to the final lymphomatous mass. It is generally accepted, on the following evidence, that neoplastically transformed T cells are present in the lymphoma: (1) the incidence of lymphomas is decreased by T-cell depletion (Sharma *et al.*, 1977), but not by B-cell depletion (Payne and Rennie, 1970; Fernando and Calnek, 1971); (2) the majority of lymphoma cells are T cells, and most of the remainder B cells (Hudson and Payne, 1973; Rouse *et al.*, 1973; Hoffmann-Fezer and Hoffmann, 1980); (3) a variable minority of lymphoma cells carry a putative tumor-specific antigen, MATSA (Powell *et al.*, 1974; Witter *et al.*, 1975b); (4) chicken lymphoblastoid-line cells and MD-derived transplantable lymphoma cells are T cells, carry MATSA, and contain multiple copies of MDV genome (see Sections V.A.2.a and V.A.2.b). Other cells in the lymphoma (the B cells, some of the T cells, and the nonlymphoid cells) are believed not to be neoplastic, but to be normal or immunocompetent cells reacting to extrinsic virus, viral or tumor antigens, or other stimuli (Payne and Roszkowski, 1972). Ross *et al.* (1981), using *in situ* hybridization techniques, observed 60–70% of lymphoma cells to contain MDV DNA.

Whether or not different strains of MDV induce lymphomas depends on an interplay between the virulence of the virus and the host's immune response. Most low-virulence strains of MDV are shown to have transforming potential when inoculated into immunologically compromised hosts (Calnek *et al.*, 1977). Only limited lymphoproliferative lesions, without gross tumor formation, have been induced by HVT in chickens, whether intact or immunodeficient, suggesting that this virus has no, or only limited, transforming ability (Witter *et al.*, 1976; Sharma *et al.*, 1980).

6. Hematological Changes

Cell-associated MDV appears in the bone marrow at 5 days after infection (Adldinger and Calnek, 1973), and this infection may lead to destruction of marrow cells and depression of red cell count, especially in chicks that lack maternal antibody against MDV (Jakowski *et al.*, 1970). In birds that develop lymphoproliferative lesions, lymphocytosis and leukemia occur (Evans and Patterson, 1971; Payne *et al.*, 1975).

In a study of sequential changes in blood-leukocyte responses in MD, absolute numbers of B cells, T cells, total lymphocytes, and heterophils increased, and those of monocytes and eosinophils decreased, during the early stage of acute infection of lymphoid tissues. During the later lymphoproliferative phase, T cells and lymphocytes increased, resulting in leukemia in some birds, and B cells, monocytes, heterophils, and basophils decreased (Payne and Rennie, 1976b).

7. Neural Lesions

Neural changes in MD occur mainly in the peripheral nerves. Lesions in the CNS, discussed later, are usually mild in MD, although they are the principal lesions in transient paralysis.

The peripheral nerves in MD are variably infiltrated by lymphoid cells, and there has been much debate about the nature of and relationships among the various neural changes (see Payne *et al.*, 1976). Two main pathological processes occur: (1) a neoplastic lymphoproliferative infiltration similar to that which give rise to lymphomas in other tissues and (2) an inflammatory demyelination that is primary, segmental, and cell-mediated, similar to that seen in experimental allergic neuritis, and characterized by penetration of nerve fibers by lymphocytes and macrophages and destruction of myelin, but usually sparing of the axon (Prineas and Wright, 1972; Fujimoto and Okada, 1977; Lampert *et al.*, 1977; Lawn and Payne, 1979) (Fig. 20). Both the neoplastic and the demyelinative processes can occur within the same bird, often in the same nerve; this is usual in infection by the more virulent strains of MDV, whereas milder strains may cause only the demyelinative disease. These processes are reflected in the several classifications that have been proposed to describe and relate the histopathological nerve lesions in MD (Tables IV and V) (Wight, 1962a; Payne and Biggs, 1967; Fujimoto *et al.*, 1971).

In the classification of Payne and Biggs (1967), three lesion types are described:

1. A-type lesion, characterized by a lymphoproliferative infiltration and demyelination (Fig. 21).
2. B-type lesion, characterized by edema, a lighter infiltration by mainly small lymphocytes and plasma cells, and demyelination (Fig. 22).
3. C-type lesion, characterized by a light infiltration by small lymphocytes and plasma cells (considered to be a mild inflammatory lesion).

The earliest change observable in peripheral nerves with the electron microscope is at 5 days after infection with MDV, consisting of infiltration mainly by macrophages, but with some small lymphocytes (Lawn and Payne, 1979). Under the light microscope, the infiltration becomes apparent at about 10 days. Macrophages and lymphoid cells increase

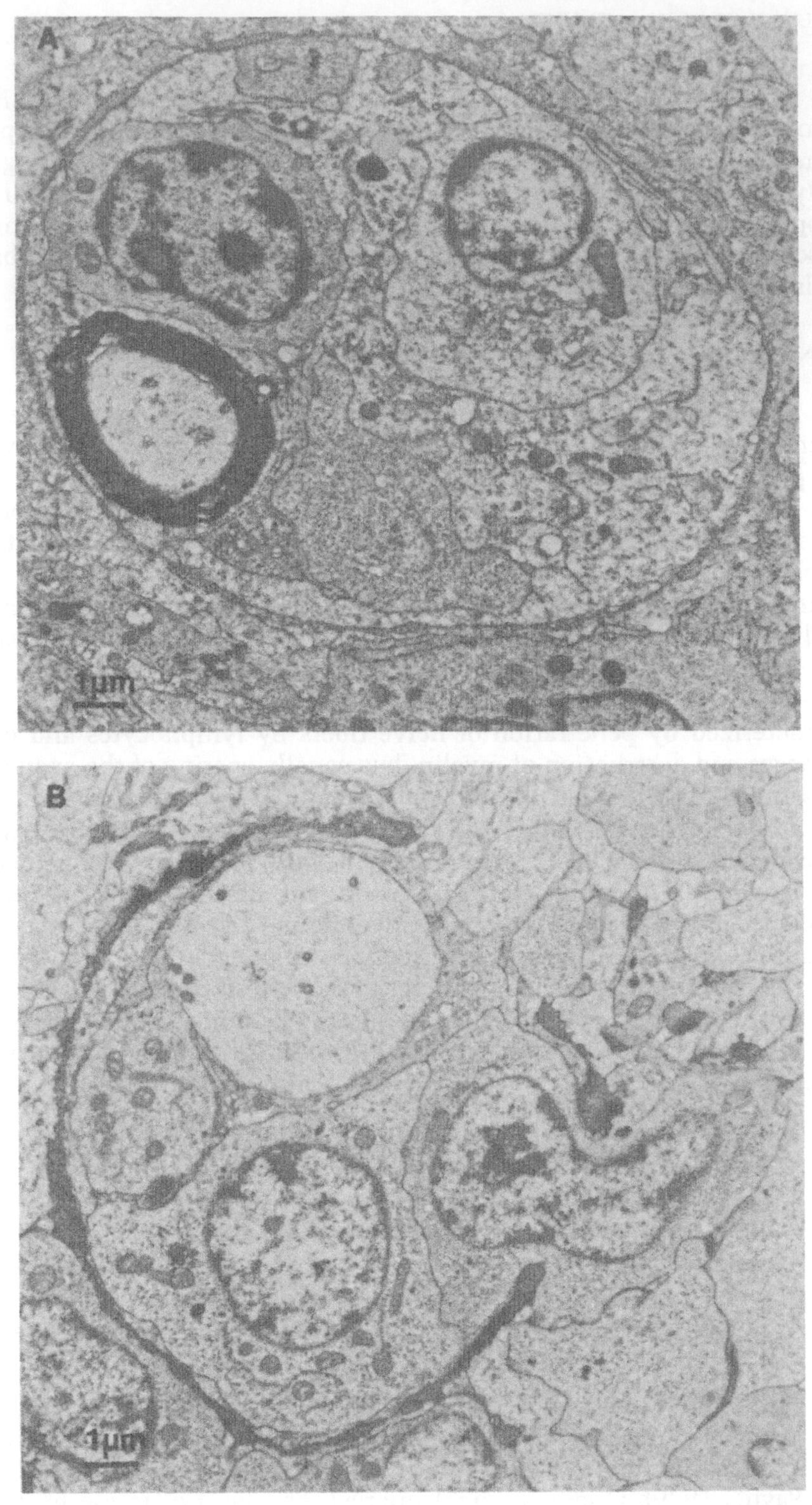

FIGURE 20. Invasion by macrophages and lymphocytes of nerve fibers (A and B) of peripheral nerve of MDV-infected chicken leading to demyelination of axon (B). Kindly provided by Dr. A. M. Lawn.

TABLE IV. Classifications of Marek's Disease Neuropathy[a]

Predominant lesion	Classification of Wight (1962a)	Payne and Biggs (1967)	Fujimoto *et al.* (1971)
Infiltration by small lymphocytes	Type I	—	T_I-type
Infiltration by mixed lymphocytes	Type I	A-type	T_{II}-type
Infiltration by lymphoblasts	Type III	—	—
Infiltration by reticular or undifferentiated mesenchymal cells	—	—	T_{III}-type
Sparse infiltration by small lymphocytes and plasma cells	—	C-type	—
Interneuritic edema, with infiltration by small lymphocytes and plasma cells	Type II	B-type	R-type

[a] From Payne *et al.* (1976), with permission.

during the 2nd and 3rd weeks, with the latter cells becoming predominant, resulting in the lymphoproliferative A-type lesion. Virus particles are rarely seen in the infiltrating cells, although cell-associated virus is present. As with lymphomas, about 75% of lymphocytes from A-type nerves are T cells, and most of the remainder B cells (Payne and Rennie,

FIGURE 21. A-type nerve lesion, showing lymphoid-cell infiltration between neurites, in peripheral nerve of 4-week-old MDV-infected chicken. From Payne (1972) by kind permission of the publishers.

TABLE V. Suggested Nature of Marek's Disease Neuropathy[a]

Suggested nature of lesions and their relationship	Classification of		
	Wight (1962a)	Payne and Biggs (1967)	Fujimoto *et al.* (1971)
Inflammatory	Types I, II	B- and C-types	R-type
Neoplastic	Type III	A-type	T_I-, T_{II}-, T_{III}-type
Relationship	Types II → I → III	A-type → B- or C-type	T-types → R-type

[a] From Payne *et al.* (1976) with permission.

1976a; Hoffmann-Fezer and Hoffmann, 1980); about 9% are MATSA-bearing (P.C. Powell and M. Rennie, unpublished results). The majority of infiltrating lymphoid cells in A-type nerves contained viral DNA (Ross *et al.*, 1981).

The initial stimulus for migration of macrophages and lymphocytes into peripheral nerves is not known. Some investigators have suggested that demyelination precedes lymphoid proliferation, resulting either from Schwann cell damage, leading to demyelination and a cellular response (Wight, 1969), or from a primary autoimmune cell-mediated demyelination (Prineas and Wright, 1972). No virions or viral antigen can

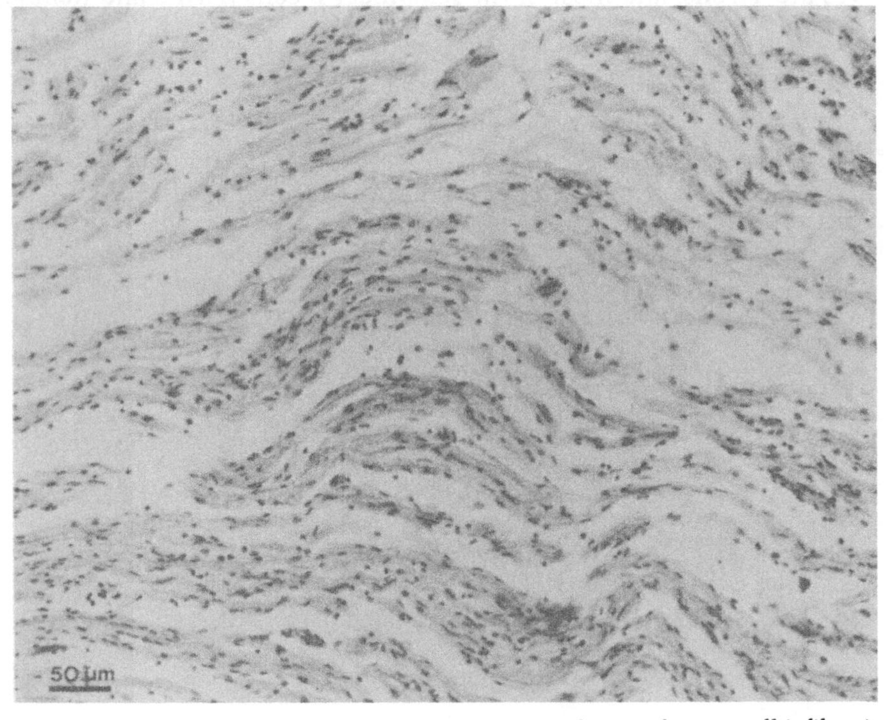

FIGURE 22. B-type nerve lesion, showing interneuritic edema and sparse cell infiltration, in peripheral nerve of 10-week-old MDV-infected chicken. From Payne and Biggs (1967).

be detected in Schwann cells, but there is evidence for a latent infection by MDV of neuronal supporting cells (Pepose *et al.*, 1981; Stevens *et al.*, 1981). On the other hand, Ross *et al.* (1981) found no viral DNA in Schwann cells and fibroblasts by hybridization techniques. A recent study, however, suggests that the cell-mediated primary demyelination is a late, secondary, event, occurring during the 4th and 5th weeks after infection, in heavily infiltrated areas and coinciding with the onset of neurological signs (Lawn and Payne, 1979), consistent with the earlier finding that the A-type lesion precedes the B-type lesion (Payne and Biggs, 1967). The demyelination is believed to result from allergic sensitization to normal nerve antigens, and is supported by the finding in MD of skin sensitivity to extracts of normal nerve and antibodies to myelin (Schmahl *et al.*, 1975; Pepose *et al.*, 1981; Stevens *et al.*, 1981). These events could be a consequence of "bystander" or other effects following an immune response to infection of infiltrating lymphoid cells (Lawn and Payne, 1979) or neuronal cells (Pepose *et al.*, 1981; Stevens *et al.*, 1981). The close similarity between the primary demyelination in MD and the Landry–Guillain–Barré syndrome in man has been noted (Payne *et al.*, 1976; Lampert *et al.*, 1977; Pepose *et al.*, 1981; Stevens *et al.*, 1981). Secondary (Wallerian) demyelination also occurs in A-type lesions (Prineas and Wright, 1972; Lampert *et al.*, 1977).

Reorganization and remyelination of damaged nerve fibers occur from the 4th week in some birds, associated with edema and sparse inflammatory cellular infiltration, which characterize the B-type lesion (Lawn and Payne, 1979).

Histopathological lesions in the brain and spinal cord in MD are usually minimal, and are comprised of perivascular lymphocytic cuffing, microgliosis, astrocytosis, and meningeal infiltration by lymphoid cells (Wight, 1962b; Fujimoto *et al.*, 1971). Primary demyelination does not occur, but secondary demyelination due to cuffing or mechanical distortion has been observed. Changes similar to these occur in the CNS in transient paralysis (Wight, 1968; see also Payne *et al.*, 1976).

8. Atherosclerosis

An association between atherosclerosis and neurolymphomatosis (MD) was first observed by Paterson and Cottral (1950). Although the arterial disease can occur in the absence of MDV infection, the incidence, distribution, and severity of the lesions are significantly increased in chickens infected with a mildly pathogenic strain of MDV, particularly when the chickens are fed a cholesterol-supplemented diet (C.G. Fabricant *et al.*, 1978; Minick *et al.*, 1979). Microscopically, the lesions, which are found in the coronary arteries, aorta, and major aortic branches, have been classified as fatty, proliferative, or fatty–proliferative, and closely resemble human atherosclerosis (Fig. 23). The role of MDV in the production of these lesions has not been clarified, but the finding of MD

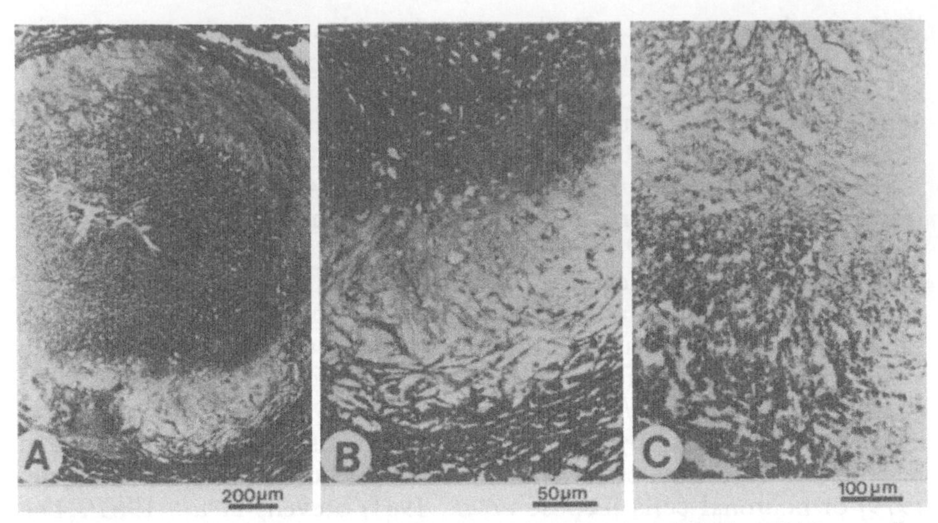

FIGURE 23. (A) Atherosclerosis of gastric artery of MDV-infected chicken. (B) Foam cells, extracellular lipid, and cholesterol clefts in media of affected artery. (C) Lipid in intima and media of affected artery. From C. G. Fabricant *et al.* (1978) by kind permission.

viral antigen in affected arterial walls suggests a direct viral effect on medial cells, possibly preceded by endothelial injury (Minick *et al.*, 1979).

9. Other Tissues

Lesions in tissues other than those described above are uncommon in MD. Necrosis of renal-tubular epithelium with intranuclear inclusion bodies, glomerulitis or glomerular necrosis, and necrosis of pancreas, proventriculus, liver, and heart have been observed (Calnek, 1972b; Ratz *et al.*, 1972). The ocular lesions sometimes associated with MD have been described (see Payne *et al.*, 1976).

VI. HOST GENETICS

A. Selection for Genetic Resistance

Fifty years ago, Asmundson and Biely (1932) showed family differences in fowl in susceptibility to neurolymphomatosis. Since then, it has become well recognized that the genotype of the host greatly influences the outcome of infection by Marek's disease virus (MDV), affecting mortality, incidence of lesions among tissues and organs, and type of lesion. These differences can be observed within and between noninbred, commercial strains of fowl, and between inbred lines. They are a consequence of either natural or artificial selection and are of importance in understanding the pathogenesis of Marek's disease (MD) and in disease control.

The resistance of older birds to MD ("age resistance") is considered to be closely related to genetic resistance (see Section VII.F.2).

The classic work on genetic resistance to MD is that of Cole and Hutt (1973). By sib and progeny testing in the presence of natural exposure to the disease over a period of some 30 years, they developed the Cornell strains susceptible (S strain) and resistant (C and K strains) to "lymphomatosis" (mainly MD) (Fig. 24). Similarly, Waters (1951) developed the East Lansing lines 6 and 7, respectively resistant and susceptible to MD (reviewed by Stone, 1975). The value of this early work is limited, though not crucially, by the presence under conditions of natural exposure of both MDV and the retroviruses, and by diagnostic confusion between the diseases they cause (see Section II). Nevertheless, these various lines are now much used in MD research.

As a basis for artificial selection for resistance, natural exposure has the disadvantages of variable and often low disease incidence, and often long latent periods. Recognition of the transmissibility of MD with cellular inocula, and of the virus responsible, encouraged the use of better-controlled, artificial exposure methods. With the use of such methods, large differences in susceptibility among lines, crosses, and families have been observed (Biggs *et al.*, 1968b; Schmittle and Eidson, 1968; Cole, 1968; Stone *et al.*, 1970; Morris *et al.*, 1970a; Crittenden *et al.*, 1972). Starting with Regional Cornell random-bred stock, with 51.1% susceptibility to MD, and by determining the response of progeny to inoculated

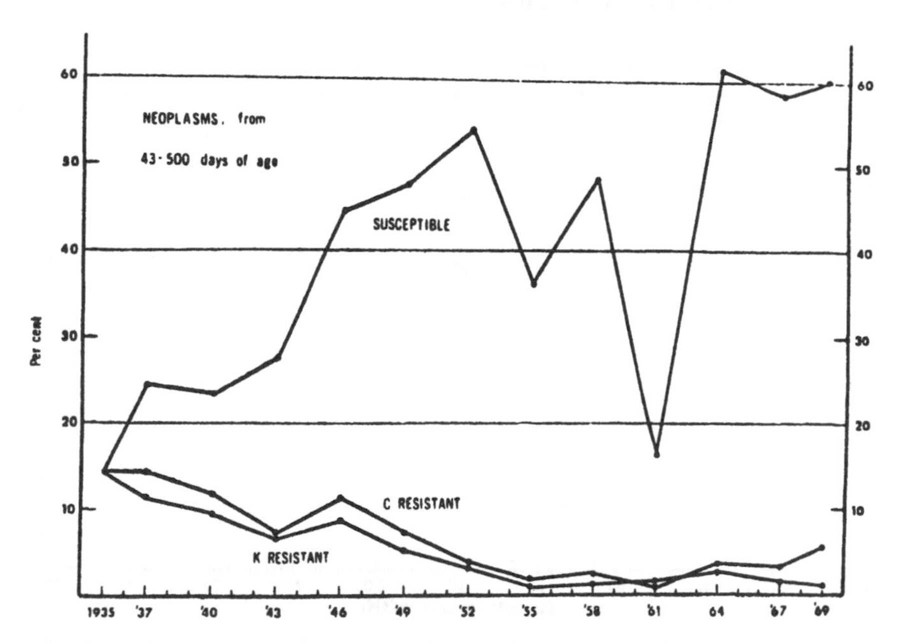

FIGURE 24. Changes in mortality, mainly from MD and lymphoid leukosis, in the Cornell S (susceptible), and C and K (resistant) strains under long-term selection. From Cole and Hutt (1973) by kind permission.

MD tumor cells, Cole (1968, 1972) selected in four generations a resistant line (N) with 6.5% susceptibility and a susceptible line (P) with 94.4% susceptibility (Fig. 25).

Controlled contact exposure to MDV has also been used to determine response, on the grounds of convenience and the possibility that resistance mechanisms operative naturally would not be bypassed. Significant correlations of responses following challenge by inoculation, contact, or field exposure have been found in various studies. Cole (1968) found that MD responses to ten commercial strains challenged by inoculation or field exposure were highly correlated ($r_s = +0.817$), and Crittenden *et al.* (1972) made similar observations in inbred lines and their crosses. Von Krosigk *et al.* (1972) reported a correlation of +0.38 for injection and contact exposure, and Hartmann and Sanz (1970) obtained correlations of +0.80 and +0.66 between injection and natural exposure in pure and crossbred commercial birds. These values suggest that selection based on injection of MDV will lead to resistance to natural exposure, but that the two selection methods may not be measuring entirely the same type of resistance. In this respect, the observation of significant statistical interactions between exposure method (inoculation or contact) and maternal influence, possibly due to maternal antibody, is of interest (Crittenden *et al.*, 1972). Of the various types of artificial exposure that can be used, the contact-exposure technique would be expected to correlate most highly with field exposure. In a test of this comparison, Grunder *et al.* (1972) obtained a correlation of +0.79.

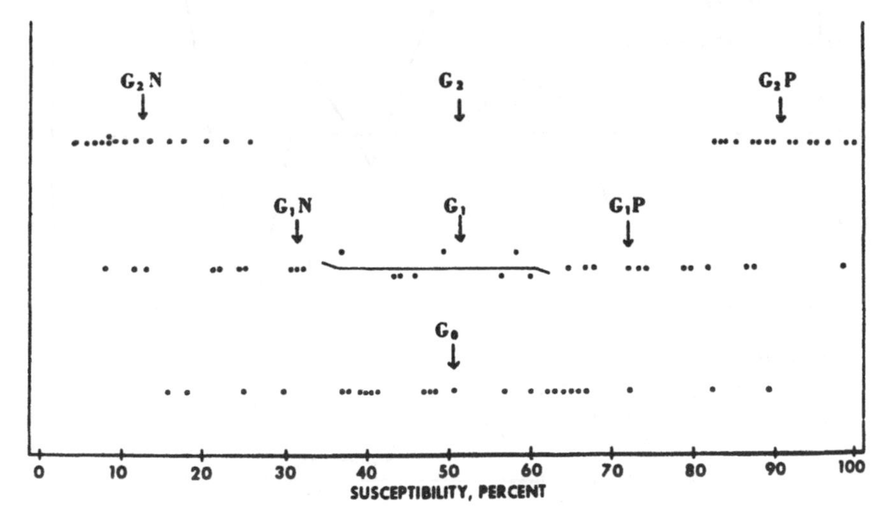

FIGURE 25. Susceptibility of sire families to MD after artificial selection for two generations. From Cole (1968) by kind permission.

Other correlations that are of practical importance are those between responses of male and female progeny (r_s = +0.867), which allows the use of surplus male chicks in progeny tests, and between total incidence of MD lesions and mortality and symptoms alone (r_s = +0.716–0.943), which allows the use of the simpler latter criteria of response (Cole, 1968). Similar observations were made in commercial lines by Morris *et al.* (1970a,b), who also found that genetic factors influenced the distribution of lesions in different tissues and organs.

Estimates of the heritability of MD resistance have been variable, perhaps because of variability of environmental factors: 0.10–0.20 (Von Krosigk *et al.*, 1972), 0.67 (Friars *et al.*, 1972) and 0.61 (Gavora *et al.*, 1974). These authors have successfully used progeny testing under commercial conditions to increase MD resistance.

B. Mode of Genetic Control

The genetic resistance observed by Cole (1972) in line N, and by Stone *et al.* (1970) in line 6, is a dominant trait; the latter authors estimated that at least three gene pairs were of importance in determining response. To date, two distinct genetic loci that play a major role in controlling genetic resistance have been identified as described below. The mechanisms involved are discussed in Section VII.F.1.

1. Involvement of the Major Histocompatibility Locus, *Ea-B*

Hansen *et al.* (1967) were the first to observe an association between resistance to MD and genes at the erythrocyte antigen group B (*Ea-B*) locus, the major histocompatibility locus of the fowl. They found that birds carrying the B^{21} haplotype were less susceptible than birds carrying B^{19}.

Subsequently, the selection of the resistant line N and the susceptible line P from common stock (Cole, 1968, 1972) on the basis of disease response was found to have been accompanied by changes in the frequency of *B* alloalleles: line N was homozygous for the allele B^{21}, whereas two other alleles existed in line P, B^{13} and B^{19}, with frequencies of 0.03 and 0.97, respectively. Test matings between the lines and their crosses confirmed the association of allele B^{21} with resistance to MD (Briles and Stone, 1975; Briles *et al.*, 1976, 1977) (Table VI). This association between B^{21} and MD resistance was also observed in studies by Pazderka *et al.* (1975) and Longenecker *et al.* (1976). The B^{21} allele was found to be widely distributed in poultry populations, including the progenitor of domestic fowl, the Red Jungle Fowl, suggesting that B^{21} may have strong survival value (Longenecker *et al.*, 1977). The association between B^{21} and resistance was influenced by other genes, either at the *B* locus or at

TABLE VI. Incidence of Marek's Disease among Backcross Progeny from the P and N Lines, Showing Influence of the B^{21} Allele[a]

	Genotype of progeny			
	$B^{19}B^{19}$		$B^{19}B^{21}$	
Mating sire and dam[b]	Number inoculated	MD (%)	Number inoculated	MD (%)
P × NP	12	58.3	17	11.8
P × PN	25	80.0	18	16.7
PN × P	32	75.0	24	8.3
PN × P	30	60.0	22	0.0
TOTALS:	99	69.7	81	8.6

[a] From Briles *et al.* (1977) by kind permission.
[b] P line = $B^{19}B^{19}$; N line = $B^{21}B^{21}$.

other loci (Longenecker *et al.*, 1976). B^2 and B^6 alleles apparently also confer some resistance to MD (Briles *et al.*, 1980) and Pevzner *et al.* (1981) have provided evidence that susceptibility maps within the immune response region of the *B* complex. An influence of other alleles (not B^{21}) at the *Ea-B* locus on transient paralysis was observed by Schierman and Fletcher (1980). Currently, there is much interest in B^{21} as a marker for resistance to MD.

2. Involvement of the *Ly-4* Locus

Although lines 6 and 7 are, respectively, highly resistant and highly susceptible to MD, both lines are identical at the *Ea-B* locus (being B^2B^2), as well as at the *Ea-A* and *Ea-C* loci (Pazderka *et al.*, 1975). A search for lymphocyte antigens that might be associated with resistance led to identification of a new locus, *Ly-4*, which controls T-lymphocyte antigens (Fredricksen *et al.*, 1977). The antigen characteristic of line 6 is designated $Ly\text{-}4^1$, determined by allele $Ly\text{-}4^a$, and that of line 7 is designated $Ly\text{-}4^2$, determined by $Ly\text{-}4^b$. An association between the presence of these alleles and response to MD was found, although other loci also were believed to be involved, because F_3 progeny homozygous for either allele did not show the extremes of response characteristic of the parental lines.

3. Involvement of Other Loci

Several other loci have been reported to have minor effects on MD susceptibility: *Akp* (alkaline phosphatase) (Grunder *et al.*, 1969; I. Okada *et al.*, 1977), hemoglobulin type (Washburn *et al.*, 1971), and *Ea-A* (blood group A) (I. Okada *et al.*, 1977).

C. Genetic Resistance and Production Traits

The genetic relationship between resistance and production traits is of importance, because the breeder does not wish, in selecting for resistance, to impair performance in economically important traits. Correlation coefficients indicate that genetically resistant birds have lower body weight and earlier sexual maturity and lay more but lighter eggs (Friars *et al.*, 1972; Von Krosigk *et al.*, 1972; Gavora *et al.*, 1974; Gavora and Spencer, 1979). Conversely, there may be a danger of increasing susceptibility to MD by selection for rapid growth (Han and Smyth, 1972) or high egg weight.

VII. IMMUNOLOGY

A. Immunosuppressive Effects of Marek's Disease Virus and Herpesvirus of Turkeys Infection

1. Depression of Immune Responses

Chickens infected with Marek's disease virus (MDV) and herpesvirus of turkeys (HVT) develop variable deficiencies in humoral and cell-mediated immune responses and in mitogen responses. These immunosuppressive effects, which depend on strain of virus and chicken, and on the stage and outcome of the infection, are of considerable interest in the understanding of events that lead to neoplasia (see Sharma, 1979; Theis, 1979).

Depressed responses to the lectin mitogens phytohemagglutinin (PHA) and concanavalin A are transiently observed at 7 days after infection in spleen cells and peripheral-blood lymphocytes, coinciding with the stage of acute cytolytic infection of lymphoid tissues (see Section V.C.2), in chickens infected with oncogenic MDV, nononcogenic MDV, or HVT (Lee *et al.*, 1978a; Schat *et al.*, 1978; Powell, 1980). Shortly after, responses are regained (although variably so) in Marek's disease (MD)-susceptible birds, and these undergo a second depression in birds developing lymphomas, but are normal or enhanced in genetically resistant birds or in those infected with nononcogenic MDV or HVT (see also Lu and Lapen, 1974) (Table VII).

The secondary depression of responsiveness coincides with deficiencies in humoral and cell-mediated immune responses: primary and secondary antibody responses to various antigens are depressed (Purchase *et al.*, 1968; Evans and Patterson, 1971; Burg *et al.*, 1971), allograft rejection is delayed, and tuberculin hypersensitivity responses are impaired (Purchase *et al.*, 1968; Payne, 1970; Schierman *et al.*, 1976). The graft-vs.-host reaction, as measured by splenomegaly, was apparently enhanced

TABLE VII. Phytohemagglutinin Response during Early, Recovery, and Late Stages of Infection by Marek's Disease Virus or Herpesvirus of Turkeys in Genetically Susceptible (Line 7_2) and Resistant (Line N) Chicks[a]

Virus	Chickens	MD lesions	PHA response[b]		
			Early	Recovery	Late
MDV	Susceptible	+	↓	±	↓
MDV	Susceptible	−	↓	+	↑
MDV	Resistant	−	↓	+	+
HVT	Susceptible	−	↓	+	↑
HVT + MDV	Susceptible	−	+[c]	+	↑

[a] From Lee *et al.* (1978a) by kind permission.
[b] (↓) Depression; (↑) (enhancement; (±) variable; (+) normal.
[c] This corrects the entry (↓) in the table as originally published; an early depression in response following MDV challenge was not observed.

(Purchase *et al.*, 1968), but virus infection may have contributed to the splenic enlargement.

Susceptibility to infection by other pathogens such as coccidia (Biggs *et al.*, 1968a) and Rous sarcoma virus (Calnek *et al.*, 1975) may also be increased in MDV-infected birds.

2. Mechanisms of Depressed Responsiveness

Lymphoid destruction, although severe during the acute cytolytic stage of infection by oncogenic MDV, is unlikely to account entirely for the early immunosuppression, because such changes are not caused by nononcogenic MDV or HVT (Calnek *et al.*, 1979), which are nevertheless immunosuppressive. These latter viruses do, however, localize in lymphoid tissues, and may therefore influence lymphocyte responsiveness. Lee *et al.* (1978a,b) attributed the early immunosuppression to suppressor macrophages present in the infected spleen. Such cells are also present in the spleen of normal chickens (Sharma, 1980). Recently, Wainberg *et al.* (1980) observed that various virus particles, either infectious or noninfectious, inhibit mitogenic responses.

In the lymphoproliferative phase of MD, Theis (1977, 1979) found a soluble suppressive factor, but no suppressor cells, in the spleen. She favored depletion of responsive cells as the cause of immunosuppression, resulting either from lymphoid atrophy or from replacement of normal lymphoid cells by nonreactive lymphoma cells (Lu and Lapen, 1974). More recently, Theis (1981) identified subpopulations of cells with suppressor activity in chickens bearing the JMV transplantable lymphoma.

These findings are consistent with the view that the early immunosuppression may be a cause of lymphoma formation and that the later immunosuppression is a consequence of this.

B. Immune Responses to Viral Antigens

1. Antibodies

Antibodies against antigens of MDV or HVT can be detected in sera from infected birds by agar–gel precipitin (AGP) tests (R.C. Chubb and Churchill, 1968), immunofluorescence (Purchase and Burgoyne, 1970), virus neutralization (VN) (Calnek, 1972a; Sharma and Stone, 1972), indirect hemagglutination (IHA) (Eidson and Schmittle, 1969), and complement-fixation (Marquardt and Newman, 1972). The first three of these techniques are mainly used.

Antibodies to three main antigens, A, B and C, are detectable by the AGP test in MD, of which those to the A antigen, partially related in MDV and HVT, have been mainly studied and used diagnostically (see Section IV.E.1). Precipitin antibodies appear 7–14 days after MDV infection, depending on genetic strain of bird, and persist; they are apparently unrelated to clinical outcome of infection (Calnek, 1972a; Higgins and Calnek, 1975). The results of IHA and AGP tests are correlated, but the former test is more sensitive (Hong and Sevoian, 1972). Precipitin responses are enhanced with less virulent strains of MDV, probably due to their milder immunosuppressive properties (M.W. Smith and Calnek, 1973).

Fluorescent immunoglobulin M (IgM) antibody is transiently detectable 5–7 days after MDV infection, at the time of first detection of neutralizing antibody, and IgG fluorescent antibodies (FAs) first appear at 7–8 days and parallel the increase in neutralizing antibodies (Higgins and Calnek, 1975).

The VN antibody response against MDV appears to be distinct from the precipitin antibody response, even though there may be a relationship between the antigens involved (Section IV.E.1). There is a strong correlation between resistance to MD and high VN antibody levels (Calnek, 1972a). However, MD-susceptible birds do not lack ability to produce VN antibodies (i.e., when infected by mild strains of virus), suggesting that the low response in birds that succumb to MD is consequential rather than causal (M.W. Smith and Calnek, 1973).

Hens infected with MDV or HVT transmit precipitin or VN antibodies to their progeny via the yolk; this passively acquired antibody persists for about 3 weeks. Such antibody against MDV has four main effects: (1) delayed onset and increased latent period to death; (2) lower MD mortality; (3) a reduction in lymphoma formation; and (4) suppression of the stage of acute cytolytic infection of lymphoid tissues (R.C. Chubb and Churchill, 1969; Spencer and Robertson, 1972; Burgoyne and Witter, 1973; Payne and Rennie, 1973). The protection afforded by passively acquired antibody can be enhanced by repeated immunization of adult breeding hens with MDV (Ball *et al.*, 1971). Maternally derived antibody exerts its effect by diminishing the severity of the initial virus infection, although the mechanism for this is unknown (see Payne *et al.*, 1976).

Maternal antibody to HVT will similarly retard the development of infection in progeny vaccinated with HVT and interfere with the onset of vaccinal immunity (Calnek and Smith, 1972; Spencer *et al.*, 1974) (see Section VII.D.2).

2. T-Cell-Mediated Responses

Delayed hypersensitivity reactions in the wattle and leukocyte migration inhibition *in vitro* against MDV and HVT antigens have been demonstrated by Byerly and Dawe (1972), Fauser *et al.* (1973), and Prasad (1978), implicating cell-mediated immunity in response to these infections. In a plaque inhibition test, Ross (1977) identified T cells in blood of MDV-infected chickens cytotoxic against leukocytes or chick kidney cells (CKC) infected with MDV. In cytotoxicity studies using effector cells from birds infected with MDV or HVT and target cells infected with either virus, responses were stronger against homologous virus than against heterologous virus (Kodama *et al.*, 1979b; Ross, 1980).

3. Antibody-Dependent Cell-Mediated Cytotoxicity (ADCC)

The specificity of the direct cell-mediated cytotoxicity for homologous virus mentioned above was lost when antisera from MDV- or HVT-infected birds were mixed with normal spleen cells or blood lymphocytes, or with blood lymphocytes from HVT-infected birds, and reacted against MDV-infected target cells (Kodama *et al.*, 1979b; Ross, 1980). It is probable that direct cell-mediated reactions, and ADCC, are important in controlling the spread of cell-associated virus in infected birds.

4. Macrophages

Although Frazier (1974) has suggested that reticulum cells, possibly macrophage precursors, are the initial targets for MDV infection *in vivo*, macrophages *in vitro* do not support virus replication (Haffer *et al.*, 1979). Evidence is increasing that macrophages play a role in immunity to MD. The suggestion that macrophages might restrict spread of MDV in the bird was first made by Higgins and Calnek (1976), who observed that silica treatment of chicks tended to increase survival of infected birds, suppressed early virus replication in the thymus and bursa, suppressed VN antibody production, and increased buffy coat macrophages, possibly thereby *enhancing* their effect. Haffer *et al.* (1979), on the other hand, *suppressed* macrophage function *in vivo* by inoculation of antimacrophage serum or trypan blue and observed elevated virus titers and increased tumor incidence, supporting an immunosurveillant role for macrophages.

Direct evidence for viral suppression by macrophages has been presented: removal of macrophages from spleens from MDV-infected chickens increased virus replication in culture (Lee, 1979), and MDV plaques in CKC cultures were inhibited by addition of peritoneal macrophages from MDV-infected birds (Kodama *et al.*, 1979a). These latter workers observed that plaque reduction by peritoneal macrophages was much less when they were derived from normal or HVT-infected birds; however, MDV plaque inhibition by normal macrophages was enhanced in the presence of MDV, but not HVT, antibody. Macrophages can also inactivate cell-free MDV in cooperation with B cells (Schat and Calnek, 1978b). These findings suggest that there are several mechanisms by which macrophages protect against MDV replication. In addition, macrophages from normal and MDV-infected birds suppress mitogenic responses by normal T cells and proliferation of MDV-transformed T cells (Lee *et al.*, 1978b; Lee, 1979; Sharma, 1980).

5. Interferon

Interferon production has been demonstrated in MDV and HVT infections; it is influenced by strain of chicken, being higher in resistant strains (Hong and Sevoian, 1971), and by strain of virus, apparently not being produced by some (Kaleta and Bankowski, 1972; Kaleta, 1977). It provides some protection against the transplantable JMV lymphoma (Vengris and Mare, 1973). The role of interferon in immunity to MD has not been clearly established, but it is considered to be minor.

C. Immune Responses to Marek's Disease Tumor-Associated Surface Antigen and Other Tumor-Associated Antigens

1. Antibodies

Antibodies against Marek's disease tumor-associated surface antigen (MATSA) have not been found in FA tests in sera from convalescent MDV or HVT-infected chickens, although they are induced by inoculation of lymphoma or MD lymphoblastoid-line cells (Witter *et al.*, 1975b; Stephens *et al.*, 1980). No cytotoxic activity against MSB-1 lymphoblastoid-line cells was detected in complement-dependent antibody cytotoxicity tests or ADCC tests in sera from MDV- or HVT-infected birds, suggesting absence of anti-MATSA activity (Sugimoto *et al.*, 1978; Kitamoto *et al.*, 1979). Humoral cytotoxic activity has been detected in sera from chickens immunized with inactivated MSB-1 cells, but its specificity for MATSA has not been shown.

2. T-Cell-Mediated Responses

Low-level cell-mediated cytotoxicity against MD lymphoblastoid-line cells has been demonstrated in ^{51}Cr-release assays using effector lymphocytes from spleen (Sharma and Coulson, 1977; Sharma *et al.*, 1978; Calnek *et al.*, 1979) or peripheral blood (Powell, 1976; Confer and Adldinger, 1980) from MDV- or HVT-infected chickens. Higher-level activity has been reported using [^{3}H]proline labeling (Dambrine *et al.*, 1980) and [^{35}S]methionine labeling of target cells (Kitamoto *et al.*, 1979). Both oncogenic and nononcogenic strains of MDV, and of HVT, induce activity (Calnek *et al.*, 1979). The effector cell has been identified as a T cell (Sharma, 1977; Kitamoto *et al.*, 1979).

The target antigen in these assays is unknown. Although MATSA, by virtue of its presence on transformed cells, has been suspected and promoted by various workers as the likely target, Schat and Murthy (1980) have shown recently that effector cells were still cytotoxic after MATSA had been blocked with specific antisera or removed by papain treatment. Furthermore, when effector cells and target cells were isogeneic, little or no cytotoxicity was observed, implicating histocompatability antigens in the response (Schat and Calnek, 1980).

Sensitized lymphocytes appear about 1 week after infection, correlating apparently with the appearance of MATSA-bearing cells. There is, however, much unexplained variation in the association of cytotoxic activity with the stage and outcome of MDV or HVT infection. The activity of effector cells from MDV- or HVT-infected birds has been variously reported as transient during the 2nd week of infection (Sharma and Coulson, 1977), prolonged but independent of lymphoma production on an individual basis (Confer and Adldinger, 1980; Confer *et al.*, 1980), present only in lymphoma-bearing birds (Dambrine *et al.*, 1980), present at 8–25 weeks in MDV-infected, but not HVT-infected, birds (Sugimoto *et al.*, 1978), and present in HVT-infected birds at 2–6 weeks (Kitamoto *et al.*, 1979). Numerous factors that may influence these findings include strain of virus and bird, source of effector cells, type of target cells, the label used in cytotoxicity assays, and the duration of the assay; more study of the effect of these is needed.

3. ADCC Responses

ADCC activity was found in serum from chickens immunized with MD lymphoblastoid cells (Powell, 1976), but was not present in sera from MDV- or HVT-infected birds (see Section VII.C.1).

4. Macrophages

Macrophages from the spleen of MDV-infected birds inhibit the DNA synthesis of MSB-1 lymphoblastoid cells *in vitro* (Lee *et al.*, 1978b) and of proliferating MATSA-bearing lymphoma cells in the spleen *in vivo*

(Lee, 1979). Splenic macrophages from normal uninfected birds will also inhibit lymphoblastoid-line cell growth (Sharma, 1980).

5. Natural Killer (NK) Cells

NK cells active against MD and other lymphoid-cell lines are present in spleens of chickens of various lines (Sharma and Coulson, 1980). They may be responsible for natural age resistance to the JMV transplantable lymphoma (Lam and Linna, 1979) and play a role in genetic and vaccinal resistance (Sharma, 1981b).

D. Vaccination

1. Types of Vaccine and Their Administration

Since 1970, MD in commercial chickens has been successfully controlled by use of live virus vaccines. These are injected into chicks at hatching to provide an early protection from pathogenic MDV, exposure to which frequently occurs in the poultry house during the first few weeks of life. The vaccine viruses are nonpathogenic and establish a permanent infection that prevents lymphoma formation following subsequent exposure to pathogenic virus.

Vaccine viruses are of three types: (1) attenuated MDV, (2) naturally apathogenic MDV, and (3) HVT.

1. *Attenuated MDV*: The first live virus vaccine to be developed and used commercially was produced by serial passage of pathogenic MDV (HPRS-16 strain) in cultured CKC, which resulted in a loss of pathogenicity but not of immunogenicity (Churchill *et al.*, 1969a,b; Biggs *et al.*, 1970). Attenuation was accompanied by changes in the antigenic and growth properties of the virus (see Section IV.D.1). Other pathogenic strains have been attenuated similarly (Nazerian, 1970; Eidson and Anderson, 1971).
2. *Naturally apathogenic MDV*: Field viruses of little or no pathogenicity occur and may be used to protect against MD. Their use as vaccines necessitates serial passage in tissue culture, which is accompanied by changes in growth rate and antigenic composition (Rispens *et al.*, 1972a,b; Blaxland *et al.*, 1972). The CVI 988 strain is a well-known naturally apathogenic MDV used as a vaccine (Rispens *et al.*, 1972a).
3. *Herpesvirus of turkeys*: HVT was first isolated from turkeys, in which it is ubiquitous, in the United States (Kawamura *et al.*, 1969; Witter *et al.*, 1970b). It is apathogenic in turkeys and chickens, antigenically related to MDV, and extensively used as an MD vaccine (Okazaki *et al.*, 1970; Purchase *et al.*, 1971b, 1972c) (Fig. 26). The FC126 strain is most widely used, but other strains have also been isolated and used as vaccines.

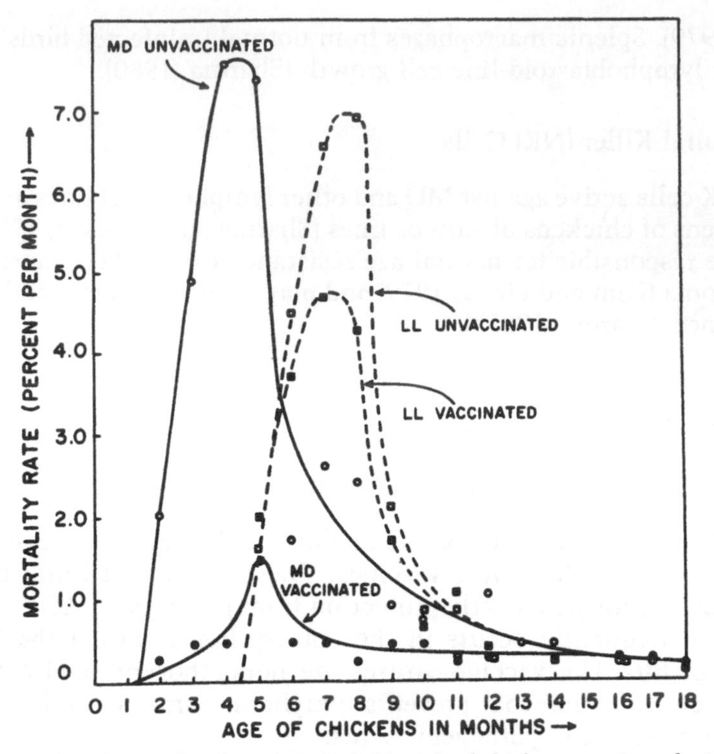

FIGURE 26. Reduced mortality from MD in field trials of chickens vaccinated with HVT. [Mortality from lymphoid leukosis (LL) was also apparently slightly reduced, although misdiagnosis may have been responsible.] From Purchase *et al.* (1972c) by kind permission.

The attenuated and naturally apathogenic strains of MDV are cell-associated and are inoculated in this form. Viable infected culture cells are stored in liquid nitrogen in medium containing dimethylsulfoxide prior to use. HVT similarly can be used in a cell-associated form, and it is also produced in a cell-free form by sonication of infected cells and lyophilization. These two forms are known colloquially as "wet" and "dry" vaccines, respectively.

In normal use, these vaccine viruses, whether cell-associated or cell-free, are inoculated in doses of 1000–2000 plaque-forming units (PFU) intramuscularly or subcutaneously into 1-day-old chicks. Generally, all three types of vaccine provide a high level of protection in the field, although in controlled trials HVT was significantly better than attenuated MDV (Eidson *et al.*, 1971). Currently, HVT is the most extensively used vaccine; the CVI 988 vaccine is mainly used in the Netherlands. Unlike the other vaccines, CVI 988 will spread in chickens by contact, but not sufficiently rapidly to avoid the need to vaccinate chicks individually. Vaccination with cell-free HVT by aerosol application is rather less effective than by inoculation (Eison and Kleven, 1977; Eidson *et al.*, 1980) and has not been adopted commercially. Recently embryonal vaccination has been accomplished (Sharma and Burmester, 1982).

The infective dose 50% (ID_{50}) and protective dose 50% (PD_{50}) of HVT is between 1 and 5 PFU (Patrascu *et al.*, 1972; Purchase *et al.*, 1972b; Witter and Burmester, 1979), although in chicks with maternally derived HVT antibodies, which have a neutralizing effect, PD_{50} values are increased, 2- to 8-fold for cell-associated virus and 15- to 80-fold for cell-free virus (Witter and Burmester, 1979). The higher doses of vaccine that are used practically are designed to provide a large margin of safety and to reduce the interval between vaccination and development of immunity to challenge. Generally, cell-free and cell-associated HVT vaccines are considered to be equally efficacious (Eidson *et al.*, 1975, 1976). Substantial protection against MD is provided 1–2 weeks after vaccination in normal circumstances. Vaccination after challenge gives no protection, and simultaneous vaccination and challenge give only partial protection (Okazaki *et al.*, 1971). Vaccinal immunity is believed to be lifelong. Waning of immunity has been considered as one cause of vaccination failures that occasionally occur (see Section VII.D.4), although an attempt to demonstrate loss of immunity was unsuccessful (Witter and Offenbecker, 1978). The mechanism of vaccinal immunity is considered in Section VII.F.3.

2. Effect of Maternal Antibody on Vaccination

Normally, dams of commercially produced chicks are vaccinated against MD, usually with HVT, and are naturally challenged with MDV. They will consequently develop antibodies to these viruses that will be passed to the chick via the yolk. The question of whether these antibodies can interfere with vaccination of the progeny is of considerable practical importance. It has been observed that such antibodies can indeed interfere, doing so more strongly against the homologous virus than against the heterologous virus. Thus, antibodies to HVT, but not to MDV, interfere with HVT vaccination, and cell-free HVT is more susceptible than cell-associated HVT (Calnek and Smith, 1972; Churchill *et al.*, 1973; Spencer *et al.*, 1974; Yoshida *et al.*, 1975; King *et al.*, 1981). The differential effect of HVT antibodies on the two types of HVT is clearly demonstrated by the higher PD_{50} value for cell-free HVT in passively immune chicks already mentioned (Witter and Burmester, 1979). The interference with vaccination is seen as a depression of HVT viremia, neutralizing antibodies, and protective effect. Although both cell-free and cell-associated HVT vaccines are usually highly effective in preventing MD (Eidson *et al.*, 1976), in some studies rather less protection was afforded by cell-free vaccine (Gavora *et al.*, 1977).

3. Effect of Strain of Chickens on Vaccination

The differences in susceptibility to MD observed in chickens of different genotypes (see Section VI.A) is also observed in vaccinated birds.

Spencer *et al.* (1972, 1974) observed that strains generally ranked in the same order of relative susceptibility whether they were vaccinated or unvaccinated. No differences were observed among strains in development of neutralizing antibodies to HVT following vaccination (although passive HVT antibodies depressed the strain responses), suggesting that strain differences in MD incidence in both vaccinated and unvaccinated birds depend on differences in response to MDV, perhaps related to immunosuppressive effects (Spencer and Gavora, 1980).

4. Failure of Vaccination

Although vaccination is highly effective in preventing MD, vaccine failures or "breaks" occur in some flocks, resulting in unacceptably high mortality from MD. Many possible causes of failure can be considered, although they are difficult to identify in practice. These possible causes include:

1. *Failure to inoculate correct dose of vaccine virus*: This may occur as a result of errors in the storage or dilution of the vaccine (Halvorson and Mitchell, 1979; Colwell *et al.*, 1975) or in chick-inoculation procedures; they are essentially trivial reasons, and the remedies are obvious. Chickens from a number of flocks investigated because of vaccination failure had a lower incidence of HVT infection than expected, suggesting that they had not been vaccinated (Okazaki *et al.*, 1973; Cho *et al.*, 1976).
2. *Challenge by MDV before normal vaccinal immunity has developed*: Since vaccinal immunity takes at least 1 week to develop, birds challenged before this time are likely to be susceptible to MD and succumb. Hygienic precautions can be taken to avoid such early exposure (Section IX), but this may be difficult to prevent, particularly on sites with birds of different ages. The failure of double vaccination to prevent losses after early exposure is to be expected (Ball and Lyman, 1977). Embryonal vaccination provides improved immunity to early challenge (Sharma and Burmester, 1982).
3. *Delayed onset of vaccinal immunity*: As discussed in Section VII.D.2, maternal antibody may interfere with HVT, particularly cell-free, reducing or preventing the establishment of HVT viremia, the development of an immune response, and onset of protection. Riddell *et al.* (1978) observed considerable variation in HVT viremia titers in different field flocks of a strain susceptible to vaccine failure. Suggested remedies against maternal antibody effects are: (a) increase the immunizing dose of HVT or revaccinate at 3 weeks when passive antibody has gone; (b) alternate HVT and attenuated MDV vaccines from generation to generation (King *et al.*, 1981); (c) do not vaccinate parent flock; or (d) use cell-associated HVT. However, controlled field trials to study these approaches have not been conducted.

Infection with infectious bursal disease virus, which is strongly immunosuppressive, will interfere with vaccinal immunity and increase losses from MD (Giambrone *et al.*, 1976).

4. *Use of nonprotective strains of vaccine virus*: A commercial HVT vaccine that induced poor viremias and protection was described by Thornton *et al.* (1975), and MDV and HVT can be overattenuated by continued passage and lose their protective properties (Witter and Offenbecker, 1979). Until recently, strains of MDV isolated from flocks with vaccination failure appeared no different from other pathogenic strains in protection trials with HVT, thus providing no evidence for antigenic variants (Okazaki *et al.*, 1973). However, highly lymphotoxic strains of MDV have now been isolated from HVT-vaccinated commercial broiler chickens experiencing excessive MD condemnations, against which HVT vaccination provided suboptimal protection in laboratory trials (Witter and Fadly, 1980; Witter *et al.*, 1980; Eidson *et al.*, 1981). An attenuated strain of a variant virus protected against the parental and related strains, and was of value in a polyvalent vaccine against various pathogenic strains (Witter, 1982).

5. Vaccination and Production Traits

Apart from reducing mortality and carcass condemnation from MD, vaccination improves economic performance in broilers and egg layers, by reducing subclinical MD and susceptibility to other infections. Benefits in vaccinated broilers include increased body weight and food conversion; in layers, increased egg production; and in all birds, decreased nonspecific mortality (Purchase, 1975, 1977).

E. Nonspecific Immunostimulation

Donahoe *et al.* (1978) reported a higher incidence of gross MD lesions, though not of mortality, in chickens inoculated with *Corynebacterium parvum*.

Levamisole slightly accelerated MD mortality, apparently by suppressing macrophage restriction of MDV replication (Kodama *et al.*, 1980; Payne and Howes, 1980), and increased the responsiveness of lymphocytes from MDV-infected chickens to PHA (Confer and Adldinger, 1981).

F. Mechanisms of Resistance

The various immune responses that have been detected in chickens infected with MDV or HVT have been reviewed (Sections VII.A–C). In this section are considered the contribution these responses make to the bird's ability to resist the pathogenic effects of virulent MDV, whether determined by genetic factors, age, or vaccination.

1. Genetic Resistance

Studies on the basis for genetic resistance to MDV (see Section VI) involve comparisons of responses between resistant and susceptible strains. The differences seen are quantitative in nature, and no mechanisms unique to resistant strains have been found.

Resistance to MDV does not lie at a general cellular level, because cultured fibroblasts or kidney cells from susceptible or resistant lines are equally susceptible to infection (Spencer, 1969; Sharma and Purchase, 1974).

Various lines of evidence point to genetically controlled differences in the lymphoid system as being the basis for differences in susceptibility. In the past, it has been usual to consider this evidence as a whole in formulating hypotheses about the nature of resistance. However, the finding of response differences among resistant lines suggested that two types of genetic resistance exist, one present at hatching, as exemplified by line 6, and the other developing during the first few weeks of life, and believed to be similar to age resistance, as exemplified by line N (Witter, 1976). Subsequently, the line 6 type of resistance has been associated (see Section VI.B.2) with the *Ly-4* locus, and is believed to depend primarily on differences in susceptibility of the presumed target cell for MDV, the T lymphocyte. The line N type of resistance, on the other hand, is associated with the B^{21} allele at the *Ea-B* locus (see Section VI.B.1) and is believed to depend on differences in immunosurveillance of transformed cells.

Compared with the susceptible line 7, the resistant line 6 develops lower levels of viremia (Sharma and Stone, 1972), and the difference in susceptibility to infection is seen even in chick embryos (Longenecker *et al.*, 1975). VN-antibody levels are higher in line 6 (Sharma and Stone, 1972), but resistance is still expressed after bursectomy, indicating that antibodies are not essential for resistance (Sharma, 1974). It is likely that immunosuppression in the susceptible line 7 was responsible for the lower VN-antibody levels. Strong evidence for the target-cell difference between lines 6 and 7 was provided by Gallatin and Longenecker (1979), and Powell *et al.* (1982), who transplanted line 7 thymuses into thymectomized line 6 chickens and made them susceptible to lymphoma formation. However, in reciprocal experiments, line 7 chickens were not made resistant. A greater susceptibility of line 7 lymphocytes to infection by MDV is believed to be partly responsible for the line differences (Gallatin and Longenecker, 1981; Powell *et al.*, 1982). Immunosurveillance also appears to have a role in line 6 resistance, because thymectomy (plus irradiation) enhanced lymphoma formation in the line (Sharma *et al.*, 1975). This finding implicates cell-mediated immunity in resistance.

Recently, higher antiviral and anti-tumor-cell cell-mediated immunity has been observed in line 6 compared with line 7 (Confer and Adldinger, 1980; Lee *et al.*, 1981) accompanied by disappearance of MATSA-bearing (possibly transformed) cells in the spleen of line 6 chickens. These results suggest that although cell-mediated immune mechanisms play a role in resistance of line 6, they are effective because of the lower level of lymphocyte infection, transformation, and immunosuppression in that line.

It may be noted that thymectomy and other adjunctive treatment in line 7 *decreased* lymphoma formation (Sharma *et al.*, 1977). Here, it may be speculated that the effect of thymectomy is to decrease the number of target cells for transformation. The differential effect of thymectomy on lymphoma formation in susceptible and resistant lines may be anticipated when T cells provide both targets for transformation and effectors for destruction of transformed cells (Payne *et al.*, 1976).

Some of the characteristics of MDV-infected chickens, such as those of line N, resistant to MD by virtue of possessing the B^{21} allele, are similar to those shown by line 6, although the mechanism is not the same. Thus, compared with susceptible line P, infected line N birds have higher VN antibodies (Calnek, 1972a), the resistance is not abolished by bursectomy (Sharma and Witter, 1975), and lymphoma incidence is increased by thymectomy (Shieh and Sevoian, 1978). The lower VN-antibody levels in line P are due to immunosuppressive effects of MDV, because both lines P and N developed similar levels of VN antibodies when infected with a nonpathogenic virus; in other words, the susceptibility of line P was not due to an inherent defect in ability to produce VN antibodies (M.W. Smith and Calnek, 1973; Higgins and Calnek, 1975).

Findings with line N differ from those with line 6 in several important respects. Levels of early infection by MDV, and severity of the acute cytolytic infection, were similar in lines N and P (J. Fabricant *et al.*, 1977), although later virus levels in line N were lower (Murthy and Calnek, 1979a). MATSA-bearing cells appeared at a similar time in the two lines, but disappeared in the resistant line (Murthy and Calnek, 1979b). Contrary to the findings of Gallatin and Longenecker (1979) with lines 6 and 7, Shieh and Sevoian (1978) reported that transfer of P line thymus (susceptible) to thymectomized N line (resistant) chicks did not influence lymphoma formation, whereas thymectomized P line chicks that received N line thymus were rendered more resistant.

These results suggest that B^{21}-associated resistance is mediated by superior ability to reject transformed lymphoid cells. A role for NK cell activity in the resistance of line N has been suggested (Sharma, 1981b).

2. Age Resistance

It is well established that older chickens are often more resistant to MD than young birds (Witter *et al.*, 1973). This resistance depends on the development of cell-mediated immunity, since it is abolished by thymectomy but not by bursectomy (Sharma *et al.*, 1975; Sharma and Witter, 1975). Age resistance was regarded by Calnek (1973) as being at least partly dependent on genetic resistance (of the B^{21}-linked type), because age had little or no influence on the tumor response of the genetically susceptible strain P, whereas chicks from resistant strain N, if free from maternal antibody, were susceptible to clinical MD when infected at 1 day old, but developed resistance over a period of several weeks. The semiproductive infection of lymphoid organs occurred in resistant birds irrespective of age, suggesting that resistance operates on a later stage of the disease. This seems to be the multiplication of transformed lymphoid cells, since age resistance depends on regression of lymphoproliferative lesions (Sharma *et al.*, 1973b), and seems to be associated with increased cytotoxic activity by lymphocytes against tumor cells (Confer *et al.*, 1980).

3. Vaccinal Immunity

Chickens vaccinated with any of the live virus vaccines (Section VII.D.1) develop resistance to lymphoma formation or other clinical disease following exposure to MDV, but become persistently superinfected by the challenge virus. Vaccinal immunity thus involves an altered relationship between oncogenic MDV and the host, and available evidence indicates that this is a consequence of immune responses: (1) vaccinal immunity may be abrogated by the immunosuppressive drug cyclophosphamide (Purchase and Sharma, 1974); (2) various immune responses against viral and tumor antigens develop in vaccinated birds (Sections VII.B and VII.C); (3) immunity against MD can be successfully induced by use of inactivated antigens, as discussed below.

Vaccination with HVT or nononcogenic MDV (SB-1 strain) is followed within 1 week by localization, but limited replication, of virus in the spleen, thymus, and bursa, unaccompanied by the acute cytolytic changes seen with oncogenic virus (Calnek *et al.*, 1979). MATSA-bearing cells, possibly transformed, appear in these organs at about the same time, and may persist for many weeks (Powell and Rennie, 1978; Calnek *et al.*, 1979). Unchallenged HVT-vaccinated birds develop a persistent viremia and mild, transient lymphoproliferative lesions in the nerves and gonads, suggesting a limited transformational event (Witter *et al.*, 1976). Various immune responses against viral and tumor antigens have been recognized in such birds, including development of VN antibodies, and of antiviral and antitumor cytotoxicity (Sections VII.B.1, VII.B.2, and VII.C.2). Following challenge, the early cytolytic infection of lymphoid tissue, and consequent immunosuppression, MDV viremia, and the lymphoproliferative response, are all suppressed (Witter *et al.*, 1976) (Figs. 27 and 28). The MDV in turn will stimulate antiviral and antitumor immune responses that can be expected to reinforce the vaccinal responses (Sections VII.B and VII.C).

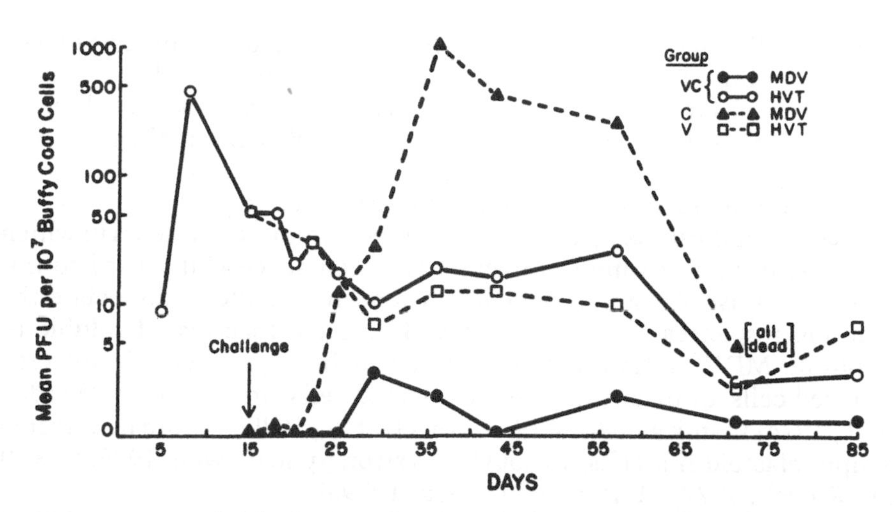

FIGURE 27. Viremia levels of HVT and MDV in chickens vaccinated and challenged (VC), only vaccinated (V), and only challenged (C), showing reduced MDV viremia in vaccinated and challenged birds. From Witter *et al.* (1976) by kind permission.

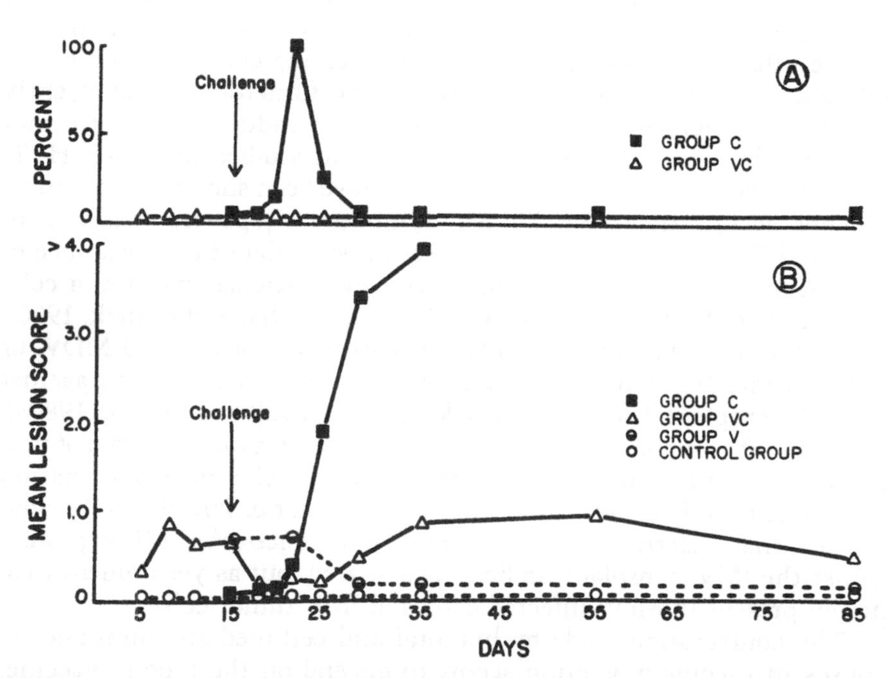

FIGURE 28. Reduction of (A) early reticuloendothelial proliferative lesions in the thymus and (B) later lymphoproliferative lesions in chickens vaccinated and challenged (VC) compared with chickens only challenged (C). Only mild lymphoproliferative lesions occurred in chickens only vaccinated (V). From Witter *et al.* (1976) by kind permission.

These findings are consistent with the "two-step" hypothesis of vaccinal immunity (Payne *et al.*, 1976), which postulates that the resistance of vaccinated birds to MD depends on (1) immune responses directed against viral antigens, which reduce the level of MDV replication, immunosuppressive effects, and malignant transformation of lymphocytes, and (2) immunological rejection of transformed cells.

The hypothesis has gained further support from experiments in which inactivated viral or tumor antigens have been successfully used to immunize against MD and in which it has been possible to dissociate the immune responses to the two types of antigen. Inactivated soluble or insoluble MDV or HVT antigens extracted from productively infected cultured cells, or inactivated whole infected cells, are protective (Kaaden *et al.*, 1974; Lesnik and Ross, 1975; Powell *et al.*, 1980b), as are inactivated lymphoblastoid-line cells, although less strongly so (Powell, 1975; Powell and Rowell, 1977; Murthy and Calnek, 1979a).

In these studies, the immunity induced by viral antigens is characterized by development of VN antibodies, lowering of MDV viremia, inhibition of virus replication, and early degenerative changes in lymphoid organs, but absence of lymphocytes cytotoxic against tumor cells. Immunization with tumor-cell antigens, on the other hand, does not influence early MDV replication and produces no lymphoid degeneration, VN antibodies, or lowering of viremia, but does stimulate development of lymphocytes cytotoxic against tumor cells. Both types of inactivated vaccine inhibit the appearance of MATSA-bearing cells, which are possibly transformed cells, and prevent lymphoma formation, but apparently by different mechanisms. Further evidence for induction of antitumor immunity by vaccine viruses has come from studies in which HVT, attenuated MDV, and nononcogenic MDV have been shown to immunize chickens against transplantable MD lymphomas (Spencer *et al.*, 1976; Schat and Calnek, 1978c). These viruses all stimulate the appearance of MATSA-bearing cells and of lymphocytes cytotoxic against tumor cells (Sharma *et al.*, 1978; Powell and Rennie, 1980b; Schat and Calnek, 1980). In addition, cells productively infected with HVT, attenuated MDV, or nononcogenic MDV are unable after inactivation to immunize against the JMV transplant (Schat and Calnek, 1978c; Powell and Rennie, 1980a), supporting the importance of tumor-specific antigens in this type of protection. The only finding to date that impugns the clear dissociation between antiviral and antitumor immunity is that of Powell (1978), who reported that inactivated cells infected with *oncogenic* MDV protect against the JMV transplant, indicating a shared, but as yet unidentified, antigen present in MDV-infected cells and JMV tumor cells.

The contribution made by humoral and cell-mediated immune responses in vaccine protection seems to depend on the type of vaccine. Else (1974) observed full protection in agammaglobulinemic, bursectomized birds vaccinated with attenuated MDV, whereas Rennie *et al.* (1980)

observed only partial protection in similar experiments with HVT. Thus, humoral immunity appears to be more important in HVT-induced immunity than in that produced by attenuated MDV. The impairment of HVT-induced vaccinal immunity by cyclophosphamide (Purchase and Sharma, 1974) was found to be transient, development of protection apparently being related to recovery of the thymic system (Payne *et al.*, 1978); however, partial recovery of the bursal system also cannot be excluded. A possible involvement of NK cells in vaccinal immunity has been suggested (Sharma, 1981b).

VIII. EPIZOOTIOLOGY

A. Spread of Infection

As discussed in Section III.B, Marek's disease virus (MDV) and herpesvirus of turkeys (HVT) are widely distributed in chickens and turkeys, respectively. Marek's disease (MD) is highly contagious, and the virus spreads from infected to uninfected chickens by direct or indirect contact (Biggs and Payne, 1967). The question of whether vertical transmission occurs was argued for many years, although confusion between MD and leukosis obscured experimental findings. Sevoian (1968) presented evidence for egg transmission, but Solomon *et al.* (1970) were unable to confirm this in extensive tests on embryos and isolator-reared chicks from infected flocks. Furthermore, no evidence of MDV infection was found in chicks hatched from eggs containing blood or meat spots (Solomon and Witter, 1973). It is common experience that progeny chicks of infected hens remain free of infection if hatched and reared in isolation, and the consensus is that vertical transmission rarely if ever occurs.

The major source of MDV in the environment is feather debris and dander from infected birds (Calnek *et al.*, 1970a), although virus has been reported to be present in feces and in oral and nasal washings (Kenzy and Biggs, 1967; Witter and Burmester, 1967; Eidson and Schmittle, 1968b). Infected birds remain carriers and shedders for life. Poultry house dust is a potent source of infection owing largely to the presence of infected epithelial cells and cell-free virus (Beasley *et al.*, 1970; Jurajda and Klimes, 1970; Carrozza *et al.*, 1973), and the infection is readily transmitted by the airborne route (Sevoian *et al.*, 1963; Colwell and Schmittle, 1968).

At room temperature, the infectivity of droppings and litter was found to persist for up to 4 months (Witter *et al.*, 1968) and of dust or feathers for 7–8 months (Calnek and Hitchner, 1973; Carrozza *et al.*, 1973). The virus was found in larval or adult darkling beetles (*Alphitobius diaperinus*), which are commonly present in poultry litter (Eidson *et al.*, 1966), but insect vectors are considered to be relatively unimportant as reservoirs of virus. The virus was also found in mosquitoes fed on infected birds, although they did not transmit the infection by biting (Brewer *et al.*, 1969). Coccidial oocysts from MDV-infected chickens did not transmit the virus (P.L. Long *et al.*, 1968).

The shedding and transmission of HVT in turkeys are similar to those of MDV in chickens. Witter and Solomon (1971) demonstrated infectivity in dirty cages previously used to rear infected turkeys and in air from isolators containing such birds. Infectivity of litter and dust could not be demonstrated. Egg transmission of HVT does not occur (Witter and Solomon, 1971; Paul *et al.*, 1972). HVT spreads by the airborne route from turkeys to turkeys, and from turkeys to chickens (Witter and Solomon, 1972), but only poorly from chickens to chickens owing to poor replication in feather-follicle epithelium of this species (Cho, 1975, 1976b).

B. Development of Flock Infection

Because of the prevalence and persistence of MDV in the environment, chicks placed in commercial poultry houses usually become infected within the first few weeks of life. In four flocks studied by Witter *et al.* (1970a), chick infection occurred as early as 9 days of age. The virus spread progressively, infecting virtually all birds by 8 weeks. Acquired precipitating antibody appeared at 6–7 weeks and also increased in incidence; the interval between 50% virus infection and 50% antibody response was 14–17 days. The incidence of microscopic lesions closely followed that of virus isolation and was high (78%) over all birds at 8 weeks (Fig. 29).

Biggs *et al.* (1972, 1973) and Jackson *et al.* (1976) found that most if not all flocks were infected by strains of MDV of varying pathogenicity during the first 8 weeks of life and that eventual mortality from MD was significantly influenced by events that occurred while birds were in the rearing pens (up to 8 weeks of age), and not by subsequent events in the production houses. They postulated that the most likely factor in the rearing pens accounting for variable MD mortality was the pathogenicity of the strain of infecting MDV and that where infection by a nonpathogenic strain preceded that by a pathogenic strain, natural vaccination occurred, reducing the incidence of clinical disease.

C. Factors That Affect Epizootiology

The major factors that influence the incidence of clinical MD, as opposed to MDV infection, were discussed earlier, and include genotype and age of the chicken (Sections VI and VII.F.2), strain of virus (Section IV.D.1), and presence of maternal antibody (Section VII.B.1). Sex of the bird has an influence, with a tendency to a higher incidence of lymphomas

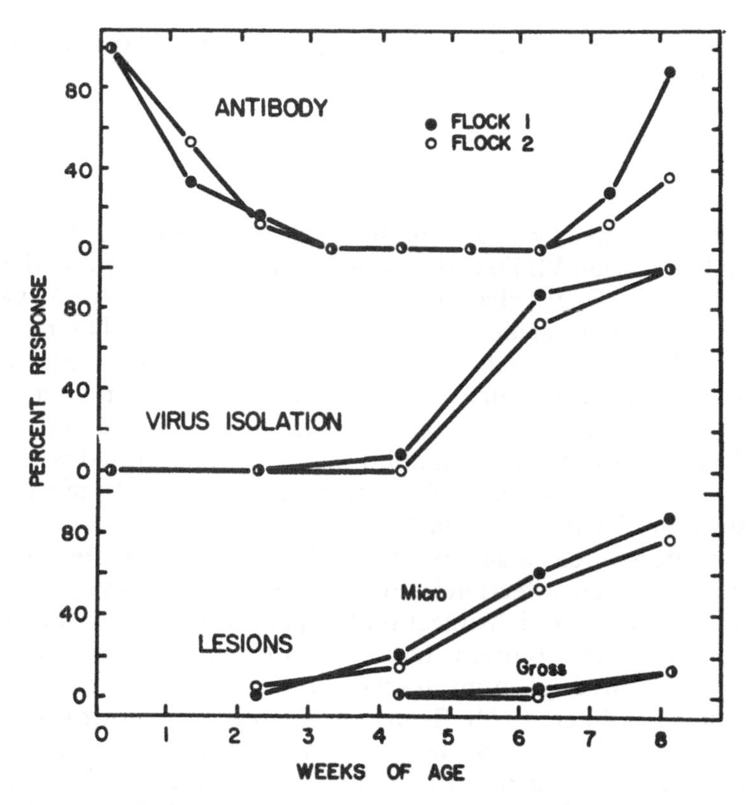

FIGURE 29. MD antibody, virus, and lesions in two flocks naturally exposed to MDV. From Witter *et al.* (1970a) by kind permission.

in females than in males, depending on the strain of virus (see Payne *et al.*, 1976; J. Fabricant *et al.*, 1978). The basis for the sex effect is not known. A number of stresses associated with commercial poultry practice, such as handling and movement of birds, vaccination against other diseases, debeaking, and dietary changes, have been anecdotally associated with increased MD incidence. Gross and Colmano (1971) exposed chickens to social stress and from them selected lines for high and low plasma corticosterone levels. When placed in MDV-contaminated high-stress and low-stress environments, the line selected for low corticosterone had MD incidences of 18 and 7%, respectively, whereas the high-corticosterone line had incidences of 41 and 40%, respectively. Thus, failure to react to stress was associated with resistance to MD. Feeding of dichlorodiphenyl-dichloroethane (DDD), which reduces the production of corticosterone by the adrenal, markedly lowered the incidence of MD (Colmano and Gross, 1971). Compared with the low-corticosterone-response line, the high-response line had a lower lymphocyte-transformation response to phytohemagglutinin and a bovine serum albumin complex, and lower mitogen-induced cellular cytotoxicity associated with increased tumor incidence (Thompson *et al.*, 1980). Metyrapone, another adrenal inhibitor, caused regression of tumors.

IX. CONTROL OF MAREK'S DISEASE

MD is controlled by a combination of three methods: (1) Vaccination (discussed in Section VII.D) is the principal means of control under commercial conditions. (2) Selection of genetically resistant stock, particularly when this can be achieved by means of easily detected markers such as the B^{21} allotype, is a valuable control method in the absence of vaccination, and may be helpful as an adjunct to vaccinal control (see Section VII.D.3). (3) Hygienic control measures, which were particularly important before the advent of vaccination, but are also required with vaccination to prevent exposure to MDV in the first 1–2 weeks of life before vaccinal immunity has developed.

Great efforts were made by the poultry industry before the use of vaccines to eliminate infection from the environment of the young chick by vigorous cleaning and disinfection by spraying or fumigation. Several common disinfection solutions inactivate MDV in dried feathers within 10 min; formaldehyde gas fumigation destroys most but not all infectivity (Calnek and Hitchner, 1973). To be successful, such control measures must eliminate all infection, the chicks must be free of infection when introduced, and subsequent introduction of infection must be prevented. These are demands difficult to meet under commercial conditions, and hygienic control had limited success. Controlled-environment houses with filtered air and positive pressure (FAPP principle) together with disinfection were used with some success by some operators to prevent infection of houses from the outside (Anderson *et al.*, 1972). The efficiency of commercial air filters against airborne MDV was studied by Burmester and Witter (1972) and Calnek and Hitchner (1973).

X. CONCLUDING REMARKS

As a result initially of agricultural needs, and more recently of the adoption of Marek's disease as a biological model system, a sizeable body of information on Marek's disease virus (MDV), and to a lesser extent herpesvirus of turkeys, now exists. Many, although by no means all, literature sources are given in this review. The disease has been brought under control, and the broad features at least of its natural history are understood. Nevertheless, much still needs to be done to provide detailed understanding of many important facets of the disease.

Application of the techniques of molecular biology (see Chapter 7) promises to provide information on virus–cell interactions, needed particularly for understanding the nature of latent or neoplastic nonproductive infections, and the basis for differences between oncogenic and nononcogenic strains of MDV. These methods will overlap with more classic approaches to disease investigation; for example, the use of *in situ* hybridization techniques for detection of viral genome in cells should provide information much needed in understanding some aspects of pathology, such as the nature of neural and atherosclerotic lesions.

Many other notable gaps in knowledge exist: transformation of lymphocytes with MDV *in vitro* cannot yet be readily achieved, viral antigens have not been fully characterized, the nature of Marek's disease tumor-associated surface antigen (MATSA) is unknown, and the immunological mechanisms involved in genetic and vaccinal immunity are known only in broadest outline. The nature of vaccination failure needs investigation, and recent work on the influence of stress on disease resistance has important theoretical and practical implications.

ACKNOWLEDGMENTS. I am grateful to Dr. A.M. Lawn and Dr. P.C. Powell for helpful comments, to Mrs. Helen Tiddy for secretarial assistance, and to Mr. A.W. Kidd and Mr. S. Hodgson for photography.

REFERENCES

Adldinger, H.K., and Calnek, B.W., 1973, Pathogenesis of Marek's disease: Early distribution of virus and viral antigens in infected chickens, *J. Natl. Cancer Inst.* **50:**1287.

Akiyama, Y., and Kato, S., 1974, Two cell lines from lymphomas of Marek's disease, *Biken J.* **17:**105.

Akiyama, Y., Kato, S., and Iwa, N., 1973, Continuous cell culture from lymphoma of Marek's disease, *Biken J.* **16:**177.

Anderson, D.P., King, D.D., Eidson, C.S., and Kleven, S.H., 1972, Filtered air positive pressure (FAPP) brooding of broiler chickens, *Avian Dis.* **16:**20.

Anonymous, 1941, Tentative pathologic nomenclature for the disease and/or disease complex variously designated as fowl leucemia, leucosis, range paralysis, fowl paralysis, iritis, lymphomatosis, lymphocytoma, neurolymphomatosis, leucotic tumors, leucemoid diseases, etc., *Am. J. Vet. Res.* **2:**116.

Asmundson, V.S., and Biely, J., 1932, Inheritance of resistance to fowl paralysis (neurolymphomatosis gallinarum). 1. Differences in susceptibility, *Can. J. Res.* **6:**171.

Ball, R.F., and Lyman, J.F., 1977, Revaccination of chicks for Marek's disease at twenty-one days old, *Avian Dis.* **21:**440.

Ball, R.F., Hill, J.F., Lyman, J., and Wyatt, A., 1971, The resistance to Marek's disease of chicks from immunized breeders, *Poult. Sci.* **50:**1084.

Bankowski, R.A., Moulton, J.E., and Mikami, T., 1969, Characterization of Cal-1 strain of acute Marek's disease agent, *Am. J. Vet. Res.* **30:**1667.

Baxendale, W., 1969, Preliminary observations on Marek's disease in ducks and other avian species, *Vet. Rec.* **85:**341.

Beasley, J.N., Patterson, L.T., and McWade, D.H., 1970, Transmission of Marek's disease by poultry house dust and chicken dander, *Am. J. Vet. Res.* **31:**339.

Benton, W.J., and Cover, M.S., 1957, The increased incidence of visceral lymphomatosis in broiler and replacement birds, *Avian Dis.* **1:**320.

Biggs, P.M., 1961, A discussion on the classification of the avian leucosis complex and fowl paralysis, *Br. Vet. J.* **117:**326.

Biggs, P.M., 1968, Marek's disease—current state of knowledge, *Curr. Top. Microbiol. Immunol.* **43**:91.

Biggs, P.M., 1975, Vaccination against oncogenic herpesviruses—a review, in: *Oncogenesis and Herpesviruses II* (G. de Thé, M.A. Epstein, and H. zur Hausen, eds.), pp. 317–322, IARC Scientific Publication No. 11, International Agency for Research on Cancer, Lyon.

Biggs, P.M. (ed.), 1980, *Resistance and Immunity to Marek's Disease*, CEC Publication EUR 6470, Luxembourg, 617 pp.

Biggs, P.M., and Milne, B.S., 1971, Use of the embryonating egg in studies on Marek's disease, *Am. J. Vet. Res.* **32**:1795.

Biggs, P.M., and Milne, B.S., 1972, Biological properties of a number of Marek's disease virus isolates, in: *Oncogenesis and Herpesviruses* (P.M. Biggs, G. de Thé, and L.N. Payne, eds.), pp. 88–94, IARC Scientific Publication No. 2, International Agency for Research on Cancer, Lyon.

Biggs, P.M., and Payne, L.N., 1963, Transmission experiments with Marek's disease (fowl paralysis), *Vet. Rec.* **75**:177.

Biggs, P.M., and Payne, L.N., 1964, Relationship of Marek's disease (neural lymphomatosis) to lymphoid leukosis, *Natl. Cancer Inst. Monogr.* **17**:83.

Biggs, P.M., and Payne, L.N., 1967, Studies on Marek's disease. I. Experimental transmission, *J. Natl. Cancer Inst.* **39**:267.

Biggs, P.M., Purchase, H.G., Bee, B.R., and Dalton, P.J., 1965, Preliminary report on acute Marek's disease (fowl paralysis) in Great Britain, *Vet. Rec.* **77**:1339.

Biggs, P.M., Long, P.L., Kenzy, S.G., and Rootes, D.G., 1968a, Relationship between Marek's disease and coccidiosis. II. The effect of Marek's disease on the susceptibility of chickens to coccidial infection, *Vet. Rec.* **83**:284.

Biggs, P.M., Thorpe, R.J., and Payne, L.N., 1968b, Studies on genetic resistance to Marek's disease in the domestic chicken, *Br. Poult. Sci.* **9**:37.

Biggs, P.M., Payne, L.N., Milne, B.S., Churchill, A.E., Chubb, R.C., Powell, D.G., and Harris, A.H., 1970, Field trials with an attenuated cell associated vaccine for Marek's disease, *Vet. Rec.* **87**:704.

Biggs, P.M., Powell, D.G., Churchill, A.E., and Chubb, R.C., 1972, The epizootiology of Marek's disease. 1. Incidence of antibody, viraemia and Marek's disease in six flocks, *Avian Pathol.* **1**:5.

Biggs, P.M., Jackson, C.A.W., and Powell, D.G., 1973, The epizootiology of Marek's disease. 2. The effect of supply flock, rearing house and production house on the incidence of Marek's disease, *Avian Pathol.* **2**:127.

Blaxland, J.D., MacLeod, A.J., Baxendale, W., and Hall, T., 1972, Observations on field trials with an experimental attenuated Marek's disease virus vaccine, *Vet. Rec.* **90**:431.

Brewer, R.N., Reid, W.M., Johnson, J., and Schmittle, S.C., 1969, Studies on acute Marek's disease. VIII. The role of mosquitoes in transmission under experimental conditions, *Avian Dis.* **13**:83.

Briles, W.E., and Stone, H.A., 1975, Relative influence of *B* isoalleles from lines N and P on Marek's mortality, *Poult. Sci.* **54**:1738.

Briles, W.E., McGibbon, W.H., and Stone, H.A., 1976, Effects of *B* alloalleles from Regional Cornell Random bred stock on mortality from Marek's disease, *Poult. Sci.* **55**:2011.

Briles, W.E., Stone, H.A., and Cole, R.K., 1977, Marek's disease: Effects of *B* histocompatibility alloalleles in resistant and susceptible chicken lines, *Science* **195**:193.

Briles, W.E., Briles, R.W., McGibbon, W.H., and Stone, H.A., 1980, Identification of *B* alloalleles associated with resistance to Marek's disease, in: *Resistance and Immunity to Marek's Disease* (P.M. Biggs, ed.), pp. 395–416, CEC Publication EUR 6470, Luxembourg.

Bülow, V.v., 1969, Marek'sche Huhnerlahmung: Reaktionen im experimentell infizierten embryonierten Ei, *Zentralbl. Veterinaermed.* **16**:97.

Bülow, V.v., 1971, Diagnosis and certain biological properties of the virus of Marek's disease, *Am. J. Vet. Res.* **32**:1275.

Bülow, V.v., 1977, Further characterization of the CVI 988 strain of Merek's disease virus, *Avian Pathol.* **6**:395.

Bülow, V.v., and Biggs, P.M., 1975a, Differentiation between strains of Marek's disease virus and turkey herpesvirus by immunofluorescence assays, *Avian Pathol.* **4**:133.
Bülow, V.v., and Biggs, P.M., 1975b, Precipitating antigens associated with Marek's disease viruses and a herpesvirus of turkeys, *Avian Pathol.* **4**:147.
Bülow, V.v., and Schmid, D.O., 1979, Antigenic characteristics of Marek's disease tumour cells, *Avian Pathol.* **8**:265.
Burg, R. W., Feldbush, T., Morris, C.A., and Maag, T.A., 1971, Depression of thymus- and bursa-dependent immune systems of chicks with Marek's disease, *Avian Dis.* **15**:662.
Burgoyne, G.H., and Witter, R.L., 1973, Effect of passively transferred immunoglobulins on Marek's disease, *Avian Dis.* **17**:824.
Burmester, B.R., and Witter, R.L., 1972, Efficiency of commercial air filters against Marek's disease virus, *Appl. Microbiol.* **23**:505.
Byerly, J.L., and Dawe, D.L., 1972, Delayed hypersensitivity reactions in Marek's disease virus infected chickens, *Am. J. Vet. Res.* **33**:2267.
Calnek, B.W., 1972a, Antibody development in chickens exposed to Marek's disease virus, in: *Oncogenesis and Herpesviruses* (P.M. Biggs, G. de Thé, and L.N. Payne, eds.), pp. 129–136, IARC Scientific Publication No. 2, International Agency for Research on Cancer, Lyon.
Calnek, B.W., 1972b, Effects of passive antibody on early pathogenesis of Marek's disease, *Infect. Immun.* **6**:193.
Calnek, B.W., 1973, Influence of age at exposure on the pathogenesis of Marek's disease, *J. Natl. Cancer Inst.* **51**:929.
Calnek, B.W., 1980, Marek's disease virus and lymphoma, in: *Oncogenic Herpesviruses* (F. Rapp, ed.), Vol. 1, pp. 103–143, CRC Press, Boca Raton, Florida.
Calnek, B.W., and Adldinger, H.K., 1971, Some characteristics of cell-free preparations of Marek's disease virus, *Avian Dis.* **15**:508.
Calnek, B.W., and Hitchner, S.B., 1973, Survival and disinfection of Marek's disease virus and the effectiveness of filters in preventing airborne dissemination, *Poult. Sci.* **52**:35.
Calnek, B.W., and Madin, S.H., 1969, Characteristics of *in vitro* infection of chicken kidney cell cultures with a herpesvirus from Marek's disease, *Am. J. Vet. Res.* **30**:1389.
Calnek, B.W., and Payne, L.N., 1976, Lack of correlation between Marek's disease tumor induction and expression of endogenous avian RNA tumor virus genome, *Int. J. Cancer* **17**:235.
Calnek, B.W., and Smith, M.W., 1972, Vaccination against Marek's disease with cell-free turkey herpesvirus: Interference by maternal antibody, *Avian Dis.* **16**:954.
Calnek, B.W., and Witter, R.L., 1978, Marek's disease, in: *Diseases of Poultry* (M.S. Hofstad, B.W. Calnek, C.F. Helmboldt, W.M.Reid, and H.W. Yoder, eds.), pp. 385–418, Iowa State University Press, Ames.
Calnek, B.W., Madin, S.H., and Kinazeff, A.J., 1969, Susceptibility of cultured mammalian cells to infection with a herpesvirus from Marek's disease and T-virus from reticuloendotheliosis of chickens, *Am. J. Vet. Res.* **30**:1403.
Calnek, B.W., Adldinger, H.K., and Kahn, D.E., 1970a, Feather follicle epithelium: A source of enveloped and infectious cell-free herpesvirus from Marek's disease, *Avian Dis.* **14**:219.
Calnek, B.W., Hitchner, S.B., and Adldinger, H.K., 1970b, Lyophilization of cell-free Marek's disease herpesvirus and a herpesvirus from turkeys, *Appl. Microbiol.* **20**:723.
Calnek, B.W., Ubertini, T., and Adldinger, H.K., 1970c, Viral antigen, virus particles, and infectivity of tissues from chickens with Marek's disease, *J. Natl. Cancer Inst.* **45**:351.
Calnek, B.W., Higgins, D.A., and Fabricant, J., 1975, Rous sarcoma regression in chickens resistant or susceptible to Marek's disease, *Avian Dis.* **19**:473.
Calnek, B.W., Fabricant, J., Schat, K.A., and Murthy, K.K., 1977, Pathogenicity of low-virulence Marek's disease viruses in normal versus immunologically compromised chickens, *Avian Dis.* **21**:346.
Calnek, B.W., Fabricant, J., Schat, K.A., and Murthy, K.K., 1978a, Rejection of a transplantable Marek's disease lymphoma in normal versus immunologically deficient chickens, *J. Natl. Cancer Inst.* **60**:623.

Calnek, B.W., Murthy, K.K., and Schat, K.A., 1978b, Establishment of Marek's disease lymphoblastoid cell lines from transplantable versus primary lymphomas, *Int. J. Cancer* **21**:100.

Calnek, B.W., Carlisle, J.C., Fabricant, J., Murthy, K.K., and Schat, K.A., 1979, Comparative pathogenesis studies with oncogenic and nononcogenic Marek's disease viruses and turkey herpesvirus, *Am. J. Vet. Res.* **40**:541.

Calnek, B.W., Shek, W.R., and Schat, K.A., 1981a, Latent infections with Marek's disease virus and turkey herpesvirus, *J. Natl. Cancer Inst.* **66**:585.

Calnek, B.W., Shek, W.R. and Schat, K.A., 1981b, Spontaneous and induced herpesvirus genome expression in Marek's disease tumor cell lines, *Infect. Immun.* **34**:483.

Campbell, J.G., 1954, Avian leucosis: A plea for clarification, Proceedings of the 10th World's Poultry Congress, (J.E. Wilson, ed.), pp. 193–197, Dept. Agriculture for Scotland, Edinburgh.

Campbell, J.G., 1956, Leucosis and fowl paralysis compared and contrasted, *Vet. Rec.* **68**:527.

Campbell, J.G., 1961, A proposed classsification of the leucosis complex and fowl paralysis, *Br. Vet. J.* **117**:316.

Campbell, W.F., and Frankel, J.W., 1979, Enhanced oncornavirus expression in Marek's disease tumors from specific-pathogen-free chickens, *J. Natl. Cancer Inst.* **62**:323.

Carrozza, J.H., Fredrickson, T.N., Prince, R.P., and Luginbuhl, R.E., 1973, Role of desquamated epithelial cells in transmission of Marek's disease, *Avian Dis.* **17**:767.

Chen, J.H., and Purchase, H.G., 1970, Surface antigen on chicken kidney cells infected with the herpesvirus of Marek's disease, *Virology* **40**:410.

Cho, B.R., 1975, Horizontal transmission of turkey herpesvirus to chickens. IV. Viral maturation in the feather follicle epithelium, *Avian Dis.* **19**:136.

Cho, B.R., 1976a, A possible association between plaque type and pathogenicity of Marek's disease herpesvirus, *Avian Dis.* **20**:324.

Cho, B.R., 1976b, Horizontal transmission of turkey herpesvirus to chickens. 5. Airborne transmission between chickens, *Poult. Sci.* **55**:1830.

Cho, B.R., 1981, A simple *in vitro* differentiation between turkey herpesvirus and Marek's disease virus, *Avian Dis.* **25**:839.

Cho, B.R., and Kenzy, S.G., 1972, Isolation and characterization of an isolate (HN) of Marek's disease virus with low pathogenicity, *Appl. Microbiol.* **24**:299.

Cho, B.R., and Kenzy, S.G., 1975, Virologic and serologic studies of zoo birds for Marek's disease virus infection, *Infect. Immun.* **11**:809.

Cho, B.R., Balch, R.K., and Hill, R.W., 1976, Marek's disease vaccine breaks: Differences in viremia of vaccinated chickens between those with and without Marek's disease, *Avian Dis.* **20**:496.

Chomiak, T.W., Luginbuhl, R.E., Helmboldt, C.F., and Kottaridis, S.D., 1967, Marek's disease. 1. Propagation of the Connecticut A (Conn-A) isolate in chicks, *Avian Dis.* **11**:646.

Chubb, L.G., and Gordon, R.F., 1957, The avian leucosis complex—a review, *Vet. Rev. Annot.* **3**:97.

Chubb, R.C., and Churchill, A.E., 1968, Precipitating antibodies associated with Marek's disease, *Vet. Rec.* **83**:4.

Chubb, R.C., and Churchill, A.E., 1969, Effect of maternal antibody on Marek's disease, *Vet. Rec.* **85**:303.

Churchill, A.E., 1968, Herpes-type virus isolated in cell culture from tumors of chickens with Marek's disease. 1. Studies in cell culture, *J. Natl. Cancer Inst.* **41**:939.

Churchill, A.E., and Biggs, P.M., 1967, Agent of Marek's disease in tissue culture, *Nature (London)* **215**:528.

Churchill, A.E., Chubb, R.C., and Baxendale, W., 1969a, The attenuation, with loss of oncogenicity, of the herpestype virus of Marek's disease (strain HPRS-16) on passage in cell culture, *J. Gen. Virol.* **4**:557.

Churchill, A.E., Payne, L.N., and Chubb, R.C., 1969b, Immunization against Marek's disease using a live attenuated virus, *Nature (London)* **221**:744.

Churchill, A.E., Baxendale, W., and Carrington, G., 1973, Viremia and antibody development in chicks following the administration of turkey herpesvirus, *Vet. Rec.* **92**:327.

Cole, R.K., 1968, Studies on genetic resistance to Marek's disease, *Avian Dis.* **12**:9.

Cole, R.K., 1972, The genetics of resistance to Marek's disease, in: *Oncogenesis and Herpesviruses* (P.M. Biggs, G. de Thé, and L.N. Payne, eds.), pp. 123–128, IARC Scientific Publication No. 2, International Agency for Research on Cancer, Lyon.

Cole, R.K., and Hutt, F.B., 1973, Selection and heterosis in Cornell White Leghorns: A review, with special consideration of interstrain hybrids, *Anim. Breeding Abstr.* **41**:103.

Coleman, R.M., and Schierman, L.W., 1980, Independence of chicken major histocompatibility antigens and tumor-associated antigen on the surface of herpes-induced lymphoma cells, *Infect. Immun.* **29**:1067.

Colmano, G., and Gross, W.B., 1971, Effect of metyrapone and DDD on infectious diseases, *Poult. Sci.* **51**:850.

Colwell, W.M., and Schmittle, S.C., 1968, Studies on acute Marek's disease. VII. Airborne transmission of the GA isolate, *Avian Dis.* **12**:724.

Colwell, W.M., Simmons, D.G., and Muse, K.E., 1974, Use of Japanese quail embryo fibroblast cells for propagation and assay of turkey herpesvirus FC-126, *Appl. Microbiol.* **27**:218.

Colwell, W.M., Simmons, D.G., Harris, J.R., Fulp, T.G., Carrozza, J.H., and Maag, T.A., 1975, Influence of some physical factors on Marek's disease vaccine virus, *Avian Dis.* **19**:781.

Confer, A.W., and Adldinger, H.K., 1980, Cell-mediated immunity in Marek's disease: Cytotoxic responses in resistant and susceptible chickens and relation to disease, *Am. J. Vet. Res.* **41**:307.

Confer, A.W., and Adldinger, H.K., 1981, The *in vivo* effect of levamisole on phytohaemagglutinin stimulation of lymphocytes in normal and Marek's disease virus inoculated chickens, *Res. Vet. Sci.* **30**:243.

Confer, A.W., Adldinger, H.K., and Buening, G.M., 1980, Cell-mediated immunity in Marek's disease: Correlation of disease-related variables with immune responses in age-resistant chickens, *Am. J. Vet. Res.* **41**:313.

Cottral, G.E., 1952, The enigma of avian leukosis, Proceedings of the 89th Annual Meeting, *Am. Vet. Med. Assoc.* 285.

Crittenden, L.B., Muhm, R.L., and Burmester, B.R., 1972, Genetic control of susceptibility to the avian leukosis complex. 2. Marek's disease, *Poult. Sci.* **51**:261.

Dambrine, G., Coudert, F., and Cauchy, L., 1980, Cell-mediated cytotoxicity in chickens infected with Marek's disease virus and the herpesvirus of turkeys, in: *Resistance and Immunity to Marek's Disease* (P.M. Biggs, ed.), pp. 320–337, CEC Publication EUR 6470, Luxembourg.

Doak, R.L., Munnell, J.F., and Ragland, W.L., 1973, Ultrastructure of tumor cells in Marek's disease virus-infected chickens, *Am. J. Vet. Res.* **34**:1063.

Doi, K., Kojima, A., Akiyama, Y., and Kato, S., 1976, Pathogenicity for chicks of line cells from lymphoma of Marek's disease, *Natl. Inst. Anim. Health Q.* **16**:16.

Donahoe, J., Kleven, S., and Eidson, C., 1978, Resistance to Marek's disease: Effect of *Corynebacterium parvum* and Marek's tumor cell vaccines on tumorigenesis in chickens, *J. Natl. Cancer Inst.* **60**:829.

Dunn, K.M., and Nazerian, K., 1977, Induction of Marek's disease virus antigens by IdUrd in a chicken lymphoblastoid cell line, *J. Gen. Virol.* **34**:413.

Eidson, C.S., and Anderson, D.P., 1971, Immunization against Marek's disease, *Avian Dis.* **15**:68.

Eidson, C.S., and Kleven, S.H., 1977, Application of the cell-free turkey herpesvirus vaccines by the aerosol route for the prevention of Marek's disease in chickens, *Poult. Sci.* **56**:1609.

Eidson, C.S., and Schmittle, S.C., 1968a, Studies on acute Marek's disease. 1. Characteristics of isolate GA in chickens, *Avian Dis.* **12**:467.

Eidson, C.S., and Schmittle, S.C., 1968b, Studies on acute Marek's disease. V. Attempted transmission of isolate GA with feces and nasal washings, *Avian Dis.* **12**:549.

Eidson, C.S., and Schmittle, S.C., 1969, Studies on acute Marek's disease. XII. Detection of antibodies with a tannic acid indirect hemagglutination test, *Avian Dis.* **13**:774.

Eidson, C.S., Schmittle, S.C., Goode, R.B., and Lal, J.B., 1966, Induction of leukosis tumors with the beetle *Alphitobius diaperinus*, *Am. J. Vet. Res.* **27**:1053.

Eidson, C.S., Anderson, D.P., Kleven, S.H., and Brown, J., 1971, Field trials of vaccines for Marek's disease, *Avian Dis.* **15**:312.

Eidson, C.S., Villegas, P., Page, R.K., and Kleven, S.H., 1975, A comparison of the efficacy against Marek's disease of cell-free and cell-associated turkey herpesvirus vaccine, *Avian Dis.* **19**:515.

Eidson, C.S., Page, R.K., Giambrone, J.J., and Kleven, S.H., 1976, Long term studies comparing the efficacy of cell-free versus cell-associated HVT vaccines against Marek's disease, *Poult. Sci.* **55**:1857.

Eidson, C.S., Giambrone, J.J., Fletcher, O.J., and Kleven, S.H., 1980, Vaccination against Marek's disease with cell-free herpesvirus of turkeys by aerosol vaccination, *Poult. Sci.* **59**:54.

Eidson, C.S., Ellis, M.N. and Kleven, S.H., 1981, Reduced vaccinal protection of turkey herpesvirus against field strains of Marek's disease herpesvirus, *Poult. Sci.* **60**:317.

Elmubarak, A.K., Sharma, J.M., Witter, R.L., Nazerian, K. and Sanger, V.L., 1981, Inductions of lymphomas and tumor antigen by Marek's disease virus in turkeys, *Avian Dis.* **25**:911.

Else, R.W., 1974, Vaccinal immunity to Marek's disease in bursectomised chickens, *Vet. Rec.* **95**:182.

Epstein, M.A., and Achong, B.G. (eds.), 1979, *The Epstein–Barr Virus*, Springer-Verlag, Berlin, Heidelberg, New York, 459 pp.

Epstein, M.A., Achong, B.G., Churchill, A.E., and Biggs, P.M., 1968, Structure and development of the herpes-type virus of Marek's disease, *J. Natl. Cancer Inst.* **41**:805.

Evans, D.L., and Patterson, L.T., 1971, Correlation of immunological responsiveness with lymphocyte changes in chickens infected with Marek's disease, *Infect. Immun.* **4**:567.

Evans, D.L., Barnett, J.W., and Dmochowski, L., 1973, Common antigens in herpesviruses from divergent species of animals, *Tex. Rep. Biol. Med.* **31**:756.

Fabricant, C.G., Fabricant, J., Litrenta, M.M., and Minick, C.R., 1978, Virus-induced atherosclerosis, *J. Exp. Med.* **148**:335.

Fabricant, J., Ianconescu, M., and Calnek, B.W., 1977, Comparative effects of host and viral factors on early pathogenesis of Marek's disease, *Infect. Immun.* **16**:136.

Fabricant, J., Calnek, B.W., Schat, K.A., and Murthy, K.K., 1978, Marek's disease virus-induced tumor transplants in development and rejection in various genetic strains of chickens, *Avian Dis.* **22**:646.

Fauser, I.S., Purchase, H.G., Long, P.A., Velicer, L.F., Mallman, H., Fauser, H.T., and Winegar, G.O., 1973, Delayed hypersensitivity and leucocyte migration inhibition in chickens with BCG or Marek's disease infection, *Avian Pathol.* **2**:55.

Fernando, W.W.D., and Calnek, B.W., 1971, The influence of the bursa of Fabricius on the infection and pathological response of chickens exposed to Marek's disease herpesvirus, *Avian Dis.* **15**:467.

Frankel, J.W., Farrow, W.M., Prickett, C.O., Smith, M.E., Campbell, W.F., and Groupé, V., 1974, Responses of isolator-derived and conventional chickens to Marek's disease herpesvirus and avian leukosis virus, *J. Natl. Cancer Inst.* **52**:1491.

Frazier, J.A., 1974, Ultrastructure of lymphoid tissue from chicks infected with Marek's disease virus, *J. Natl. Cancer Inst.* **52**:829.

Frazier, J.A., and Powell, P.C., 1975, The ultrastructure of lymphoblastoid cell lines from Marek's disease lymphomata, *Br. J. Cancer* **31**:7.

Fredericksen, T.L., Longenecker, B.M., Pazderka, F., Gilmour, D.G., and Ruth, R.F., 1977, A T-cell antigen system of chickens: Ly-4 and Marek's disease, *Immunogenetics* **5**:535.

Friars, G.W., Chambers, J.R., Kennedy, A., and Smith, A.D., 1972, Selection for resistance to Marek's disease in conjunction with other economic traits in chickens, *Avian Dis.* **16**:2.

Fujimoto, Y., and Okada, K., 1977, Pathological studies of Marek's disease. III. Electron microscopic observation of demyelination of the peripheral nerves, *Jpn. J. Vet. Res.* **25**:59.

Fujimoto, Y., Nakagawa, M., Okada, K., Okada, M., and Matsukawa, K., 1971, Pathological studies of Marek's disease. I. The histopathology on field cases in Japan, *Jpn. J. Vet. Res.* **19**:7.

Fujimoto, Y., Okada, K., Kakihata, K., Matsui, T., Narita, M., Onuma, M., and Mikami, T., 1974, Initial lesions in chickens infected with JM strain of Marek's disease virus, *Jpn. J. Vet. Res.* **22**:80.

Gallatin, W.M., and Longenecker, B.M., 1979, Expression of genetic resistance to an oncogenic herpesvirus at the target cell level, *Nature* (*London*) **280**:587.

Gallatin, W.M. and Longenecker, B.M., 1981, Genetic resistance to herpesvirus-induced lymphoma at the level of the target cell determined by the thymic microenvironment, *Int. J. Cancer*, **27**:373.

Gavora, J.S., and Spencer, J.L., 1979, Studies on genetic resistance of chickens to Marek's disease—a review, *Comp. Immun. Microbiol. Infect. Dis.* **2**:359.

Gavora, J.S., Grunder, A.A., Spencer, J.L., Gowe, R.S., Robertson, A., and Speckmann, G.W., 1974, An assessment of effects of vaccination on genetic resistance to Marek's disease, *Poult. Sci.* **53**:889.

Gavora, J.S., Gowe, R.S., and McAllister, A.J., 1977, Vaccination against Marek's disease: Efficacy of cell-associated and lyophilised herpesvirus of turkeys in nine strains of leghorns, *Poult. Sci.* **56**:846.

Giambrone, J.J., Eidson, C.S., Page, K.K., Fletcher, O.J., Barger, B.O., and Kleven, S.H., 1976, Effect of infectious bursal agent on the response of chickens to Newcastle disease and Marek's disease vaccination, *Avian Dis.* **20**:534.

Goodchild, W.M., 1969, Some observations of Marek's disease (fowl paralysis), *Vet. Rec.* **84**:87.

Goto, N., Fujimoto, Y., Okada, K., Mikami, T., Kodama, H., and Ichijo, K., 1979, Effect of thymectomy on the initial cytolytic lesions and nerve demyelination of Marek's disease, *Zentralbl. Veterinaermed. Reike B* **26**:61.

Gross, W.B., and Colmano, G., 1971, Effect of infectious agents on chickens selected for plasma corticosterone response to social stress, *Poult. Sci.* **50**:1213.

Grunder, A.A., Dickerson, G.E., Robertson, A., and Morin, E., 1969, Incidence of Marek's disease as related to phenotypes of serum alkaline phosphatase, *Poult. Sci.* **48**:1608.

Grunder, A.A., Jeffers, T.K., Spencer, J.L., Robertson, A., and Speckman, G.W., 1972, Resistance of strains of chickens to Marek's disease, *Can. J. Anim. Sci.* **52**:1.

Haffer, K., Sevoian, M., and Wilder, M., 1979, The role of the macrophage in Marek's disease: *In vitro* and *in vivo* studies, *Int. J. Cancer* **23**:648.

Hahn, E.C., Ramos, L., and Kenyon, A.J., 1977, Lymphoproliferative diseases of fowl: JM-V leukemic lymphoblasts in cell culture (brief communication), *J. Natl. Cancer Inst.* **59**:267.

Hahn, E.C., Jakowski, R.M., Ramos, L., and Kenyon, A.J., 1978, Lymphoproliferative disease of fowl: Characterization of transplantable G-B1 Marek's disease tumor cells in culture, *Avian Dis.* **22**:409.

Haider, S.A., Lapen, R.F., and Kenzy, S.G., 1970, Use of feathers in a gel precipitation test for Marek's disease, *Poult. Sci.* **69**:1654.

Halvorson, D.A., and Mitchell, D.O., 1979, Loss of cell-associated Marek's disease vaccine titer during thawing, reconstitution, and use, *Avian Dis.* **23**:848.

Hamdy, F., Sevoian, M., and Holt, S.C., 1974, Biogenesis of Marek's disease (type II leukosis) virus *in vitro*: Electron microscopy and immunological study, *Infect. Immun.* **9**:740.

Han, P.F.-S, and Smyth, J.R., 1972, The influence of growth rate on the development of Marek's disease in chickens, *Poult. Sci.* **51**:975.

Hansen, M.P., Van Zandt, J.N., and Law, G.R.J., 1967, Differences in susceptibility to Marek's disease in chickens carrying two different *B* locus blood group alleles, *Poult. Sci.* **46**:1268.

Hartmann, W., and Sanz, M., 1970, Resistance to Marek's disease in broiler strains under natural and artificial exposure, *Proc. XIVth World's Poult. Congr.* **3**:309.

Hein, M.B., and Mora, E.C., 1980, Distribution of herpesvirus in turkeys, *Poult. Sci.* **59**:494.

Higgins, D.A., and Calnek, B.W., 1975, Fowl immunoglobulins: Quantitation and antibody activity during Marek's disease in genetically resistant and susceptible birds, *Infect. Immun.* **11**:33.

Higgins, D.A., and Calnek, B.W., 1976, Some effects of silica treatment on Marek's disease, *Infect. Immun.* **13**:1054.

Hinshaw, V.L., and Mora, E.C., 1980, Localization of herpesvirus of turkeys in blood cells, *Poult. Sci.* **59**:258.

Hložánek, I., and Sovova, V., 1974, Lack of pathogenicity of Marek's disease herpesvirus and herpesvirus of turkeys for mammalian hosts and mammalian cell cultures, *Folia Biol. (Prague)* **20**:51.

Hložánek, I., Mach, O., and Jurajda, V., 1973, Cell-free preparations of Marek's disease virus from poultry dust, *Folia Biol. (Prague)* **19**:118.

Hoffmann-Fezer, G., and Hoffman, R., 1980, Anatomical distribution of T- and B-lymphocytes in Marek's disease—an immunohistochemical study, *Vet. Immunol. Immunopathol.* **1**:113.

Hong, C.C., and Sevoian, M., 1971, Interferon production and host resistance to type II avian (Marek's) leukosis virus (JM strain), *Appl. Microbiol.* **22**:818.

Hong, C.C., and Sevoian, M., 1972, Comparison of indirect hemagglutination and immunodiffusion tests for detecting type II leukosis (Marek's) infection in S- and K-line chickens, *Appl. Microbiol.* **23**:449.

Hudson, L., and Payne, L.N., 1973, An analysis of the T and B cells of Marek's disease lymphomas of the chicken, *Nature (London) New Biol.* **241**:52.

Ikuta, K., Akiyama, Y., Konobe, T., and Kato, S., 1976a, A lymphoblastoid cell line dually infected with Marek's disease virus and avian leukosis virus, *Biken J.* **19**:33.

Ikuta, K., Kato, S., Ando, T., and Konobe, T., 1976b, Lack of expressions of endogenous (gs, chick helper factor) and exogenous avian RNA tumor viruses in the MOB-1 and MSB-1 lines derived from Marek's disease lymphomas, *Biken J.* **19**:39.

Ikuta, K., Nishi, Y., Kato, S. and Hirai, K., 1981, Immunoprecipitation of Marek's disease virus-specific polypeptides with chicken antibodies purified by affinity chromatography, *Virology*, **114**:277.

Inage, F., Kodama, H., and Mikami, T., 1979, Differences between early and late membrane antigens on cultured cells infected with herpesvirus of turkeys, *Avian Pathol.* **8**:23.

Inoue, M., Mikami, T., Kodama, H., Onuma, M., and Izawa, H., 1980, Antigenic differences between intracellular and membrane antigens induced by herpesvirus of turkeys, *Arch. Virol.* **63**:23.

Ishikawa, T., Naito, M., Osafune, S., and Kato, S., 1972, Cell surface antigen on quail cells infected with herpesvirus of turkey or Marek's disease virus, *Biken J.* **15**:215.

Jackson, C.A.W., Biggs, P.M., Bell, R.A., Lancaster, F.M., and Milne, B.S., 1976, The epizootiology of Marek's disease. 3. The interrelationship of virus pathogenicity, antibody and the incidence of Marek's disease, *Avian Pathol.* **5**:105.

Jakowski, R.M., Fredrickson, T.N., Chomiak, T.W., and Luginbuhl, R.E., 1970, Hematopoietic destruction in Marek's disease, *Avian Dis.* **14**:374.

Jakowski, R.M., Fredrickson, T.N., and Schierman, L.W., 1974, A transplantable lymphoma induced with Marek's disease virus, *J. Natl. Cancer Inst.* **53**:783.

Johnson, E.A., Burke, C.N., Fredrickson, T.N., and DiCapua, R.A., 1975, Morphogenesis of Marek's disease virus in feather follicle epithelium, *J. Natl. Cancer Inst.* **55**:89.

Jurajda, V., and Klimes, B., 1970, Presence and survival of Marek's disease agent in dust, *Avian Dis.* **14**:188.

Kaaden, O.R., Dietzschold, B., and Ueberschar, S., 1974, Vaccination against Marek's disease: Immunizing effect of purified turkey herpesvirus and cellular membranes from infected cells, *Med. Microbiol. Immunol. (Berlin)* **159**:261.

Kaleta, E.F., 1977, Vermehrung, Interferenz und Interferoninduktion aviarer Herpesvirusarten: Beitrag zur Schutzimpfung gegen die Marieksche Krankheit, *Zentralbl. Veterinärmed. Reihe B* **24**:406.

Kaleta, E.F., and Bankowski, R.A., 1972, Production of interferon by the Cal-1 and turkey herpesvirus strains associated with Marek's disease, *Am. J. Vet. Res.* **33**:567.

Kaschka-Dierich, C., Thomssen, R., and Nazerian, K., 1979, Studies on the temperature-dependent DNA replication of herpesvirus of the turkey in chicken embryo fibroblasts, *J. Gen. Virol.* **45**:253.

Kato, S., and Akiyama, Y., 1975, Lymphoid cell lines from lymphomas of Marek's disease, in: *Oncogenesis and Herpesviruses II* (G. de Thé, M.A. Epstein, and H. zur Hausen, eds.), pp. 101–108, IARC Scientific Publication No. 11, International Agency for Research on Cancer, Lyon.

Kato, S., Ono, K., Naito, M., Doi, T., Iwa, N., Mori, Y., and Onoda, T., 1970, Isolation of herpes-type virus from chickens with Marek's disease using duck embryo fibroblast cultures, *Biken J.* **13**:193.

Kawamura, H., King, D.J., and Anderson, D.P., 1969, A herpesvirus isolated from kidney cell culture of normal turkeys, *Avian Dis.* **13**:853.

Kawamura, H., Yamaguchi, S., Matsuda, H., and Kato, S., 1977, Studies on two new lymphoma cell lines established from Marek's disease (H. Kawamura, ed.), English Summaries of the Studies on Poultry Diseases in Japan, pp. 8–9, Japanese Veterinary Poultry Association.

Kenzy, S.G., and Biggs, P.M., 1967, Excretion of the Marek's disease agent by infected chickens, *Vet. Rec.* **80**:565.

Kenzy, S.G., and Cho, B.R., 1969, Transmission of classical Marek's disease by affected and carrier birds, *Avian Dis.* **13**:211.

Kenzy, S.G., Cho, B.R., and Kim, Y., 1973, Oncogenic Marek's disease herpesvirus in avian encephalitis (temporary paralysis), *J. Natl. Cancer Inst.* **51**:977.

Khare, M.L., Grun, J., and Adams, E.V., 1975, Marek's disease in Japanese quail—a pathological, virological and serological study, *Poult. Sci.* **54**:2066.

King, D., Page, D., Schat, K.A., and Calnek, B.W., 1981, Differences in influence between homologous and heterologous maternal antibodies on response to serotype-2 and serotype-3 Marek's disease vaccines, *Avian Dis.* **25**:74.

Kitamoto, N., Ikuta, K., Kato, S., and Yamaguchi, S., 1979, Cell-mediated cytotoxicity of lymphocytes from chickens inoculated with herpesvirus of turkey against a Marek's disease lymphoma cell line (MSB-1), *Biken J.* **22**:11.

Kodama, H., Mikami, T., Inoue, M., and Izawa, H., 1979a, Inhibitory effects of macrophages against Marek's disease virus plaque formation in chicken kidney cell cultures, *J. Natl. Cancer Inst.* **63**:1267.

Kodama, H., Sugimoto, C., Inage, F., and Mikami, T., 1979b, Anti-viral immunity against Marek's disease virus infected chicken kidney cells, *Avian Pathol.* **8**:33.

Kodama, H., Mikami, T., and Izawa, H., 1980, Effects of levamisole on pathogenesis of Marek's disease, *J. Natl. Cancer Inst.* **65**:155.

Lam, K.M. and Cho, Y., 1981, *In vitro* transformation of chicken lymphocytes with Marek's disease virus. Proc. Xth International Symposium for Comparative Research on Leukemia and Related Diseases, Los Angeles, 92 (abstract).

Lam, K.M., and Linna, T.J., 1979, Transfer of natural resistance to Marek's disease (JMV) with non-immune spleen cells. I. Studies of cell population transferring resistance, *Int. J. Cancer* **24**:662.

Lampert, P., Garrett, R., and Powell, H., 1977, Demyelination in allergic and Marek's disease virus induced neuritis: Comparative electron microscopic studies, *Acta Neuropathol. (Berlin)* **40:**103.

Lapen, R.F., Piper, R.C., and Kenzy, S.G., 1970, Cutaneous changes associated with Marek's disease of chickens, *J. Natl. Cancer Inst.* **45:**941.

Lapen, R.F., Kenzy, S.G., Piper, R.C., and Sharma, J.M., 1971, Pathogenesis of cutaneous Marek's disease in chickens, *J. Natl. Cancer Inst.* **47:**389.

Lawn, A.M., and Payne, L.N., 1979, Chronological study of ultrastructural changes in the peripheral nerves in Marek's disease, *Neuropathol. Appl. Neurobiol.* **5:**485.

Lee, L.F., 1979, Macrophage restriction of Marek's disease virus replication and lymphoma cell proliferation, *J. Immunol.* **123:**1088.

Lee, L.F., Sharma, J.M., Nazerian, K., and Witter, R.L., 1978a, Suppression and enhancement of mitogen response in chickens infected with Marek's disease virus and the herpesvirus of turkeys, *Infect. Immun.* **21:**474.

Lee, L.F., Sharma, J.M., Nazerian, K., and Witter, R.L., 1978b, Suppression of mitogen-induced proliferation of normal spleen cells by macrophages from chickens inoculated with Marek's disease virus, *J. Immunol.* **120:**1554.

Lee, L.F., Powell, P.C., Rennie, M., Ross, L.J.N., and Payne, L.N., 1981, Studies on nature of genetic resistance to Marek's disease in chickens, *J. Natl. Cancer Inst.* **66:**789.

Lerche, F., and Fritzsche, K., 1934, Histopathologie und Diagnostik der Geflügellahme, *Z. Infektsionskr. Hanstiere,* **45:**89.

Lesnik, F., and Ross, L.J.N., 1975, Immunization against Marek's disease using Marek's disease virus-specific antigens free from infectious virus, *Int. J. Cancer* **16:**153.

Long, P.A., Clark, J.L., and Velicer, L.E., 1975, Marek's disease herpesviruses. II. Purification and further characterization of Marek's disease herpesvirus A antigen, *J. Virol.* **15:**1192.

Long, P.L., Kenzy, S.G., and Biggs, P.M., 1968, Relationship between Marek's disease and coccidiosis. I. Attempted transmission of Marek's disease by avian coccidia, *Vet. Rec.* **83:**260.

Longenecker, B.M., Pazderka, F., Stone, H.S., Gavora, J.S., and Ruth, R.F., 1975, *In ovo* assay for Marek's disease virus and turkey herpesvirus, *Infect. Immun.* **11:**922.

Longenecker, B.M., Pazderka, F., Gavora, J.S., Spencer, J.L., and Ruth, R.F., 1976, Lymphoma induced by herpesvirus: Resistance associated with a major histocompatibility gene, *Immunogenetics* **3:**401.

Longenecker, B.M., Pazderka, F., Gavora, J.S., Spencer, J.L., Stephens, E.A., and Witter, R.L., 1977, Role of the major histocompatibility complex in resistance to Marek's disease: Restriction of the growth of JMV-MD tumor cells in genetically resistant birds, *Adv. Exp. Med. Biol.* **88:**287.

Lu, Y.S., and Lapen, R.F., 1974, Splenic cell mitogenic response in Marek's disease: Comparison between noninfected tumor-bearing and nontumor-bearing infected chickens, *Am. J. Vet. Res.* **35:**977.

Marek, J., 1907, Multiple Nervenentzündung bei Hühnen, *Dtsch. Tieraerztl. Wochenschr.* **15:**417.

Marquardt, W.W., and Newman, J.A., 1972, A modified direct complement-fixation test for detection of Marek's disease antibodies in chick serum, *Avian Dis.* **16:**986.

Matsuda, H., Ikuta, K., Miyamoto, H., and Kato, S., 1976a, Demonstration of Marek's disease tumor-associated surface antigen (MATSA) on six cell lines derived from Marek's disease lymphomas, *Biken J.* **19:**119.

Matsuda, H., Ikuta, K., and Kato, S., 1976b, Detection of T-cell surface determinants in three Marek's disease lymphoblastoid cell lines, *Biken J.* **19:**29.

Matthews, R.E.F., 1979, Classification and nomenclature of viruses, *Intervirology* **12:**133.

Mikami, T., Onuma, M., and Hayashi, T.T.A., 1973, Membrane antigens in arginine-deprived cultures infected with Marek's disease herpesvirus, *Nature (London) New Biol.* **246:**211.

Mikami, T., Onuma, M., Hayashi, T.T.A., Narita, M., Okada, K., and Fujimoto, Y., 1975,

Pathogenic and serologic studies of Japanese quail infected with JM strain of Marek's disease herpesvirus, *J. Natl. Cancer Inst.* **54**:607.

Mikami, T., Inoue, M., Kodama, H., Inage, F., Onuma, M., and Izawa, H., 1980a, Relation between the neutralization of herpesvirus of turkeys and the antibody to late-appearing membrane antigen induced by the virus, *J. Gen. Virol.* **47**:221.

Mikami, T., Suzuki, K., Kodama, H., Onuma, M., Izawa, H. and Okada, I., 1980b, Antigenic difference between Marek's disease tumor-associated surface antigens of MSB-1 and RPL-1 cell lines derived from Marek's disease lymphoma, in: *Viruses in Naturally Occurring Cancers*, Cold Spring Harbor Conferences on Cell Proliferation, **7**:199.

Minick, C.R., Fabricant, C.G., Fabricant, J., and Litrenta, M.M., 1979, Atheroarteriosclerosis induced by infection with a herpesvirus, *Am. J. Pathol.* **96**:673.

Morris, J.R., Ferguson, A.E., and Jerome, F.N., 1970a, Genetic resistance and susceptibility to Marek's disease, *Can. J. Anim. Sci.* **50**:69.

Morris, J.R., Ferguson, A.E., and Jerome, F.N., 1970b, Distribution of Marek's disease lesions in tissues and organs of Columbian Plymouth Rocks selected for resistance and susceptibility to induced Marek's disease infection, *Can. J. Anim. Sci.* **50**:83.

Munch, D., Hohlstein, L., and Sevoian, M., 1978, *In vitro* establishment of Marek's disease herpesvirus transformed productive and non-productive lymphoblastoid cell lines, *Infect. Immun.* **20**:315.

Murthy, K.K., and Calnek, B.W., 1978, Pathogenesis of Marek's disease: Early appearance of Marek's disease tumor-associated surface antigen in infected chickens, *J. Natl. Cancer Inst.* **61**:849.

Murthy, K.K., and Calnek, B.W., 1979a, Pathogenesis of Marek's disease: Effect of immunization with inactivated viral and tumor-associated antigen, *Infect. Immun.* **26**:547.

Murthy, K.K., and Calnek, B.W., 1979b, Marek's disease tumor-associated surface antigen (MATSA) in resistant versus susceptible chickens, *Avian Dis.* **23**:831.

Murthy, K.K., Dietert, R.R., and Calnek, B.W., 1979, Demonstration of chicken fetal antigen (CFA) on normal splenic lymphocytes, Marek's disease lymphoblastoid cell lines and other neoplasms, *Int. J. Cancer* **24**:349.

Naito, M., Ono, K., Doi, T., and Kato, S., 1971, Antibodies in human and monkey sera to herpes-type virus from a chicken with Marek's disease and to EB virus detected by the immunofluorescence test, *Biken J.* **14**:161.

Nazerian, K., 1970, Attenuation of Marek's disease virus and study of its properties in two different cell cultures, *J. Natl. Cancer Inst.* **44**:1257.

Nazerian, K., 1971, Further studies on the replication of Marek's disease virus in the chicken and in cell culture, *J. Natl. Cancer Inst.* **47**:207.

Nazerian, K., 1972, Virology and immunology of Marek's disease—a review, in: *Oncogenesis and Herpesviruses* (P.M. Biggs, G. de Thé, and L.N. Payne, eds.), pp. 59–73, IARC Scientific Publication No. 2, International Agency for Research on Cancer, Lyon.

Nazerian, K., 1973, Studies on intracellular and membrane antigens induced by Marek's disease virus, *J. Gen. Virol.* **21**:193.

Nazerian, K., 1974, DNA configuration in the core of Marek's disease virus, *J. Virol.* **13**:1148.

Nazerian, K., 1979, Marek's disease lymphoma of chicken and its causative herpesvirus, *Biochim. Biophys. Acta* **560**:375.

Nazerian, K., 1980, Marek's disease: A herpesvirus-induced malignant lymphoma of the chicken, in: *Viral Oncology* (G. Klein, ed.), pp. 665–682, Raven Press, New York.

Nazerian, K., and Burmester, B.R., 1968, Electron microscopy of a herpesvirus associated with the agent of Marek's disease in cell culture, *Cancer Res.* **28**:2454.

Nazerian, K., and Chen, J.H., 1973, Immunoferritin studies of Marek's disease virus directed intracellular and membrane antigens, *Arch. Gesainte Virusforsch.* **41**:59.

Nazerian, K., and Lee, L.F., 1974, Deoxyribonucleic acid of Marek's disease virus in a lymphoblastoid cell line from Marek's disease tumours, *J. Gen. Virol.* **25**:317.

Nazerian, K., and Lee, L.F., 1976, Selective inhibition by phosphonoacetic acid of MDV DNA replication in a lymphoblastoid cell line, *Virology* **74**:188.

Nazerian, K., and Purchase, H.G., 1970, Combined fluorescent antibody and electron microscopy study of Marek's disease virus infected cell culture, *J. Virol.* **5**:79.

Nazerian, K., and Sharma, J.M., 1975, Detection of T-cell surface antigens in a Marek's disease lymphoblastoid cell line, *J. Natl. Cancer Inst.* **54**:277.

Nazerian, K., and Witter, R.L., 1970, Cell-free transmission and *in vivo* replication of Marek's disease virus, *J. Virol.* **5**:388.

Nazerian, K., Solomon, J.J., Witter, R.L., and Burmester, B.R., 1968, Studies on the etiology of Marek's disease. II. Finding of a herpesvirus in cell culture, *Proc. Soc. Exp. Biol. Med.* **127**:177.

Nazerian, K., Lee, L.F., Witter, R.L., and Burmester, B.R., 1971, Ultrastructural studies of a herpesvirus of turkeys antigenically related to Marek's disease virus, *Virology* **43**:442.

Nazerian, K., Lee, L.F., and Sharma, J.M., 1976, The role of herpesvirus in Marek's disease lymphoma of chickens, *Prog. Med. Virol.* **22**:123.

Nazerian, K., Stephens, E.A., Sharma, J.M., Lee, L.F., Gailitis, M., and Witter, R.L., 1977, A nonproducer T lymphoblastoid cell line from Marek's disease transplantable tumor (JMV), *Avian Dis.* **21**:69.

Nazerian, K., Neiman, P., Okazaki, W., Smith, E.J., and Crittenden, L.B., 1978, Status of endogenous avian RNA tumor virus in Marek's disease lymphoblastoid cell lines and susceptibility of those lines to exogenous RNA tumor viruses, *Avian Dis.* **22**:732.

Nazerian, K., Elmubarak, A. and Sharma, J.M., 1982, Establishment of B-lymphoblastoid cell lines from Marek's disease virus-induced tumors in turkeys, *Int. J. Cancer*, **29**:63.

Neumann, U., 1980, Lack of serological homology between chicken alphafoetoprotein, chickenfoetal red blood cell antigen and Marek's disease tumour-associated surface antigen, *Avian Pathol.* **9**:597.

Okada, I., Yamada, Y., Akiyama, M., Nishimura, I., and Kano, N., 1977, Changes in polymorphic gene frequencies in strains of chickens selected for resistance to Marek's disease, *Br. Poult. Sci.* **18**:237.

Okada, K., Fujimoto, Y., Mikami, T., and Yonehara, K., 1972, The fine structure of Marek's disease virus and herpesvirus of turkey in cell culture, *Jpn. J. Vet. Res.* **20**:57.

Okada, K., Fujimoto, Y., Nakanishi, Y.H., Onuma, M., and Mikami, T., 1980, Cohelical arrangement of the DNA strain in the core of Marek's disease virus particles, *Arch. Virol.* **64**:81.

Okazaki, W., Purchase, H.G., and Burmester, B.R., 1970, Protection against Marek's disease by vaccination with a herpesvirus of turkeys, *Avian Dis.* **14**:413.

Okazaki, W., Purchase, H.G., and Burmester, B.R., 1971, The temporal relationship between vaccination with the herpesvirus of turkeys and challenge with virulent Marek's disease virus, *Avian Dis.* **15**:753.

Okazaki, W., Purchase, H.G., and Burmester, B.R., 1973, Vaccination against Marek's disease: Possible causes of failure of herpesvirus of turkeys (strain FC126) to protect chickens against Marek's disease, *Am. J. Vet. Res.* **34**:813.

Ono, K., Tanabe, S., Naito, M., Doi, T., and Kato, S., 1970, Antigen common to a herpes type virus from chickens with Marek's disease and EB virus from Burkitt's lymphoma cells, *Biken J.* **13**:213.

Ono, K., Doi, T., Ishikawa, T., Iwa, N., Naito, M., Kato, S., Koyama, K., Konobe, T., and Takaku, K., 1974, Studies on herpesvirus of turkeys and its virological characteristics (in Japanese), *Jpn. J. Vet. Sci.* **36**:407.

Onoda, T., Ono, K., Konobe, T., Naito, M., Mori, Y., and Kato, S., 1970, Propagation of herpes-type virus isolated from chickens with Marek's disease in Japanese quail embryo fibroblasts, *Biken J.* **13**:219.

Onuma, M., Mikami, T., and Hayashi, T.T.A., 1974, Properties of the common antigen associated with Marek's disease herpesvirus and turkey herpesvirus infections, *J. Natl. Cancer Inst.* **52**:805.

Onuma, M., Mikami, T., Hayashi, T.T.A., Okada, K., and Fujimoto, Y., 1975, Studies of Marek's disease herpesvirus and turkey herpesvirus specific common antigen which stimulates the production of neutralizing antibodies, *Arch. Virol.* **48**:85.

Onuma, M., Mikami, T., and Hayashi, T.T.A., 1976, Relationship between common antigen and membrane antigens associated with Marek's disease herpesvirus and turkey herpesvirus infections, *Arch. Virol.* **50**:305.

Pappenheimer, A.M., Dunn, L.C., and Cone, V., 1926, A study of fowl paralysis, *Conn. Storrs Agric. Exp. Stn. Bull.* **143**:186.

Pappenheimer, A.M., Dunn, L.C., and Cone, V., 1929a, Studies on fowl paralysis (neurolymphomatosis gallinarum). I. Clinical features and pathology, *J. Exp. Med.* **49**:63.

Pappenheimer, A.M., Dunn, L.C., and Seidlin, S.M., 1929b, Studies on fowl paralysis (neurolymphomatosis gallinarum). II. Transmission experiments, *J. Exp. Med.* **49**:87.

Paterson, J.C., and Cottral, G.E., 1950, Experimental coronary sclerosis. III. Lymphomatosis as a cause of coronary sclerosis in chickens, *Arch. Pathol.* **49**:699.

Patrascu, I.V., Calnek, B.W., and Smith, M.W., 1972, Vaccination with lyophilized turkey herpesvirus (HVT): Minimum infective and protective doses, *Avian Dis.* **16**:86.

Paul, P., Larsen, C.T., Kumar, M.C., and Pomeroy, B.S., 1972, Preliminary observations on egg transmission of turkey herpesvirus (HVT) in turkeys, *Avian Dis.* **16**:27.

Payne, L.N., 1970, Immunosuppressive effects of avian oncogenic viruses, *Proc. R. Soc. Med.* **63**:16.

Payne, L.N., 1972, Pathogenesis of Marek's disease—a review, in: *Oncogenesis and Herpesviruses* (P.M. Biggs, G. de Thé, and L.N. Payne, eds.), pp. 21–37, IARC Scientific Publication No. 2, International Agency for Research on Cancer, Lyon.

Payne, L.N., 1973, Marek's disease: A possible model for herpesvirus-induced neoplasms in man, in: *Analytic and Experimental Epidemiology of Cancer* (W. Nakahara, T. Hirayama, K. Nishioka, and H. Sugano, eds.), pp. 235–257, University of Tokyo Press, Tokyo.

Payne, L.N., 1979, Marek's disease in comparative medicine, *J. R. Soc. Med.* **72**:635.

Payne, L.N., and Biggs, P.M., 1967, Studies on Marek's disease. II. Pathogenesis, *J. Natl. Cancer Inst.* **39**:281.

Payne, L.N., and Howes, K., 1980, Lack of beneficial effect of levamisole on Marek's disease, *Avian Pathol.* **9**:525.

Payne, L.N., and Rennie, M., 1970, Lack of effect of bursectomy on Marek's disease, *J. Natl. Cancer Inst.* **45**:387.

Payne, L.N., and Rennie, M., 1973, Pathogenesis of Marek's disease in chicks with and without maternal antibody, *J. Natl. Cancer Inst.* **51**:1559.

Payne, L.N., and Rennie, M., 1976a, The proportions of B and T lymphocytes in lymphomas, peripheral nerves and lymphoid organs in Marek's disease, *Avian Pathol.* **5**:147.

Payne, L.N., and Rennie, M., 1976b, Sequential changes in the numbers of B and T lymphocytes and other leukocytes in the blood in Marek's disease, *Int. J. Cancer* **18**:510.

Payne, L.N., and Roszkowski, J., 1972, The presence of immunologically uncommitted bursa- and thymus-dependent lymphoid cells in the lymphomas of Marek's disease, *Avian Pathol.* **1**:27.

Payne, L.N., Powell, P.C., and Rennie, M., 1975, Response of B and T lymphocyte and other blood leukocytes in chickens with Marek's disease, *Cold Spring Harbor Symp. Quant. Biol.* **39**:817.

Payne, L.N., Frazier, J.A., and Powell, P.C., 1976, Pathogenesis of Marek's disease, *Int. Rev. Exp. Pathol.* **16**:59.

Payne, L.N., Rennie, M.C., Powell, P.C., and Rowell, J.G., 1978, Transient effect of cyclophosphamide on vaccinal immunity to Marek's disease, *Avian Pathol.* **7**:295.

Payne, L.N., Howes, K., Rennie, M., Bumstead, J.M. and Kidd, A.W., 1981, Use of an agar culture technique for establishing lymphoid cell lines from Marek's disease lymphomas, *Int. J. Cancer*, **28**:757.

Pazderka, F., Longenecker, B.M., Law, G.R.J., Stone, H.A., and Ruth, R.F., 1975, Histocompatibility of chicken populations selected for resistance to Marek's disease, *Immunogenetics* **2**:93.

Pepose, J.S., Stevens, J.G., Cook, M.L. and Lampert, P.W., 1981, Marek's disease as a model for the Landry-Guillain-Barré syndrome, *Am. J. Pathol.* **103**:309.

Peters, W.P., Kufe, D., Schlom, J., Frankel, J.W., Prichett, C.O., Groupé, V., and Spiegelman,

S., 1973, Biological and biochemical evidence for an interaction between Marek's disease herpesvirus and avian leukosis virus *in vivo, Proc. Natl. Acad. Sci. U.S.A.* **70**:3175.

Pevzner, I.Y., Kujdych, I. and Nordskog, A.W., 1981, Immune response and disease resistance in chickens. II. Marek's disease and immune response to GAT, *Poult. Sci.* **60**:927.

Phillips, P.A., and Biggs, P.M., 1972, Course of infection in tissues of susceptible chickens after exposure to strains of Marek's disease virus and turkey herpesvirus, *J. Natl. Cancer Inst.* **49**:1367.

Powell, P.C., 1975, Immunity to Marek's disease induced by glutaraldehyde-treated cells of Marek's disease lymphoblastoid cell lines, *Nature (London)* **257**:684.

Powell, P.C., 1976, Studies on Marek's disease lymphoma-derived cell lines, *Bibl. Haematol.* **43**:348.

Powell, P.C., 1978, Protection against the JMV Marek's disease-derived transplantable tumour by Marek's disease virus-specific antigens, *Avian Pathol.* **7**:305.

Powell, P.C., 1980, *In vitro* stimulation of blood lymphocytes by phytohaemagglutin during the development of Marek's disease, *Avian Pathol.* **9**:471.

Powell, P.C., 1981, Immunity to Marek's disease, in: *Avian Immunology* (M.E. Rose, L.N. Payne, and B.M. Freeman, eds.), pp. 263–283, British Poultry Science, Edinburgh.

Powell, P.C., and Rennie, M., 1978, Marek's disease tumour-specific antigen induced by herpesvirus of turkeys in vaccinated chickens, *Vet. Rec.* **103**:232.

Powell, P.C., and Rennie, M., 1980a, Failure of attenuated Marek's disease virus and herpesvirus of turkey antigens to protect against the JMV Marek's disease-derived transplantable tumour, *Avian Pathol.* **9**:193.

Powell, P.C., and Rennie, M., 1980b, Marek's disease tumour-associated surface antigens (MATSA) induced in inoculated chickens by vaccines against Marek's disease, in: *Resistance and Immunity to Marek's Disease* (P.M. Biggs, ed.), pp. 341–349, CEC Publication EUR 6470, Luxembourg.

Powell, P.C., and Rowell, J.G., 1977, Dissociation of antiviral and antitumor immunity in resistance to Marek's disease, *J. Natl. Cancer Inst.* **59**:919.

Powell, P.C., Payne, L.N., Frazier, J.A., and Rennie, M., 1974, T lymphoblastoid cell lines from Marek's disease lymphomas, *Nature (London)* **251**:79.

Powell, P.C., Lawn, A.M., Payne, L.N., Rennie, M., and Ross, L.J.N., 1980a, The effect of virus dose on the development of Marek's disease in two strains of chickens, *Avian Pathol.* **9**:567.

Powell, P.C., Rennie, M., Ross, L.J.N. and Mustill, B.M., 1980b, The effect of bursectomy on the adoptive transfer of resistance to Marek's disease, *Int. J. Cancer*, **26**:681.

Powell, P.C., Lee, L.F., Mustill, B.M. and Rennie, M., 1982, The mechanism of genetic resistance to Marek's disease, *Int. J. Cancer*, **29**:169.

Prasad, L.B.M., 1978, Induction of delayed hypersensitivity to turkey herpesvirus in fowls, *Res. Vet. Sci.* **24**:258.

Prasad, L.B.M., 1979, Turkey herpesvirus and Marek's disease virus: A comparative appraisal, *Comp. Immun. Microbiol. Infect. Dis.* **2**:335.

Prasad, L.B.M., and Spradbrow, P.B., 1977, Isolation and characterization of herpesvirus from turkeys, *Aust. Vet. J.* **53**:582.

Priester, W.A., and Mason, T.J., 1974, Human cancer mortality in relation to poultry populations, by county, in 10 south eastern states, *J. Natl. Cancer Inst.* **53**:45.

Prineas, J.W., and Wright, R.G., 1972, The fine structure of peripheral nerve lesions in a virus-induced demyelinating disease in fowl (Marek's disease), *Lab. Invest.* **26**:548.

Purchase, H.G., 1969, Immunofluorescence in the study of Marek's disease. 1. Detection of antigen in cell culture and an antigenic comparison of eight isolates, *J. Virol.* **3**:557.

Purchase, H.G., 1970, Virus-specific immunofluorescent and precipitin antigen and cell-free virus in tissues of birds infected with Marek's disease, *Cancer Res.* **30**:1898.

Purchase, H.G., 1972, Recent advances in the knowledge of Marek's disease, *Adv. Vet. Sci. Comp. Sci.* **16**:223.

Purchase, H.G., 1975, Progress in the control of Marek's disease, *Am. J. Vet. Res.* **36**:587.

Purchase, H.G., 1976, Prevention of Marek's disease: A review, *Cancer Res.* **36**:696.

Purchase, H.G., 1977, The etiology and control of Marek's disease of chickens and the economic impact of a successful research program, in: *Beltsville Symposia in Agricultural Research*, Vol. 1, *Virology in Agriculture*, pp. 63–81, Allanheld, Osmum, Montclair, New Jersey.

Purchase, H.G., and Biggs, P.M., 1967, Characterization of five isolates of Marek's disease, *Res. Vet. Sci.* **8**:440.

Purchase, H.G., and Burgoyne, G.H., 1970, Immunofluorescence in the study of Marek's disease: Detection of antibody, *Am. J. Vet. Res.* **31**:117.

Purchase, H.G., and Sharma, J.M., 1974, Amelioration of Marek's disease and absence of vaccine protection in immunologically deficient chicks, *Nature (London)* **248**:419.

Purchase, H.G., Chubb, R.C., and Biggs, P.M., 1968, Effect of lymphoid leukosis and Marek's disease on the immunologic responsiveness of the chicken, *J. Natl. Cancer Inst.* **40**:583.

Purchase, H.G., Burmester, B.R., and Cunningham, C.H., 1971a, Responses of cell cultures from various avian species to Marek's disease virus and herpesvirus of turkeys, *Am. J. Vet. Res.* **32**:1811.

Purchase, H.G., Okazaki, W., and Burmester, B.R., 1971b, Field trials with the herpes virus of turkeys (HVT) strain FC 126 as a vaccine against Marek's disease, *Poult. Sci.* **50**:775.

Purchase, H.G., Mare, C.J., and Burmester, B.R., 1972a, Antigenic comparison of avian and mammalian herpesviruses and protection tests against Marek's disease, *Proc. 76th Annu. Meet. U.S. Anim. Health Assoc.* **1972**:484.

Purchase, H.G., Okazaki, W., and Burmester, B.R., 1972b, The minimum protective dose of the herpesvirus turkeys vaccine against Marek's disease, *Vet. Rec.* **91**:79.

Purchase, H.G., Okazaki, W., and Burmester, B.R., 1972c, Long-term field trials with the herpesvirus of turkeys vaccine against Marek's disease, *Avian Dis.* **16**:57.

Ratz, F., Széky, A., and Ványi, A., 1972, Studies on the histopathologic changes of acute Marek's disease, *Acta Vet. Acad. Sci. Hung.* **22**:349.

Rennie, M., and Powell, P.C., 1979, Serological characterisation of Marek's disease tumour-associated surface antigens on Marek's disease lymphoma cells and on cell lines derived from Marek's disease lymphomas, *Avian Pathol.* **8**:173.

Rennie, M., Powell, P.C., and Mustill, B.M., 1980, The effect of bursectomy on vaccination against Marek's disease with the herpesvirus of turkeys, *Avian Pathol.* **9**:557.

Riddell, C., Milne, B.S., and Biggs, P.M., 1978, Herpesvirus of turkey vaccine: Viraemias in field flocks and in experimental chickens, *Vet. Rec.* **102**:123.

Rispens, B.H., von Vloten, H., Mastenbroek, N., Maas, H.J.L., and Schat, K.A., 1972a, Control of Marek's disease in the Netherlands. I. Isolation of an avirulent Marek's disease virus (strain CVI 988) and its use in laboratory vaccination trials, *Avian Dis.* **16**:108.

Rispens, B.H., von Vloten, H., Mastenbroek, N., Maas, H.J.L., and Schat, K.A., 1972b, Control of Marek's disease in the Netherlands. II. Field trials on vaccination with an avirulent strain (CVI 988) of Marek's disease virus, *Avian Dis.* **16**:126.

Roizman, B., 1972, The biochemical features of herpesvirus-infected cells, particularly as they relate to their potential oncogenicity—a review, in: *Oncogenesis and Herpesviruses* (G. de Thé, W. Henle, and F. Rapp, eds.), p. 1079, IARC Scientific Publication No. 2, International Agency for Research on Cancer, Lyon.

Ross, L.J.N., 1977, Antiviral T cell-mediated immunity in Marek's disease, *Nature (London)* **268**:644.

Ross, L.J.N., 1980, Mechanism of protection conferred by HVT, in: *Resistance and Immunity to Marek's Disease* (P.M. Biggs, ed.), pp. 289–297, CEC Publication EUR 6470, Luxembourg.

Ross, L.J.N., Biggs, P.M., and Newton, A.A., 1973, Purification and properties of the "A" antigen associated with Marek's disease virus infections, *J. Gen. Virol.* **18**:291.

Ross, L.J.N., Powell, P.C., Walker, D.J., Rennie, M., and Payne, L.N., 1977, Expression of virus-specific, thymus-specific and tumour-specific antigens in lymphoblastoid cell lines derived from Marek's disease lymphomas, *J. Gen. Virol.* **35**:219.

Ross, L.J.N., DeLorbe, W., Varmus, H.E., Bishop, J.M., Brahic, M. and Haase, A., 1981, Persistence and expression of Marek's disease virus DNA in tumour cells and peripheral nerves studied by *in situ* hybridization, *J. Gen. Virol.* **57**:285.

Rouse, B.T., Wells, R.J.H., and Warner, M.L., 1973, Proportion of T and B lymphocytes in lesion of Marek's disease: Theoretical implications for pathogenesis, *J. Immunol.* **110**:534.

Schat, K.A., 1981, Role of the spleen in the pathogenesis of Marek's disease, *Avian Pathol.* **10**:171.

Schat, K.A., and Calnek, B.W., 1978a, Characterization of an apparently nononcogenic Marek's disease virus, *J. Natl. Cancer Inst.* **60**:1075.

Schat, K.A., and Calnek, B.W., 1978b, *In vitro* inactivation of cell-free Marek's disease herpesvirus by immune peripheral blood lymphocytes, *Avian Dis.* **22**:693.

Schat, K.A., and Calnek, B.W., 1978c, Protection against Marek's disease-derived tumor transplants by the non-oncogenic SB-1 strain of Marek's disease virus, *Infect. Immun.* **22**:225.

Schat, K.A., and Calnek, B.W., 1978d, Demonstration of Marek's disease tumour-associated surface antigen in chickens infected with nononcogenic Marek's disease virus and herpesvirus of turkeys, *J. Natl. Cancer Inst.* **61**:855.

Schat, K.A., and Calnek, B.W., 1980, *In vitro* cytotoxicity of spleen lymphocytes against Marek's disease tumour cells: Induction by SB-1, an apparently non-oncogenic Marek's disease virus, in: *Resistance and Immunity to Marek's Disease* (P.M. Biggs, ed.), pp. 301–316, CEC Publication EUR 6470, Luxembourg.

Schat, K.A., and Murthy, K.K., 1980, *In vitro* cytotoxicity against Marek's disease lymphoblastoid cell lines after enzymatic removal of Marek's disease tumor associated surface antigen, *J. Virol.* **34**:130.

Schat, K.A., Schultsz, R.D., and Calnek, B.W., 1978, Marek's disease: Effect of virus pathogenicity and genetic susceptibility on responses of peripheral blood lymphocytes to concanavalin-A, in: *Advances in Comparative Leukemia Research* (P. Benvelzen, J. Hilgers, and D.S. Yohn, eds.), pp. 183–186, Elsevier/North-Holland, Amsterdam.

Schat, K.A., Calnek, B.W. and Fabricant, J., 1981, Influence of the bursa of Fabricius on the pathogenesis of Marek's disease, *Infect. Immun.* **31**:199.

Schidlovsky, G., Ahmed, M., and Jensen, K.E., 1969, Herpesvirus in Marek's disease tumors, *Science* **164**:959.

Schierman, L.W., and Fletcher, O.J., 1980, Genetic control of Marek's disease virus-induced transient paralysis: Association with the major histocompatibility complex, in: *Resistance and Immunity to Marek's Disease* (P.M. Biggs, ed.), pp. 429–440, CEC Publication EUR 6470, Luxembourg.

Schierman, L.W., Theis, G.A., and McBride, R.A., 1976, Preservation of a T cell-mediated immune response in Marek's disease virus-infected chickens by vaccination with a related virus, *J. Immunol.* **116**:1497.

Schmahl, W., Hoffman-Fezer, G., and Hoffman, R., 1975, Zur Pathogenese der Nervenläsionen bei Marekscher Krankheit des Huhnes. I. Allergische Lantreaktion gegen Myelin peripherer Nerven, *Z. Immuntaetsorsch. Exp. Ther.* **150**:175.

Schmittle, S.C., and Edison, C.S., 1968, Studies on acute Marek's disease. IV. Relative susceptibility of different lines and crosses of chickens to GA isolate, *Avian Dis.* **12**:571.

Sekiya, Y., Sakata, M., Mima, K., Namiki, T., Yoshizawa, A., and Chikatisune, M., 1977, Properties of JMV Marek's disease tumour cells (JMV-S), (H. Kawamura, ed.), English Summaries of the Studies on Poultry Diseases in Japan, p. 10, Japanese Veterinary Poultry Association.

Sevoian, M., 1968, Egg transmission studies of type II leukosis infection (Marek's disease), *Poult. Sci.* **47**:1644.

Sevoian, M., Chamberlain, D.M., and Counter, F.T., 1962, Avian lymphomatosis. I. Experimental reproduction of the neural and visceral forms, *Vet. Med.* **57**:500.

Sevoian, M., Chamberlain, D.M., and Larose, R.N., 1963, Avian lymphomatosis. V. Airborne transmission, *Avian Dis.* **7**:102.

Sevoian, M., Larose, R.N., and Chamberlain, D.M., 1964, Avian lymphomatosis: Increased pathogenicity of JM virus, *Proc. 101st Annu. Meet. Am. Vet. Med. Assoc.* 342.

Sharma, J.M., 1974, Resistance to Marek's disease in immunologically deficient chickens, *Nature (London)* **247:**117.

Sharma, J.M., 1975, Marek's disease, in: *Isolation and Identification of Avian Pathogens* (S.B. Hitchner, C.H. Domermuth, H.G. Purchase, and J.E. Williams, eds.), pp. 235–250, Arnold Printing, Ithaca, New York.

Sharma, J.M., 1977, Cell-mediated immunity to tumor antigen in Marek's disease: Susceptibility of effector cells to antithymocyte serum and enhancement of cytotoxic activity by *Vibrio cholerae* neuraminidase, *Infect. Immun.* **18:**46.

Sharma, J.M., 1978, Immunopathology of Marek's disease, in: *Animal Models of Comparative and Developmental Aspects of Immunity and Disease* (M.E. Gershwin and E.L. Cooper, eds.), pp. 132–142, Peramon Press, Oxford.

Sharma, J.M., 1979, Immunosuppressive effects of lymphoproliferative neoplasms of chickens, *Avian Dis.* **23:**315.

Sharma, J.M., 1980, *In vitro* suppression of T-cell mitogenic response and tumor cell proliferation by spleen macrophages from normal chickens, *Infect. Immun.* **28:**914.

Sharma, J.M., 1981a, Fractionation of Marek's disease virus induced lymphoma by velocity sedimentation and association of infectivity with cellular functions with and without tumor antigen expression, *Am. J. Vet. Res.* **42:**483.

Sharma, J.M., 1981b, Natural killer cell activity in chickens exposed to Marek's disease virus: inhibition of activity in susceptible chickens and enhancement of activity in resistant and vaccinated chickens, *Avian Dis.* **25:**882.

Sharma, J.M. and Burmester, B.R., 1982, Resistance to Marek's disease at hatching in chickens vaccinated as embryos with the turkeys herpesvirus, *Avian Dis.* **26:**134.

Sharma, J.M., and Coulson, B.D., 1977, Cell-mediated cytotoxic response to cells bearing Marek's disease tumor-associated surface antigen in chickens infected with Marek's disease virus, *J. Natl. Cancer Inst.* **58:**1647.

Sharma, J.M., and Coulson, B.D., 1980, Natural immunity against Marek's disease tumour cells: Cytotoxicity of spleen cells from specific pathogen free chickens against MSB-1 target cells, in: *Resistance and Immunity to Marek's Disease* (P.M. Biggs, ed.), pp. 223–229, CEC Publication EUR 6470, Luxembourg.

Sharma, J.M., and Purchase, H.G., 1974, Replication of Marek's disease virus in cell cultures derived from genetically resistant chickens, *Infect. Immun.* **9:**1092.

Sharma, J.M., and Stone, H.A., 1972, Genetic resistance to Marek's disease: Delineation of the response of genetically resistant chickens to Marek's disease virus infection, *Avian Dis.* **16:**894.

Sharma, J.M., and Witter, R.L., 1975, The effect of B-cell immunosuppression on age-related resistance of chickens to Marek's disease, *Cancer Res.* **35:**711.

Sharma, J.M., Kenzy, S.G., and Rissberger, A., 1969, Propagation and behaviour in chicken kidney cultures of the agent associated with classical Marek's disease, *J. Natl. Cancer Inst.* **43:**907.

Sharma, J.M., Davis, W.C., and Kenzy, S.G., 1970, Etiologic relationship of skin tumors (skin leukosis) of chickens to Marek's disease, *J. Natl. Cancer Inst.* **44:**901.

Sharma, J.M., Witter, R.L., Shramek, G., Wolfe, L.G., Burmester, B.R., and Deinhardt, F., 1972, Lack of pathogenicity of Marek's disease virus and herpesvirus of turkeys in marmoset monkeys, *J. Natl. Cancer Inst.* **49:**1191.

Sharma, J.M., Witter, R.L., Burmester, B.R., and Landon, J.C., 1973a, Public health implications of Marek's disease virus and herpesvirus of turkeys: Studies on human and subhuman primates, *J. Natl. Cancer Inst.* **51:**1123.

Sharma, J.M., Witter, R.L., and Burmester, B.R., 1973b, Pathogenesis of Marek's disease in old chickens: Lesion regression as the basis for age-related resistance, *Infect. Immun.* **8:**715.

Sharma, J.M., Witter, R.L., and Purchase, H.G., 1975, Absence of age resistance in neonatally thymectomized chickens as evidence for cell-mediated immune surveillance in Marek's disease, *Nature (London)* **253:**477.

Sharma, J.M., Coulson, B.D., and Young, E., 1976, Effect of *in vitro* adaptation of Marek's disease virus on pock induction on the chorioallantoic membrane of embryonated chicken eggs, *Infect. Immun.* **13**:292.

Sharma, J.M., Nazerian, K., and Witter, R.L., 1977, Reduced incidence of Marek's disease gross lymphomas in T-cell-depleted chickens, *J. Natl. Cancer Inst.* **58**:689.

Sharma, J.M., Witter, R.L., and Coulson, B.D., 1978, Development of cell-mediated immunity to Marek's disease tumor cells in chicks inoculated with Marek's disease vaccines, *J. Natl. Cancer Inst.* **61**:1273.

Sharma, J.M., Lee, L.F., and Witter, R.L., 1980, Effect of neonatal thymectomy on pathogenesis of herpesvirus of turkeys in chickens, *Am. J. Vet. Res.* **40**:761.

Shearman, P.J., and Longenecker, B.M., 1980, Selection for virulence and organ-specific metastasis of herpesvirus-transformed lymphoma cells, *Int. J. Cancer* **25**:363.

Shearman, P.J. and Longenecker, B.M., 1981, Clonal variation and functional correlation of organ-specific metastasis and an organ-specific metastasis-associated antigen, *Int. J. Cancer* **27**:387.

Shearman, P.J., Gallatin, W.M., and Longenecker, B.M., 1980, Detection of a cell-surface antigen correlated with organ-specific metastasis, *Nature (London)* **286**:267.

Shieh, H.K., and Sevoian, M., 1974, Antibody response of susceptible and resistant chickens (type II leukosis) infected with JM and JM-V leukosis strains and herpesvirus of turkeys (HVT), *Avian Dis.* **18**:318.

Shieh, H.K., and Sevoian, M., 1978, The influence of humoral and cellular immune system on chickens with Marek's disease, *J. Chinese Vet. Sci.* **4**:3.

Siegmann, O., Kaleta, E.F., and Schindler, P., 1980, Short- and long-term stability studies on four lyophilized and one cell-associated turkey herpesvirus vaccines against Marek's disease of chickens, *Avian Pathol.* **9**:21.

Smith, M.E., Campbell, W.F., Farrow, W.M., and Frankel, J.W., 1975, Enhancement and interference in chickens inoculated with Marek's disease herpesvirus and oncornaviruses, *Proc. Soc. Exp. Biol. Med.* **150**:574.

Smith, M.W., and Calnek, B.W., 1973, Effect of virus pathogenicity on antibody production in Marek's disease, *Avian Dis.* **17**:727.

Smith, M.W., and Calnek, B.W., 1974, Comparative features of low-virulence and high-virulence Marek's disease virus infections, *Avian Pathol.* **3**:229.

Smith, T.W., Albert, D.M., Robinson, N., Calnek, B.W., and Schwabe, O., 1974, Ocular manifestations of Marek's disease, *Invest. Ophthalmol.* **13**:586.

Solomon, J.J., and Witter, R.L., 1973, Absence of Marek's disease in chicks hatched from eggs containing blood or meat spots, *Avian Dis.* **17**:141.

Solomon, J.J., Witter, R.L., Nazerian, K., and Burmester, B.R., 1968, Studies on the etiology of Marek's disease. I. Propagation of the agent in cell culture, *Proc. Soc. Exp. Biol. Med.* **127**:173.

Solomon, J.J., Witter, R.L., Stone, H.A., and Champion, L.R., 1970, Evidence against embryo transmission of Marek's disease virus, *Avian Dis.* **14**:752.

Spencer, J.L., 1969, Marek's disease herpesvirus: *In vivo* and *in vitro* infection of kidney cells of different genetic strains, *Avian Dis.* **13**:753.

Spencer, J.L., and Calnek, B.W., 1967, Storage of cells infected with Rous sarcoma virus or JM strain avian lymphomatosis agent, *Avian Dis.* **11**:274.

Spencer, J.L., and Gavora, J.S., 1980, Influence of genotype of chickens and immune status of dams on response to vaccination with turkey herpesvirus, in: *Resistance and Immunity to Marek's Disease* (P.M. Biggs, ed.), pp. 519–535, CEC Publication EUR 6470, Luxembourg.

Spencer, J.L., and Robertson, A., 1972, Influence of maternal antibody on infection with virulent or attenuated Marek's disease herpesvirus, *Am. J. Vet. Res.* **33**:393.

Spencer, J.L., Grunder, A.A., Robertson, A., and Speckmann, G.W., 1972, Attenuated Marek's disease herpesvirus: Protection conferred on strains of chickens varying in genetic resistance, *Avian Dis.* **16**:94.

Spencer, J.L., Gavora, J.S., Grunder, A.A., Robertson, A., and Speckmann, G.W., 1974, Immunization against Marek's disease: Influence of strain of chickens, maternal antibody and type of vaccine, *Avian Dis.* **18**:33.

Spencer, J.L., Gavora, J.S., Hare, W.C.D., Grunder, A.A., Robertson, A., and Speckmann, G.W., 1976, Studies on genetic resistance and on vaccination-induced resistance of chickens to lymphoid tumor transplants. I. Marek's disease tumor transplant (JMV), *Avian Dis.* **20**:268.

Stephens, E.A., Witter, R.L., Lee, L.F., Sharma, J.M., Nazerian, K., and Longenecker, B.M., 1976, Characteristics of JMV Marek's disease tumor: A non-productively infected transplantable cell lacking in rescuable virus, *J. Natl. Cancer Inst.* **57**:865.

Stephens, E.A., Witter, R.L., Nazerian K., and Sharma, J.M., 1980, Development and characterization of a Marek's disease transplantable tumor in inbred, line 7_2 chickens homozygous at the major (B) histocompatibility locus, *Avian Dis.* **24**:358.

Stevens, J.G., Pepose, J.S. and Cook, M.L., 1981, Marek's disease: A natural model for the Landry-Guillain-Barré syndrome, *Ann. Neurol.* suppl. **9**:102.

Stone, H.A., 1975, Use of highly inbred chickens in research, *Agric. Res. Serv. U.S. Dept. Agric. Tech. Bull.* **1514**:22.

Stone, H.A., Holly, E.A., Burmester, B.R., and Coleman, T.H., 1970, Genetic control of Marek's disease, *Poult. Sci.* **49**:1441.

Sugimoto, C., Kodama, H., and Mikami, T., 1978, Anti-tumor immunity against Marek's disease-derived lymphoblastoid cell line (MSB-1), *Jpn. J. Vet. Res.* **26**:57.

Sugiyama, M., Hagiwara, S., Ichinose, M., Horiuchi, T., and Isoda, M., 1973, Pathological changes of spontaneous cases of Marek's disease. I. Lesions classified by location in the peripheral and central nervous system, *Bull. Nippon Vet. Zootech. Coll. (Nihon Jui Chikusan Daigaku Kiyo)* **22**:9.

Takagi, N., Sasaki, M., Ikuta, K., and Kato, S., 1977, Chromosomal characteristics of six cultured lymphoblastoid cell lines originating from Marek's disease lymphomas, *Biken J.* **20**:21.

Tanaka, A., Joel, M., Silver, S., and Nonoyama, N., 1978, Biochemical evidence of the nonintegrated status of Marek's disease virus DNA in virus-transformed lymphoblastoid cells of chicken, *Virology* **88**:19.

Theis, G.A., 1977, Effects of lymphocytes from Marek's disease infected chickens on mitogenic responses of syngeneic normal chicken spleen cells, *J. Immunol.* **118**:887.

Theis, G.A., 1979, Immune reactivity in Marek's disease during viral oncogenesis and protective vaccination, in: *Immune Mechanisms and Disease* (D.B. Amos, R.S. Schwartz, and B.W. Janicki, eds.), pp. 323–338, Academic Press, New York.

Theis, G.A., 1981, Subpopulations of suppressor cells in chickens infected with cells of a transplantable lymphoblastic leukemia, *Infect. Immun.* **34**:526.

Theis, G.A., Schierman, L.W., and McBride, R.A., 1974, Transplantation of a Marek's disease lymphoma in syngeneic chickens, *J. Immunol.* **113**:1710.

Thompson, D.L., Elgert, K.D., Gross, W.B., and Siegel, P.B., 1980, Cell-mediated immunity in Marek's disease virus-infected chickens genetically selected for high and low concentrations of plasma corticosterone, *Am. J. Vet. Res.* **41**:91.

Thornton, D.H., Hinton, M.H., and Muskett, J.C., 1975, Efficacy of Marek's disease vaccines: Protection and viraemia studies with turkey herpes virus vaccines, *Avian Pathol.* **4**:97.

Thurston, R.J., Hess, R.A., Biellier, H.V., Adldinger, H.K., and Solorzano, R.F., 1975, Ultrastructural studies of semen abnormalities and herpesvirus associated with cultured testicular cells from domestic turkeys, *J. Reprod. Fertil.* **45**:235.

Velicer, L.F., Yager, D.R., and Clark, J.L., 1978, Marek's disease herpesvirus. III. Purification and characterization of Marek's disease herpesvirus B antigen, *J. Virol.* **27**:205.

Vengris, V.E., and Mare, C.J., 1973, Protection of chickens against Marek's disease virus JM-V strain with statolon and exogenous interferon, *Avian Dis.* **17**:758.

Von Krosigk, C.M., McClary, C.F., Vielitz, E., and Zander, D.V., 1972, Selection for resistance to Marek's disease and its expected effects on other important traits in White Leghorn Strain crosses, *Avian Dis.* **16**:11.

Wainberg, M.A., Beiss, B., and Israel, E., 1980, Virus-mediated abrogation of chicken lymphocyte responsiveness to mitogenic stimulus, *Avian Dis.* **24**:580.

Washburn, K.W., Eidson, C.S., and Lowe, R.H., 1971, Association of hemoglobulin type with resistance to Marek's disease, *Poult. Sci.* **50**:90.

Waters, N.F., 1951, Mortality from lymphomatosis and other causes among inbred lines of White Leghorns, *Poult. Sci.* **30**:531.

Weiss, R.A., and Biggs, P.M., 1972, Leukosis and Marek's disease viruses of feral Red Jungle Fowl and domestic fowl in Malaya. *J. Natl. Cancer Inst.* **49**:1713.

Wight, P.A.L., 1962a, Variations in peripheral nerve histopathology in fowl paralysis, *J. Comp. Pathol.* **72**:40.

Wight, P.A.L., 1962b, The histopathology of the central nervous system in fowl paralysis, *J. Comp. Pathol.* **72**:348.

Wight, P.A.L., 1966, Histopathology of the skeletal muscles in fowl paralysis (Marek's disease), *J. Comp. Pathol.* **76**:333.

Wight, P.A.L., 1968, The histopathology of transient paralysis of the domestic fowl, *Vet. Rec.* **82**:749.

Wight, P.A.L., 1969, The ultrastructure of sciatic nerves affected by fowl paralysis (Marek's disease), *J. Comp. Pathol.* **79**:563.

Withell, J., 1973, A focus forming technique for the assay of HVT vaccines against Marek's disease, *Dev. Biol. Stand.* **25**:287.

Witter, R.L., 1971, Marek's disease research—History and perspectives, *Poult. Sci.* **50**:333.

Witter, R.L., 1972, Turkey herpesvirus: Lack of oncogenicity for turkeys, *Avian Dis.* **16**:666.

Witter, R.L., 1976, Natural mechanisms of controlling lymphotropic herpesvirus infection (Marek's disease) in the chicken, *Cancer Res.* **36**:681.

Witter, R.L., 1982, Protection by attenuated and polyvalent vaccines against highly virulent strains of Marek's disease virus, *Avian Pathol.*, **11**:49.

Witter, R.L., and Burmester, B.R., 1967, Transmission of Marek's disease with oral washings and feces from infected chickens, *Proc. Soc. Exp. Biol. Med.* **124**:59.

Witter, R.L., and Burmester, B.R., 1979, Differential effect of maternal antibodies on efficacy of cellular and cell-free Marek's disease vaccines, *Avian Pathol.* **8**:145.

Witter, R.L., and Fadly, A.M., 1980, Characteristics of some selected Marek's disease virus field isolates, in: *Resistance and Immunity to Marek's Disease* (P.M. Biggs, ed.), pp. 181–191, CEC Publication EUR 6470, Luxembourg.

Witter, R.L., and Offenbecker, L., 1978, Duration of vaccinal immunity against Marek's disease, *Avian Dis.* **22**:396.

Witter, R.L., and Offenbecker, L., 1979, Non-protective and temperature-sensitive variants of Marek's disease vaccine viruses, *J. Natl. Cancer Inst.* **62**:143.

Witter, R.L., and Sharma, J.M., 1974, Transient infectivity and heterokaryon formation in hamster cell cultures inoculated with cell-associated stocks of Marek's disease virus and herpesvirus of turkeys, *J. Natl. Cancer Inst.* **53**:1731.

Witter, R.L., and Solomon, J.J., 1971, Epidemiology of a herpesvirus of turkeys: Possible sources and spread of infection in turkey flocks, *Infect. Immun.* **4**:356.

Witter, R.L., and Solomon, J.J., 1972, Experimental infection of turkeys and chickens with a herpesvirus of turkeys (HVT), *Avian Dis.* **16**:34.

Witter, R.L., Burgoyne, G.H., and Burmester, B.R., 1968, Survival of Marek's disease agent in litter and droppings, *Avian Dis.* **12**:522.

Witter, R.L., Solomon, J.J., and Burgoyne, G.H., 1969, Cell culture techniques for primary isolation of Marek's disease-associated herpesvirus, *Avian Dis.* **13**:101.

Witter, R.L., Moulthrop, J.I., Burgoyne, G.H., and Connell, H.C., 1970a, Studies on the epidemiology of Marek's disease herpesvirus in broiler flocks, *Avian Dis.* **14**:255.

Witter, R.L., Nazerian, K., Purchase, H., and Burgoyne, G.H., 1970b, Isolation from turkeys of a cell-associated herpesvirus antigenically related to Marek's disease virus, *Am. J. Vet. Res.* **31**:525.

Witter, R.L., Purchase, H.G., and Burgoyne, G.H., 1970c, Peripheral nerve lesions similar to those of Marek's disease in chickens inoculated with reticuloendotheliosis virus, *J. Natl. Cancer Inst.* **45**:567.

Witter, R.L., Solomon, J.J., Champion, L.R., and Nazerian, K., 1971, Long-term studies of Marek's disease infection in individual chickens, *Avian Dis.* **15**:346.

Witter, R.L., Nazerian, K., and Solomon, J.J., 1972, Studies on the *in vivo* replication of turkey herpesvirus, *J. Natl. Cancer Inst.* **49**:1121.

Witter, R.L., Sharma, J.M., Solomon, J.J., and Champion, L.R., 1973, An age-related resistance of chickens to Marek's disease: Some preliminary observations, *Avian Pathol.* **2**:43.

Witter, R.L., Solomon, J.J., and Sharma, J.M., 1974, Response of turkeys to infection with virulent Marek's disease viruses of turkey and chicken origins, *Am. J. Vet. Res.* **35**:1325.

Witter, R.L., Lee, L.F., Okazaki, W., Purchase, H.G., Burmester, B.R., and Luginbuhl, R.E., 1975a, Oncogenesis by Marek's disease herpesvirus in chickens lacking expression of endogenous (gs, chick helper factor, Rous-associated virus—O) and exogenous avian RNA tumor viruses, *J. Natl. Cancer Inst.* **55**:215.

Witter, R.L., Stephens, E.A., Sharma, J.M., and Nazerian, K., 1975b, Demonstration of a tumor-associated surface antigen in Marek's disease, *J. Immunol.* **115**:177.

Witter, R.L., Sharma, J.M., and Offenbecker, L., 1976, Turkey herpesvirus infection in chickens: Induction of lymphoproliferative lesions and characterization of vaccinal immunity against Marek's disease, *Avian Dis.* **20**:676.

Witter, R.L., Kato, S., Calnek, B.W., Powell, P.C., et al., 1979, A proposed method for designating avian cell lines and transplantable tumours, *Avian Pathol.* **8**:487.

Witter, R.L., Sharma, J.M., and Fadly, A.M., 1980, Pathogenicity of variant Marek's disease virus isolants in vaccinated and unvaccinated chickens, *Avian Dis.* **24**:210.

Wyn-Jones, A.P., and Kaaden, O.-R., 1979, Induction of virus-neutralising antibody by glycoproteins isolated from chicken cells infected with a herpesvirus of turkeys, *Infect. Immun.* **25**:54.

Yoshida, I., Yuasa, N., Horiuchi, T., and Tsubahara, H., 1975, Effect of maternal immunity against development of Marek's disease and protective ability of vaccine, *Natl. Inst. Anim. Health Q.* **15**:1.

Zander, D.V., 1959, Experiences with epidemic tremor control, *Proc. 8th Annu. Western Poult. Dis. Conf.* 18–23.

Zanella, A., Bettoni, V., and Mambelli, N., 1974, Lyophilised turkey herpesvirus (HVT): Stability of vaccine and minimum protective dose, *Avian Pathol.* **3**:15.

Zygraich, N., and Huygelen, C., 1972, Inoculation of one-day-old chicks with different strains of turkey herpesvirus. II. Virus replication in tissues of inoculated animals, *Avian Dis.* **16**:793.

Index

CPSIA information can be obtained at www.ICGtesting.com
Printed in the USA
LVOW02s2359091113

360695LV00005B/405/P